IP OVER WDM

IP OVER WDM

Building the Next-Generation Optical Internet

Edited by

Sudhir Dixit

WILEY-
INTERSCIENCE

A JOHN WILEY & SONS PUBLICATION

Library of Congress Cataloging-in-Publication Data:

Dixit, Sudhir.
    IP over WDM : building the next-generation optical internet / Sudhir Dixit.
      p. cm.
    Includes bibliographical references.
    ISBN 0-471-21248-2 (cloth : alk. paper)
    1. Wavelength division multiplexing. 2. TCP/IP (Computer network protocol) 3.
Optical communications. I. Title

    TK5103.592.W38 D59 2003
    621.382'7–dc21
                                        2002027394

Printed in the United States of America

10 9 8 7 6 5 4 3 2 1

*With gratitude to my wonderful parents for their undying love and inspiration,
and for laying a solid foundation of my life—I could not be luckier.*

CONTENTS

7   Multiprotocol over WDM in the Access and Metropolitan Networks   181

Maurice Gagnaire

8   Ethernet Passive Optical Networks   229

Glen Kramer, Biswanath Mukherjee, and Ariel Maislos

13   Internetworking Optical Internet and Optical Burst Switching   397

Myungsik Yoo and Younghan Kim

14   IP-over-WDM Control and Signaling   421

Chunsheng Xin, Sudhir Dixit, and Chunming Qiao

CONTRIBUTORS

JAVIER ARACIL, Departamento de Automática y Computación, Universidad Pública de Navarra, Campus Arrosadía, 31006 Pamplona, Spain

RICHARD BARRY, Sycamore Networks, 150 Apollo Drive, Chelmsford, MA 01824

YANG CAO, Sycamore Networks, 150 Apollo Drive, Chelmsford, MA 01824

SUDHIR DIXIT, Nokia Research Center, 5 Wayside Road, Burlington, MA 01803

MAURICE GAGNAIRE, Computer Science and Networks Department, École Nationale Supérieure des Télécommunications, 46 Rue Barrault, 75634 Paris Cedex 13, France

RANJAN GANGOPADHYAY, Department of Electrical and Communication Engineering, Indian Institute of Technology, Kharagpur 721302, India

MIKEL IZAL, Departamento de Automática y Computación, Universidad Pública de Navarra, Campus Arrosadía, 31006 Pamplona, Spain

SHIVKUMAR KALYANARAMAN, Department of ECSE, Rensselaer Polytechnic Institute, 110 8th Street, Troy, NY 12180-3590

YOUNGHAN KIM, School of Electronic Engineering, Soongsil University, Seoul, Korea

GLEN KRAMER, Department of Electrical Engineering, University of California–Davis, Davis, CA 95616

ARIEL MAISLOS, Passave Technologies, AMGAR House, 7 Rival Street, Tel-Aviv, Israel

DANIEL MORATO, Departamento de Automática y Computación, Universidad Pública de Navarra, Campus Arrosadía, 31006 Pamplona, Spain

BISWANATH MUKHERJEE, Department of Electrical Engineering, University of California–Davis, Davis, CA 95616

CHUNMING QIAO, Department of Computer Science and Engineering, State University of New York, Buffalo, NY 14260-2000

BYRAV RAMAMURTHY, Department of Computer Science and Engineering, University of Nebraska–Lincoln, 215C Ferguson Hall, Lincoln, NE 68588-0115

GEORGE N. ROUSKAS,   Department of Computer Science, North Carolina State University, Raleigh, NC 27695-7534

BIPLAB SIKDAR,   Department of ECSE, Rensselaer Polytechnic Institute, 110 8th Street, Troy, NY 12180-3590

ARI TERVONEN,   Nokia Research Center, Itämerenkatu 11-13, FIN-00180, Helsinki, Finland

TI-SHIANG WANG,   Nokia Research Center, 5 Wayside Road, Burlington, MA 01803

CHUNSHENG XIN,   Nokia Research Center, 5 Wayside Road, Burlington, MA 01803

SHUN YAO,   Department of Electrical Engineering, University of California–Davis, Davis, CA 95616

YINGHUA YE,   Nokia Research Center, 5 Wayside Road, Burlington, MA 01803

MYUNGSIK YOO,   School of Electronic Engineering, Soongsil University, Seoul, Korea

PREFACE

The primary objective of this book is to provide, in an understandable style, comprehensive and practical information on the transport of the Internet protocol (IP) over the optical/wavelength-division-multiplexed (WDM) layer. This is accomplished by providing sufficient background information on the enabling technologies, followed by an in-depth discussion on the various aspects of IP over WDM. The vision for IP over WDM is that when it becomes a reality it will revolutionize networking as we know it today. It is already presenting new major business opportunities and could potentially affect the future of Synchronous Optical Network/Synchronous Digital Hierarchy (SONET/SDH), asynchronous transfer mode (ATM), and other technologies. The motivation for writing this book came from the realization that there was a shortage of technical material in a single place on IP-over-WDM technology.

Although this book could have been written by a single author, I felt that different perspectives from different experts, who have confirmed their work in their respective specialties with their peers, would bring tremendous value and make the contents more interesting. It has been our intention to encourage authors to be concerned about the adjoining layers and technologies even though their primary interest has been to focus deeply on only one aspect of the technology spectrum. To meet this objective of interoperability I felt all along that it is good to have some overlaps in the subject material since the context often differs from one focus area to another, and how a certain issue or problem is perceived can vary widely.

Publication of this book is very timely because information on the topic is scattered. Although there are many books in the area of optical networking focusing on the optical layer and the enabling devices and technologies, they do scant justice to the field of intelligent optical Internet, where IP is mapped directly over the optical layer. This book is intended to be a single authoritative source of information to both industry and academia. The primary audience for the book are practicing engineers and designers and engineering managers. In addition, the book is written in a format and style that makes it suitable as an undergraduate (senior year) or graduate-level textbook in a one-semester (or with omission of some topics) one-quarter format. It is our belief that this book will provide readers with sufficient exposure to the knowledge, with a good mix of theory and practice, to lay a solid foundation in IP-over-WDM and related technologies.

ACKNOWLEDGMENTS

I thank my wife, Asha, and my children, Sapna and Amar, for their support, understanding, and sacrifice of time spent while I worked long hours on this project. They were just wonderful.

I would also like to thank Reijo Juvonen, head of Communication Systems Laboratory, Nokia Research Center (NRC), Helsinki, Finland, and Dr. Pertti Karkkainen, head of NRC, USA, for their continued support and encouragement throughout the course of writing and editing this book.

I am grateful to the contributing authors, who were diligent in their submissions and were always willing and prompt in revising their material. Without their help and cooperation, this project would not have been possible—my heartfelt thanks to them. I also thank all the reviewers, who provided me with thorough reviews in a timely manner. Last, but not least, I thank the wonderful editorial staff of John Wiley, who were always accessible and helpful in guiding me though the labyrinth of this task.

Finally, the authors and I have tried our best to make each chapter complete in itself and their contents as accurate as possible, but some errors and omissions may remain. Any feedback to correct errors and to improve the book would be sincerely appreciated.

Boston, Massachusetts   SUDHIR DIXIT
November 2002
(Email: sudhir.dixit@ieee.org)

IP OVER WDM

1 IP-over-WDM Convergence

SUDHIR DIXIT

Nokia Research Center, Burlington, Massachusetts

1.1 INTRODUCTION

In the new millennium, the telecommunications industry is abuzz with megabits per second (Mbps), gigabits per second (Gbps), terabits per second (Tbps), and so on. Those working in the forefront of the technologies involved are getting used to the power of the exponent, as major advances are happening in silicon, routing, switching, radio frequencies, and fiber transport. Whereas the bandwidth, or computing power, has been doubling per Moore's law, and price has been decreasing by half every 18 months or so, that threshold has been exceeded in the optical arena. The bandwidth distance multiplier has been doubling every year. Nowadays, 40 wavelengths, each wavelength running at OC-48 (2.5 Gbps), are common over a pair of fibers. Much newer systems being deployed are capable of carrying 160 10-Gbps channels (i.e., 1.6 Tbps) over a single pair of fibers. Systems capable of running at 80 Gbps on each channel have already been demonstrated in the laboratory environment. At this rate, 160 channels would equal 12.8 Tbps of capacity, which is much more than the entire bandwidth of the voice network in the world today. The rate of growth in fiber capacity is simply phenomenal, with plenty of room still left for growth (the theoretical limit being about 10^{15} bps or 1 Pbps from ideal physics).

A critical question arises as to what impact this capacity would have on the network infrastructure of the future [e.g., quality of service (QoS), digital loop carriers, protocol stacks, core network, metro network, and the access network]. It is important to look at why *wavelength-division multiplexing* (WDM) is so important and whether Internet protocol (IP) over WDM is here to stay. It is important to look at the historical reasons why there are currently so many layers in the network. The industry must examine whether all of those layers will be needed in the future. If a protocol stack were built from scratch today, would all those layers still be included? If not, the industry must consider how to evolve toward establishing this infrastructure and what challenges lie ahead. Issues such as QoS (including

IP over WDM: Building the Next-Generation Optical Internet, Edited by Sudhir Dixit.
ISBN 0-471-21248-2 © 2003 John Wiley & Sons, Inc.

reliability), time to provision (or to set up connections on demand), bandwidth granularity, cost, and management issues must also be considered before this vision becomes a reality. In the remainder of this book the terms *WDM* and *DWDM* (*dense WDM*) are used interchangeably and refer to the same thing.

Before we delve more into IP over WDM, it is worthwhile to review the today's infrastructure, shown in Fig. 1.1. It is usually built with three tiers of rings (local, regional, or metro) and core or backbone (long haul). The rings connect with each other via either optical cross-connects (OXCs) for high-throughput junction points or add–drop multiplexers (ADMs) for low-throughput junction points. The access network typically consists of copper drops using analog channels, xDSL, ISDN, fiber drops, or hybrid fiber coax (HFC). The OXCs and ADMs in today's architecture rely on O/E/O conversion since only the interfaces are optical, but all processing in the node is done in the electrical domain. The metro and local rings mostly use Synchronous Optical Network/Synchronous Digital Hierarchy (SONET/SDH) in the transport. The backbone rings or meshes use DWDM with multiple wavelengths in each fiber pair. The signal regeneration is also done in the electrical domain after every 200 to 500 km (depending on the type of the fiber). This architecture dates back to the time when telephony was supreme and the network was optimized for the basic granularity of 64-kbps voice channels. Present-day architecture suffers from the problems of high costs due to O/E/O conversion that needs to happen in the regenerators, OXCs, and ADMs. It also makes it difficult for the operators to upgrade the network and it takes too long to provision the circuits through multiple rings, which by and large is still a manual process.

For these reasons there is a lot of push to use mesh networks as much as possible, with IP/optical routers (with interconnecting DWDM links) setting up the lightpaths dynamically on-demand, where all signal regeneration/amplification and routing will be done in the optical domain without any O/E/O conversion. Dynamic connection setups which may encompass LSPs from the IP layer all the way down to fibers would slash provisioning times to milliseconds or seconds from days or months. The costs will be slashed by eliminating O/E/O conversions and reducing signal regenerations except in long-haul networks. The new Internet architecture of tomorrow can be illustrated as in Fig. 1.2. The key attributes of such a network can be summarized as follows:

1. DWDM will be deployed all over the infrastructure, including in the access. Passive optical networks (PONs), especially Ethernet PONs, would provide another alternative to serve lots of customers at low cost by sharing bandwidth resources. The access networks will be more ring based than point to point, which would also enhance reliability.

2. ADMs and OXCs will be replaced by optical switches and routers that would set up end-to-end lightpaths in response to client signaling. This would eliminate O/E/O conversions and expensive regeneration.

3. Since rings are inherently poor in bandwidth utilization, these will be replaced by mesh in the local, metro, and backbone regions.

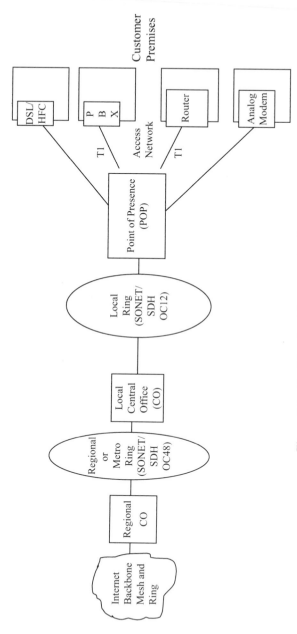

Figure 1.1 Traditional architecture as it exists today.

3

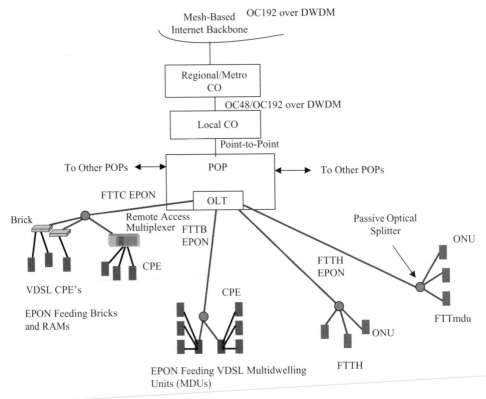

Figure 1.2   Future architecture with increasing use of DWDM and fiber in the access.

4. Gigabit Ethernet will be used increasingly in the carrier networks, replacing SONET/SDH and ATM wherever possible.
5. Most traffic will be packetized and QoS differentiation made by using the IP DiffServ (differentiated services) and MPLS (multiprotocol label switching) approaches.

Some technologies to enable the new infrastructure are already mature and deployed in a limited way, whereas others are at an early phase of development. The reader will learn more about the state of the emerging packet network and the associated technologies in the remainder of this book.

1.2   WHY IP AND WHY WDM?

A definite transition from circuit-to packet-switched networks has unquestionably occurred, and the packet mindset has already set in. In volume terms, the data traffic has already surpassed the voice traffic today, and in about five years' time, it will increase five- to tenfold. Investment in IP solutions is natural since it is the common revenue-generating integration layer that is transparent to different layer 2 and physical layer technologies. This is despite the lack of IP layer support for real-time services or the services that require QoS guarantees. While many QoS solutions for IP are still awaiting maturity and commercial success, fortunately, this problem has been alleviated to some extent, due to the rapid explosion in the availability of bandwidth that the optical networks provide. Because of this, real-time multimedia services are already becoming possible even with the best-effort Internet. In the near future we will see more deployment of fiber in the metro and access networks that will transform the Internet into a real end-to-end optical Internet, with copper, coax, or wireless in the final drop to customer premises still playing a major role.

Packet-based public networks require massive, reliable, and scalable routers, add–drop multiplexers, cross-connects, and so on. If the bandwidth capability continues to increase over fiber, the network elements noted above must be able to process the data at the line rate, preferably all optically. To realize this vision, the network architecture must be changed. The public-switched network must also be scalable and quality of service enabled. It must support network-based servers, applications, and Web farms. All this needs to happen at as least a cost as possible. Fortunately, the optical networks provide much lower cost per bit than the other nonfiber media that were used in the past. Broadband access is critical for optical backbone capacity (which is at present overbuilt) to be fully utilized and economic benefits to be derived.

Increasing the efficiency of the network will require fewer network layers and integrated support for IP. That requires developing an adaptation layer [or an IP-aware media access control (MAC) layer] that goes (as a shim) between the IP and WDM layers. The natural switching granularity in the backbone will be 2.5/ 10 Gbps. If the rate reaches 2.5 or 10 Gbps on each channel in the optical fiber, the optical transport layer should be in that range as well, since switching a packet at a lower rate is much more expensive in the optical domain than in the electrical domain. Finally, concatenated 155-Mbps, 622-Mbps, and 2.5-Gbps interfaces with lightweight SONET/SDH framing will make performance monitoring easy with standard chip sets.

1.3   WHAT DOES WDM OFFER?

WDM offers massive bandwidth multiplication in existing fibers. It solves the problem of best-effort quality of service in the short term. While QoS standards (e.g., MPLS, DiffServ) mature and are implemented in production networks, enormous bandwidth capacity already exists that could potentially solve the problem of quality of service by overprovisioning. WDM is like an expandable highway, where one can simply turn on a different color of light in the same fiber to achieve more capacity. The upgrade is much less expensive than the forklift upgrade that is necessary with a SONET/SDH solution. WDM offers a potentially low-cost solution; three times the cost will amount to 30 times the capacity. It offers a secure network, needs low power, and requires low maintenance.

WDM also provides the ability to establish lightpaths. This could lead to minimizing the number of hops in the network. Numerous applications requiring QoS would benefit from having a single hop from one edge to the other edge of the network or from one major city to another major city just by having an optical cut-through (without E/O/E conversion) lightpath. WDM also offers improved rerouting and protection switching and transparency in signals. It can carry a mix of analog and digital signals and protocols. It is protocol transparent; it does not matter what a wavelength is carrying in its payload. Although initial deployments of WDM were point to point for transport, mesh topologies are now preferred with more and more intelligence residing in the optical layer. Thus, the WDM is evolving from a dumb tactical transport to an intelligent on-demand reconfigurable strategic network layer technology.

1.4 CAPACITY, INTERFACE SPEEDS, AND PROTOCOLS

An increasing amount of data transport is in high-capacity tubes. Previously, the industry had 155-Mbps interfaces, and now the demand is for 620-Mbps interfaces. There is also a demand for OC-48 (2.5 Gbps) to the end user. The new services are increasingly based on IP, which leads to the evolution of IP-optimized high-capacity networks today. Both asynchronous transfer mode (ATM) switches and IP routers have direct SONET/SDH interfaces, so there is no need for lower-order granularity. The ATM switches use SONET/SDH as an interface because it was available at the time. The WDM in the current solution basically carries the SONET/SDH signal on a wavelength. In cost-optimized networks, transport network granularity should be in a proper relationship with network capacity. Over time, the granularity has been increasing from OC-3 to OC-12 to OC-48 (Fig. 1.3). At

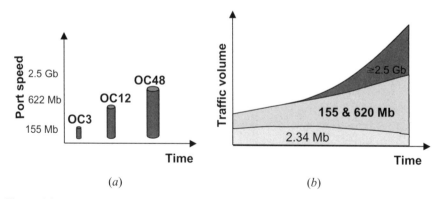

Figure 1.3 Optimization of capacity, interface speeds, and protocols: (*a*) interface speed of data services; (*b*) data volumes in different capacity levels.

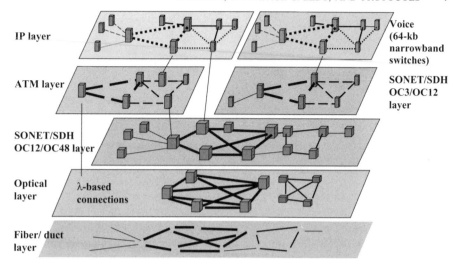

Figure 1.4   Network layers.

higher interface speeds, it becomes natural to use one of the wavelengths of the optical fiber. Market forecast shows that the traffic volume for DS1/E1 (1.5/2.0 Mbps) to DS3/E3 (45/34 Mbps) will decrease in the future, whereas the traffic over higher-speed ports will increase in the future.

Typically, the network consists of multiple layers, from the IP to the ATM to the SONET/SDH to the optical to the fiber duct (Fig. 1.4). Each of those layers offers its own flexibility, advantages, and features. The goal is to optimize the network to take advantage of the capabilities of a particular layer in a certain part of the network. For example, it does not make sense to do IP layer forwarding with traffic from New York to San Francisco at two or three routers in between. It would make much more sense to carry the traffic that is originating in New York to San Francisco in one single hop in the optical domain completely. That will provide higher performance, protocol transparency, and cost optimization.

Currently, there are too many layers in the protocol stack. IP is running over a frame relay and/or ATM layers. SONET/SDH, which is quite dominant, is also included here. Now, the emergence of WDM results in IP over ATM over WDM or IP over SDH over WDM. Figure 1.5 shows the kind of evolution the industry is experiencing today. Multiplicity of these layers is really a historical coincidence and is generally the result of the need to be backward compatible. If we were to devise the protocol layers from scratch today, many of the layers (e.g., ATM, SONET/SDH) would probably not be there. Let's discuss the various layers briefly.

IP over ATM   ATM was developed in the late 1980s, and it was fully specified by 2000. ATM envisioned that the entire infrastructure (end user to end user) would be ATM based and will support all types of traffic classes with varying QoS requirements. The fixed packet size of 53 bytes would enable fast switching and

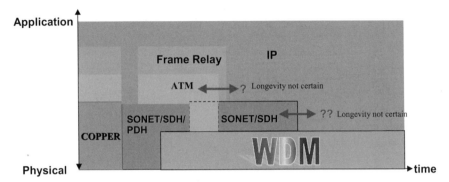

Figure 1.5 Too many layers?

would run over SONET/SDH. Initially, there was no effort to support IP. However, the huge success of the Internet changed all that, and the ATM developers began to develop IP over ATM interoperability specifications. Unfortunately, IP and ATM differ greatly since both were developed independently with different objectives and visions. For example, they use different addressing structures; IP relies on hop-by-hop routing, whereas ATM is a point-to-point solution; in IP the reliability is provided by TCP and relies on packet loss to determine the bit rate, whereas ATM proactively works to eliminate packet loss; and IP is best-effort connectionless, whereas ATM is connection oriented and capable of providing hard guarantees. So what is going to happen in the future? Although ATM has failed to reach the terminal, the service providers are continuing to deploy ATM in the backbone and IP as the service layer. ATM in the backbone allows setting up of virtual connections of varying granularity on demand. There is also a lot of interest in deploying packet (or IP) over SONET/SDH.

ATM over SONET/SDH The original vision of ATM was to provide a single intelligent packet-switched infrastructure for voice and data, but the new genera-tions of IP routers provided faster and cheaper processing at the line rate of the optical interface. The question then becomes why ATM is needed when it failed to reach the desktop. Similarly, the vision of SONET/SDH was that the world's infrastructure will be circuit switched, and therefore it was optimized for a granularity of 64 kbps. This was mainly for two reasons. First, the bit rate of uncompressed voice is 64 kbps. Second, due to the limited WAN capacity, the bandwidth was an expensive resource, which made it sensible to dole it out in smaller chunks. However, today, more bits are sent in packets than in circuits. The packet data world benefits tremendously when the pipes are bigger (which gives a higher level of statistical resource sharing) and the resources are shared. Thus, both the ATM and SONET/SDH have fallen short of reaching their intended visions. ATM switches were developed with SONET/SDH interfaces since that was the only high-speed physical-layer interface available at the time, but this role is likely to be taken over by the WDM interface.

SONET/SDH over WDM   Typically, WDM supports OC-48 (2.5 Gbps) or OC-192 (10 Gbps) in each wavelength, and in the absence of any other framing layer it has been reasonable to run SONET/SDH over WDM, but it is highly unrealistic to assume that these channels would be filled up with only the circuit-switched voice. For one thing, these channels would be OC-48c or OC-192c to avoid the unnecessary overhead, and second to provide the entire continuous channel more suited for multiplexed packetized data. SONET/SDH and WDM were designed for entirely different purposes in very different times in the history of telecommunications. It is safe to assume that the new infrastructure would be built on WDM, and it would support many of the SONET/SDH functions.

Packet over SONET/SDH   Recently, there has been a lot of interest and preference for running IP directly over SONET/SDH while eliminating completely the ATM layer in between. Running the connectionless IP over a connection-oriented SONET/SDH is rather pointless since there are no virtual tributaries (connections) carrying TDM voice where everything is in packets. Packet over SONET/SDH implementations are just point-to-point links between routers rather than ring topologies. Some operators would, however, prefer to have an intervening layer to enable some form of virtual connections [with varying levels of granularities, e.g., ATM, frame relay, PPP (point-to-point protocol) over MPLS] to make it easier for traffic engineering, network design, troubleshooting, and so on.

IP over WDM   In the future, IP-over-WDM layering will depend on the ability of the IP layer to provide quality of service. Also, it would depend on whether the WDM network would be survivable and as robust as the SONET/SDH. The rate of evolution or adoption would also depend on whether the interfaces would be inexpensive enough for the IP over WDM to be commercially attractive. ATM over WDM as backbone will depend on the overhead efficiency and the presence of native ATM applications, which is rather unlikely to happen, due to the huge success of Ethernet and IP technologies to the desktop.

1.5   WHY IP OVER WDM?

WDM exists because the fiber is in place. WDM provides a lot of bandwidth, and IP offers great convergence, which is a winning combination. IP will be in the network because it is a revenue-generating layer, whereas WDM is a cost-reducing layer that also provides tons of capacity. ATM and SONET/SDH do not add much value for data transport because SONET/SDH was designed for the circuit-switched world, where the basic bandwidth granularity was 64 kbps, when data were not as predominant as they are today. The value of ATM and SONET/SDH will also be minimized because IP will potentially deliver the quality of service. WDM will support the rich features of SONET/SDH, and nodes will be able to process information at line rates. Having said this, it must be noted that ATM, SONET/SDH, and WDM continue to get deployed in the Internet core and metro networks

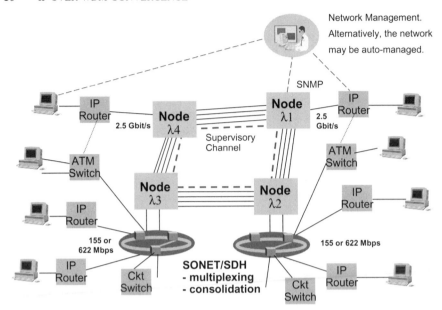

Figure 1.6   Coexistence of SONET/SDH and high-capacity IP traffic in the optical layer, where high-capacity IP routers (e.g., 2.5/10-Gbps interfaces) may connect directly with the optical layer. (*Note*: Nodes could be cross-connects, OADMs, or optical/wavelength routers/ switches.)

for reasons of various levels of granularity, protection/reliability, and capacity that they offer.

The question is not whether IP over WDM is a long-term solution, but how long it would take for it to get deployed at large scale. Closely associated with this trend is the generalized multiprotocol lable switching (GMPLS), which is emerging as a common control and signaling protocol to take care of switching and routing at the fiber, wavelength, packet, and even the slot level. While this evolution takes place, SONET/SDH and high-capacity IP traffic will coexist in the optical WDM backbone (Fig. 1.6). The SDH or SONET traffic could be multiplexed in the WDM network in individual wavelengths, so wavelengths could be added and/or dropped in the optical nodes. Similarly, an IP router with a WDM interface with tunable laser(s) could be connected directly with an optical switch or an optical router in the backbone.

Conceptually, it is quite possible to build a network node that will switch at different layers of the protocol stack. For example, at the lowest layer, one could have a fiber cross-connect. At a higher layer, it would be possible to cross-connect the wavelengths from one incoming fiber to an outgoing fiber. At an even higher layer, one could do ATM switching with the SONET/SDH interfaces. Some of the wavelengths could carry the SONET/SDH traffic, which will have ATM in the pay-loads. Another layer up could have an IP switch. This type of platform, which could switch at multiple layers of the protocol stack, can indeed be built today (Fig. 1.7).

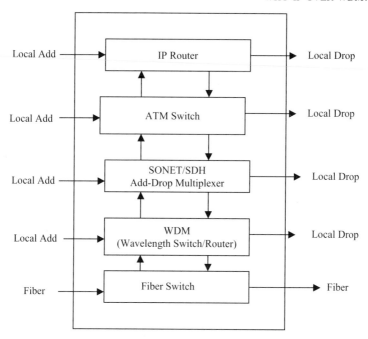

Figure 1.7 Multilayer network node. Some layers may be bypassed or omitted altogether depending on the node function and architecture.

IP over WDM involves higher-layer IP protocol integration issues. For example, one challenge is to find techniques to do wavelength routing and wavelength assignment to build an optical lightpath-based backbone networks. This may require a measurement-based approach that will relate to the open shortest-path-first (OSPF) protocols. Another consideration is lightpath assignment and routing. That is, can there be cut-through based on the lightpaths? Network reliability and protection switching are also major issues for WDM networks. If SDH/SONET will be replaced, WDM has to provide network reliability and protection switching that the carriers are accustomed to in the SONET/SDH world that has worked so well so far.

Another issue is how to access the channels. SONET/SDH can be used for channel access, but the overall goal is to minimize electrical-to-optical (E/O) and optical-to-electrical (O/E) conversion. There are a number of proposals for a generic WDM MAC, which brings about issues of framing at the WDM MAC layer. The extent of optical transparency is also still unclear. That is, the extent to which one should do O-E or E-O conversion depends on the amount of the traffic load and the ability to carry optically transparent traffic as much as possible. One thing is clear: that the O ↔ E conversions and signal regeneration or amplification due to loss of power add significantly to the network costs and reduction in overall performance.

Another issue is managed versus unmanaged network bandwidth. Would there be a need to manage the bandwidth, the optical layer, and the WDM layer? Or could it remain unmanaged because of the tremendous capacity that will be available at negligible cost per bit? Finally, a number of interoperability specifications are currently being developed by organizations such as the Optical Internetworking Forum (OIF), the Internet Engineering Task Force (IETF), and the ITU (International Telecommunications Union).

1.6 BOOK OUTLINE

The major focus of this book is on the why, how, and when of issues associated with the support of an IP-over-optical/WDM layer. To lay the foundation to address effectively the topic of IP over WDM, Chapters 1 through 9 have been devoted to reviewing the background material on the technologies involved in enabling IP over WDM. Chapters 10 through 16 specifically target the issues, requirements, solutions, and standards efforts pertaining to IP over WDM. Chapter 1 reviews the motivation for IP-over-WDM networking with a historical perspective and discusses some of the key challenges ahead to make this architecture a reality.

Chapter 2: Protocol Design Concepts, TCP/IP, and the Network Layer   This chapter covers the concept of layering in the Internet protocol stack and the position and role of the IP layer in relation to the other layers (e.g., TCP, ATM, link layer, and the physical layer) from the viewpoint of end-to-end bit transfer. The chapter focuses specifically on TCP/IP, IP over ATM, IP switching, MPLS, and Internet QoS architectures. The chapter concludes with the observation that the intelligent optical networking functions can be placed inside the Internet provided that it offers a favorable economic trade-off in achieving the basic goal of the Internet, which is end-to-end connectivity.

Chapter 3: Optical Enabling Technologies for WDM Systems   As the title implies, this chapter covers the critical technologies at the device level and the characteristics of fibers that are making the intelligent optical networking possible. The propagation issues in optical fibers, the physics and performance characteristics, and the implementation issues in such devices as WDM transmitters, filters, and amplifiers are described. This is followed by a discussion of dispersion compensation, optical attenuators, optical switching devices, and wavelength converters. Finally, how the various enabling technologies have come together to enable the complete WDM systems capable of transporting bit streams from one end of the network to the other is discussed. The chapter concludes with some projections for new developments in technologies that will potentially bring forth the tremendous improvements in fiber transmission, signal quality, switching speeds, and bandwidth capacities.

Chapter 4: Electro-optic and Wavelength Conversion   Electro-optic and wavelength conversion are the cornerstones of optical networking. Therefore, a complete chapter is devoted to this topic. The E/O/E conversion allows electronic equipment to interface with the optical components and networks, whereas the wavelength converters enable much better utilization of network, reduced blocking rate, and superior network survivability. This chapter provides an overview of the need for electro-optic and wavelength conversion, enabling technologies that have made them possible, design methods, and analytical models used in wavelength-convertible networks. Finally, the advantages of wavelength conversion are described.

Chapter 5: Contention Resolution in Optical Packet Switching   It is well known that in packet-switched networks, contention in all elements of the network is resolved by buffering in the electronic domain. In all optical networks, where packets are routed and/or switched in the optical domain, resolution for contention among packets inside the node or enroute to the same output port necessitates some type of buffering. Buffering in the optical domain is very limited and rather complex, if not nonexistent. This chapter provides an overview of the various optical contention resolution schemes in time, wavelength, and space domains. This is followed by a comparison of performance of these individual approaches and hybrid (including electro-optical) approaches. The transport control protocol (TCP) performance of the hybrid approach is also studied.

Chapter 6: Advances toward Optical Subwavelength Switching   The current optical switching/routing solutions (e.g., optical cross-connects) support only the granularity at the wavelength level unless electronic grooming switch fabrics are used. However, a future-proof network not only needs high-capacity circuit switching but also needs high-performance switching with finer granularities. Therefore, this chapter focuses on various subwavelength switching techniques, such as optical packet switching, photonic slot routing, and optical burst switching.

Chapter 7: Multiprotocol over WDM in the Access and Metropolitan Networks
Although there is ample bandwidth in the backbone of the Internet, the bottleneck is still in the access of the Internet, which is limiting the success of many multimedia applications. In fact, the backbone is already overbuilt and not being tapped to its limit, denying revenues to operators. The role and viability of multiprotocols, especially IP over WDM in the access and metropolitan networks are discussed in this chapter. After discussing the geographical coverage of carrier networks into three subdomains—access networks, metropolitan area networks (MANs), and core networks—the chapter highlights some of the key characteristics of these subdomains. A number of requirements in the access and in metro networks are pointed out, together with several broadcast and select network architectures presented to meet those requirements. Both the passive optical star and active optical ring architectures have been considered in the discussion. The chapter also delves into the two variants of the optical metropolitan rings called dynamic packet transport (DPT) and resilient packet ring (RPR). On the access side, two variants of passive optical networks (PONs) are discussed: ATM PONs or APONs and broadband PONs or BPONs. The topic of Ethernet PONs is covered in great detail in Chapter 8.

Chapter 8: Ethernet Passive Optical Networks   Ethernet passive optical networks (EPONs), an emerging local subscriber access architecture that combines low-cost point-to-multipoint fiber infrastructure with inexpensive and proven Ethernet technology, are described in this chapter. EPONs use a single trunk fiber that extends from a central office to a passive optical splitter, which then fans out to multiple optical drop fibers connected to subscriber nodes. Other than the end terminating equipment, no component in the network requires electrical power, hence the term *passive*. Local carriers have long been interested in passive optical networks for the benefits they offer: minimal fiber infrastructure and no powering requirement in the outside plant. With Ethernet now emerging as the protocol of choice for carrying IP traffic in the metro and access network, EPON has emerged as a potential optimized architecture for fiber to the building and fiber to the home. This chapter provides an overview of EPON enabling technologies, investigates EPON design challenges, and evaluates its performance.

Chapter 9: Terabit Switching and Routing Network Elements   Data-centric traffic on the Internet has increased, although circuit-switching requirements remain and will persist well into the future. To keep pace with the growing demand, transmission speeds have increased to the level of gigabits or even terabits per second. To use this huge bandwidth efficiently as well as satisfy the increasing Internet bandwidth demand, carriers are working to define scalable system architectures that are able to provide multigigabit capacity now and multiterabit capacity in the near future. Both constituencies also realize that these terabit-class switches and routers will need to do much more than pass cells and packets at significantly higher aggregate line rates. In this chapter, an overview of related activities in terabit switching and routing (TSR) is provided. Key issues and requirements for TSR networks are outlined. In addition, to understand the functionality that TSR supports, some key building blocks are described and their implementations presented. To pave the way for an all IP-based network, the implications of IP on the design of terabit switching and routing nodes are also introduced. Finally, the impact of MPLS and generalized MPLS on terabit switching and routing is discussed.

Chapter 10: Optical Network Engineering   The application of traffic engineering and planning methodologies to the design and operation of optical networks employing WDM technology is described in this chapter. The routing and wavelength assignment (RWA) problem is introduced as the fundamental control problem in WDM networks. The RWA problem is then studied in two different but complementary contexts. The static RWA arises in the network design phase, where the objective is capacity planning and the sizing of network resources. The dynamic RWA is encountered during the real-time network operation phase and involves the dynamic provisioning of optical connections. Both the static and dynamic variants of the problem are explored in depth, with an emphasis on solution approaches and algorithms. The chapter concludes with a discussion of control plane issues and related standardization activities in support of traffic engineering functionality.

Chapter 11: Traffic Management for IP-over-WDM Networks   Traffic management is an important issue in the Internet and the IP. The same applies for the WDM layer: how to manage lightpath resources at the wavelength and subwavelength level. Mapping IP over WDM brings an additional set of issues, which differ from other kinds of traffic such as voice traffic. This chapter describes why network dimensioning and provisioning deserve special treatment. This is followed by a description of the impact of IP traffic in WDM networks in first-generation (static lightpath) and second-generation (dynamic lightpath/optical burst switching) WDM networks. Concerning the former, the chapter covers the scenario and the traffic grooming problem, where the use of overflow bandwidth to absorb traffic peaks is described. Concerning second-generation networks, analyses of the trade-off between burstiness and long-range dependence and the queuing performance implications are presented. Then the signaling aspects of IP encapsulation over WDM and label switching solutions are described. Finally, the end-to-end issues in IP-over-WDM networking are discussed. Specifically, the performance of TCP (with high-speed extensions and split-TCP) in the IP-over-WDM scenario is presented.

Chapter 12: IP- and Wavelength-routing Networks   This chapter focuses mainly on IP- and wavelength-routing networks. A brief introduction to the current routing protocols (e.g., IP) for datagram networks is presented. Then wavelength-routing network architectures and routing and wavelength assignment (RWA) policy in WDM networks are presented in detail. Both the layered graph and virtual wavelength path approach are described in depth. This provides a straightforward vehicle for the design and analysis of WDM networks with and without wavelength conversion. Finally, the concepts of MPLS/GMPLS, IP-over-WDM integration, integrated routing, and waveband routing are described.

Chapter 13: Internetworking Optical Internet and Optical Burst Switching   Recently, there has been much interest in optical burst switching (OBS) since it allows for all optical transmission of packets in a bufferless optical infrastructure while offering a mechanism to support differentiated quality of service. This chapter covers the various technical issues related with the OBS. First, the OBS technology is reviewed from the protocol and architectural aspects. Then the routing and signaling interface issues between IP and OBS domains are addressed in the context of internetworking in the optical Internet. To provide differentiated service in the optical Internet, the QoS provisioning problem in the OBS domain is defined. Finally, survivability issues to ensure that the optical Internet is reliable are described.

Chapter 14: IP-over-WDM Control and Signaling   The optical layer is evolving to a fully functional optical network to support on-demand bandwidth provisioning. In this evolution, network control and signaling are essential to address the critical challenge of lightpath provisioning in real time and network survivability. First, network control in two different directions rooted from the telecommunication and Internet is introduced. Due to the consensus that IP traffic will be dominant in future data networks, this chapter focuses on IP-centric control and signaling. The MPLS control plane and its extensions to optical networks, called MPλS/GMPLS, are discussed in detail in terms of the functional requirements and issues raised by the optical network. Two families of IP signaling protocols, LDP/CR-LDP/CR-LDP GMPLS extension and RSVP/RSVP-TE/RSVP-TE GMPLS extension, are introduced and examples given to illustrate the LSP/lightpath setup. Finally, optical internetworking and signaling across the optical network boundary are examined, and a sample control plane is presented.

Chapter 15: Survivability in IP-over-WDM Networks   IP-over-WDM networks have been acknowledged extensively as next-generation network infrastructure and have to stand up to the same level of survivability as the multilayer network architecture. To understand effective survivability solutions for IP-over-WDM networks, this chapter covers network architecture, survivability strategies and algorithms, and detailed survivability capabilities of each layer. It also addresses traffic grooming problems in meshed optical network and introduces multilayer survivability mechanisms. Then fault detection and notification approaches are described, elaborating on the operational requirements of the corresponding signaling protocol and survivability techniques for future packet-based optical networks. The chapter concludes with the notion that successful deployment of survivability should consider the requirements of recovery time scale, flexibility, resource efficiency, and reliability for multiservices supported by IP-over-WDM networks.

Chapter 16: Optical Internetworking Models and Standards Directions   One of the most critical aspects of a high-bandwidth network is how to internetwork the network layer with the intelligent optical layer. In the traditional architecture, intelligence has resided primarily at the electrical layer, and the optical layer has functioned as the dumb inflexible transport. Increasingly, as time progresses, the optical layer is becoming more intelligent, allowing dynamic and automated control of the optical layer. The chapter begins with a review of the traditional networking view and introduces the concept of intelligent optical networking. A number of internetworking models that efficiently emulate the interaction between the electrical and intelligent optical layers are described. This is followed by a description of an overlay model that is commonly referred to as the user network interface (UNI) model, where a client device requests a connection from the optical layer and a description of a peer model where the electrical layer systems and the optical layer systems act as peers. Finally, the status of many optical internetworking and Ethernet standards activities, such as IETF, OIF, ITU-T, and IEEE, are described. The discussion of IEEE standards activities focuses mainly on gigabit Ethernet standards and resilient packet rings (RPRs).

1.7   CONCLUDING REMARKS

The key challenges to IP over WDM are to build multilayer, multiservice modular network nodes and to develop the supporting optical infrastructure. Interlayer communication and management are also key challenges. For example, the WDM may have a client layer that will access some of the services of the WDM layer and provide the requirements from the client layer to set up the lightpaths. The industry is already developing techniques for topology design, routing and wavelength assignment, wavelength conversion, wavelength cut-through, channel access, fast packet switching, network survivability, scheduling and buffering, and quality of service in the optical network. Another issue is managed versus unmanaged network bandwidth. Would there be a need to manage the bandwidth, the optical layer, and the WDM layer? Or could it remain unmanaged because of the tremendous capacity that will be available at a very low cost per bit? For the new vision of the optical Internet to become a reality, the cost of optical components must also decrease, which has fortunately been declining by a factor of approximately 50 to 75% every year. In addition, the standards organizations and industry forums must standardize the solutions, define the interoperability requirements, and set the vision of the emerging end-to-end optical Internet. To find answers to some of these intriguing questions, the reader is strongly encouraged to read the remainder of the book.

2 Protocol Design Concepts, TCP/IP, and the Network Layer

SHIVKUMAR KALYANARAMAN and BIPLAB SIKDAR

Rensselaer Polytechnic Institute, Troy, New York

2.1 INTRODUCTION

A central idea in the design of protocols is that of layering; and a guiding principle of Internet protocols is the end-to-end principle. In this chapter we review these ideas and describe the transport and network layers in the Internet stack. In particular, we start with notions of protocols, layering, and the end-to-end principle. We then discuss transport control protocol (TCP), Internet protocol (IP), and routing algorithms in some detail. This is followed by a discussion of asynchronous transfer mode (ATM) networks, IP over ATM, IP switching, multiprotocol label switching (MPLS), and Internet quality of service (QoS) architectures (IntServ and DiffServ). The purpose of this chapter is to give a broad overview of the internetworking context into which optical networking fits.

2.1.1 Protocols and Layering

Protocols are complex, distributed pieces of software. Abstraction and modular design are standard techniques used by software engineers to deal with complexity. By *abstraction*, we mean that a subset of functions is carefully chosen and set up as a black box or module (see Fig. 2.1). The module has an interface describing its input/output behavior. The interface outlives implementation of the module in the sense that the technology used to implement the interface may change often, but the interface tends to remain constant. Modules may be built and maintained by different entities. The software modules are then used as building blocks in a larger design. Placement of functions to design the right building blocks and interfaces is a core activity in software engineering.

Protocols have an additional constraint of being distributed. Therefore, software modules have to communicate with one or more software modules at a distance.

IP over WDM: Building the Next-Generation Optical Internet, Edited by Sudhir Dixit.
ISBN 0-471-21248-2 © 2003 John Wiley & Sons, Inc.

Figure 2.1 Abstraction of functionality into modules.

Such interfaces across a distance are termed *peer-to-peer interfaces*; and the local interfaces are termed *service interfaces* (Fig. 2.2). Since protocol function naturally tends to be a sequence of functions, the modules on each end are organized as a (vertical) sequence called *layers*. The set of modules organized as layers is also commonly called a *protocol stack*. The concept of layering is illustrated in Fig. 2.3.

Over the years, some layered models have been standardized. The ISO Open Systems Interconnection (ISO/OSI) layered model has seven layers and was developed by a set of committees under the auspices of the International Standards Organization (ISO). The TCP/IP has a four-layered stack that has become a de facto standard of the Internet. The TCP/IP and ISO/OSI stacks and their rough correspondences are illustrated in Fig. 2.4. The particular protocols of the TCP/IP stack are also shown.

The physical layer deals with getting viable bit transmission out of the underlying medium (fiber, copper, coax cable, air, etc.). The data-link layer then converts this bit transmission into a framed transmission on the link. *Frames* are used to

Figure 2.2 Communicating software modules.

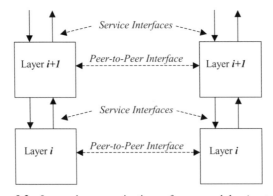

Figure 2.3 Layered communicating software modules (protocols).

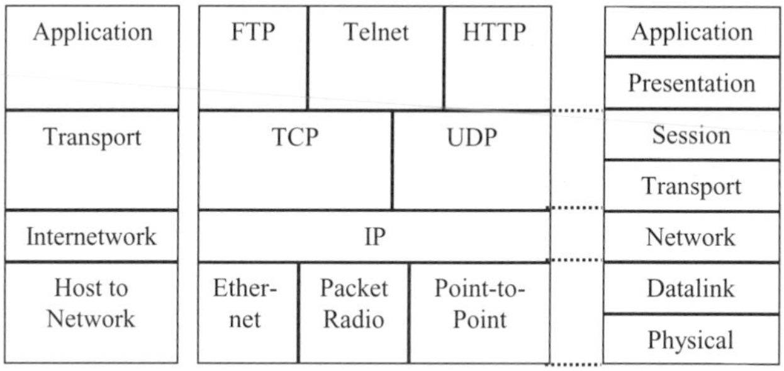

Application	FTP	Telnet	HTTP	Application
				Presentation
Transport	TCP		UDP	Session
				Transport
Internetwork	IP			Network
Host to Network	Ether-net	Packet Radio	Point-to-Point	Datalink
				Physical

Figure 2.4 TCP/IP versus ISO/OSI protocol stack.

allow multiplexing of several streams and define the unit of transmission used in error detection and flow control. In the event that the link is actually shared (i.e., multiple access), the data link layer defines the medium access control (MAC) protocol as well. The network layer deals with packet transmission across multiple links from the source node to the destination node. The functions here include routing, signaling, and mechanisms to deal with the heterogeneous link layers at each link.

The transport layer protocol provides end-to-end communication services (i.e., it allows applications to multiplex the network service and may add other capabilities, such as connection setup, reliability, and flow/congestion control). Examples of communication abstractions provided by the transport layer include a reliable byte-stream service (TCP) and an unreliable datagram service (UDP). These abstractions are made available through application-level programming interfaces (APIs) such as the BSD socket interface. The application layers (session, presentation, application) then use the communication abstractions provided by the transport layer to create the basis for such interesting applications as email, Web, file transfer, multimedia conference, and peer-to-peer applications. Examples of such protocols include SMTP, HTTP, DNS, H.323, and SIP.

2.1.2 Internet Protocol Design: The End-to-End Principle

A key principle used in the design of TCP/IP protocols is the *end-to-end* (E2E) *principle* [4], which guides the placement of functionality in a complex distributed system. The principle suggests that "functions placed at the lower levels may be redundant or of little value when compared to the cost of providing them at the lower level. . . ." In other words, a system (or subsystem level) should consider only functions that can be implemented within it *completely* and *correctly*. All other functions are best moved to the system level, where they can be implemented completely and correctly.

In the context of the Internet, the implication is that several functions, such as reliability, congestion control, and session/connection management, are best moved to the end systems (i.e., performed on an end-to-end basis), and the network layer focuses on functions that it can fully implement (i.e., routing and datagram delivery). As a result, the end systems are intelligent and in control of the communication while the forwarding aspects of the network are kept simple. This leads to a philosophy diametrically opposite to the telephone world which sports dumb end systems (the telephone) and intelligent networks.

The end-to-end principle argues that even if the network layer did provide the functions in question (i.e., connection management and reliability), transport levels would have to add reliability to account for the interaction at the transport-network boundary or if the transport needs more reliability than the network provides. Removing these concerns from the lower-layer packet-forwarding devices stream-lines the forwarding process, contributing to systemwide efficiency and lower costs. In other words, the costs of providing the "incomplete" function at the network layers would arguably outweigh the benefits.

It should be noted that the end-to-end principle emphasizes function placement vis-à-vis correctness, completeness, and overall system costs. The argument does say that "sometimes an incomplete version of the function provided by the communication system may be useful as a performance enhancement. . . ." In other words, the principle does allow a cost–performance trade-off, and incorporation of economic concerns. However, it cautions that the choice of such "incomplete versions of functions" to be placed inside the network should be made very prudently.

One issue regarding the incomplete network-level function is the degree of state maintained inside the network. Lack of state removes any requirement for the network nodes to notify each other as endpoint connections are formed or dropped. Furthermore, the endpoints are not, and need not be, aware of any network components other than the destination, first-hop router(s), and an optional name resolution service. Packet integrity is preserved through the network, and transport checksums and any address-dependent security functions are valid end to end. If the state is maintained only in the endpoints, in such a way that the state can only be destroyed when the endpoint itself breaks (also termed *fate sharing*), then as networks grow in size, the likelihood of component failures affecting a connection becomes increasingly frequent. If failures lead to loss of communication, because the key state is lost, the network becomes increasingly brittle, and its utility degrades. However, if an endpoint itself fails, there is no hope of subsequent communication anyway. Therefore, one interpretation of the end-to-end model is that only the endpoints should hold the critical state.

In the situation of ISPs providing quality of service (QoS) and charging for it, the trust boundary has to be placed inside the network, not in end systems. In other words, some part of the network has to participate in decisions of resource sharing, and billing, which cannot be entrusted to end systems. A correct application of the end-to-end principle in this scenario would allow placement of these functions in the network, in consideration of the economic and trust model issues. Applications may be allowed to participate in the decision process, but the control belongs to the network, not the end system, in this matter. The differentiated services architecture discussed later in the chapter has the notion of the network edge, which is the repository of these functions.

In summary, the end-to-end principle has guided a vast majority of function placement decisions in the Internet and it remains relevant today even as the design decisions are intertwined with complex economic concerns of multiple ISPs and vendors. A recent paper examines the end-to-end principle in today's complex networking design context [2].

2.2   TRANSPORT LAYER AND TCP*

The transport layer provides a mechanism for end-to-end communication between various applications running on different hosts. The transport layer allows different applications to communicate over heterogeneous networks without having to worry about different network interfaces, technologies, and so on, and isolates the applications from the details of the actual network characteristics. The transport layer offers some typical *services* to its upper layers. It is important to understand the difference between a *service* and a *protocol* that may offer these services. A transport layer service refers to the set of functions that are provided to the upper layers by the transport layer. A protocol, on the other hand, refers to the details of how the transport layers at the sender and receiver interact to provide these services. We now take a look at the services provided by the transport layer.

2.2.1   Service Models at the Transport Layer

There are two basic services provided by the transport layer: connection oriented and connectionless. A *connection-oriented service* has three phases of operation: connection establishment, data transfer, and connection termination. When a host wants to send some data to another user by means of a connection-oriented service, the sending host's transport layer first explicitly sets up a connection with the transport layer at the other end. Once the connection is established, the hosts exchange the data and at the end of the transfer the sender's transport layer explicitly asks the transport layer of the other end to terminate the connection. While all connection-oriented services go through these three phases, they have two variations, depending on how the data given by the application layer is broken down by the transport layer before transmission: message oriented and byte stream. In *message-oriented services*, the sender's messages have a specified maximum size and the message boundaries are preserved. For example, if the sender gets two 1-kB messages to be transmitted, they are delivered as the same two distinct 1-kB messages. The transport layer will not combine them into one 2-kB or four 500-byte messages. In the *byte-stream service* mode, the data from the sender are viewed as an unstructured sequence of bytes that are transmitted in the same order as they arrive. The data in this case are not treated as messages, and any data given to the transport layer is appended to the end of the byte stream. One example of a connection-oriented service is TCP, which we look at in detail in Section 2.2.3.

*Some material in this section is based on [5], [7], and [14].

A *connectionless service*, on the other hand, has only one phase of operation, data transfer, and there is no connection establishment and termination phase. All data to be transmitted are given directly to the transport layer, and a message-oriented service model is used to transfer the packets. The user datagram protocol (UDP) is the most commonly used connectionless transport layer service, and we take a detailed look at it in Section 2.2.2.

2.2.1.1 Reliable and Unreliable Transport

Before we move on to a more detailed study of TCP and UDP, we first discuss briefly the reliability of data transferred by the transport protocol. Reliable data transfer involves four different features: no loss, no duplicates, ordered form, and data integrity. A no-loss service guarantees that either all the data are delivered to the receiver or the sender is notified in case some of the data are lost. It ensures that the sender is never under the misconception of having delivered its data to the receiver when in fact it was not. TCP provides a lossless delivery mechanism, whereas UDP does not guarantee a lossless operation. Any data submitted to an UDP sender may fail to get delivered without the sender being informed.

A no-duplicates service (e.g., TCP) guarantees that all the data given to the sender's transport layer will be delivered at the receiver at most once. Any duplicate packets that arrive at the receiver will be discarded. On the other hand, UDP does not provide this guarantee, and duplicates may occur at the receiver. The third property of a reliable transport protocol is ordered data transport. An ordered service such as TCP delivers data to the receiver in the same order in which data were submitted at the sender. In case data packets arrive at the receiver out of order due to network conditions, the transport layer at the receiver sorts them in the right order before giving them to the upper layers. An unordered service such as UDP, on the other hand, may try to preserve the order but does not make any guarantees.

Finally, data integrity implies that the data bits delivered to the receiver are identical to those submitted at the sender. Both TCP and UDP provide integrity to the data they transfer by including error-detection mechanisms using coding mechanisms. Note that the four aspects of reliability are orthogonal in the sense that ensuring one does not imply the presence of any of the other three functions. We take a further look at these features and how they are achieved in TCP and UDP in the next section.

2.2.1.2 Multiplexing and Demultiplexing

Another critical set of services that are provided by the transport layer is that of application multiplexing and demultiplexing. This feature allows multiple applications to use the network simultaneously and ensure that the transport layer can differentiate the data it receives from the lower layer according to the application or process to which the data belong. To achieve this, the data segment headers at the transport layer have a set of fields to determine the process to which the segment's data are to be delivered. At the receiver, these fields are examined to determine the process to which the data segment belongs, and the segment is then directed to that process. This functionality at the receiver's transport layer that delivers the data it receives to the correct

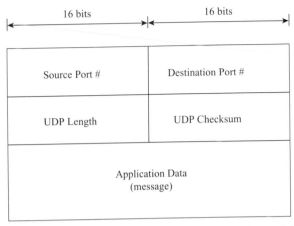

Figure 2.5   Source and destination port numbers from multiplexing and demultiplexing.

application process is called *demultiplexing*. The process at the sender where information about the various active processes is collected and the corresponding information is incorporated in the segment headers to be given to the lower layers is called *multiplexing*.

Both TCP and UDP provide multiplexing and demultiplexing by using two special fields in the segment headers: *source port number field* and the *destination port number field*. These two fields are shown in Fig. 2.5. Each application running at the host is given a unique port number and the range of the port numbers is between 0 and 65535. When taken together, the sources and destination port fields uniquely identify the process running on the destination host.

Having taken a look at the services provided by the transport layer and the differences in there functionalities, we now move on to take a more detailed look at two of the most widely used transport layer protocols: UDP and TCP. We divide the discussion on these two protocols in terms of their connectionless and connection-oriented nature, respectively.

2.2.2   UDP and Connectionless Transport

UDP is the most basic transport layer protocol and does not provide any features other than multiplexing/demultiplexing and data integrity services. Being a connectionless service, a data transfer using UDP does not involve connection setup or teardown procedures, nor does it provide reliability, ordered delivery, or protection against duplication. UDP takes the messages from the upper layers, attaches the source and destination port numbers along with some other fields to the header, and passes the resulting segment to the network layer.

Why is UDP an interesting protocol? Even though it does not provide reliability or in-order delivery, it is a very lightweight protocol, enhancing its suitability for various applications. First, the absence of the connection establishment phase allows UDP to start transferring data without delays. This makes it suitable for applications that transfer of very small amounts of data, such as DNS, and the delay in connection establishment may be significant compared to the time taken for data transfer. Also, unlike TCP, UDP does not maintain any information on connection state parameters such as send and receive buffers, congestion control parameters, and sequence and acknowledgment number. This allows UDP to support many more active clients for the same application as does TCP. Also, while TCP tries to throttle its rate according to its congestion control mechanism in the presence of losses, the rate at which UDP transmits data is limited only by the speed at which it gets the data from the application. The rate control in TCP flows can have a significant impact on the quality of real-time applications, while UDP avoids such problems, and consequently, UDP is used in many of multimedia applications, such as Internet telephony, streaming audio and video, and conferencing.

Note that the fact that UDP does not react to congestion in the network is not always a desirable feature. Congestion control is needed in the network to prevent it from entering a stage in which very few packets get to their destinations and to limit packet delays. To address this problem, researchers have proposed UDP sources with congestion control mechanisms [16,17]. Also, note that although UDP itself does not provide reliability, the application itself may have reliability. In fact, many current streaming applications have some form of reliability in their application layers, while they use UDP for the transport layer.

2.2.2.1 *UDP Segment Structure*

The UDP segment structure along with the various fields in its header is shown in Fig. 2.5. In addition to the port numbers described earlier, the UDP header has two additional fields: length and checksum. The *length field* specifies the length of the UDP segment, including the headers in bytes. The *checksum* is used at the receiving host to check if any of the bits were corrupted during transmission and covers both the headers and the data.

The checksum is calculated at the sender by computing the 1's complement of the sum of all the 16-bit words in the segment. The result of this calculation is again a 16-bit number that is put in the checksum field of the UDP segment. At the receiver, the checksum is added to all the 16-bit words in the segment. If no errors have been introduced in the segment during transmission, the result should be 1111111111111111, and if any one of the bits is a zero, an error is detected. If on receipt of a segment the receiver detects a checksum error, the segment is discarded and no error messages are generated. With this we conclude our discussion of UDP and move on to TCP and reliable connection-oriented transport protocols.

2.2.3 TCP and Connection-Oriented Transport

Although TCP provides multiplexing/demultiplexing and error detection using means similar to UDP, the fundamental differences in them lies in the fact that TCP is connection oriented and reliable. The connection-oriented nature of TCP

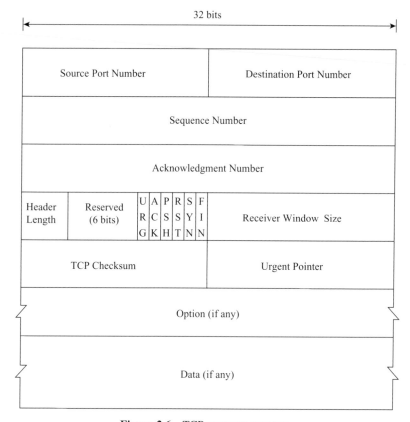

Figure 2.6 TCP segment structure.

implies that before a host can start sending data to another host, it first has to set up a connection using a handshaking mechanism. Also, during the connection setup phase, both hosts initialize many TCP *state variables* associated with the TCP connection. We now look at TCP in detail, beginning with its segment structure.

2.2.3.1 TCP Segment Structure   Before going into the details of TCP operation, we look at TCP's segment structure, details of which are shown in Fig. 2.6. In contrast to the 8-byte header of UDP, TCP uses a header that is typically 20 bytes long. The source and destination port numbers are each 2 bytes long and are similar to those used in UDP. The *sequence number* and *acknowledgment number* fields are used by the TCP sender and receiver to implement reliable transport services. The sequence number identifies the byte in the stream of data from the sender that the first byte of this segment represents. The sequence number is a 32-bit unsigned number which wraps back to 0 after reaching $2^{32}-1$. Since TCP is a byte-stream protocol and every byte is numbered, the acknowledgment number represents the next sequence number that the receiver is expecting. The sequence number field is initialized during the initial connection establishment, and this initial value is called the *initial sequence number*. The 4-bit *header length* field gives the length of the header in terms of 32-bit words. This is required since TCP can have a variable-length options field. The use of 4 bits for the header length limits the maximum size of a TCP header to 60 bytes. The next 6 bits in the header are reserved for future use, and the next 6 bits constitute the *flag field*. The *ACK* bit is used to indicate that the current value in the acknowledgment field is valid. The *SYN* bit is set to indicate that the sender of the segment wishes to set up a connection, and the *FIN* bit is set to indicate that the sender is finished sending data. The *RST* bit is set in case the sender wants to reset the connection and the use of the *PSH* bit indicates that the receiver should pass the data to the upper layer immediately. Finally, the *URG* bit is used to indicate that there are some data in the segment that the sender's upper layers have deemed to be urgent. The location of the last byte of this urgent data is indicated by the 16-bit *urgent pointer* field. Note that the contents of this field are valid only if the URG bit is set. The 16-bit *receiver window size* is used for flow control and indicates the number of bytes that the receiver is willing to accept. This field is provided to limit the rate at which the sender can transmit data so that it does not overwhelm a receiver with slower speed or link rates. The *checksum* field is similar to that in UDP and is used to preserve the integrity of the segment.

2.2.3.2   TCP Connection Establishment and Termination   A TCP connection is set up using a three-way handshake. Figure 2.7 shows the details of this handshake. For a connection to be established, the requesting end (also called a *client*) sends a

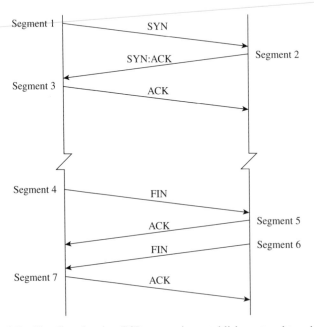

Figure 2.7   Timeline showing TCP connection establishment and termination.

SYN segment specifying the port number of the receiver (also called the *server*) that it wants to connect to along with the initial sequence number. In the second step, the server responds to this message with its own SYN segment containing the server's initial sequence number. In addition, the server acknowledges the client's SYN in the same segment by sending an acknowledgment for the client's initial sequence number plus one. In the third step, the client acknowledges the SYN from the server by sending an acknowledgment for the server's initial sequence number plus one. The exchange of these three segments constitutes the three-way hand-shake and completes the connection establishment.

While the connection establishment phase of TCP involves the exchange of three segments, it takes four segments for connection termination. This is because a TCP connection is full duplex (i.e., data can flow in either direction independently of the data in the other direction). When either side is done with sending its data, it sends a FIN packet to indicate that it wants the connection to be terminated. On receiving a FIN segment, TCP notifies the application that there are no more data to come and sends back an acknowledgment packet for the received sequence number plus one. The receipt of a FIN implies only that there will be no more data in that direction. TCP can still send data in the other direction after receiving a FIN. Once the other side is also done with sending its data, it sends a FIN segment of its own, to which the other side responds with an acknowledgment, thereby formally closing the connection in both directions. In TCP parlance, the host that sends the first FIN performs the *active close* and the other host performs the *passive close*. The segments exchanged during TCP connection termination are shown in Fig. 2.7.

2.2.3.3   *Reliability of TCP*   Apart for the fact that TCP is connection oriented, that main difference that sets it apart from UDP flows is that it is reliable. TCP provides reliability to the data it transfers, in the following way. Before looking at TCP's reliability in detail, we look at how TCP uses the sequence and acknowl-edgment numbers since that will be crucial to understanding TCP's reliability.

TCP breaks up the data passed on to it by the application layer into what it considers the best-sized chunks or segments. Note that this is different from UDP, where the datagram size is determined by the application. TCP views the data passed on to it by the upper layers as an ordered stream of bytes and uses the sequence numbers to count the number of bytes it transmits rather than the number of segments. As mentioned before, the sequence number of a TCP segment is the byte-stream number of the first byte in the segment. While sending the data it receives from the upper layers, TCP inherently counts the sequence number in terms of bytes. On receipt of a segment, TCP sends out an acknowledgment informing the sender about the amount of data correctly received and the sequence number of the next byte of data that it is expecting. For example, if the receiver gets a segment with the bytes 0 to 1023, it is now expecting the sender to give it byte number 1024 next. The segment that the receiver sends in response to the receipt of bytes 0 to 1023 thus contains 1024 in the acknowledgment number field. It is important to note here the response of TCP when it receives an out-of-order segment. Consider again the case where the receiver gets a segment containing bytes 0 to 1023. Now, for some reason, the next segment sent by the sender (which contained bytes 1024 to 2047) is lost and the receiver instead gets a segment with bytes 2048 to 3071. Since the receiver is still waiting for byte number 1024, the acknowledgment sent in response to this segment again has 1024 in the acknowledgment number field. Since TCP only acknowledges bytes up to the first missing byte in the stream, TCP is said to provide *cumulative acknowledgments*. Note that on the receipt of out-of-order segments, TCP has two options: (1) immediately discard out-of-order segments, or (2) keep them in the buffer and wait until the missing bytes turn up before giving the bytes to the upper layers. In practice, the latter option is usually implemented. Finally, in full-duplex operation, acknowledgments for client-to-server data can be carried in a segment server to client data instead of having a separate segment being sent just for the acknowledgment. Such acknowledgments are termed *piggybacked acknowledgments*.

Let us now see how TCP provides reliability and uses sequence and acknowledgment numbers for this purpose. When TCP sends the segment, it maintains a timer and waits for the receiver to send an acknowledgment on receipt of the packet. When TCP receives data from the other end of the connection, it sends an acknowledgment. This acknowledgment need not be sent immediately; usually, there is a delay of a fraction of a second before it is sent. If an acknowledgment is not received by the sender before its timer expires, the segment is retransmitted. This mode of error recovery, which is triggered by a timer expiration, is called a *timeout*.

Another way in which TCP can detect losses during transmission is through *duplicate acknowledgments*. Duplicate acknowledgments arise due to the cumulative acknowledgment mechanism of TCP wherein if segments are received out of order, TCP sends an acknowledgment for the next byte of data that it is expecting. Duplicate acknowledgments refer to those segments that reacknowledge a segment for which the sender received an earlier acknowledgment. Now let us see how TCP can use duplicate acknowledgments to detect losses. Note that in the case of a first-time acknowledgment, the sender knows that all the data sent so far have been received correctly at the receiver. When TCP receives a segment with a sequence number greater than the next-in-order packet that is expected, it detects a gap in the data stream or, in other words, a missing segment. Since TCP uses only cumulative acknowledgments, it cannot send explicit negative acknowledgments informing the sender of the lost packet. Instead, the receiver simply sends a duplicate acknowledgment reacknowledging the last-in-order byte of data that it has received. If the TCP sender receives three duplicate acknowledgments for the same data, it takes this as an indication that the segment following the segment that has been acknowledged three times has been lost. In this case the sender now retransmits the missing segment without waiting for its timer to expire. This mode of loss recovery is called *fast retransmit*.

2.2.3.4   TCP Congestion Control   TCP has in-built controls which allow it to modulate its transmission rate according to what it perceives to be the current network capacity. This feature of end-to-end congestion control involving the source and destination is provided since TCP cannot rely on network-assisted congestion control. Since the IP layer does not provide any explicit information to TCP about the congestion on the network, it has to rely on its stream of acknowledgment packets and the packet losses it sees to infer congestion in the network and take preventive action. We now take a more detailed look at the TCP congestion control mechanism.

The congestion control mechanism of TCP is based on limiting the number of packets that it is allowed to transmit at a time without waiting for an acknowledgment for any of these packets. The number of these unacknowledged packets is usually called TCP *window size*. TCP's congestion control mechanism works by trying to modulate window size as a function of the congestion that it sees in the network. However, when a TCP source starts transmitting, it has no knowledge of the congestion in the network. Thus it starts with a conservative, small window and begins a process of probing the network to find its capacity. As the source keeps getting acknowledgments for the packets it sends, it keeps increasing its window size and sends more and more packets into the network. This process of increasing the window continues until the source experiences a loss that serves as an indication that at this rate of packet transmission, the network is becoming congested and it is prudent to reduce the window size. The TCP source then reduces its window and again begins the process of inflating its window to slowly use up the bandwidth available in the network. This process forms the basis of TCP's congestion control mechanism. We now look at the details of the increase/decrease process for the window size.

To implement TCP's congestion control mechanism, the source and destinations have to maintain some additional variables: the *congestion window*, cwnd, and the *threshold*, ssthresh. The congestion window, together with the window size advertised by the receiver in the "receiver window size" field (which we denote rwin) in the TCP header, serve to limit the amount of unacknowledged data that the sender can transmit at any given point in time. At any instant, the total unacknowledged data must be less than the minimum of cwnd and rwin. While rwin is advertised by the receiver and is determined only by factors at the receiver such as processor speed and available buffer space, the variations in cwnd are more dynamic and affected primarily by the network conditions and the losses the sender experiences. The variable ssthresh also controls the way in which cwnd increases or decreases.

Usually, while controlling the rate at which a TCP source transmits, rwin plays a very minor role and is not considered a part of the congestion control mechanism since it is controlled solely by the conditions at the receiver, not by network congestion. To get a detailed look at how TCP's congestion control works, we consider a source which has data to transfer and takes data from the application, breaks them up into segments and transmits them. For the time being, forget about the size limitation due to rwin. The size of the packets is determined by the maximum allowable length for a segment on the path followed by the segment, called the *maximum segment size* (MSS). Being a byte-stream-oriented protocol, TCP maintains its cwnd in bytes and once the connection establishment phase is over, it begins transmitting data starting with an initial cwnd of one MSS. Also, while the connection is set up, ssthresh is initialized to a default value (65,535 bytes). When the segment is transmitted, the sender sets the retransmission timer for this packet. If an acknowledgment for this segment is received before the timer expires, TCP takes that as an indication that the network has the capacity to transmit more than one segment without any losses (i.e., congestion) and thus increases its cwnd by one MSS, resulting in cwnd = 2MSS. The sender now transmits two segments, and if these two packets are acknowledged before their retransmission timer expires, cwnd is increased by one MSS for each acknowledgment received, resulting in a cwnd of 4MSS. The sender now transmits four segments and the reception of the four acknowledgments from these segments leads to a cwnd of 8MSS. This process of increasing the cwnd by one MSS for every acknowledgment received results in an exponential increase in the cwnd, and this process continues until cwnd < ssthresh, and the acknowledgments keep arriving before their corresponding retransmission timer expires. This phase of TCP transmission is termed the *slow-start phase* since it begins with a small initial window size and then grows bigger.

Once cwnd reaches ssthresh, the slow-start phase ends and now, instead of increasing the window by one MSS for every acknowledgment received, cwnd is increased by 1/cwnd every time an acknowledgment is received. Consequently, the window increases by one MSS when the acknowledgments for all the unacknowledged packets arrive. This results in a linear increase of the cwnd as compared to the exponential increase during slow start, and this phase is called the *congestion avoidance phase*. This form of additive increase results in cwnd increasing by one MSS every round-trip time regardless of the number of acknowledgments received in that period. An example of the slow-start and congestion avoidance phases from a TCP flow is shown in Fig. 2.8, where ssthresh was assumed to be 16 segments. Thus the exponential increase in the slow-start phase continues until cwnd becomes

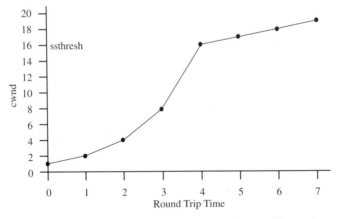

Figure 2.8   Example of the slow-start and congestion avoidance phases. (From [14].)

16 MSS and then enters the congestion avoidance phase, where `cwnd` increases linearly.

On the successful reception of acknowledgments, the window keeps increasing whether the flow is in the slow-start or congestion avoidance phase. While the window increases, the sender puts more and more packets into the network and at some point exceeds the network's capacity, leading to a loss. When a loss is detected, `ssthresh` is set to max{`cwnd`/2,2MSS} (i.e., set to half of the current value of `cwnd` or 2MSS, whichever is higher). Also, `cwnd` is reduced to a smaller value; the exact value to which it is set depends on the version of TCP. In older versions of TCP (called TCP Tahoe), on the detection of a loss (whether through timeout or through three duplicate acknowledgments), `cwnd` is set to 1 MSS and the sender goes into slow start. In the case of TCP Reno, a more recent version of TCP and the one most commonly implemented versions of TCP today, depending on whether we have a timeout or three duplicate acknowledgments, `cwnd` is set to different values. In the case of a timeout, TCP Reno sets `cwnd` to 1 MSS and begins slow start again. In case the loss is detected through three duplicate acknowledgments, Reno does a fast retransmit and goes into what is known as *fast recovery*. When we have a fast retransmit, TCP Reno sets its `cwnd` to half its current value (i.e., `cwnd`/2) and instead of doing a slow start, goes into congestion avoidance directly. In both TCP Tahoe and Reno, however, detection of a loss by either means results in `ssthresh` to be set at max {`cwnd`/2,2MSS}. Note that now if the flow goes into slow start, it enters the congestion avoidance phase much earlier than before since `ssthresh` has now been halved. This reduction of TCP's transmission rate in the presence of losses and its attempt to modulate its window increase pattern based on network congestion forms TCP's congestion avoidance mechanism. For more details on how TCP operates, the reader is referred to [14].

2.3 NETWORK LAYER

The network layer in the TCP/IP stack deals with internetworking and routing. The core problems of internetworking are heterogeneity and scale. *Heterogeneity* is the problem of dealing with disparate layer 2 networks to create a workable forwarding and addressing paradigm and the problem of providing meaningful service to a range of disparate applications. *Scale* is the problem of allowing the Internet to grow without bounds to meet its intended user demands. The Internet design applies the end-to-end principle to deal with these problems.

2.3.1 Network Service Models

One way of dealing with heterogeneity is to provide translation services between the heterogeneous entities when forwarding across them is desired. Examples of such design include multiprotocol bridges and multiprotocol routers. But this gets too complicated and does not allow scaling because every new entity that wishes to join the Internet will require changes in all existing infrastructure. A preferable requirement is to be able to upgrade the network incrementally. In an alternative strategy called an *overlay model*, a new protocol (IP), with its own packet format and address space, is developed, and the mapping is done between all protocols and this intermediate protocol.

By necessity, IP has to be simple, so that the mapping between IP and lower-layer protocols is simplified. As a result, IP opts for a best-effort, unreliable data-gram service model, where it forwards datagrams between sources and destinations situated on, and separated by, a set of disparate networks. IP expects a minimal link-level frame forwarding service from lower layers. The mapping between IP and lower layers involves address mapping issues (e.g., address resolution) and packet format mapping issues (e.g., fragmentation/reassembly). Experience has shown that this mapping is straightforward in many subnetworks, especially those that are not too large and those that support broadcasting at the LAN level. Address resolution can be a complex problem on nonbroadcast multiple-access (NBMA) subnetworks; and the control protocols associated with IP (especially BGP routing) can place other requirements on large subnetworks (e.g., ATM networks) which make the mapping problems difficult. Hybrid technologies such as MPLS are used to address these mapping concerns and to enable new traffic engineering capabil-ities in core networks.

For several applications, it turns out that the simple best-effort service provided by IP can be augmented with end-to-end transport protocols such as TCP, UDP, and RTP to be sufficient. Other applications having stringent performance expectations (e.g., telephony) need either to adapt and/or use augmented QoS capabilities from the network. Although several mechanisms and protocols for this have been developed in the last decade, a fully QoS-capable Internet is still a holy grail for the Internet community. The difficult problems surround routing, interdomain/multiprovider issues, and the implications of QoS on a range of functions (routing, forwarding, scheduling, signaling, application adaptation, etc.).

In summary, the best-effort, overlay model of IP has proved to be enormously successful, it has faced problems in being mapped to large NBMA subnetworks, and continues to faces challenges in the interdomain/multiprovider and QoS areas.

2.3.2   Internet Protocol: Forwarding Paradigm

The core service provided by IP is datagram forwarding over disparate networks. This itself is a nontrivial problem. The end result of this forwarding service is to provide connectivity. The two broad approaches to getting connectivity are: direct connectivity and indirect connectivity. *Direct connectivity* refers to the case where the destination is only a single link away (this includes shared and unshared media). *Indirect connectivity* refers to connectivity achieved by going through intermediate components or intermediate networks. The intermediate components (bridges, switches, routers, network address translator (NAT) boxes, etc.) are dedicated to functions to deal with the problem of scale and/or heterogeneity. Indeed, the function of providing indirect connectivity through intermediate networks can be thought of as a design of a large virtual intermediate component, the Internet. These different forms of connectivity are shown in Figs. 2.9 to 2.11.

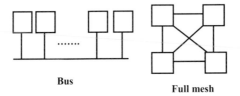

Bus

Full mesh

Figure 2.9   Direct connectivity architectures.

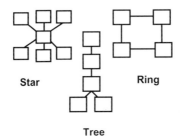

Star

Ring

Tree

Figure 2.10   Indirect connectivity through intermediate components.

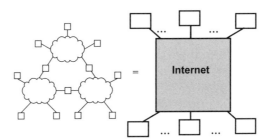

Figure 2.11   Indirect connectivity through intermediate neworks and components.

The problem of scaling with respect to a parameter (e.g., number of nodes) is inversely related to the efficiency characteristics of the architecture with respect to the same parameter. For example, direct connectivity architectures do not scale because of the finite capacity of a shared medium, finite interface slots, or the high cost of provisioning a full mesh of links. A way to deal with this is to build a switched network, where the intermediate components (*switches*) provide filtering and forwarding capabilities to isolate multiple networks to keep them within their scaling limits while providing scalable interconnection. In general, the more efficient the filtering and forwarding of these components, the more scalable is the architecture. Layer 1 hubs do pure broadcast, and hence no filtering, but can forward signals. Layer 2 bridges and switches can filter to an extent using forwarding tables learned by snooping; but their default to flooding on a spanning tree when the forwarding table does not contain the address of the receiver. This default behavior of flooding or broadcast is inefficient, and hence limits scalability. This behavior is also partially a result of the flat addressing structure used by L2 networks.

In contrast, layer 3 (IP) switches (i.e., routers) *never* broadcast across subnetworks and rely on a set of routing protocols and a concatenated set of local forwarding decisions to deliver packets across the Internet. IP addressing is designed hierarchically, and address assignment is coordinated with routing design. This enables intermediate node (or hosts) to do a simple determination: whether the destination is connected *directly* or *indirectly*. In the former case, simple layer 2 forwarding is invoked; and in the latter case, a layer 3 forwarding decision is made to determine the next hop, an intermediate node on the same subnetwork, and then the layer 2 forwarding is invoked.

Heterogeneity is supported by IP because it invokes only a minimal forwarding service of the underlying L2 protocol. Before invoking this L2 forwarding service, the router has to (1) determine the L2 address of the destination (or next hop), an address resolution problem; and (2) map the datagram to the underlying L2 frame format. If the datagram is too large, it has to do something: fragmentation/reassembly. IP does not expect any other special feature in lower layers and hence can work over a range of L2 protocols.

In summary, the IP forwarding paradigm naturally comes out of the notions of direct and indirect connectivity. The "secret sauce" is in the way addressing is designed to enable the directly/indirectly reachable query and in the scalable design of routing protocols to aid determination of the appropriate next hop if the destination is connected indirectly. Heterogeneity leads to mapping issues, which are simplified because of the minimalist expectations of IP from its lower layers (only an forwarding capability expected). All other details of lower layers are abstracted out.

2.3.3   Internet Protocol: Packet Format, Addressing, and Fragmentation/Reassembly

In this section we explore the design ideas in IP.

2.3.3.1   IP Packet Format   The IP packet format is shown in Fig. 2.12. The biggest fields in the header are the source and destination 32-bit IP *address* fields. The second 32-bit line (*ID, flags, fragment offset*), related to fragmentation/reassembly, is explained later. The length field, which indicates the length of the entire datagram, is required because IP accepts variable-length payloads. The *checksum* field covers only the header, not the payload, and is used to catch any header errors to avoid misrouting garbled packets. Error detection in the payload is the responsibility of the transport layer (both UDP and TCP provide error detection). The *protocol* field allows IP to demultiplex the datagram and deliver it to a higher-level protocol. Since it has only 8 bits, IP does not support application multiplexing. Providing port number fields to enable application multiplexing is another required function in transport protocols on IP.

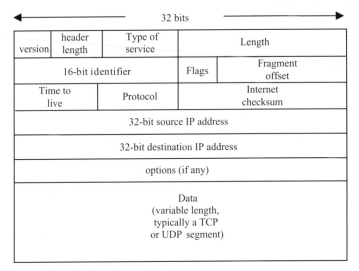

◄─────────────── 32 bits ───────────────►

version	header length	Type of service	Length	
16-bit identifier			Flags	Fragment offset
Time to live		Protocol	Internet checksum	
32-bit source IP address				
32-bit destination IP address				
options (if any)				
Data (variable length, typically a TCP or UDP segment)				

Figure 2.12 IP packet format.

The *time-to-live* (TTL) *field* is decremented at every hop, and the packet is discarded if the field is 0; this prevents packets from looping forever in the Internet. The TTL field is also used in a simple way to scope the reach of the packets and can be used in conjunction with ICMP, multicast, and so on, to support administrative functions. The *type-of-service* (TOS) *field* was designed to allow optional support for differential forwarding but has not been used extensively. Recently, the differentiated services (DiffServ) WG in IETF renamed this field the *DS byte* to be used to support DiffServ. The version field indicates the version of IP and allows extensibility. The current version of IP is version 4. IPv6 is the next generation of IP that may be deployed over the next decade to support a larger 128-bit IP address space. *Header length* is a field used because *options* can be variable in length. But options are rarely used in modern IP deployments, so we will not discuss them further.

2.3.3.2 IP Addressing and Address Allocation   An address is a *unique computer-understandable identifier*. Uniqueness is defined in a domain. Outside that domain one needs to have either a larger address space or do translation. Ideally, an address should be valid regardless of the location of the *source*, but may change if the *destination* moves. Fixed-size addresses can be processed faster. The concept of addresses is fundamental to networking. There is no (nontrivial) network without addresses. Address space size also limits the scalability of networks. A large address space allows a large network (i.e., it is required fundamentally for network scalability). A large address space also makes it easier to assign addresses and minimize configuration. In connectionless networks, the most interesting differences revolve around addresses. After all, a connectionless net basically involves

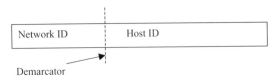

Figure 2.13   Hierarchical structure of an IP address.

putting an address in a packet and sending it, hoping that it will get to the destination. IPv4 uses 32-bit addresses, whereas IPv6 uses 128-bit addresses. For convenience of writing, a dotted decimal notation became popular. Each byte is summarized as a base 10 integer with dots placed between these numbers (e.g., 128.113.40.50).

IP addresses have two parts: a network part (prefix) and a host part (suffix). This is illustrated in Fig. 2.13. Recall that the intermediate nodes (or hosts) have to make a determination whether the destination is connected directly or indirectly. Examining the network part of the IP address allows us to make this determination. If the destination is connected directly, the network part matches the network part of an outgoing interface of the intermediate node. This hierarchical structure of addressing, which is fundamental to IP scaling, is not seen in layer 2 (IEEE 802) addresses. The structure has implications on address allocation because all interfaces on a single subnetwork have to be assigned the same network part of the address (to enable the forwarding test mentioned above).

Unfortunately, address allocation was not well thought out during the early days of IP, and hence it has followed a number of steps of evolution. Part of the evolution was forced because of the then unforeseen sustained exponential growth of the Internet. The evolution largely centered around placement of the conceptual demarcator between the network ID and host ID, as shown in Fig. 2.13. Initially, the addressing followed a *classful scheme*, where the address space was divided into a few blocks and static demarcators were assigned to each block. Class A has an 8-bit demarcator; class B has a 16-bit demarcator; class C has a 24-bit demarcator. Class D was reserved for multicast and class E for future use. This scheme is shown in Fig. 2.14. This scheme ran into trouble in the early 1980s for two reasons: (1) class

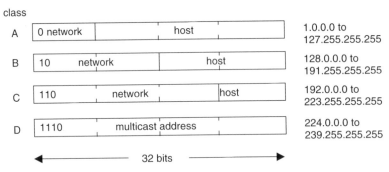

Figure 2.14   Initial classful addressing for IPv4.

B was popular (class C was largely unallocated) and (2) the host space in classes A and B was largely unused because no single subnetwork (e.g., Ethernet) was large enough to utilize the space fully. The solution to these problems is simple—allow the host space to be subdivided further, and allow demarcators to be placed more flexibly rather than statically.

These realizations led to the development of subnet and supernet masking, respectively. A mask is a 32-bit pattern, the 1's of which indicate the bits belonging to the network ID and the zeros, the host ID bits. For simplicity, the 1's in the masks are contiguous. For example, a subnet mask 255.255.255.0 applied to IP address 128.113.40.50 indicates that the network ID has been extended from 16 bits (since this is a class B address) to 24 bits. Supernet masks are used between autonomous systems to indicate address allocations or to advertise networks for routing. For example, the notation 198.28.29.0/18 indicates an 18-bit address space. The supernet mask written as /18 is actually 255.255.192.0. Observe that the 198.28. 29.0 belonged to the class C space according to the earlier classful scheme, and class C admits only of /24 networks (i.e., with a host space of 8 bits).

Since these class boundaries are no longer valid with the supernet masks, this allocation scheme is also called *classless allocation*; and the routing scheme that accompanied this development is called *classless interdomain routing* (CIDR). One effect of CIDR and supernet masking is that it is possible for a destination address to match multiple prefixes of different lengths. To resolve this, CIDR prescribes that the longest-prefix match be chosen for the L3 forwarding decision. As a result, all routers in the mid-1980s had to replace their forwarding algorithms. Similarly, when subnet masking was introduced, hosts and routers had to be configured with subnet masks and had to apply the mask in the forwarding process to determine the true network ID. Recall that the network ID is used to determine if the destination is connected directly or indirectly. These evolutionary changes are examples of how control-plane changes (CIDR and address allocation) could also affect the data-plane (IP forwarding) operation.

In modern networks, two other schemes are also used to further conserve public address space: the dynamic host configuration protocol (DHCP) and NAT. DHCP was originally a network booting protocol that configured essential parameters to hosts and routers. Now it is used primarily to *lease* a pool of scarce public addresses among hosts who need it for connecting to the Internet. Observe that the leasing model means that host interfaces no longer "own" IP addresses.

The network address translator (NAT) system enables the use of private address spaces within large enterprises. The Internet Assigned Numbers Authority (IANA) has reserved the following three blocks of the IP address space for private internets:

10.0.0.0–10.255.255.255 (10/8 prefix)
172.16.0.0–172.31.255.255 (172.16/12 prefix)
192.168.0.0–192.168.255.255 (192.168/16 prefix)

The NAT boxes at the edge of these private networks then translate public addresses to private addresses for all active sessions. Since early applications (e.g., FTP) overloaded the semantics of IP addresses and included them in application-level fields, NAT has to transform these addresses as well. NAT breaks certain security protocols, notably IPSEC. IPSEC protects (through per-packet encryption or authentication) both the IP payload and some fields of the IP header, notably the IP addresses. Therefore, a NAT box cannot change IP addresses without breaking IPSEC.

The combination of these techniques has delayed the deployment of IPv6 that proposes a more long-lived solution to address space shortage. IETF and the IPv6 Forum have been planning the deployment of IPv6 for over a decade now, and it remains to be seen what will be the major catalyst for IPv6 adoption. The potential growth of 3G wireless networks and/or the strain on interdomain routing due to multihoming have been cited as possible catalysts. ISPs project that the IPv4 address space can be prolonged for another decade using the foregoing techniques.

2.3.3.3 ARP, Fragmentation, and Reassembly

Recall that the overlay model used by IP results in two mapping problems: address mapping and packet format mapping. Address mapping is resolved by a sub-IP protocol called the *address resolution protocol* (ARP); packet mapping is done within IP by fragmentation/ reassembly procedures. The mapping problems in IP are far simpler than those of other internetworking protocols in the early 1980s because IP has minimal expectations from the lower layers.

The address mapping problem occurs once the destination or next hop is determined at the IP level (i.e., using the L3 forwarding table). The problem is as follows: The node knows the IP address of the next hop (which by definition is connected directly (i.e., accessible through layer 2 forwarding). But now to be able to use L2 forwarding, it needs to find the next hop's L2 address. Since the address spaces of L2 and L3 are assigned independently, the mapping is not a simple functional relationship (i.e., it has to be discovered dynamically). ARP is used to discover L3 address-to-L2 address mapping.

ARP at the node sends out a link-level broadcast message requesting the mapping. Since the next hop is on the same layer-2 *wire,* it will respond with a unicast ARP reply to the node giving its L2 address. Then the node uses this L2 address and encloses the IP datagram in the L2 frame payload and *drops* the frame on the L2 wire. ARP then uses caching (i.e., an ARP mapping table) to avoid the broadcast request response for future packets. In fact, other nodes on the same L2 wire also snoop and update their ARP tables, thus reducing the need for redundant ARP broadcasts. Since the mapping between L3 and L2 addresses could change (because both L2 and L3 addresses can be assigned dynamically), the ARP table entries are aged and expunged after a timeout period.

The packet-mapping problem occurs when the IP datagram to be forwarded is larger than the maximum transmission unit (MTU) possible in the link layer. Every link typically has a MTU, for reasons such as fairness in multiplexing, error detection efficiency, and so on. For example, Ethernet has an MTU of 1518 bytes. The solution is for the IP datagram to be fragmented such that each fragment fits the L2 payload. Each fragment now becomes an independent IP datagram; hence the IP header is copied over. However, it also needs to indicate the original datagram, the position (or offset) of the fragment in the original datagram and whether it is the last datagram. These pieces of information are filled into the fragmentation fields in the IP header (ID, flags, frag offset), respectively. The reassembly is then done at the IP layer in the ultimate destination. Fragments may come out of order or be delayed. A reassembly table data structure and a timeout per datagram are maintained at the receiver to implement this function. Reassembly is not attempted at intermediate routers because all fragments may not be routed through the same path.

In general, although fragmentation is a necessary function for correctness, it has severe performance penalties. This is because any one of the fragments lost leads to the entire datagram being discarded at the receiver. Moreover, the remaining fragments that have reached the receiver (and are discarded) have consumed and effectively wasted scarce resources at intermediate nodes. Therefore, modern transport protocols try to avoid fragmentation as much as possible by first discovering the minimum MTU of the path. This procedure is also known as *path-MTU discovery*. Periodically (every 6 seconds or so), an active session will invoke the path-MTU procedure. The procedure starts by sending a maximum-sized datagram with the *"do not fragment"* bit set in the flags field. When a router is forced to consider fragmentation due to a smaller MTU than the datagram, it drops the datagram and sends an ICMP message indicating the MTU of the link. The host then retries the procedure with the new MTU. This process is repeated until an appropriately sized packet reaches the receiver, the size of which is used as the maximum datagram size for future transmissions.

In summary, the mapping problems in IP are solved by ARP (a separate protocol) and fragmentation/reassembly procedures. Fragmentation avoidance is a performance imperative and is carried out through path-MTU discovery. This completes discussion of the key data-plane concepts in IP. The missing pieces now are the routing protocols used to populate forwarding tables such that a concatenation of local decisions (forwarding) leads to efficient global connectivity.

2.3.4   Routing in the Internet

Routing is the magic enabling connectivity. It is the control-plane function, which sets up the local forwarding tables at the intermediate nodes such that a concatenation of local forwarding decisions leads to global connectivity. The global connectivity is also efficient in the sense that loops are avoided in the steady state. Internet routing is scalable because it is hierarchical. There are two categories of routing in the Internet: interdomain routing and intradomain routing. *Interdomain routing* is performed between autonomous systems (ASs). An autonomous system defines the locus of single administrative control and is connected internally (i.e., employs appropriate routing so that two internal nodes need not use an external route to reach each other). The internal connectivity in an AS is achieved through *intradomain routing* protocols.

Once the nodes and links of a network are defined and the boundary of the routing architecture is defined, the routing protocol is responsible for capturing and condensing the appropriate global state into a local state (i.e., the forwarding table). Two issues in routing are completeness and consistency. In the steady state, the routing information at nodes must be *consistent* [i.e., a series of independent local forwarding decisions must lead to connectivity between any (source, destination) pair in the network]. If this condition is not true, the routing algorithm is said not to have *converged* to steady state (i.e., it is in a transient state). In certain routing protocols, convergence may take a long time. In general, a part of the routing information may be consistent while the rest may be inconsistent. If packets are forwarded during the period of convergence, they may end up in loops or traverse the network arbitrarily without reaching the destination. This is why the TTL field in the IP header is used. In general, a faster convergence algorithm is preferred and is considered more stable; but this may come at the expense of complexity. Longer convergence times also limit the scalability of the algorithm, because with more nodes, there are more routes, and each could have convergence issues independently.

Completeness means that every node has sufficient information to be able to compute all paths in the entire network locally. In general, with more complete information, routing algorithms tend to converge faster, because the chances of inconsistency are reduced. But this means that a more distributed state must be collected at each node and processed. The demand for more completeness also limits the scalability of the algorithm. Since both consistency and completeness pose scalability problems, large networks have to be structured hierarchically (e.g., as areas in OSPF), where each area operates independently and views the other areas as a single border node.

2.3.4.1   *Distance Vector and Link-State Algorithms and Protocols*   In packet-switched networks, the two main types of routing are link state and distance vector. *Distance vector protocols* maintain information on a *per-node* basis (i.e., a vector of elements), where each element of the vector represents a distance or a path to that node. *Link-state protocols* maintain information on a *per-link* basis, where each element represents a weight or a set of attributes of a link. If a graph is considered as a set of nodes and links, it is easy to see that the link-state approach has complete information (information about links also implicitly indicates the nodes which are the endpoints of the links), whereas the distance vector approach has incomplete information.

The basic algorithms of the distance vector (Bellman–Ford) and the link state (Dijkstra) attempt to find the shortest paths in a graph in a fully distributed manner, assuming that distance vector or link-state information can only be exchanged between immediate neighbors. Both algorithms rely on a simple recursive equation. Assume that the shortest-distance path from node i to node j has distance $D(i,j)$, and it passes through neighbor k to which the cost from i is $c(i,k)$; then we have the equation

$$D(i,j) = c(i,k) + D(k,j) \qquad (2.1)$$

In other words, the subset of a shortest path is also the shortest path between the two intermediate nodes.

The *distance vector (Bellman–Ford) algorithm* evaluates this recursion iteratively by starting with initial distance values:

$D(i,i) = 0$;
$D(i,k) = c(i,k)$ if k is a neighbor (i.e., k is one hop away); and
$D(i,k) =$ infinity for all other nonneighbors k.

Observe that the set of values $D(i,*)$ is a *distance vector at node i*. The algorithm also maintains a next-hop value for every destination j, initialized as

next-hop$(i) = i$;
next-hop$(k) = k$ if k is a neighbor; and
next-hop$(k) =$ unknown if k is a nonneighbor.

Note that the next-hop values at the end of every iteration go into the forwarding table used at node i.

In every iteration, each node i exchanges its distance vectors $D(i,*)$ with its immediate neighbors. Now each node i has the values used in equation (2.1) [i.e., $D(i,j)$ for any destination and $D(k,j)$ and $c(i,k)$ for each of its neighbors k]. Now if $c(i,k) + D(k,j)$ is smaller than the current value of $D(i,j)$, then $D(i,j)$ is replaced with $c(i,k) + D(k,j)$, as per (2.1). The next-hop value for destination j is set now to k. Thus after m iterations, each node knows the shortest path possible to any other node which takes m hops or less. Therefore, the algorithm converges in $O(d)$ iterations, where d is the maximum diameter of the network. Observe that each iteration requires information exchange between neighbors. At the end of each iteration, the next-hop values for every destination j are output into the forwarding table used by IP.

The *link-state (Dijkstra) algorithm* pivots around the link cost $c(i,k)$ and the destinations j rather than the distance $D(i,j)$ and the source i in the distance-vector approach. It follows a greedy iterative approach to evaluating (2.1), but it collects all the link states in the graph *before* running the Dijkstra algorithm *locally*. The Dijkstra algorithm at node i maintains two sets: set N, which contains nodes to which the shortest paths have been found so far, and set M, contains all other nodes. Initially, set N contains node i only, and the next hop $(i) = i$. For all other nodes k, a value $D(i,k)$ is maintained which indicates the current value of the path cost (distance) from i to k. Also, a value $p(k)$ indicates what is the predecessor node to k on the shortest known path from i [i.e., $p(k)$ is a neighbor of k]. Initially,

$D(i,i) = 0$ and $p(i) = i$;
$D(i,k) = c(i,k)$ and $p(k) = i$ if k is a neighbor of i;
$D(i,k) =$ infinity and $p(k) =$ unknown if k is *not* a neighbor of i.
Set N contains node i only, and the next hop $(i) = i$.
Set M contains all other nodes j.

In each iteration, a new node j is moved from set M into set N. Such a node j has the minimum distance among all current nodes in M [i.e., $D(i,j) = \min_{\{l \in M\}} D(i,l)$]. If multiple nodes have the same minimum distance, any one of them is chosen as j. Node j is moved from set M to set N, and the next hop (j) is set to the neighbor of i on the shortest path to j. Now, in addition, the distance values of any neighbor k of j in set M is reset as

If $D(i,k) < c(j,k) + D(i,j)$, then $D(i,k) = c(j,k) + D(i,j)$, and $p(k) = j$.

This operation, called *relaxing* the edges of j, is essentially the application of equation (2.1). This defines the end of the iteration. Observe that at the end of iteration p the algorithm has effectively explored paths, which are p hops or smaller from node i. At the end of the algorithm, set N contains all the nodes, and knows all the next hop (j) values, which are entered into the IP forwarding table. Set M is empty upon termination. The algorithm requires n iterations, where n is the number of nodes in the graph. But since the Dijkstra algorithm is a *local* computation, they are performed much quicker than in the distance vector approach. The complexity in the link-state approach is largely due to the need to wait to get all the link states $c(j,k)$ from the entire network.

The protocols corresponding to the distance-vector and link-state approaches for intradomain routing are RIP and OSPF, respectively. In both these algorithms, if a link or node goes down, the link costs or distance values have to be updated. Hence information needs to be distributed and the algorithms need to be rerun. RIP is used for fairly small networks, due primarily to a convergence problem called *count to infinity*. The advantage of RIP is simplicity (25 lines of code!). OSPF is a more complex standard that allows hierarchy and is more stable than RIP. Therefore, it is used in larger networks (especially enterprise and ISP internal networks). Another popular link-state protocol commonly used in ISP networks is IS-IS, which came from the ISO/OSI world but was adapted to IP networks.

BGP-4 is the interdomain protocol of the Internet. It uses a vectoring approach but uses full AS paths instead of the distances used in RIP. BGP-4 is designed for policy-based routing between autonomous systems, and therefore it does not use the Bellman–Ford algorithm. BGP speakers announce routes to certain destination prefixes expecting to receive traffic that they then forward along. When a BGP speaker receives updates from its neighbors that advertise paths to destination prefixes, it assumes that the neighbor is actively using these paths to reach that destination prefix. These route advertisements also carry a list of *attributes* for each prefix.

BGP then uses a list of tiebreaker rules (as in a tennis tournament) to determine which of the multiple paths available to a destination prefix is chosen to populate the forwarding table. The tiebreaker rules are applied to attributes of each potential path to destination prefixes learned from neighbors. The AS path length is one of the highest-priority items in the tiebreaker process. Other attributes include local preference, multiexit discriminator (i.e., LOCAL_PREF, MED: used for redundancy/load balancing), ORIGIN (indicates how the route was injected into BGP), and

NEXT-HOP (indicates the BGP-level next hop for the route). For further details on these routing protocols, see the reference list at the end of the chapter (e.g., [11], [12], [13], and [15]).

2.4 ASYNCHRONOUS TRANSFER MODE

The asynchronous transfer mode (ATM) technology was a culmination of several years of evolution in leased data networks that grew out from the telephony world. ATM was developed as a convergence technology where voice and data could be supported on a single integrated network. Its proponents had a grander vision of ATM taking over the world (i.e., becoming the dominant networking technology). However, it is fair to observe after over a decade of deployment that ATM has had its share of successes and failures. ATM switching started appearing in the late 1980s and is recognized as a stable technology that is deployed in several carrier and RBOC networks to aggregate their voice, leased lines, and frame relay traffic.

However, ATM lost out in the LAN space because of the emergence of Fast Ethernet in the early 1990s and Gigabit Ethernet in the mid to late 1990s. ATM never became the basis for end-to-end transport (i.e., native ATM applications never got off the ground). This was primarily because the Web, email, and FTP became critical entrenched applications which ensured that TCP/IP would stay. Until the mid-1990s, ATM still offered performance advantages over IP routers. Therefore, ATM was deployed as IP backbones, and a lot of development went into internetworking IP and ATM. However, since IP and ATM had completely different design choices, the mapping problem of IP and routing protocols such as BGP onto ATM led to several complex specifications (NHRP, MPOA, etc.). MPLS was then developed as a hybrid method of solving this problem, with a simple proposition: Take the IP control plane and merge it with the ATM data plane. MPLS has been developed considerably since then, and ISPs are slowly moving away from ATM backbones to MPLS.

2.4.1 ATM Basics

ATM has a fixed cell size of 53 bytes (48-byte payload). The fixed size was chosen for two reasons: (1) queuing characteristics are better with fixed service rates (and hence delay and jitter can be controlled for voice/video), and (2) switches in the late 1980s could be engineered to be much faster for fixed-length packets and simple label-based lookup. ATM uses virtual circuits (VCs) that are routed and established before communication can happen. It is similar to circuit switching in the sense that its virtual circuits are established prior to data transmission, and data transmission takes the same path. But there the similarity stops. Telephone networks use time-division multiplexing (TDM) and operate on a fundamental time frame of 125 μs, which translates into 8000 samples/s, or a minimum bit rate of 8 kbps; all rates are multiples of 8 kbps. In fact, since the base telephone rate is 64 kbps (corresponding

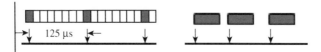

Figure 2.15 Synchronous versus asynchronous transfer mode.

to 8 bits/sample × 8000 samples/s), the TDM hierarchy (e.g., T1, T3, OC-3, OC-12, OC-48, OC-192, etc.) are all multiples of 64 kbps. However, in ATM this restriction and associated multiplexing hierarchy aggregation requirements are removed (as illustrated in Fig. 2.15). One can signal an ATM VC and operate at any rate. Since telephony networks are synchronous (periodic), ATM was labeled *asynchronous transfer mode*. Moreover, several aspect of leased-line setup are manual. ATM virtual circuit setup is automatic. ATM differs from frame relay in that ATM cells are of fixed size, 53 bytes, and offers a richer set of services and signaling/ management plane support.

ATM offers five service classes: constant bit rate (CBR), variable bit rate (VBR-rt and VBR-nrt), available bit rate (ABR), guaranteed frame rate (GFR) and unspecified bit rate (UBR). CBR was designed for voice, VBR for video, and ABR, GFR, and UBR for data traffic. ATM offers adaptation layers for different applications to be mapped to lower-level transmissions. The adaptation layer also does fragmentation and reassembly of higher-level payload into ATM cells. Of the five adaptation layers, AAL1 and AAL2 are used for voice/video and AAL5 is used for data. AAL 3/4, which was developed for data, has been deprecated.

The ATM layer then provides services including transmission/switching/reception, congestion control/buffer management, cell header generation/removal at the source/destination, cell address translation and sequential delivery. The ATM header format is illustrated in Fig. 2.16. ATM uses 32-bit labels (VCI and VPI) instead of addresses in its cells. ATM addresses are 20 bytes long and are used in the signaling phase alone. During signaling, each switch on the path assigns an outgoing label from its local label space and accepts the label assignment of the prior switch on the path for the inbound link. The ATM forwarding table entries each has four fields: inbound label, inbound link, outbound link, and outbound

GFC/VPI		VPI	
VPI		VCI	
VCI			
VCI		PTI	CLP
Header Error Check (HFC)			
Payload			

Figure 2.16 ATM header format.

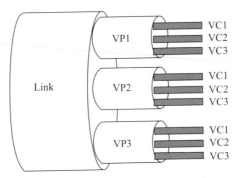

Figure 2.17 ATM virtual paths (VPs) versus virtual channels (VCs).

label. The label in the ATM header is hence swapped at every switch, unlike IP addresses, which do not change en route.

ATM paths are actually set up in two granularities: virtual path and virtual circuit. The distinction between these two concepts is illustrated in Fig. 2.17. *Virtual paths* are expected to be long-lived, and *virtual circuits* could be *switched* (i.e., set up and torn down upon demand). The VP label is called the VPI and the VC label the VCI in Fig. 2.16. To keep our notation simple, we refer to both as "VC" in what follows. ATM VC signaling involves specification of the traffic class and parameters, routing through the network (using a scalable QoS-enabled link-state protocol, PNNI) and establishing a VC. Then cells flow through the path established. Traffic is policed, and flow controlled or scheduled in the data plane according to the service specification.

The ATM Forum has specified several interfaces to ATM:

- *User-to-network interface* (UNI, with public and private flavors)
- *Network-to-node interface* [NNI, the private flavor is called PNNI; the public flavor, called the interswitching system interface (ISSI), allows intra- or inter-LATA communication]
- *Broadband intercarrier interface* (B-ICI) between carriers
- *Data exchange interface (DXI)* between routers and ATM digital service units (DSUs)

Of these, UNI and P-NNI are fairly popular and well implemented. Now we move on to understanding IP–ATM internetworking issues.

2.4.2 IP over ATM

ATM was developed as a competitor to IP and Ethernet (i.e., as a ubiquitous end-to-end technology). But it had its share of successes and failures. ATM had a strong backing from carriers who were in the leased line and frame relay businesses. Essentially, these technologies provided reliable capabilities for enterprises to create their own private networks. Private networks based on leased line and frame relay were popular for good reasons. First, they were very reliable (carrier class). Second, they were opaque to the payload offered, which was possible because leased line operated at layer 1 and frame relay operated at layer 2. Moreover, they guaranteed bandwidth (i.e., data could "burst" up to the allowed peak rates, with a minimum rate guaranteed). Both technologies could carry a mix of voice and data traffic. Voice could be mapped directly to leased lines because it conforms to voice trunk aggregation rules, and voice could be packetized and transmitted in frames with a minimum frame rate in FR. ATM was proposed as a core technology to integrate and evolve from these two high-margin technologies.

But ATM also lost certain crucial battles: Its adaptation for LANs (LAN emulation) lost to 100-Mbps and 1-Gbps Ethernet; and it was very difficult to internetwork IP and ATM in a scalable manner. The LAN operators' familiarity with Ethernet and the emergence of faster/cheaper forms of the well-understood Ethernet sealed the fate of ATM in the LAN area. We summarize some of the internetworking attempts here because it is relevant to the growth of MPLS technology.

One problem with ATM was that did not provide a good internetworking solution (i.e., one could not map ATM to arbitrary link layers, due primarily to its complexity and VC-oriented behavior. Therefore, the ATM Forum (and IETF in parallel) decided to develop a set of standards that would allow ATM to coexist with IP and Ethernet. These standards include LAN emulation (LANE), IP over ATM (RFC 1483: encapsulation and RFC 1577: architecture), next-hop routing protocol (NHRP), and multiprotocol over ATM (MPOA).

The key idea in LANE was to make ATM a sub-802.3 technology at the LAN level to provide a migration path. In this standard, assuming that the destination Ethernet address is known, ATM-LANE could accept Ethernet frames but transport it over ATM VCs. In IP over ATM (RFC 1483 and 1577), assuming that the next-hop IP address is known, IP packets would be mapped to a LLC frame of MTU = 8096 bytes and would be carried over ATM VCs. The central problem in both these technologies was that both Ethernet and IP had their own addressing, and ATM had its own addressing.

There was hence an *address resolution* problem in both cases, and ATM was a nonbroadcast multiple-access (NBMA) medium. The general approach taken was to have a server (LANE server or a RFC 1577 server) to which all nodes set up VCs and to send the address resolution request. Once the reply was obtained, a VC would be established and packets transmitted. This resolution was required only for the first packet. It is easy to see that this approach has two problems: the server would become a central point of failure and would not scale to support many VCs; and the setup and teardown of VCs had to be managed. This resolution procedure in RFC 1577 that converted IP addresses to ATM addresses was called ATMARP. In LANE, a LAN emulation server (LES) would translate Ethernet addresses to ATM addresses.

Moreover, in the case of IP over ATM, the IP routing protocols (especially iBGP) required peering sessions between every pair of routers and hence required a full mesh of VCs. Actually, iBGP requires only a full mesh of logical adjacencies, but in ATM, adjacencies had to be implemented as VCs. BGP tends to exchange large amounts of information, and hence this combination of large routing traffic and $O(N^2)$ VCs to be maintained did not scale.

RFC 1577 broke up ATM clouds into subnets called logical IP subnets (LISs). Within each LIS, an ATMARP server provided address resolution service. Nodes would have a VC with the ATMARP server, resolve their next-hop IP address to an ATM address, and then set up a VC to the next hop (if not set up already) before transmission. A problem with RFC 1577 was that if the ATM network was large, a packet may need to cross several LIS boundaries, which meant crossing several routers on the path. Also, since routers were slow in those days and had to reassemble cells, this was undesirable. A desirable alternative was to have the source set up a shortcut VC to the destination.

The next-hop routing protocol (NHRP) was developed in IETF to deal with this problem. A source would contact an NHRP server for the ATM address of the last hop on the ATM network (which would be several subnet hops away). The NHRP server would then contact other NHRP servers to resolve the query into an ATM address. A cut-through or shortcut VC would then be established for the transmission. NHRP was considered very complex, and it also led to cases where stable routing loops emerged.

Multiprotocol over ATM (MPOA) was an effort of the ATM Forum to integrate and evolve LANE and NHRP standards and support non-IP network layers. In essence, it would have NHRP at layer 3 and LANE at layer 2. The idea was to integrate the multiple resolution and next-hop servers into a single framework. Without going into the details of these protocols, history has shown that they were too complex for implementers and operators, not to mention the long standardization process.

Despite IP-ATM internetworking problems, operators still liked ATM because it was a stable technology and provided carrier-class traffic engineering features. Unlike dynamic IP routing, the provisioning of virtual circuits could be well controlled by operators. Moreover, until the mid-1990s it was the only alternative to get speeds of 155 Mbps and above in both the LAN and WAN areas. Also, it was the only technology that allowed for quality-of-service (QoS) guarantees to be associated with VCs. IP did not have an answer in any of these categories until the mid-1990s.

Two key developments led to the overshadowing of ATM: (1) the emergence of IP switching and MPLS, and (2) the development of fast lookup, forwarding, and switching for IP leading to gigabit routers. We examine these developments in the following section.

2.5   IP SWITCHING*

The attractive features of ATM that have led to its popularity and deployment are its high capacity, scalability, and ability to support various classes of traffic, each with

*Some material in this section is based on [8], [9], and [14].

different requirements. However, while ATM is connection oriented, IP and a majority of the network protocols are connectionless. Using connectionless protocols over an inherently connection oriented service such as ATM results in a lot of undesirable complexity rising out of the mismatch. While ATM is quite popular at the data link layers, IP has grown in popularity at the network layer and is the most dominant network protocol in the Internet today, capable of supporting such applications as real time and multimedia, traditionally considered as strong points of ATM. Thus it becomes imperative to find a solution for using an IP-based network layer on top of an ATM-based data link layer without the complexity and inefficiency arising out of using a connectionless protocol over a connection-oriented protocol.

In [8], Newman et al. proposed a solution for implementing IP directly on top of ATM hardware while preserving the connectionless model of IP. To achieve this, the connection-oriented nature of the ATM protocol stack was discarded and ATM switching hardware was connected directly to IP. This arrangement has the advantage of not requiring a signaling or address resolution protocol and requires only the standard IP routing protocols. It also has the additional advantage of being able to support IP multicast that is not compatible with ATM's implementation of multicast. This structure of removing ATM's connection-oriented nature and using IP directly on ATM hardware is called *IP switching*. We now look at IP switching in more detail.

2.5.1 Connectionless Services over ATM

The main motivation for having connectionless services over ATM is to take advantage of the speed and capacity of ATM switching hardware without sacrificing the scalability and flexibility that come from the connectionless nature of IP. In fact, IP switching is not the first attempt at a solution to this problem, and classical IP can be used over ATM [18]. In the classical IP over ATM, the shared-medium LAN is replaced with an ATM subnet. However, the LAN-based paradigm of classical IP preserved and logical IP subnets, for example, still need to communicate with each other via router, regardless of whether direct connection via ATM is possible. Other popular approaches for providing IP over ATM include LAN emulation and multiprotocol over ATM. In LAN emulation, ATM emulates the physical shared medium by establishing an ATM multicast group between all of the end stations that belong to the ATM LAN segment and translating the 48-bit MAC addresses into an ATM address [6]. Multiprotocol over ATM (MPOA) is a collection of all these approaches, including LAN emulation, classical IP over ATM, and the next-hop resolution protocol (NHRP) combined into one [1]. All these approaches have various drawbacks. First, there is duplication of functionality at the ATM and IP layers, such as routing protocols and management functions. Also, network management is complicated by the need to manage separately the legacy components of the upper layers and the ATM switches. Additionally, while LAN emulation and classical IP over ATM work well for smaller networks, they do not scale well for larger networks.

2.5.2 IP Switching Architecture

IP switching tries to avoid the drawbacks of other approaches to using IP and ATM together and combine the advantages of IP (simplicity, scalability, and robustness) with the speed, capacity, and multiservice capabilities of ATM. IP switching is based on the concept of flows, which can be thought of as a sequence of packets sent from a particular source to a particular destination that are related in terms of their routing and any local handling policy they may require [3]. Flows can be thought of being the connectionless network equivalent of connections in connection-oriented networks.

The basic structure of an IP switch is shown in Fig. 2.18, which is essentially a router, with the hardware having the additional capacity to cache routing decisions in the switching hardware. An IP switch can be constructed on a standard ATM switch with its hardware left unchanged. However, all the control software in the switch above AAL-5 is removed and replaced by IP routing software along with a *flow classifier* and a driver to control the switch hardware. The flow classifier is mainly responsible for determining whether a particular source–destination pair is a long- or short-duration flow. This architecture is called an IP switch because it allows the flows to be *switched* without going into the router once the routing information has been cached in the switch. For longer, high-volume flows, the IP switch maps the flow onto ATM virtual channels established across the ATM switch. Once a virtual circuit is established for a flow, all further traffic from that flow is switched directly through the ATM switch without having to go to the IP layer for routing decisions. This greatly reduces the load on the forwarding engine. IP switches use the ipsilon flow management protocol (IFMP), as specified in RFC

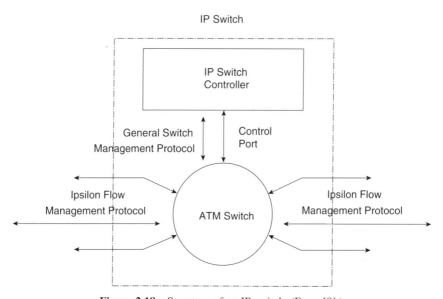

Figure 2.18   Structure of an IP switch. (From [9].)

1953, to propagate the mapping between flows and VCI to upstream routers and set up the path for the flow. While longer flows are switched directly at the ATM switch, it is inefficient to set up virtual channels for smaller flows such as name server queries and short transactions. Consequently, for shorter flows, the forwarding of packets is handled directly by the forwarding engine. The flow classifier dynamically selects flows to be forwarded in the switch fabric while the remaining flows are forwarded hop by hop. The flow classification feature allows the IP switch to intrinsically support multicast, quality of service differentiation, firewalls, and policy-based routing decisions.

Note that IP switching requires that each packet entering the network be labeled according to its flow, which is computationally more intensive than simple forwarding because more fields need to be examined for flow classification than for packet forwarding. On the other hand, this labeling needs to be done only once while the packet enters the switched network, unlike the forwarding operation, which needs to be done at each router. This labeling operation is done at the edges of the network. Also, flow classification is done based on local policies, and the classifier makes the decisions based on the local policy expressed in a table.

2.6   QoS, INTEGRATED SERVICES, AND DIFFERENTIATED SERVICES

Quality of service (QoS) is a broad term that means different things to different people. It could span human factors (e.g., customer relations) to network/link availability to specific guarantees on parameters, such as bandwidth, delay, and jitter. Moreover, every networking layer has its own interpretation of QoS. For example, application-level QoS parameters would be response time, throughput, and so on, whereas TCP-level parameters could be timeouts, mean and variance of per-flow throughput, and so on. Also, QoS has implications on several layers of the network stack—it spans from application-level parameter specification and mapping to network layer routing to scheduling and buffer management algorithms. Indeed, the functions need to be coordinated to achieve QoS objectives. For our treatment we focus on the simpler QoS problem of achieving bandwidth and delay guarantees from a network.

Consider a server with fixed capacity C. The sum of all bandwidth allocations (which are guaranteed a priori) cannot exceed C—this is a zero-sum game in the bandwidth dimension. More subtly, the conservation law of queuing theory says that for a stable queuing system, irrespective of the scheduling and buffer management scheme used, the weighted sum of delays experienced by users is a constant— another *zero-sum game in terms of delay*. In other words, at some point, the fact that some user is allocated bandwidth and delay guarantees means that someone else must not get the guarantee. This argues for limiting the number of users to whom the guarantees are given (i.e., to do admission control). It turns out that when users keep their traffic's statistical profile within a given envelope of parameters, it is possible to increase the degree of multiplexing (i.e., the maximal number of users admitted). The goal of traffic management and QoS is to develop a combination of shapers/policers, buffer management, and scheduling schemes that can allow a mix of guarantees and maximize the multiplexing gains.

The underlying assumption of QoS schemes is that capacity is costly to upgrade (which is why we assume it to be fixed). This assumption has been challenged in recent times because raw capacity has become cheaper, due to the growth of optical networking. However, on an end-to-end basis or on a site-to-site basis, provisioned path bandwidth (as opposed to raw link bandwidth) is still a challenging problem, especially when the path crosses multiple provider boundaries. The two primary efforts in the IETF regarding QoS were the integrated services (IntServ) and differentiated services (DiffServ) standards.

2.6.1 Integrated Services and RSVP

The IntServ working group (WG) aimed to provide an overall architecture for QoS on the Internet. It resulted in the specification of a number of service classes designed to meet the requirements of a number of applications. A related group worked on resource reservation (i.e., signaling) protocol, called RSVP. The IntServ architecture was built on pioneering theoretical work by Parekh and Gallagher [19], which said that delay guarantees can be achieved in a network, provided that careful shaping and policing was done at the edges and that weighted fair queueing (WFQ), a scheduling scheme, was deployed at interior nodes with appropriate bandwidth reservations. The traffic specification in IntServ is called the *TSpec*, which leads to a resource request called the *RSpec*. The IntServ components then act on these specifications according to Parekh–Gallagher theory, to signal these requirements to routers (using RSVP in the control plane), admit flows (using RSVP), and then police, classify, and schedule the flows in the data plane.

The WG defined two service classes: guaranteed service and controlled load. *Guaranteed service* is intended to serve the needs of applications that require a hard guarantee of bandwidth and delay. *Controlled load* refers to a service that emulates a lightly loaded network—useful for adaptive applications, which do not have graceful degradation modes when multiplexed at high loads.

The RSVP protocol was also designed bottom up to support the IP connectionless paradigm and IP multicast. Therefore, it was decoupled from IP routing (unlike ATM), and was receiver-driven (although source-driven RSVP is now available and is more popular). RSVP has two types of messages: a PATH message, which "marks" the forward path, and an RESV message, which carries the *TSpec* and *RSpec* and traverses the path from the receiver to the sender, establishing reservations along the way. RSVP also introduced the notion of *soft state*, where state (e.g., reservations) is not maintained permanently and therefore has to be refreshed periodically.

2.6.2 Differentiated Services

IntServ is widely agreed to have scalability issues because of its operation at very fine granularity (i.e., with microflows). Moreover, ISPs wanted more freedom in defining the services rather than have the IETF define it for them. In 1997 the IETF started working on a coarser-grained model of QoS, dubbed *differentiated services architecture*. The simple concept in DiffServ is to apply the end-to-end principle, move more complexity to edge routers of a cloud, and simplify core routers to be either stateless or maintain a coarser-grained state. It is easy to see that this would simply scale core routers. Work by David Clark and Van Jacobson [20] demonstrated that a near-equivalent of a frame-relay service (with a committed rate and a burstable peak rate) and a packetized leased line service (with a delay, jitter, and bandwidth guarantee) could be achieved using such a model.

Because the DiffServ core is expected to support a few classes, a packet's class can be marked directly into the packet. This contrasts with the IntServ model, in which a signaling protocol was required to tell the routers which of the flow's packets required special QoS treatment. The packet field, called a *differentiated services code point* (DSCP), is carried in a 6-bit subfield of the differentiated services (DS) byte, which was formerly the TOS octet in the IP header. Therefore, in theory, a forwarding path could support 64 different classes simultaneously.

Formally, the DSCP is just an encoding, which identifies a per-hop behavior (PHB), which is an abstraction of differential forwarding, classification, buffer management, packet dropping, and scheduling services offered to that class at any hop of the router. The particular implementation of PHBs at routers is left to individual vendors to decide. ISPs can design their own services, define PHBs, and map it to DSCPs. These are examples of ISP flexibility in the design of services. However, three standard PHBs have also been defined to promote development of standard interoperable services:

- *Default*: no special treatment (i.e., equivalent to best effort).
- *Expedited forwarding* (EF): forward with minimal delay and low loss. Implemented by aggregating EF packets into a single queue, perfectly smoothing the arrival process, and ensuring that the service rate is larger than the arrival rate. This is a building block to develop a packetized leased-line service.
- *Assured forwarding* (AF): a set of PHBs that effectively generalize the popular frame-relay CIR/PIR (committed information rate, peak information rate) service where the flow is assured the committed rate, but can burst up to the peak rate.

The DSCP is set upon arriving packets at the trust boundary (ingress edge node) based on preconfigured policy. As the packets traverse the network, the PHB at each hop is determined from the DSCP in the packet header. When enforced through a combination of lower-level implementation mechanisms, the PHB determines the QoS treatment received by a flow.

As the industry evolves, ISPs are deploying a mix of MPLS and DiffServ into their networks, coexistent with current OSPF or IS/IS-based domains. Virtual private networks (VPNs) are expected to be a key driver of QoS mechanisms on ISP infrastructures.

2.7   MULTIPROTOCOL LABEL SWITCHING*

Another approach to combining the specific strengths of ATM and IP is through multiprotocol label switching (MPLS) [10]. MPLS is a hybrid technology aimed at enabling very fast forwarding at the cores and conventional routing at the edges. It is based on assigning labels to packets based on *forwarding equivalent classes* (FEC) as they enter the network. All packets that belong to a particular FEC at a given node follow the same path and the same forwarding decision is applied to all. Packets are then switched through the MPLS domain through simple label lookups. Labels are used to differentiate the type of service that each packet gets and are valid as long as they are within the MPLS domain. At each hop, the routers and switches do not have to use any complex search algorithms but use the packet labels to index the routing table to determine the next-hop router and a new value for the label. The old label gets replaced with this new label and the packet is forwarded to the next hop. As each packet exits the network, it is stripped of the label at the egress router and then routed using conventional mechanisms. The term *multiprotocol* comes from the fact that this technique can be used with any network layer protocol and is not restricted to IP. MPLS also allows for ensuring that packets follow a predefined path. A router that uses MPLS is called a *label switching router* (LSR), and the path taken by a packet after being switched by LSRs through a MPLS domain is called a *label-switched path* (LSP). We now look at MPLS in greater detail.

2.7.1   Labels

The main strength of MPLS comes from the fact that the IP packet header needs to be examined once as it enters the MPLS domain by the *label edge routers* (LERs), which are responsible for assigning the labels to the packets. Once labels are assigned, forwarding of the packets is done sorely on the basis of the labels used by the label-switched path followed by the packet. A label is defined as a short, fixed-length locally significant identifier which is used to identify a forwarding equivalence class [10]. The label that is put on a particular packet represents the forwarding equivalence class to which that packet is assigned and the packet's class of service. The term *locally significant* implies that any two routers may decide on a choice of labels to signify a particular forward equivalence class for traffic flowing between them. The same label can be used by other routers inside the MPLS domain to represent different FECs. Routers use a label distribution protocol to inform others of the label to FEC bindings that it has made.

Usually, instead of carrying just a single label, a packet carries multiple labels, which allows MPLS to implement hierarchical routing and tunneling. The set of labels attached to a packet, called the *label stack*, is arranged in a last in/first out manner. As the packet traverses through the network, forwarding decisions are made on the basis of the topmost label in the stack. Forwarding at each node is

---

*Some material in this section is based on [10].

based on the *next-hop label forwarding entry* (NHLFE), which determines the packet's next hop and updates the label with a new label or pops the label stack.

2.7.2   Route Selection

The route selection problem in a MPLS network refers to the problem of mapping a FEC to a label-switched path. A label-switched path is the predetermined route that packets belonging to a particular FEC are forwarded through as they traverse the MPLS domain. A LSP is unidirectional and the return traffic has to use a different LSP. LSPs can be chosen for a given FEC in either of the following two ways: (1) *hop-by-hop routing*, where the routes are chosen at each router as in conventional IP forwarding, or through (2) *explicit routing*, where the boundary ingress router specifies the particular LSR to be taken by the packets of that FEC.

While hop-by-hop routing allows a node independently to choose the next hop for a packet, in the case of explicit routing, a single router specifies most or all of the LSRs in the LSP. The sequence of LSRs in an explicitly routed LSP can be chosen based either on the configuration or dynamically selected by a node (usually, the ingress node). Explicit routing is similar to source routing in IP networks and can be used for policy-based routing and traffic engineering. Although MPLS may have some similarities to IP source routing, it is more efficient since the explicit route needs to be specified only at the time that labels are assigned and does not have to be specified with each IP packet.

2.8   SUMMARY

This chapter has provided a broad overview of core concepts at the network and transport layer. The coverage extends from architectural principles such as end-to-end argument, to transport and network layer issues and a short overview of popular protocols. The purpose of this chapter has been to lay out a broad context in which optical networking fits. In particular, the control plane of optical networks is an extension of the MPLS architecture, known as G-MPLS (generalized MPLS). Optical networks deal with large aggregates of packet traffic, but it is important to ensure that such aggregates preserve the integrity of end-to-end microflows (e.g., TCP flows are affected by reordered packets). The discussion of routing algorithms provides a counterpoint to the style of control-plane algorithms used in optical networks. Optical networks also change the trade-off with respect to problems such as QoS. Complex QoS mechanisms need to be placed in the network only when network capacity is scarce. With optical networks, it is likely that for several years the core network may have more capacity than the sum of all its feeder or metro-area networks. Moreover, the fast provisioning and reconfigurability of such networks gives new traffic engineering tools for a network operator to meet the dynamic QoS needs of its customers. The discussion of the end-to-end principle shows that functions (e.g., intelligent optical networking) can be placed inside the Internet, provided that it offers a favorable economic trade-off in achieving the basic goal of the Internet: connectivity.

REFERENCES

1. C. Brown, Baseline text for MPOA, *ATM Forum/95-0824r4.txt*, May 1995.

2. M. Blumenthal and D. Clark, Rethinking the design of the Internet: The end to end arguments vs. the brave new world, *ACM Transactions on Internet Technology*, 1(1): 70–109, Aug. 2001.

3. S. Deering and R. Hinden, Internet protocol, version 6 (IPv6), IETF draft, *draft-ietf-ipngwg-ipv6-spec-02.txt*, June 1995.

4. J. Saltzer, D. Reed, and D. Clark, End-to-end arguments in system design, *ACM Transactions on Computer Systems*, 2(4):195–206, 1984.

5. S. Iren, P. D. Amer, and P. T. Conrad, The transport layer: Tutorial and survey, *ACM Computing Surveys*, 31(4), Dec. 1999.

6. N. Kavak, Data communication in ATM networks, *IEEE Network Magazine*, pp. 28–37, May 1995.

7. J. F. Kurose and K. Ross, *Computer Networking: A Top-Down Approach Featuring the Internet*, Addison-Wesley, Reading, MA, 2001.

8. P. Newman, T. Lyon, and G. Minshall, Flow labelled IP: A connectionless approach to ATM, *Proceedings of IEEE INFOCOM*, San Francisco, pp. 1251–1260, Mar. 1996.

9. P. Newman, G. Minshall, T. Lyon, and L. Huston, IP switching and gigabit routers, *IEEE Communications Magazine*, Jan. 1997.

10. E. Rosen, A. Viswanathan, and R. Callon, Multiprotocol label switching architecture, *IETF RFC 3031*, Jan. 2001.

11. B. Davie and Y. Rekhter, *MPLS: Technology and Applications*, Morgan Kauffman, San Francisco, 2000.

12. J. Moy, *OSPF: Anatomy of an Internet Routing Protocol*, Addison-Wesley, Reading, MA, 1998.

13. C. Huitema, *Routing in the Internet*, 2nd ed., Prentice Hall, Upper Saddle River, NJ, 2000.

14. W. R. Stevens, *TCP/IP Illustrated*, Vol. 1, Addison-Wesley, Reading, MA, 1994.

15. J. Stewart, *BGP4 Inter-domain Routing in the Internet*, Addison-Wesley, Reading, MA, 1999.

16. J. Mahdavi and S. Floyd, TCP-friendly unicast rate-based flow control, unpublished note, Jan. 1997. http://www.psc.edu/networking/papers/tcp_friendly.html.

17. S. Floyd, M. Handley, J. Padhye, and J. Widmer, Equation-based congestion control for unicast applications, *Proceedings of ACM SIGCOMM*, Stockholm, Sweden, Aug. 2000.

18. M. Laubach, Classical IP and ARP over ATM, IETF RFC 1577, Jan. 1994.

19. A. K. Parekh and R. G. Gallager, A generalized processor sharing approach to flow control in integrated services networks: The multiple node case, *IEEE/ACM Transactions on Networking*, Vol. 2, no. 2, pp. 137–150, April 1994.

20. V. Jacobson and D. Clark, Flexible and efficient resource management for datagram networks, unpublished manuscript, April 1991.

3 Optical Enabling Technologies for WDM Systems

ARI TERVONEN

Nokia Research Center, Helsinki, Finland

3.1 INTRODUCTION

Optical communication had its beginning some 30 years ago, with the development of semiconductor laser diodes and silica-based optical fibers, making available both the practical signal source and the transmission medium. Extensive developments during the two successive decades led to higher bit rates and longer transmission distances, perfecting the long-wavelength single-longitudinal-mode laser sources and very low loss single-mode optical fiber. However, the optics in field-deployed systems was quite limited to receivers and detectors at the two ends of the optical fiber transmission link, until the introduction of wavelength-division multiplexing (WDM) technology into practical systems that happened after 1990.

WDM systems introduced several new optical component technologies into field use. The key enabler was erbium-doped fiber amplifier (EDFA), built using many types of optical components. It made possible the amplification of optical signals without conversion back to the electrical level, thus removing the attenuation limitations for the distance span of optical transmission. Longer optical transmission spans also necessitated taking into use external modulation, to produce signals having optimum tolerance against the effects of fiber dispersion. A key advantage of EDFAs was their ability to amplify multiple optical signals at different wavelengths within their roughly 30-nm amplification bandwidth. As distributed feedback (DFB) laser sources with accurately defined wavelengths, as well as filter components to combine and separate closely spaced wavelengths, could be developed, the time of WDM technology came, leading to further development in transmission capacity and length.

Figure 3.1 illustrates the taking into use of WDM in an optical link. WDM is more cost-efficient, just taking into account the total equipment cost in the long-span transmission link. The most important cost advantage comes because instead

IP over WDM: Building the Next-Generation Optical Internet, Edited by Sudhir Dixit.
ISBN 0-471-21248-2 © 2003 John Wiley & Sons, Inc.

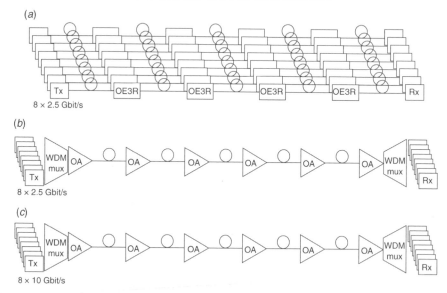

Figure 3.1 (a) Space-division-multiplexed link using 2.5-Gbps transmission in eight separate fibers and consisting of a large number of optoelectronic 3R regenerators (OE3R) has the same total capacity that can be achieved with (b) an eight-wavelength WDM system, introducing WDM filter multiplexers (WDM mux) and optical amplifiers (OA). When higher bit rates are taken into use (c), additional increase in WDM system capacity is achieved.

of a large number of electro-optic regenerators, a smaller number of optical amplifiers is needed. Also, if a sufficient number of fibers is not available for carrying optical signals separately, WDM is the choice that gives additional cost advantages, as there is no need to install new fiber. In further upgrading to higher bit rates, WDM will continue to be used.

To understand WDM technologies, it is first necessary to take a look at the transmission properties and limitations of optical fiber, which are first reviewed, including some basic formulas. Equations given in this connection are mainly to illustrate in mathematical terms the physical optics basis of propagation phenomena, and also to give some rough quantities and basic rules of thumb concerning the relevant phenomena. After this, the actual key technologies for WDM systems are introduced and described. Proceeding from this, the development of WDM systems is discussed, including current trends and recent developments in high-capacity optical transmission. Finally, a concluding summary takes a look at some future directions in fiber-optic technologies.

3.2   TRANSMISSION LIMITATIONS IN OPTICAL FIBER

3.2.1   Propagation in Optical Single-Mode Fiber

Light in an optical fiber propagates confined in the fiber core, which has a refractive index value n_{core} higher than the surrounding cladding refractive index $n_{cladding}$.

Optical fiber has translation symmetry along the axial propagation direction, meaning that the refractive index distribution does not vary along this direction. Due to this, propagation in the fiber can be described in terms of guided modes. We consider monochromatic light, optical fields consisting of a single frequency, ω. If the fiber axis is along the z-coordinate direction and we consider wave propagation in the direction of positive z-axis, this means that electromagnetic field solutions for electric and magnetic field $\bar{E}$ and $\bar{H}$, respectively, are of the form

$$\bar{E}(\bar{r}) = \bar{E}(x,y)e^{-i(\beta z - \omega t)} \qquad \bar{H}(\bar{r}) = \bar{H}(x,y)e^{-i(\beta z - \omega t)} \tag{3.1}$$

Fields can be separated into the transverse-mode field distribution in the x, y-plane and the axial z-dependence that oscillates in phase, thus having the nature of a propagating wave. Here, β, called the *propagation constant* of the mode, can also be represented as the effective index $n_{\mathrm{eff}} = \beta/k_0$, where $k_0 = \omega/c = 2\pi/\lambda$—with speed of light c and optical wavelength λ, both in vacuum. The guided modes have their transverse-mode field distributions localized in the vicinity of the fiber core and thus propagate with their energy confined to the core region. The effective index value of the guided modes fulfills the condition

$$n_{\mathrm{cladding}} < n_{\mathrm{eff}} < n_{\mathrm{core}} \tag{3.2}$$

For single-mode fiber, core size and refractive index contrast are so low that only a single guided mode exists. The cutoff condition is written for single-mode operation with a step-index fiber having a round core with radius a and index contrast $\Delta = n_{\mathrm{core}} - n_{\mathrm{cladding}}$, defining the parameter normalized frequency

$$V = \frac{2\pi a}{\lambda} n_{\mathrm{cladding}} \sqrt{2\Delta} < 2.405 \tag{3.3}$$

This cutoff condition gives the minimum wavelength (or maximum frequency) value allowed for single-mode propagation. For wavelength values below the cutoff limit, more than one guided mode exists. A typical single-mode fiber has $a = 4$ μm and $\Delta = 0.003$, with n_{cladding} around 1.465. Single-mode operation is achieved for both 1.3- and 1.55-μm wavelength regions with a value of V around 2, and the mode-field distribution can be well approximated by a suitable Gaussian that has a width value slightly larger than that of the fiber core. This well-confined mode field has a tolerance for fiber bending and other perturbations.

Also taking into account the polarization of the optical field, it is to be noted that single-mode fibers actually support two modes with orthogonal linear polarizations. In an ideal fiber, with a perfect cylindrical symmetry, these modes are degenerate, differing only in the direction of field polarization. In a real fiber, perfect symmetry does not exist, so the degeneracy of the polarization modes is broken and there is birefringence, which is nonuniform and random in nature. Because of this, the light launched into a single-mode fiber has an arbitrary state of polarization after propagating a relatively short distance, and this polarization state also typically varies with time.

To include fiber loss, we write the transmission equation in a simplified form, representing the scalar optical field amplitude E only, in terms of distance z along the fiber, and leaving out the harmonic time dependence

$$E(z) = E(0)e^{-(\alpha/2+i\beta)z} \tag{3.4}$$

Here α is the transmission loss coefficient. As the power carried by the mode is the integrated intensity over the fiber cross section, and intensity is proportional to the square of the field amplitude, the power decrease in transmission through a fiber of length L is given, with input launch power P_{in} and output power P_{out}, as

$$P_{out} = P_{in}e^{-\alpha L} \tag{3.5}$$

Fiber transmission loss expressed in units of dB/km is obtained for fiber of length L (in kilometers) as

$$\alpha_{dB} = -\frac{10\log_{10}(P_{out}/P_{in})}{L} \tag{3.6}$$

Main factors causing transmission loss in fibers are absorption and Rayleigh scattering, the latter of which is dominating at telecommunication wavelengths. The two telecommunication wavelength windows are the 1.3 μm window (approximately 80 nm width at 0.4 dB/km typical loss) and the 1.55-μm window (approximately 180 nm width at 0.25 dB/km typical loss).

3.2.2 Chromatic Dispersion

Equation (3.1) assumed a single optical frequency component, which is strictly correct only in the case of continuous-wave light carrying no modulated signal. Optical signals have bandwidths that are small compared to the frequency of the optical carrier, so more realistic signals can be described by expanding the optical field to include frequency components around this carrier frequency. Optical pulses in a single-mode fiber propagate at a group velocity $1/\beta_1$, with $\beta_1 = d\beta/d\omega$. Pulses with finite spectral width change shape in propagation due to the spectral dependence of β_1, which causes different spectral components making up the pulse to propagate at different velocities. $\beta_2 = d^2\beta/d\omega^2$ is the *group velocity dispersion parameter*. For single-mode fibers, *dispersion parameter D* is typically specified as a derivative of β_1 with respect to wavelength λ instead of frequency ω: $D = -2\pi c\beta_2/\lambda^2$. Chromatic dispersion in single-mode fibers consists of material and waveguide dispersion, the former due to material optical properties and the latter due to the wavelength dependence of the existing guided-mode solution and its propagation constant.

With a simple heuristic model, in a signal having spectral width $\Delta\lambda$, after traveling through a fiber of length L with dispersion parameter D, the differential group delay of pulse is $\Delta t = DL\,\Delta\lambda$. The dispersion causes signal deterioration through the effect called *intersymbol interference*: pulses representing 1-bits spread to overlap neighboring bit slots. Allowable bit-spreading is usually described by limiting the differential group delay to a given fraction of the bit period $t_{bit} = 1/B$, where B is the signal bit rate. This fraction of bit period varies depending on the specific signal type and allowable deterioration; typically, a fraction around $e = \frac{1}{4}$ is used. This gives the dispersion limit

$$|D|L\,\Delta\lambda < et_{bit} \Rightarrow B|D|L\,\Delta\lambda < e \qquad (3.7)$$

However, depending on the type of the source and modulation, the spectral character of signal varies and three cases of importance are:

1. A source with a spectral linewidth that is wide compared with the modulation bandwidth. Typically, this is a Fabry–Perot laser diode source, described quite well by equation (3.7), although some special characteristics need to be taken into account. Anyway, this is less relevant in connection with WDM technology, so it need not to be discussed in more detail here.

2. A source with a modulation-induced frequency chirping. Typically, this is a narrow optical linewidth DFB laser modulated directly through the driving current. Direct modulation causes chirping, variation in the source optical frequency with the modulation, so that the dynamic linewidth is much widened. Dispersion limitation can be estimated from equation (3.7), keeping in mind that the description of chirped pulse spreading is somewhat more complicated. This is because pulse waveform has limited linewidth at a specific time, but optical frequency is changing with time, so different parts of the original pulse actually propagate at different velocities. Directly modulated lasers have negative chirp, meaning that instantaneous optical frequency of pulses increases with time, or from leading to trailing edge. With anomalous dispersion, $D > 0$, as in standard single-mode fibers in the 1.55-μm window; the leading edge of the pulse thus travels faster than the trailing edge.

3. A source with a spectral linewidth narrow in comparison with the modulation bandwidth. Typically, this is a narrow optical linewidth DFB laser with external zero-chirp intensity modulation. Equation (3.7) can still be used, but the signal is now transform limited, meaning that the spectral width is now due to modulation, not from the source chirping. This is the minimum spectral width that a given signal can have, making it quite tolerant to dispersion. Roughly, spectral width in frequency is the modulation bit rate B. Changing this to wavelength value, $\Delta\lambda = B\lambda^2/c$, thus giving, from equation (3.7), a dispersion limit

$$\frac{B^2|D|L\lambda^2}{c} < e \qquad (3.8)$$

It is to be noted that transmission distance limited by dispersion now is proportional to the inverse of the squared bit rate, thus decreasing fast with increasing bit rate.

With a special external modulator, positive chirp can be impressed on modulated pulses, which would thus compress first to a certain minimum width before starting to spread. With careful design, this can be used to extend transmission span to some distance beyond the limit for a transform-limited signal.

3.2.3   Polarization-Mode Dispersion

An additional dispersion effect seen in optical fiber cables is polarization-mode dispersion (PMD). Actual fiber is always somewhat nonideal, leading to the two polarization modes traveling at slightly different group velocities, due to the ellipticity of the fiber core and the strain-induced birefringence. The two principal states of polarization, as well as their group velocity difference, vary along the fiber. Also, the coupling between polarization states is random in nature, as signal travels long distances in fiber. This PMD effect causes spreading of the pulses in signal, as photons making up the pulse are rather randomly distributed between the polarization states during propagation. Thus PMD is statistical in nature, group delay difference Δt increasing proportional to the square root of fiber link length L, as a random-walk process:

$$\Delta t = D_{PMD}\sqrt{L} \tag{3.9}$$

Here D_{PMD} is the fiber PMD parameter, in units of ps/$\sqrt{km}$. Actually, PMD is also statistical in nature in the sense that the value of D_{PMD} in cable varies with time, for example due to mechanical or thermal disturbances. It is expected to obey the Maxwellian probability distribution. To have an additional margin for such statistical variation, typically as a rule of thumb, the maximum average Δt due to PMD is allowed to be 0.1 times the bit period. PMD can be critical in old fibers having large values of D_{PMD}. New fiber cables have very low PMD, and typically, the effects of the chromatic dispersion are more severe. However, as chromatic dispersion is deterministic in nature, it can be compensated by methods that are quite simple in principle. In dispersion-compensated links operating at high bit rates, PMD can be a critical effect. Technology for PMD compensation is under development, based on dynamic polarization control and delay methods.

3.2.4   Nonlinear Optical Effects

Examination is now extended to include the nonlinear propagation phenomena in fiber [1]. These can generally be expressed by including a nonlinear term in propagation equation (3.4), for long distance $z = L$ along fiber, and with initial power P_{in} launched into fiber, this can be written

$$E(L) = E(0)\exp\left[-\left(\frac{\alpha}{2} + i\beta\right)L + \frac{\gamma P_{in}L_{eff}}{2A_{eff}}\right] \tag{3.10}$$

Here an additional power-dependent factor γ appears to describe the nonlinearity. Like equation (3.4), this is for continuous wave propagation, but due to the introduction of nonlinearity, it can no longer be linearly expanded for time-varying signals. This equation is given here mainly in an illustrative form, giving some basic insight into the nature of nonlinearities without mathematical complication.

Due to the attenuation, the magnitude of nonlinearity decreases along propagation. This effect is included by integrating the nonlinearity across propagation distance L, using effective length L_{eff} as a factor:

$$L_{\text{eff}} = \frac{1 - e^{-\alpha L}}{\alpha} \qquad (3.11)$$

So equation (3.10) is valid for distances L that exceed L_{eff}. As an example, in a low-loss fiber having $\alpha_{\text{dB}} = 0.22\,\text{dB/km}$, $L_{\text{eff}} \leq 20\,\text{km}$.

Also, the magnitude of nonlinearity depends not directly on the power carried by the mode, but on the intensity, power density over area. For this reason, there appears in equation (3.10) the effective area of fiber A_{eff}, describing the cross-sectional area of the mode field. The value of A_{eff} is about $80\,\mu\text{m}^2$ in standard single-mode fiber, and A_{eff} is about $50\,\mu\text{m}^2$ in dispersion-shifted fiber. To minimize nonlinear effects in special fibers, large effective areas are usually the aim.

Examining equation (3.10), two different groups of nonlinear effects can be seen:

1. *Stimulated scattering* (γ has real value, nonlinear effect leads to attenuation or amplification of propagating field); the effects are stimulated Brillouin scattering (SBS) and stimulated Raman scattering (SRS).
2. *Nonlinear refractive index* (γ is imaginary, nonlinear effect leads to development of nonlinear phase factor for field); three main effects are: self-phase modulation (SPM), cross-phase modulation (XPM), and four-wave mixing (FWM).

Phonons or lattice vibrations in the silica material of optical fiber are seen by light as refractive index nonuniformities, which can cause scattering of light. Such scattering events are known as Brillouin and Raman scattering, and scattering from acoustic and optical phonons, respectively. Both stimulated Brillouin scattering and stimulated Raman scattering are processes in which intense optical field induces lattice vibrations, causing scattering, so that in the process the incident (pump) photon is scattered to another lower-energy (Stokes) photon with the excitation of a phonon. The nature of scattering is ruled by the dispersion relation of involved phonons, based on the conservation of energy and momentum in the process.

In stimulated Brillouin scattering, acoustic phonons participate. The scattered Stokes light is propagating backward in fiber (in comparison with the incident pump light), with a frequency downshift of about 11 GHz (at 1.55-μm wavelength). When the power of optical signal in fiber is above SBS threshold, there is depletion of signal power and generation of a strong backward-propagating Stokes wave. As there is quite random variation in the level of SBS scattering, signal also becomes noisy. SBS is the lowest-threshold optical nonlinearity, and the threshold for a narrow-linewidth continuous-wave source may be as low as about 1 mW. The gain bandwidth of SBS is about 20 MHz. As optical sources often have wider linewidths than this, and particularly, modulated optical signals have larger bandwidths, the SBS threshold is increased. For directly modulated lasers, due to the large dynamic linewidth from chirping, SBS is typically no issue. However, for transmitters based on narrow-linewidth lasers and external modulation, signals can be nearly transform-limited and threshold is increased only by a factor of 2 to 4. Additional dithering of the source (modulation with a low frequency outside the receiver bandwidth) can be used to broaden the signal linewidth and to increase SBS threshold. Finally, due to the narrow gain bandwidth of SBS, in WDM systems the threshold power for each wavelength channel is determined independently. Thus SBS sets the limitation for launched power per WDM channel, but it is possible to increase this threshold to a noncritical level.

In stimulated Raman scattering, optical phonons participate. Different from SBS, SRS has a very large gain bandwidth. The frequency downshift of Stokes light can vary up to about 15 THz (corresponding to a wavelength upshift of more than 100 nm), and SRS gain coefficient increases quite linearly with a Stokes frequency difference, thus favoring larger frequency differences between pump and Stokes light. Also, pump and Stokes waves may propagate in the same or opposite directions in fiber. For a single-wavelength channel, the SRS threshold is roughly at the level of 1 W, so it is no issue in single-wavelength systems. However, in a WDM system, SRS causes depletion of optical power from shorter wavelength channels, while wider wavelength channels are amplified. Due to signal modulation, this depletion/amplification is not by a factor that is constant in time, but has a crosstalk-like characteristic, causing signal distortion and eye closure. SRS can be seen as setting a kind of upper limit for the capacity of WDM systems. This limitation can be expressed roughly as

$$P_{tot}\Delta f < 1000 \text{ W} \cdot \text{GHz} \tag{3.12}$$

Thus the product of the total optical power P_{tot} in all wavelength channels and the total occupied optical frequency bandwidth Δf must be below a limit to maintain the effects of SRS at an acceptable level. The limit can vary to some extent from the value given here, depending on the dispersion, modulation, and fiber type, but it gives a rough general limit.

The nonlinear refractive index leads to a modification of the optical signal phase. An optical pulse, representing a single 1-bit in the signal, when propagating through a fiber at high optical power, is connected with an instantaneous perturbation of the refractive index that is propagating with the pulse. This is caused by the material susceptibility having nonlinear terms dependent on the local optical field E, observed as a nonlinear, intensity-dependent refractive index:

$$n(E) = n_0 + n_2 E^2 \tag{3.13}$$

The refractive index perturbation has a consequence that different parts of the optical pulse experience different values of the refractive index on propagation through the fiber. As the phase change for the optical wave is proportional to the integrated product of the refractive index over the distance traveled, differential phase changes accumulate over the pulse. As frequency is the first derivative of the phase, this means that the pulse develops chirping, or frequency shifts, in various parts of the pulse. Particularly, pulse acquires a positive chirp. This does not directly cause the distortion in power distribution or shape of the pulse. However, from an examination of the effects of dispersion, we know that chirp causes distortion of the pulse through dispersion when propagating in fiber. This effect is known as *self-phase modulation*, and its effect is enhanced dispersion-caused distortion of short optical pulses. Actually, anomalous dispersion present in standard single-mode fiber in the 1.55-μm window leads to initial compression for pulses with positive chirp, and after compression leads to enhanced spreading of pulses in further propagation. The total chirp acquired by the pulse is proportional to the product of launch power and effective length. With careful planning, SPM can be used to extend the dispersion-limited link span somewhat, but it also limits the launch powers that can be used for optical signals.

In WDM systems, an additional effect can be observed, wherein the phase of the optical pulse is modulated not only by the pulse itself but also by the pulses in all other wavelength signals propagating together in the fiber. This phase perturbation of pulses by the signals at other wavelengths is called *cross-phase modulation* (CPM). In the first observation, for two pulses at different wavelengths but otherwise similar, aligned in time, the phase modulation produced by CPM is seen to be twice as large as that produced by SPM. So, with multiple WDM signals, CPM would be expected to dominate. However, pulses in WDM signals do not have correlated alignment, and as they are traveling at different group velocities in a dispersive fiber, any occasional alignment of pulses will not be maintained. The pulses walk through each other. Due to this, substantial phase differences cannot be expected to accumulate across the pulse through CPM, and unlike in SPM, pulses will not acquire significant chirp in propagation through long distances. The phase perturbations are averaged out as long as there are substantial group delay differences between different wavelengths, as is the case in dense WDM systems using standard single-mode fiber.

There is one additional effect from the nonlinear part of material susceptibility, which was observed as the nonlinear refractive index. When optical frequencies ω_i, ω_j, and ω_k (in a general case, these need not be different frequencies) are present with high powers in a fiber, new waves are generated at the frequencies $\omega_i \pm \omega_j \pm \omega_k$. This phenomenon, known as *four-wave mixing* (FWM), is a generation of the fourth frequency wave from the existing three waves. Like all other nonlinear effects, long propagation distances are needed to accumulate a substantial effect. For this reason, FWM can generally be observed only in cases in which all the waves have only small differences in frequency. This is since the optical phase of the wave generated depends on the phases of the original waves. Because of fiber dispersion, the particular phase relationship between different frequency waves will be maintained only through a propagation distance that is inversely proportional to the frequency difference of the waves. After this distance, there is no longer phase matching between the FWM-generated new wave and the previously accumulated FWM product wave at the same frequency—when they are off-phase, no more energy can be transferred from the original waves to the FWM product. (It is to be noted here that this phase matching is not critical in stimulated Brillouin or Raman scattering phenomena, in which phonons are also involved, making these processes automatically phase matched.) So, for FWM to occur, the frequencies ω_i, ω_j, and ω_k need to be closely spaced (as in dense WDM systems) and the FWM products can only be observed at frequencies $\omega_i + \omega_j - \omega_k$ ($\omega_j \neq \omega_k$) that are also closely spaced to the original frequencies. The effect of four-wave mixing on signal depletion is usually rather small, but the new mixing products in WDM systems with equal wavelength spacing coincide with the existing signal, causing nonlinear crosstalk between channels. It is obvious that the maintenance of phase matching depends critically also on the value of dispersion in fiber. FWM effects in WDM systems are critical primarily at dispersion that is close to zero, in which case it makes WDM transmission practically infeasible.

3.2.5   Types of Single-Mode Optical Fiber

Currently, three main single-mode optical fiber types specified by ITU-T are in use:

1. *Standard single-mode fiber* (SSMF), having dispersion zero inside the 1.3-μm window and dispersion parameter value D around 17 ps/nm/km in the 1.55-μm window [2]. SSMF dispersion is good for WDM application, but severely limits the achievable transmission spans at bit rates of 10 Gbps and higher.

2. *Dispersion-shifted single-mode fiber* (DSF), having dispersion zero inside the 1.55-μm window. This suits well for single-channel high-bit-rate transmission in the minimum-loss window of fiber, but due to four-wave mixing, is unsuitable for WDM application [3].

3. *Nonzero dispersion single-mode fiber* (NZDF), having small but finite values for dispersion within the bandwidth for WDM use [4]. The absolute value of D is typically between 4 and 8 ps/nm per kilometer, and it may be positive or negative. Bandwidth for use is defined to be within conventional Er-doped optical amplifier bandwidth (1535 to 1565 nm). The purpose of NZDF is to find a balance between the effects of dispersion and nonlinearities.

3.3   KEY TECHNOLOGIES FOR WDM SYSTEMS

The primary standardization recommendation related to WDM technologies is the ITU-T definition of used wavelengths [5]. WDM wavelengths are positioned in a grid having exactly 100 GHz (about 0.8 nm) spacing in optical frequency, with a reference frequency fixed at 193.10 THz (1552.52 nm). The main grid is placed inside the optical fiber amplifier bandwidth, but it can be extended to wider bandwidths. With progress into more dense WDM, a natural extension was to take into use 50- and 25-GHz-spaced subgrids. The importance of defined standard wavelengths is central to component manufacturing, since fiber-optic WDM components can be manufactured to fit a single common standard.

Not only developments in optical technologies are driving the evolution of fiber-optic communication systems. To justify taking into use the relatively expensive WDM technology was the fact that the traditional method to increase transmission capacity, by using higher bit rates, was becoming more and more difficult and costly. The fast electronics close to the high-bit-rate optical interfaces uses high-end semiconductor processes.

There are some generic functionalities carried out by high-speed electronics in optical systems. These include optical/electrical conversion, clock and data recovery, serialization/deserialization of data in the electrical multiplexing/demultiplexing process, and framing/deframing. Framing adds to the optical signal the digital overhead particular to the specific protocol used in the transmission. In addition, high-bit-rate systems are increasingly using forward error-correction algorithms implemented on the digital level, with additional overheads and fast processing.

The 2.5-Gbps systems have been using high-speed semiconductor processes such as GaAs MESFET and silicon bipolar technology, but later silicon CMOS has been able to meet these speed requirements. With 10-Gbps bit rates, GaAs, SiGe, and to some extent CMOS are in use. Emerging 40-Gbps systems are again stretching the limits of electronics. Although SiGe has been used to some extent in early technology pilots, InP heterojunction bipolar transistors may represent the most potential process to be used, owing to its high speed, low voltage requirements, and high breakdown voltages.

In the following sections we describe the key optical technologies used in WDM systems. We have already seen that the WDM breakthrough was enabled by wavelength-specific sources, external modulation technologies, narrowband filters, and erbium-doped fiber amplifiers. All these established technologies are discussed. Dispersion compensation technologies, which are becoming very important as higher transmission bitrates are introduced into systems in large scale, are described. Additional emerging technologies for optical networks—optical variable attenuators, fiber-optic switches, and wavelength converters—are also included.

3.3.1 WDM Transmitters

InGaAsP compound semiconductor DFB laser diodes are single-longitudinal-mode narrow-linewidth sources. They were developed for use within the minimum-loss window of the single-mode fiber at 1.55 µm, in which there is substantial dispersion. With these sources, attenuation-limited transmission at a 2.5-Gbps rate could be achieved. To serve as WDM sources, the main requirement was to have them at defined wavelengths. Combining processing targeted at specific wavelengths and component selection, they could be produced as specified for the ITU-T grid of WDM wavelengths. DFB lasers have temperature dependence of wavelength of about 0.1 nm/°C. Thus they need thermoelectric coolers to maintain specified wavelengths. The fine adjustment of wavelength exactly to ITU-T grid is achieved with the set temperature for operation. Tolerances in WDM technology currently allow wavelength control in a 100-GHz spaced system at a 2.5-Gbps channel rate using only temperature control of sources. Often, better wavelength accuracy has been a requirement, and therefore optical wavelength locking techniques have been used. These use a small portion of laser output tapped out to a wavelength-selective element, which provides two signals to detectors, based on which a feedback loop accurately maintains the output wavelength. For some time, WDM lasers have also been available with wavelength lockers integrated into the laser package [6].

With linewidths of a few megahertz, DFB lasers can achieve optical transmission spans longer than 500 km in standard single-mode fiber at a 2.5-Gbps rate—far longer than for attenuation-limited spans. This made them well suited for WDM application with optical amplifiers. However, due to wavelength chirping, such long transmission spans cannot be achieved using direct modulation. So, from the beginning, external modulation techniques were essential for WDM systems. Also, modulation bandwidths of DFB lasers are limited, and it has been very challenging to achieve even 10-Gbps rates using direct modulation.

LiNbO$_3$ external modulators [7,8] were the first technology taken into use for external modulation and are still currently in use for achieving the highest performance. They are an example of integrated optics applied to component manufacturing. Optical waveguides are fabricated into LiNbO$_3$ crystal substrates with either Ti indiffusion or proton exchange, patterned onto the substrate surface using lithography. Light propagates in the waveguides laterally confined, similar to the optical fiber core. There is a strong electro-optic effect in LiNbO$_3$, so the modulation of refractive index in the waveguide areas can be achieved with the electric field produced by surface thin-film electrodes processed close to the waveguides. The modulation speed is limited by the electrode structure. Using a traveling-wave electrode structure, into which the driving radio-frequency (RF) signal is applied, modulation rates of 40 Gbps can be achieved. The modulation rate is then basically limited by the velocity matching of RF signal in electrodes to the optical signal in waveguides. This needs to be maintained through the propagation length needed to achieve sufficient electro-optic modulation. The phase modulation of the signal is converted into intensity modulation in a Mach–Zehnder waveguide interferometer configuration. With the symmetric waveguide geometry for the two waveguide arms, and with push-pull electrodes—having opposite but equally strong modulations for the two arms—the phase of the signal produced by the interference is not modulated, and thus no frequency chirping is connected with the intensity modulation. Alternatively, different geometries can be utilized in these modulators for controlling and tailoring chirp properties. The large polarization-sensitivity of LiNbO$_3$ devices is of no importance here, since the modulated wave is from a single-polarization continuous-wave DFB laser.

Currently, integrated transmitter-modulators [9] have also been widely taken into use for long-distance transmission at 2.5- and 10-Gbps rates. In these, DFB lasers have been monolithically integrated with electroabsorption modulators, using semiconductor manufacturing processes similar to those used for discrete laser diodes. Multi-quantum-well electroabsorption modulators, illustrated in Fig. 3.2,

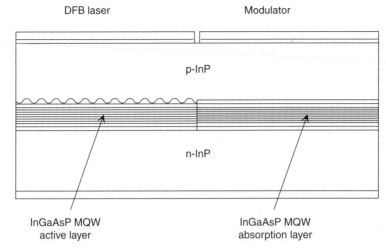

Figure 3.2   Structure of an integrated electroabsorption modulator source. DFB laser source and modulator are controlled by currents through separate top electrodes.

are based on a quantum confined Stark effect, which allows highly efficient, high-speed modulation. They have also demonstrated 40 Gbps rates [10].

3.3.2   WDM Filter Components

There are certain general requirements for optical wavelength filters in WDM systems. First, they have to be designed for a certain number of WDM channels having a given frequency spacing (200 GHz, 100 GHz, etc.) in the ITU-T grid. Other necessary requirements are:

- Over the passband, around the nominal center frequency, small loss and small variation of loss (spectral flatness) are needed, as the signal wavelength can drift and at high bit rates the signal will occupy a bandwidth that is a substantial fraction of the frequency spacing.
- Over adjacent signal passbands, crosstalk isolation from neighboring signal channels should be high (typically, >25 dB). Also, isolation for total cumulative crosstalk from all other channels should be low (typically, >21 dB). With a large number of channels, this sets quite large crosstalk isolation requirements, at least for nonadjacent channels.
- Low polarization dependence and high thermal stability are required.
- For high-bit-rate signals, dispersion and polarization mode dispersion values must be sufficiently low.

Dielectric thin-film multilayer passband filters have been the most popular choice for WDM filters. These are manufactured by growing alternate layers of high- and low-refractive-index dielectric films (usually, tantalum oxide and silica, respectively) on top of each other in a multilayer stack. Very large numbers of such filters for 200-GHz-spaced WDM have been installed in the field. To achieve wide, flat passbands, three cavity filters are needed. This consists of three layers having half-wavelength optical thickness, separated and surrounded by reflectors consisting of multilayer stacks of alternating low- and high-index quarter-wavelength layers. Thin-film growth of such filters was more challenging than that of any previous optical-coating applications, due to the very low tolerances for film thickness and refractive index variations. Also, high-quality dense structure for films is required for good thermal and environmental stability. For 100-GHz spacing, four or five cavity structures are needed. The total number of layers in such a passband filter is 100 to 200. Electron-beam evaporation and ion-beam sputtering techniques, with constant optical monitoring of growth, are used. Filter growth can last one to two days, and passband wavelengths vary across different areas of substrate, so they are characterized and mapped before the substrate is cut into small discrete filters. Different filters need to be grown for each wavelength. Manufacturing costs for dielectric passband filters still remain rather high, but a more effective, high-quality thin-film growth process may change this.

Application of dielectric passband filters in multiplexers uses regular micro-optic hybrid assembly techniques. Input and output coupling between the fiber and the filter is done using cylindrical microlenses to expand and collimate the optical beam. For a multichannel WDM multiplexer, several filters are cascaded. Figure 3.3

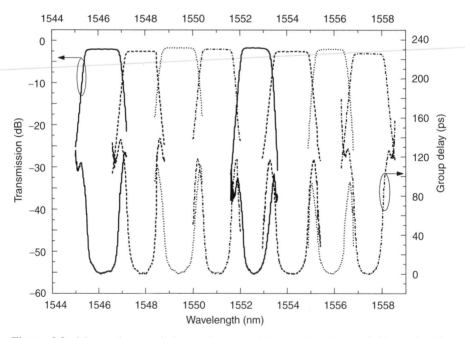

Figure 3.3 Measured transmission and group delay spectra of an eight-wavelength dielectric multilayer bandpass filter WDM demultiplexer. (Measurement results provided by T. Niemi, Helsinki University of Technology.)

shows measured transmission and group delay spectra of an eight-channel multi-plexer. These filters have low losses, high flatness over wide passbands, crosstalk isolation that is sufficient for adjacent channels and is very low for nonadjacent channels, low polarization-dependent loss, and low-temperature dependence of wavelength (<5 pm/$°$C). When several filters are cascaded, increased losses for some channels are, of course, the result.

Alternatively, fiber Bragg gratings [11] (FBGs) can be used as narrowband filters. The possibility of using the photosensitivity of optical fiber at short wavelengths for writing permanent gratings was discovered accidentally in 1978 [12]. Exposure to ultraviolet light causes changes in refractive index, particularly in fiber cores with germanium doping. A grating pattern can be achieved with two-beam interference or, more flexibly, using a phase mask in the exposure process. The refractive index perturbation is quite small (typically, about 10^{-4}), so the gratings are rather long (millimeters and even centimeters in length). FBGs are band-reflection filters, reflecting a narrow resonant bandwidth in a backward direction and transmitting the rest of the spectrum. With proper apodization of the grating modulation profile, spectral reflection profiles having a very flat top and a steep drop at the band edges can be achieved, with very low loss and good crosstalk isolation. Using advanced FBG design methods, filters have been demonstrated that are spectrally flat and dispersion-free over more than 75% of the 25-GHz channel spacing [13]. FBGs have a temperature coefficient for wavelength drift around 0.1 nm/$°$C. They can be packaged athermally, so that difference in thermal expansion coefficients induces stress in fiber that compensates for this, achieving negligible thermal drift.

Fiber Bragg gratings can achieve excellent performance, and they are, as such, very simple and compact, being essentially short pieces of fiber with permanent modulation written in. The disadvantage is, however, that separate gratings are needed for each wavelength, and they need to be cascaded for multiwavelength operation. Implementation of FBGs in WDM multiplexers has some complications, since as two-port pieces of fiber, additional components are needed to separate the backward and forward waves at the fiber end, as shown in Fig. 3.4. A simple, moderate-loss solution is to use an optical circulator. An optical isolator is a nonreciprocal component based on the magneto-optic Faraday effect. An optical isolator has two optical ports, and nonreciprocal operation is low-loss connection between the ports in one direction and high attenuation in the opposite direction. An optical circulator has at least three ports, and connections between these are nonreciprocal. The circulator symbol in Fig. 3.4*a* illustrates these connections by having an arrow. Light input in any given port is coupled out from the port that is next to it in the direction given by the arrow. Optical circulators are complicated and still rather expensive micro-optic devices. Alternatively, a fiber-optic Mach–Zehnder interferometer, having two identical FBGs on both interferometer paths between two 3-dB couplers, can be used as WDM add–drop devices (Fig. 3.4*b*). These require, however, very accurate control of optical pathlengths. An attractive variation of this structure is realized in a single piece of twin-core fiber [14].

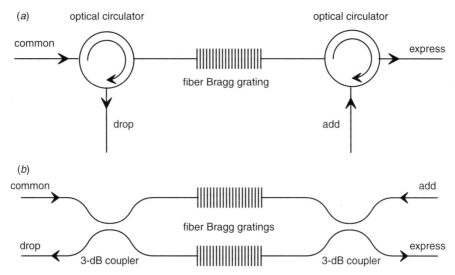

Figure 3.4   Fiber Bragg grating–based single-wavelength add-drop filters. A complete set of wavelengths is input at the common port, the wavelength in the FBG reflection band is coupled to the drop port, while the rest are passed to the express port. From the add port, a new signal at the dropped wavelength can be added to the output. (*a*) Construction with optical circulators. (*b*) Mach–Zehnder interferometer, having two identical FBGs on both interferometer paths between two 3-dB couplers.

Phased-array WDM devices [15–20] [also known as arrayed waveguide gratings (AWGs)] are integrated optics waveguide devices that can be used for multi-wavelength filtering. These devices have proved to offer several advantages that make them most important, particularly for dense, multichannel wavelength-division multiplexing applications. Since they can be realized with a basic fabrication process for single-mode waveguides, fabrication of these devices is relatively straightforward. Typically, silica-on-silicon waveguide technology is used. Doped-silica waveguides are fabricated by deposition with various techniques of silica layers on substrates, which are typically silicon wafers [21–23]. A phased-array waveguide device, shown in Fig. 3.5, can be considered as a generalization of Mach–Zehnder interferometer with multiple paths, or as a diffraction grating built from channel waveguides—it has both focusing and dispersive properties. Figure 3.5 illustrates the phased-array waveguide configuration: an array of single-mode waveguides coupled to both input and output waveguides radiatively through planar waveguide regions. Alternatively, it can be looked at as two star couplers connected by uncoupled waveguides of unequal lengths. It is a good example of a fundamental integrated optics circuit and may be further integrated with additional waveguide devices.

In a phased-array structure, both input waveguides and arrayed waveguides on the input side are arranged on a circle (Rowland circle configuration): The circle of arrayed waveguides is centered at the center input waveguide end, and the radius of

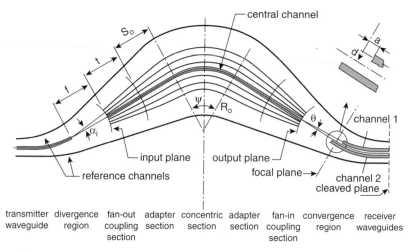

Figure 3.5   Waveguide geometry for a phased-array demultiplexer; for clarity, only a single input waveguide and two output waveguides are drawn. (From [15], © 1991 IEEE.)

this circle is called the focal length at array input. Light from the input waveguide diffracts horizontally in the planar waveguide region and couples into arrayed waveguides, which are first separated to become uncoupled in a fan-out section. The waveguides have unequal lengths, so that there is a constant length difference for adjacent waveguides. For the center wavelength of array, this length difference is a multiple of optical wavelength in the channel waveguide, so that the wavefront at array input is reproduced at the array output and focused into a center output waveguide. Thus a symmetric phased array forms a 1:1 image-forming structure. At a wavelength different from the center wavelength, additional phase difference between adjacent waveguides is accumulated while propagating through an array, thus producing a linear phase modulation or tilting of the array output wavefront. This results in off-center focusing of light, so that different wavelength channels are coupled into different output waveguides. If there are also off-center input waveguides, light is coupled from these into a phased array with a tilted wavefront, so that the output waveguide is selected by both the input waveguide position and the wavelength in combination.

The advantage of a phased array is that in comparison to the main competitor technologies, it has essentially parallel operation. Thus increasing the number of wavelength channels adds quite a little to the excess loss or cost of the device. Presently, the number of channels at which phased arrays become superior is around 16, a number that is already quite widely in use. Another advantage of phased-array filtering properties is that for very dense—100- or 50-GHz spaced—WDM, they can readily be designed and also have very low dispersion at passband edges. This dispersion can become critical in WDM networks, where a large number of filters can be cascaded in the optical path.

Initially, the performance of phased-array devices was in many respects non-ideal, so they were not at once attractive. Intense effort has led to several important advances in phased-array devices. They have been demonstrated with very low insertion loss (below 1 dB), 400 channels, and 25-GHz spacing [24]. Crosstalk was long considered a limitation to the number of useful channels, since total cumulative crosstalk from dozens of channels can easily come to a large number, but it has been possible to reduce this to a very low level [25]. Solutions have been found for polarization dependence [26,27], flattened wide passband response, temperature stabilization [28], and even athermal operation [29]. Typical commercial devices now available have 32 or 40 channels with 100-GHz spacing, about 5 dB insertion loss, and total cumulative crosstalk isolation better than 21 dB. For temperature stabilization, a heater or thermoelectric cooler element is used, since silica-on-silicon phased arrays have a temperature coefficient for wavelength drift of around 0.1 nm/°C. Flat-top passbands can readily be achieved, but at the cost of 2 to 3 dB of excess loss. Recently, a configuration to achieve flat-top passbands without inherent excess loss was demonstrated by integrating a first-stage Mach–Zehnder interferometer with a phased-array demultiplexer [30]. Still, bulk-optic demultiplexers using free-space gratings can achieve performance comparable to that of phased arrays. For example, such a demultiplexer with 16 100-GHz spaced channels showed minimal temperature dependence without temperature control [31].

One approach to upgrading to narrower WDM channel spacing is to use interleaver techniques. For example, in a 50-GHz-spaced WDM, the wavelength set is divided into two complementary sets with 100-GHz spacing. An WDM interleaver is the component that separates and combines the two wavelength sets having double-channel spacing, and other WDM filtering technologies are used at the double-channel spacing level. Thus interleavers are filters that have a spectral response that is periodic at a period of twice the channel spacing; otherwise, they have requirements quite similar to those of ordinary WDM filters. Several interleaver technologies exist. Figure 3.6 shows one of these, Fourier-expansion filters based on integrated optics waveguides, composed of cascaded directional couplers and delay lines, forming a Mach–Zehnder interferometer cascade [32]. Complex spectral transfer functions can be synthesized based on such filters. By integrating such an interleaver with two phased-array devices using silica-on-silicon technology, a 102-channel 50-GHz spacing WDM filter device with a 1-dB passband of 30 GHz width was demonstrated [33].

3.3.3 Erbium-Doped Fiber Amplifiers

The first demonstrations of Er-doped fiber [34] as a gain medium for optical signals were done in 1987 [35,36]. After that, the main remaining step in the way toward field-deployable optical amplifiers was the development of suitable semiconductor laser diode for pump sources. InGaAsP pump laser diodes at 1.48 μm were applied for Er-doped fiber amplifier pumping in 1989 [37]. It took some more time to develop pump sources at 980 nm, which have advantages in achieving lower noise properties and needing less electrical power. For some time it appeared that the

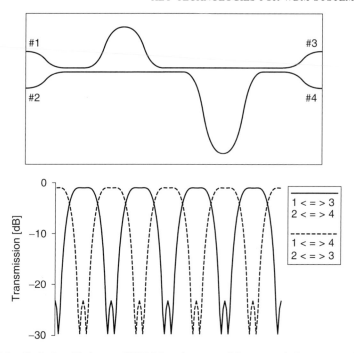

Figure 3.6 Optical guided-wave WDM interleaver and its transmission spectra between various ports.

most promising approach to achieving reliable 980-nm pump sources was to use Al-free GaInP/GaInAsP/InGaAs structures [38]. However, the catastrophic optical failure problem of AlGaInAs lasers could be solved, and in mid-1990, reliable 980-nm pump laser diodes were widely available [39]. Since then, much of the development in EDFAs has been led by achievements of increased output powers in pump laser diodes.

Optical amplification can be achieved in a fiber that has its core doped with erbium. Figure 3.7 is a simplified illustration of the three lowest energy levels for Er^{3+} and transitions between them. The energy levels are actually relatively wide energy bands: Each level is first split into several sublevels by Stark splitting caused by local crystal fields. As local environment in the amorphous glass host material varies for different Er ions, crystal fields and Stark splitting varies between them. This is known as *inhomogeneous broadening*. Also, there is additional phonon broadening (or *homogeneous broadening*). Photons having suitable wavelengths (e.g., 980 nm or 1480 nm) can be absorbed by Er^{3+} ions in the glass host material, leading to excitation of Er^{3+} from the ground state to an excited state. Absorption of a 1480-nm photon leads directly to excitation at the $^4I_{13/2}$ level. Absorption of a 980-nm photon ends up fast on this same energy level, through excitation first at the $^4I_{11/2}$ level and then nonradiative transition with a lifetime of several microseconds. The $^4I_{13/2}$ level is a metastable level, having a long lifetime (around 10 ms) for

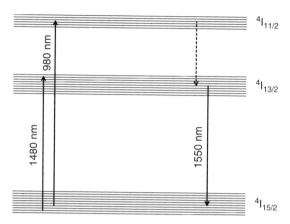

Figure 3.7   Energy levels and transitions for Er^{3+}. Radiative transitions with associated photon wavelengths shown with solid lines, noradiative transition with dashed line.

deexcitation through radiative transition, in which photons in the 1520- to 1570-nm wavelength region are emitted. Thus high-intensity 980- or 1480-nm light can excite most of the Er^{3+} ions in the glass host to this state, causing population inversion. Optical pump sources at either of these two wavelengths are used to achieve population inversion in Er-doped fiber, which is needed for optical signal amplification. It is to be noted, that thermalization within the various sublevels of the $^4I_{13/2}$ band leads to a very fast distribution of energies across the band.

Er^{3+} ions work actually as a three-level laser system. In this system three processes involved with transitions between the two lowest levels occur:

1. *Spontaneous emission* from higher level to the ground state at a rate ruled by the relatively long metastable lifetime.
2. Transition from a lower to a higher level through *absorption*; the absorption rate is decreased in strong population inversion, as there are not many Er^{3+} ions remaining at the ground level.
3. *Stimulated emission*, in which a photon having an energy corresponding to the excitation energy of the Er^{3+} ion stimulates the emission of a photon that is coherent and in phase with the stimulating photon.

All three processes occur at signal wavelengths (1520 to 1570 nm). For a signal propagating through a fiber, absorption and stimulated emission are competing processes: With population inversion, stimulated emission rate is higher and the signal experiences net gain. Spontaneous emission can be seen as a competing process consuming the energy pumped into the inverted material, but in addition the spontaneously emitted photons that are guided by the fiber core are also amplified; they thus consume even more of the energy in the fiber. Also, since these photons have random phases, this leads to an accumulation of optical background noise that

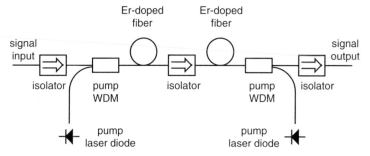

Figure 3.8   Erbium-doped fiber amplifier structure.

spectrally overlaps with the signal. As spontaneous and stimulated emission rates are connected by the Einstein coefficients relationship, there is a fundamental level of this amplified spontaneous emission relative to the signal power that is always present. This sets a lower limit to the noise in EDFAs.

An example of an EDFA structure is shown in Fig. 3.8. Er-doped fiber can be pumped codirectionally, contradirectionally, or as shown here, bidirectionally. Optical isolators are used to allow transmission through the EDFA in only one direction, to avoid amplified multiple reflection, which can deteriorate performance and even lead to laser oscillation of the amplifier. Optical tap couplers are typically used to couple small portions of input and output for signal monitoring and EDFA control purposes. EDFA optics is assembled by fusion splicing the various fiber-optic components together, and packaging this into a gain block package. Additional electronics is needed to drive the pump laser, to monitor its optical output, to control the pump laser temperature using a thermoelectric cooler, to monitor the optical input and output, and generally, to control the EDFA operation.

EDFAs have many good properties, which have led to using them in large numbers in optical transmission systems:

- Operation in the 1.55-μm minimum-transmission-loss window of standard single-mode fibers.
- Availability of high gain, low noise, and high optical output powers.
- Naturally applicable for WDM transmission, as operation is achieved across a wide optical bandwidth (typically, 30 nm).
- Due to the long excited-state lifetime, the dynamics of population inversion is slow, and thus there is insignificant crosstalk between various WDM signals at the high bit rates used in optical communication.
- High power conversion efficiency achievable.
- Low polarization dependence and temperature sensitivity.

Optical signals experience gain in transmission through an Er-doped fiber coil that has been pumped to high inversion. EDFA performance and properties depend on the details of the design, which includes Er-doped fiber type and length, pump powers, pump wavelengths, and pumping scheme. Modeling of EDFAs in detail has some complexities, as a detailed model has to include several interrelated optical intensity components across a wide spectrum and in two propagation directions. With the known inversion distribution through fiber, the total gain for signals at different wavelengths can be calculated from the average inversion through the fiber. However, to obtain the inversion distribution, some iterative type of calculation is needed. Inversion is achieved by absorption of pump photons propagating in either or both directions in the fiber, pump propagation and absorption depending on the local inversion. In most cases, the inversion profile particularly close to the output end is also affected by the signal, which has gained substantial power. In addition, there is backward and forward propagation of amplified spontaneous emission (ASE), which accumulates relatively high power levels at the two ends of the doped fiber, affecting inversion while competing with signals for pump energy.

Key parameters of optical amplifiers are gain and noise figure as a function of wavelengths for given input conditions. Gain is defined for signal input power P_{in} and signal output power P_{out} of signal at wavelength λ as

$$G(\lambda) = \frac{P_{out}}{P_{in}} \qquad (3.14)$$

Noise effects are characterized in terms of signal-to-noise ratio (SNR) at the electrical level. *Noise figure* is defined as the ratio of SNR at signal input to SNR at signal output. However, it can be shown that the noise figure of EDFA is related directly to optical quantities and can be expressed in terms of them. This makes it possible to examine noise accumulation in EDFA systems at the optical level, and to calculate the impact of it on signal quality.

The noise contribution of EDFA comes from the ASE component it adds to the amplified signal. ASE spectral density can be expressed as

$$\frac{P_{ASE}}{\Delta f} = 2n_{sp}(G - 1)hf \qquad (3.15)$$

Here expressed as P_{ASE} is the ASE power in the narrow frequency bandwidth Δf, G is the EDFA gain, hf is the photon energy (product of Planck's constant h and optical frequency f). The factor of 2 is due to the presence of two polarization modes in the single-mode fiber, both having the same ASE density. Spontaneous emission factor n_{sp} can also be expressed as the inversion parameter, relating it to inversion of the fiber:

$$n_{sp} = \frac{\sigma_e(\lambda)N_2}{\sigma_e(\lambda)N_2 - \sigma_a(\lambda)N_1} \qquad (3.16)$$

Here $\sigma_e(\lambda)$ and $\sigma_a(\lambda)$ are the spectral emission and absorption cross sections of Er^{3+}, and N_1 and N_2 are the populations of the Er^{3+} ground state and excited state, respectively. It is to be noted that as the inversion parameter is a local quantity and equation (3.16) describes the total ASE density at the fiber output, the spontaneous emission factor and inversion parameter are strictly the same only when fiber has a constant inversion through it. However, it can be observed that to achieve a low noise level, high inversion is needed. Also, there is always a minimum noise level for given gain, as $n_{sp} > 1$.

It can be shown that the noise figure N_F of EDFA is

$$N_F = \frac{2n_{sp}(G-1)}{G} + \frac{1}{G} \tag{3.17}$$

Thus relating it to the ASE power density yields

$$N_F = \frac{(P_{ASE}/\Delta f)}{hfG} + \frac{1}{G} \tag{3.18}$$

For high gain, $G \gg 1$, it is a good approximation to simplify $N_F = 2n_{sp}$, and

$$\frac{P_{ASE}}{\Delta f} = N_F Ghf \tag{3.19}$$

It is observed that the theoretical minimum noise figure expressed in decibels is 3 dB. Noise figures close to this theoretical minimum have been demonstrated for EDFAs. Practical EDFAs typically have noise figures in the range 4 to 6 dB.

Finally, quick estimates for ASE power density can be made from equation (3.19). ASE power of EDFA in dBm in the 0.1-nm bandwidth is, with gain G and noise figure N_F in decibels,

$$P_{ASE} = -57\,\text{dBm} + G + N_F \tag{3.20}$$

To characterize the signal quality at the optical level, an optical signal-to-noise ratio may be used. *Optical SNR* is defined as the ratio of signal power to ASE power in a selected bandwidth; typically, 0.1-nm bandwidth is used. For different types of signals, minimum optical SNR limits can be set.

EDFAs can be divided into three main functional categories: power amplifiers used to boost signal power at the transmitter end of a link; line amplifiers to compensate for transmission losses in midspan of a long fiber link; preamplifiers used to increase sensitivity at the receiver end of a link.

Power amplifiers are used to increase signal power by adding some gain for signals that are at a moderate level. Typical WDM transmitter output powers are at the 0-dBm level. In a WDM system, there is additional loss from multiplexing the various wavelength signals in the same fiber. Typically, this is below 10 dB but can be higher, when combiners are used for multiplexing instead of filters. So a power amplifier may also compensate for multiplexing loss, but its general purpose is to achieve the high launch powers necessary to maintain sufficient input power levels at other amplifiers in the link. It is seen that the relative level of ASE noise added by an EDFA with a given gain and noise figure decreases with the increase in input power. Thus the total noise contribution of the link can be minimized using a power amplifier. The main requirement for a power amplifier is a sufficiently high output power, particularly in multichannel WDM systems, in which total power is divided between a multitude of wavelengths. Output power can be 15 to 20 dBm or even higher, with 10 to 15 dB gain. Because of the moderate input power, the noise figure of the power amplifier is not very critical, and it can be 7 to 8 dB. Backward pumping configurations are typically used, since they can support maximum inversion at the EDFA output end, where the signal level is at a maximum and pump energy use is maximized to maintain gain. For high output powers, high-powered pumps are needed; 1480-nm pumps are often used, since they are available at higher powers than those of 980-nm pumps.

Line amplifiers are used to compensate for transmission losses in the typically 80- to 120-km span between the amplifiers. Thus they need large gains (20 to 30 dB) and high output powers. Additionally, due to the lower input levels, low noise figure is also a requirement. Bidirectional pumping is often used. Many line amplifiers are in fact of two-stage design, cascading two EDFAs. Division into two stages is mainly through isolation of the backward second-stage ASE from the first stage, achieved by an optical isolator between the stages, but it also involves independent pumping of the stages. Simple examination of signal and ASE noise levels contributed by the cascade of two amplifier stages gives the result, using noise figures $N_{F,1}$ and $N_{F,2}$ and gains G_1 and G_2 of the first and second stage, respectively:

$$N_{F,\text{total}} = N_{F,1} + \frac{N_{F,2}}{G_1} \qquad G_{\text{total}} = G_1 \times G_2 \qquad (3.21)$$

Thus the noise figure of the first stage dominates. A low-noise first stage can be combined with a second stage that has high output power, with the total gain that is simply the product of gains for them. To achieve low noise, a 980-nm pump is used, since 1480-nm pumps cannot achieve maximum levels of inversion. Either forward pumping is used, so that inversion is maximum at the EDFA input side, leading to a lower forward ASE—or backward pumping is carried out using relatively high pump power, so that high inversion is maintained throughout the Er-doped fiber loop. Although this latter approach leads to a large fraction of unused remnant pump power, the advantage is that there is no pump multiplexing component at the EDFA input, with its added insertion loss. It is observed that every decibel of additional loss at the EDFA input, before the gain-producing Er-doped fiber, is added to the noise figure and subtracted from the gain of the total EDFA unit. If there is a need to place additional functionality at the line amplifier location, such as dispersion compensation or WDM add–drop, the optimum position is between the stages of the two-stage EDFA. The insertion loss of added optical elements then has the lowest total impact on the link noise.

Optical preamplifiers can substantially increase receiver sensitivity. Ordinary direct-detection receivers have sensitivity limited by the thermal and shot noise in the electronic circuitry of the receiver. Using an optical preamplifier, optical signals are amplified before detection to a level where such circuit noise components become insignificant, and the noise resulting from ASE dominates. At the detector, photons are converted into electrons in a square-law process (electronic signal electrical field component is proportional to the squared optical signal electrical field component). From the interference of signal and ASE photons sharing the same optical frequency band, but having random phase relationships, beat frequency components are generated which appear as electronic noise. Beat frequencies that are inside the receiver frequency band contribute to the electronic current SNR. This is known as signal-spontaneous beat noise. There is an additional, smaller contribution from ASE photons beating between themselves, known as spontaneous–spontaneous beat noise. An optical filter that filters out the ASE noise outside signal optical bandwidth can substantially decrease the latter component. Most dense WDM systems have sufficient ASE filtering in demultiplexing the WDM signals. This is one more respect in which EDFAs are WDM amplifiers, since to minimize ASE effects, optically narrowband operation is required in the system.

3.3.4 Dispersion Compensation

As chromatic dispersion is a fully deterministic effect, it is in principle possible to compensate for it fully [40]. This is achieved for a signal having gone through a fiber having a known total dispersion by putting the signal through an additional element having a dispersion with equal magnitude and opposite sign. The total cumulative dispersion will then be equal to zero. In practice, it is sufficient to reduce the total dispersion to a sufficiently low value by adding positive and negative dispersion components in the total transmission span. As a dispersion management technique, this has a strong advantage, since the local dispersion can remain high enough to keep the four-wave mixing effect insignificant while avoiding the dispersion limitations for the transmission.

In WDM systems this is complicated by the spectral dependence of the fiber dispersion, since the dispersion experienced by different wavelength channels varies. If large amounts of dispersion need to be compensated, matching of total positive and negative dispersion components needs to be done accurately, and there are two basic approaches to achieving this. Either dispersion compensation is done for each wavelength separately, to adjust dispersion for each, or if compensation is done for the full WDM bandwidth, it is necessary not only to match opposite magnitudes of dispersion but also spectral slopes of dispersion. There is some preference for the latter approach, as a smaller number of compensating elements is needed.

Currently, dispersion-compensating modules applied in the field are all based on special fibers having tailored values of dispersion. Typically, high-negative-dispersion fibers have been made for this application that can compensate for 100 km of standard single-mode fiber in a length of 20 km. Also, more recently, these dispersion-compensating fibers have become available with the ratio of dispersion to dispersion slope matching various commercial fiber types, so accurate, wideband dispersion compensation can be achieved.

The main alternative technology for dispersion compensation has been chirped fiber Bragg gratings, manufactured the same way as FBGs for WDM filtering. The grating constant of these varies linearly through the length of grating, so that effectively, different wavelengths are reflected at different distances and experience variable group delays. Typically, only single-channel dispersion compensation is achieved, since very long gratings would be required to compensate across a wide EDFA band. Sufficient perfection of these FBG compensators has not yet been achieved, the main limitation being the ripple of group delays.

3.3.5 Variable Optical Attenuators

Integrated, easily controllable compact variable optical attenuators find a number of important uses in optical networks. They are used to tune WDM signal powers for optimum performance and to allow control of network operation, which is seen quite essential to manage WDM networks. For example, it is often seen as necessary to equalize WDM channel powers at each network node. This means having attenuators at each node for each WDM channel, so the number of attenuators can be large indeed. Clearly, integrated attenuator arrays are of interest here. Use of guided-wave variable optical attenuators in this context was introduced in [41]. Variable attenuator structures were tunable Mach–Zehnder interferometers (MZIs) in silica-on-silicon waveguides, similar to optical modulators but slower, for semistatic operation. Operation is based on a thermo-optic effect driven by small heater electrodes. System requirements were defined as: attenuation dynamic range, 20 dB; minimum loss; below 2 dB; and response time, milliseconds. A similar advanced variable attenuator device was also published [42]. Currently, attenuators and particularly attenuator array devices are available from several vendors.

3.3.6 Optical Switching Devices

Moving from pure WDM transmission to other WDM network functionalities, optical switching has a key role. The critical performance issues for switching components are [43]:

- Transmission characteristics: loss, crosstalk, polarization dependence, and wavelength dependence
- Switching time
- Power consumption
- Temperature sensitivity
- Requirements related to implementation: packaging, manufacturing yield, size, reliability, and manageability
- Scalability properties, including number of ports and blocking properties

Crosstalk is a particularly critical issue in networks. Routing signals through many cascaded switching stages leads to a crosstalk accumulation, and particularly critical is crosstalk within the signal wavelength band, originating either from other signals in the network having the same wavelength or from the signal itself from traveling through alternative crosstalk paths. This in-channel crosstalk cannot be filtered out and can cause coherent beat noise at the receiver even at relatively low levels. The tolerable switch component crosstalk level varies depending on the node architecture and on the number of nodes, but typically, -40 dB is considered safe in many architectures. For circuit switching, switching times of around 1 ms or even several milliseconds is quite adequate.

There are numerous switching technologies, but not all of them have the potential for achieving large-scale switching fabrics. Here, only technologies based on integrated optics and optical MEMS (micro-optoelectromechanical systems) are discussed, as two of the most promising approaches. An additional potential future technology is based on the use of semiconductor optical amplifiers (SOAs), semiconductor laser diode structures with low-reflectance facets, acting as optical amplifiers instead of sources. They can be switched with electrical current in a nanosecond time scale between an on state with gain and an off state with high isolation. Such SOA gate switching has been demonstrated [44], but challenges remain in integrating them into large-scale switching fabrics.

In addition to the purely passive functions, the thermo-optic effect allows some dynamic functions to be realized with silica-on-silicon technology, as discussed in connection with the variable optical attenuators. For example, a large number of 2×2 switch elements with a Mach–Zehnder structure have been integrated to realize a compact 16×16 switch matrix [45]. The limitation is mainly modest speed (around milliseconds of switching time) achievable with thermo-optic effects and high electrical power (typically, 0.5 to 1 W for a single switch element) drawn by microresistor heating elements deposited on top of the waveguide devices. To achieve larger-scale switch matrices, a 1×128 switch was demonstrated, having 4 dB of insertion loss and a 40-dB extinction ratio [46]. By interconnecting such switch modules by fiber cross-wiring, a 128×128 optical switch can be achieved, but this requires 256 switch chips and a very large number of fiber interconnections.

Polymer materials have allowed tailoring of waveguide properties to a larger extent. Particularly for thermo-optic switching operation, better performance than using silica-on-silicon in terms of switching time and electrical power consumption has been achieved, due to the availability of a higher thermo-optic coefficient and lower thermal conductivity. Although switching times are still in the millisecond regime, electrical switching powers for single switch elements that are below 100 mW have been achieved. A thorough reliability assurance program has indicated that polymer optical switch technology can meet the requirements for telecommunication applications [47]. Integration capability has been demonstrated up to an 8×8 switch matrix [48]. Instead of a Mach–Zehnder interferometer-type switch used in silica waveguide technology, a high thermo-optic coefficient allows use of a digital optical switch structure, which does not require precise electrical control. However, loss and birefringence properties of waveguides have difficulty competing with glass waveguides, and much additional work is needed to develop materials and technology for use in passive device operation. It has been suggested to use hybrid polymer–silica integration technology, combining the advantages of both materials for integrated optics circuits [49].

As an alternative to thermo-optics, similar guided-wave structures can be fabricated using electro-optic $LiNbO_3$ waveguide technology, which was the basis for external modulators. Very fast switching in nanosecond regime can be achieved, and up to 16×16 switch matrices [50] have been demonstrated, but there are issues with polarization dependence, and complications with manufacturing such large-scale waveguide circuits in this material system have kept the manufacturing at quite a limited scale. It must be said that more recently, micromechanical solutions have emerged as a technology that is more advantageous than integrated optics for realization of large-scale compact switch matrices. An interesting hybrid solution combines silica planar lightwave circuits with switching based on total internal reflection from a bubble generated in the fluid-containing trench by a thermal actuator. A very compact 32×32 crossbar switch fabric with good performance has been demonstrated [51].

As an example of the potential of optical MEMS, a 112×112 matrix (112 input fibers connected to 112 output fibers) was fabricated using 224 micromirrors on a silicon wafer. These mirrors can be tilted around two axes using electrostatic actuation, and they can switch free-space beams. Nonblocking switching is achieved with the number of mirrors scaling as $2N$, where N is the number of input/output fibers: two mirrors in each switching path. Figure 3.9 illustrates the construction of

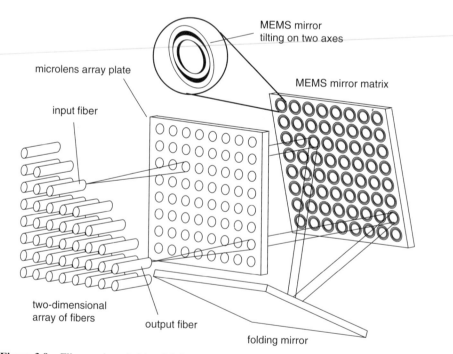

Figure 3.9   Fiber-optic switching fabric construction based on a MEMS micromirror matrix.

such switching fabric. Recently, a wavelength-selective 72×72 cross-connect switch based on this MEMS technology and on phased-array multiplexers was published [52]. The reliability performance of optical cross-connects based on MEMS technology was analyzed [53]. Although it appears to be more reliable than other proposed technologies, there are still some questions, such as whether adequate reliability is achievable for very large scale optical cross-connects.

3.3.7 Wavelength Converters

Wavelength conversion is needed to more fully exploit the wavelength dimension in WDM networks. Wavelength converters perform wavelength adaptation at network interfaces. They are used to avoid wavelength blocking in optical cross-connects, to facilitate allocation of wavelength paths for each link, generally to allow simpler management of network resources and scalability of architecture, and to perform routing functions in combination with optical filters.

Several approaches have been proposed for wavelength conversion, the simple optoelectronic conversion as in WDM transponders being one. Today, the only all-optical alternative approaching maturity is utilization of cross-phase modulation in semiconductor optical amplifier (SOA) structures [54]. Interferometric structures are used to convert cross-phase modulation to intensity modulation. Compared with a Michelson interferometer, a Mach–Zehnder interferometer (MZI) has become more popular due to separation of input and output signals by counterpropagation (see Fig. 3.10). These interferometric wavelength converters operate typically with fixed-input wavelength and variable-output wavelength or with fixed-output wavelength and variable-input wavelength. Output wavelength is determined by the

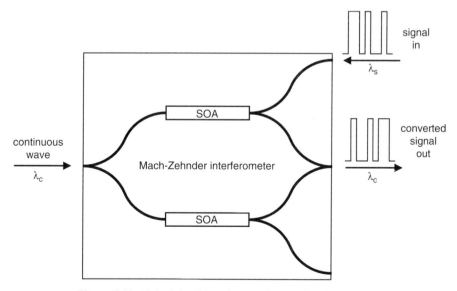

Figure 3.10   Principle of interferometric wavelength converter.

external continuous-wave (CW) source modulated at the SOA structure. Such wavelength converters are good examples of high-level monolithic optical integration of waveguide devices. Carrier dynamics of SOAs allows high-bit-rate operation: above 10 Gbps and up to 20 Gbps or even 40 Gbps. A low chirp of output signal allows long-span transmission. The nonlinear transfer function of the device is utilized for more complete regeneration of signal; not only is the signal power regenerated but there is also reshaping, regenerating the extinction ratio, and thus the device acts as an optical 2R-regenerator. This compensates partially for signal impairments during transmission and routing.

3.4 DEVELOPMENT OF WDM SYSTEMS

First-generation WDM systems were using 2.5-Gbps channel rates. Channel plans were limited up to 32 channels at 100-GHz spacing, based on the available total EDFA gain bandwidth and WDM filter technologies combined with the wavelength stability of sources. At the time, existing terrestrial fiber plant consisted of standard single-mode fiber, excluding dispersion-shifted fiber, which was not suitable for WDM use, due to the four-wave mixing problem. As dispersion-limited transmission spans for standard single-mode fiber at 2.5 Gbps are 500 to 800 km, dispersion was no issue with externally modulated sources. So, in these first-generation optically amplified WDM systems, the transmission limitations came from the ASE noise accumulation and from the optical nonlinearities.

Sufficient signal-to-noise ratios in EDFA cascades can be achieved by shortening amplifier spacing or by increasing launch powers. EDFA spacing was limited typically to 80- to 100-km minimum for long-span links. For launch powers, compromise had to be made between deterioration due to noise and nonlinearities (mainly self-phase modulation). Commercial systems are currently available having up to 160 and even 320 WDM channels. Channel spacing down to 25 GHz is available using present technology. Beyond the conventional erbium-doped fiber amplifier band, the C-band at 1530 to 1565 nm, an additional L-band at 1565 to 1570 nm has been taken into use. Gain in the L-band is also provided by EDFA technology. Special L-band EDFAs are based on the fact that Er-doped fibers can also provide more modest gain in this longer-wavelength side when pumped to lower levels of inversion. Longer Er-fiber coils as well as different engineering for pumping is used, and amplification in the C and L bands is done using separate parallel amplifiers for the two bands and with band-separating wideband optical filters.

Also, the single-channel bit rates have seen the jump from 2.5 Gbps to the next level, provided by electronic time-division multiplexing of 10 Gbps. With this, the dispersion again became an issue. Maximum 10-Gbps transmission span in standard single-mode fiber, limited by dispersion, is around 100 km. In nonzero dispersion fiber, several types of which have been developed as lower-dispersion fibers for WDM use, about 500-km spans can be achieved. So, for WDM systems, use of dispersion compensation became a necessity. Terrestrial fiber links are usually limited to using an existing fiber plant, consisting of a single type of optical fiber. Dispersion-managed transmission cable spans, consisting of subsequent sections of positive- and negative-dispersion fiber with average dispersion close to zero, cannot be used. The technique to be used consists of adding sections of suitable dispersion-compensating fiber at the optical amplifier sites. The disadvantage is that these dispersion-compensation modules contribute additional loss without adding to transmission spans. Use of nonzero dispersion fiber would still be advantageous, since with it there is smaller material need and loss contribution from dispersion compensation.

Also, with a higher bit rate and different dispersion map, the effect of nonlinearities needs to be reconsidered. The effect of cross-phase modulation could, for example, accumulate over several optically amplified spans, if accurate dispersion compensation reproduces exactly the initial alignment of WDM channel bit streams at the beginning of each span. Also, enhanced nonlinearity of dispersion-compensating fiber, with its small effective area, could be a problem, particularly as the optimum position of dispersion-compensating modules is between the stages of a double-stage EDFA.

Finally, it is to be noted that to maintain the acceptable bit-error rates, when upgrading from 2.5 to 10 Gbps, an increase of 6 dB in receiver power and in optical signal-to-noise ratios in the system is needed. In upgrading, it is typically not feasible to change amplifier spacing, so higher output powers from EDFAs are needed. Optical nonlinearities can hinder such an addition of launching power, and then other approaches to maintain bit-error rates are needed.

One possibility is to take into use forward error correction coding at the digital electronic level, to achieve sufficient bit-error rates [55]. Such coding can typically add an effective 5 to 6 dB into the system power budget. Another approach is to use Raman amplification. Raman amplifiers, utilizing stimulated Raman scattering, were actually studied well before EDFAs. Through SRS, a high-energy pump wave propagating in fiber can provide gain for optical signals at longer wavelengths. The advantage of Raman amplification is that standard telecommunication fiber can be used as a gain medium, and it can be used at any wavelength, provided that the suitable lower-wavelength pump lasers are available. Mostly, Raman amplifiers are used in combination with EDFAs, so that at EDFA sites, backward Raman pumping into preceding transmission fiber section is used for distributed gain over a rather long distance transmission link. The advantage of backward pumping is that relative intensity noise of pump lasers is averaged over the total gain length. As an instantaneous process, SRS can duplicate forward pump noise to signal. The disadvantages of Raman amplification include the high pump power levels (around 1 W) required and the need for careful engineering to avoid too much penalty from amplified multiple Rayleigh scattering. Pump laser diodes have wavelengths between 1400 and 1500 nm and optical powers around 0.5 W, and they are quite similar to high-power, 1480-nm pumps for EDFAs. Polarization multiplexing of two pumps is used, effectively combining the output of two lasers into different optical polarizations of a single fiber. This gives higher pump power and polarization-independent gain. Additionally, WDM multiplexing of several pump wavelengths is used when designed carefully to achieve a sufficient level of gain that is flat over the optical bandwidth required. Added Raman amplification can substantially improve the effective noise figure of amplifiers, since it adds gain a considerable distance before the actual amplifier site, thus preventing the signal from falling to a low power level. Also, the distributed Raman gain does not amplify signals to such high power levels that nonlinear effects would become critical. An additional balancing between nonlinearities and amplification comes from the fact that Raman amplification uses a nonlinear process in fiber. Fibers that are more effective for utilization of pump power in distributed Raman amplification also have lower thresholds for nonlinear effects in general.

The development of WDM systems deployed in the field over the past five years is seen to have achieved increases by a factor of 4 to 5 in both achievable WDM channel number and channel density, and also in channel bit rates. It is to be noted that this is not simply a quantitative development but can clearly be observed as a qualitative evolution. The 100-GHz spaced systems at 2.5-Gbps rates were not yet representing real dense WDM. Spectral efficiency can be measured as a rate of channel capacity to channel spacing, and present state-of-art 50-GHz spaced systems at 10-Gbps rates have spectral efficiency of 0.2 bit/s per hertz. With 25-GHz spaced technology also available at commercial system level, spectral efficiency is 0.4 bit/s per hertz. It will also be as a standard in the emerging next-generation 100-GHz spaced systems at 40-Gbps rates [56]. Achieving such spectral efficiencies was no small challenge using straightforward WDM technology, as the signal bandwidth occupies a substantial part of the total filter bandwidth, and all wavelength tolerances of filters and sources need to be accounted.

Already, dense WDM systems are approaching the limits for total system capacity set by the stimulated Raman scattering. From the above-mentioned 6 dB need of increase in optical signal power, when upgrading from 2.5 to 10 Gbps (and from 10 to 40 Gbps), it is observed that this SRS limit basically depends only on the total system capacity and spectral efficiency. At the time of first-generation WDM systems, some five years ago, record laboratory achievements for single-fiber transmission capacity were at the 1-Tbps level. Currently, above-10-Tbps capacities have been achieved. Systems of 40 Gbps per channel are expected to readily achieve total WDM capacity around 3 Tbps, using 100-GHz spacing in C and L bands. Presently achieved transmission capacity record experiments show possible routes for higher capacities [57,58].

Using vestigial sideband (VSB) modulation, with 128×40 Gbps WDM channels and alternative 75- and 50-GHz channel spacing, a 5.12-Tbps transmission experiment for a 300-km total span of nonzero dispersion fiber was achieved. In VSB filtering, optical filtering at the receiver end was done with a filter having a narrow 30-GHz bandwidth tuned 15 GHz from the center of the signal bandwidth to the 75-GHz spacing side. This isolates one of the two redundant signal sidebands that has the smaller adjacent crosstalk component. Spectral efficiency was 0.64 bit/s per hertz. This experiment is illustrated in Fig. 3.11. The 128 different wavelengths are divided into sets of 32 wavelenghts, two interleaved sets in the C-band and two in the L-band. Fiber-optic 1- to-32 combiners are used to couple

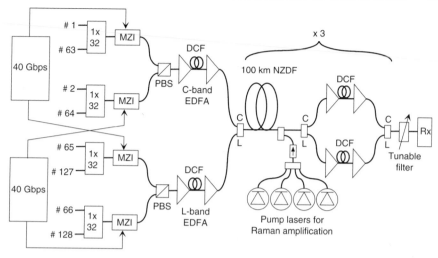

Figure 3.11 A 5.12-Tbps experiment. MZI, Mach–Zehnder interferometer modulator; PBS, polarization beam splitter [57].

each set into a LiNbO$_3$ modulator, modulating the same 40-Gbps signal for all 32 wavelengths. Polarization beam splitters can combine two interleaved sets into the same fiber, as modulators have single-polarization outputs. Two-stage C- and L-band amplifiers with dispersion compensation boost the signals to launch level before combining the C and L bands into transmission fiber. The link consists of three similar 100-km spans. Distributed Raman amplification in the NZDF transmission fiber, and dispersion-compensating fiber (DCF) modules in two-stage line amplifiers were used. Typical of this type of record-breaking laboratory experiments, it was not a completely built up transmission system, but one aimed to verify such transmission capacity to be achievable. Therefore, the experiment did not include unnecessary redundancy in the equipment. For example, only two different pseudorandom-bit sequences were actually transmitted, so that only adjacent wavelength signals, as primary sources of crosstalk, were completely independent. Of course, dispersion would also substantially unsynchronize the other bit streams. Also, only a single channel at a time was received, for measuring the bit error rate.

This experiment was extended to double the total capacity to 10.2 Tbps [59] with 100 km of transmission distance, in a nonzero dispersion fiber, further using polarization division multiplexing. Another set of signals at the same wavelengths but having orthogonal polarizations was added to the transmission link. Polarization demultiplexing was carried out at the receiver end using a polarization controller and polarization beam splitter. FEC coding was used, FEC overhead increasing the optical bit rates to 42.7 Gbps. Spectral efficiency was 1.28 bit/s per hertz.

An alternative route toward increased capacity, in addition to the higher spectral efficiency, is through enhanced optical bandwidths. S-band amplifiers in the

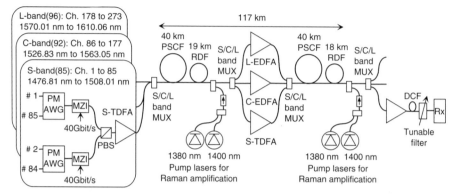

Figure 3.12 A 10.92-Tbps transmission experiment. C-EDFA, C-band EDFA; L-EDFA, L-band EDFA; S-TDFA, S-band thulium-doped fiber amplifier [60].

1490-nm regions have been demonstrated using Raman amplification and thulium-doped fiber amplifiers. The latter alternative has recently been demonstrated as a workable technology, and it was used in a 10.92-Tbps transmission experiment [60], illustrated in Fig. 3.12. This experiment used 50-GHz-spaced 273 × 40 Gbps transmission in S, C, and L bands, with spectral efficiency of 0.8 bit/s per hertz. Polarization multiplexing was again used in a form of polarization interleaving: two sets of 100-GHz spaced channels, offset from each other being at 50 GHz of optical frequency, which were combined at orthogonal polarizations. Transmission span was 117 km and used successive sections of pure-silica core fiber (PSCF) with reduced nonlinearity, and reverse-dispersion fiber (RDF), partially compensating for PSCF dispersion and acting as an effective medium for distributed Raman amplification. Demultiplexing at the receiver combines frequency and polarization filtering. As in the experiment described above, polarization multiplexing is rather simple to achieve, as transmitter outputs have typically known linear polarization. Due to the random state of polarization after transmission, polarization controller with feedback is required at the output. With optical PMD compensators, polarization control may become included in high-capacity WDM systems in the future. This kind of polarization interleaving is also sensitive to PMD, since PMD is related directly to the spectral dependence of polarization rotation; with too high PMD, polarization states of adjacent wavelengths will no longer be orthogonal, causing additional crosstalk penalty.

In all the multiterabit transmission experiments described above, SRS effects were observed as a transition of power from shorter to longer wavelengths, and spectrally nonflat gain profiles were used to partially compensate for this. An additional consideration of extending gain bandwidth to the S band or other regions is that, with Raman amplification in the system, there may be a conflict between signals and Raman pumps trying to occupy the same spectral region.

The 10-Tbps capacities demonstrated are already very impressive and are efficient in using about 10 THz out of the total maximum optical fiber bandwidth, which is around 50 THz. Using direct detection or binary modulation techniques, as in these experiments, it is not possible to achieve much higher spectral efficiencies. Classical Shannon limits on spectral efficiency depend on the noise present in the information channel. In fiber-optic systems, increase of signal power to achieve a higher signal-to-noise ratio leads to limitations set by optical nonlinearities. One of the most thorough attempts to define limits for optical fiber transmission capacities considers the limiting nonlinear effect to be cross-phase modulation [61]. With coherent optical detection, achievable spectral efficiency values were then calculated to be around 4 bit/s per hertz. However, even this work has not yet defined the ultimate limits, having for example left polarization multiplexing out of consideration [62].

3.5 SUMMARY

It has been pointed out that fiber-optic transmission capacities, doubling every 16 months, have been outrunning the well-known Moore's law describing the development of semiconductor integrated-circuit technology development. However, it would be a mistake to conclude that fiber-optic technologies are developing at a rate faster than that for integrated circuits. Single-fiber transmission capacity increases are not coming only from increased performance of related technologies, but typically also from adding more technologies, more components with larger footprints, and more equipment, lined up in parallel at the end of fiber cable. They are much more costly and consume increasing amounts of power.

It is obvious that there is high diversity and fragmentation in fiber-optic technologies, involving numerous material systems, fabrication processes and manufacturing methods. Experience from the field exhibits the manual-labor-intensive nature of fiber-optic component production: the need for accurate optical alignments, fiber splicing and pigtailing, micro-optics assembly and packaging, and time-consuming optical measurements and characterization. Many of the technologies lack the potential for integration and for effective automation to achieve better scalability of production. Although present technology can well serve the high-end core networking market, economy of scale would be needed to achieve volume production, bringing the price of optical bandwidth down to the level that would enable it to penetrate farther from core to metropolitan and access network markets. The diversity of photonic technologies, lacking standards and universal manufacturing practices even inside specific technologies, are parts of this picture. It remains to be seen how the evolution of optical technologies can meet this anticipated challenge of making cost-effective volume production. Some of the promising technology areas that may have important roles in such a future are pointed out below.

It is expected that gigabit Ethernet transceivers as well as emerging 10-gigabit Ethernet transceivers, will be high-volume optical communication products, bringing cost down to a low level. These modules will mainly be using semiconductor electronics and photonic devices with quite limited and basic functionality in optics. Optoelectronic integrated circuits can be developed for high-speed optical interfaces, where InP is utilized both for optics and electronics in receivers and transmitters. Among the advantages of such monolithic integration are effective packaging, reduced footprint, and parasitics. Optical MEMS technologies provide wide scalability and a high level of integration in large-scale switching fabrics, as discussed earlier in the chapter.

Integrated optics promises enhanced functionality and performance, compactness, and cost-effective volume production capability. Silica-on-silicon phased-array WDM multiplexers are good examples of parallel functionality in a compact module, with impressive advances in performance and volume production, and lower total costs. Integration of phased-array WDM devices with a number of other guided-wave devices using silica-on-silicon platform has been demonstrated in various examples: for example, to make an optical add–drop multiplexer with individually controlled switches for each wavelength channel [63]. Recently, vendors have developed new products integrating variable attenuators for each channel of phased-array demultiplexer, which is an example of the emergence of complicated integrated optical circuits as commercial products. In addition, highly functional components can be realized using hybrid integration where optical waveguide circuits are used as a platform for mounting active semiconductor components such as lasers and detectors [64]. Examples of these components are multiwavelength sources and receivers. An attractive goal is the low-cost manufacturing of integrated optics bidirectional single-fiber transceivers, which can be key enablers for optical access networks. Substantial progress has also been made in the development of supporting technologies for integrated optics manufacturing, including packaging, testing, and component design.

Finally, InP waveguide technology monolithically integrated with active semiconductor photonic devices and even electronics has long been expected as a technology capable of combining all the photonic functionalities. Currently, instead of this comprehensive level of integration, research has concentrated on two main approaches to more limited levels of integration. These are optoelectronic integrated circuits (OEICs)—integrating optical and electronic devices—and photonic integrated circuits (PICs)—integrating waveguides with active optical devices such as lasers and detectors. In PICs, phased-array devices are of interest. InP-based phased arrays have dimensions that are an order-of-magnitude smaller than those of silica-on-silicon devices. Due to this smaller size, their insertion losses are also at the same level, although other performance, including crosstalk, has not yet reached quite as good a level. The advantage for InP PICs is integration with active devices, various demonstrations of which have been made [65]. For example, integrated WDM receiver combines phased array with detectors, and optical cross-connect devices integrate a phased array with switches. However, a vast amount of work is still needed to develop such photonic integrated circuits. To integrate different devices, compromises need to be made in materials, process quality and simplicity, and in performance in order to achieve compatible manufacturing processes.

REFERENCES

1. A. R. Chraplyvy, Limitations on lightwave communications imposed by optical-fiber nonlinearities, *Journal of Lightwave Technology*, 8(10):1548–1557, 1990.

2. *Characteristics of a Single-Mode Optical Fibre Cable*, ITU-T Recommendation G.652 (04/97).

3. *Characteristics of a dispersion-shifted single-mode optical fibre cable*, ITU-T Recommendation G.653 (04/97).

4. *Characteristics of a non-zero dispersion shifted single-mode optical fibre cable*, ITU-T Recommendation G.655 (10/96).

5. *Optical interfaces for multichannel systems with optical amplifiers*, ITU-T Recommendation G.692 (10/98).

6. B. Villeneuve, M. Cyr, and H. B. Kim, High-stability wavelength-controlled DFB laser sources for dense WDM application, in *Optical Fiber Communication Conference*, Vol. 2, OSA Technical Digest Series, Optical Society of America, Washington, DC, 1998, pp. 381–382.

7. R. A. Becker, Traveling wave electro-optic modulator with maximum bandwidth-length product, *Applied Physics Letters*, 45:1168–1170, 1984.

8. E. L. Wooten, K. M. Kissa, A. Yi-Yan, E. J. Murphy, D. A. Lafaw, P. F. Hallemeier, D. Maack, D. V. Attanasio, D. J. Fritz, G. J. McBrien, and D. E. Bossi, A review of lithium niobate modulators for fiber-optic communications systems, *IEEE Journal on Selected Topics in Quantum Electronics*, 6:69–82, 2000.

9. A. Ramdane, Integrated laser-modulators for high speed transmission, *Proc. ECIO'97 8th European Conference on Integrated Optics*, Stockholm, Apr. 2–4, 1997, pp. 112–117.

10. H. Takeuchi, K. Tsuzuki, K. Sato, M. Yamamoto, Y. Itaya, A. Sano, M. Yoneyama, and T. Otsuji, Very high-speed light-source module up to 40 Gb/s containing an MQW electroabsorption modulator integrated with a DFB laser, *IEEE Journal on Selected Topics in Quantum Electronics*, 3:336–343, 1997.

11. R. Kashyap, *Fiber Bragg Gratings*, Academic Press, San Diego, CA, 1999.

12. K. O. Hill, Y. Fujii, D. C. Johnson, and B. S. Kawasaki, Photosensitivity in optical waveguides: Application to reflection filter fabrication, *Applied Physics Letters,* 32:647, 1978.

13. M. Ibsen, P. Petropoulos, M. N. Zervas, and R. Feced, Dispersion-free fibre Bragg gratings, *OSA Trends in Optics and Photonics*, Vol. 54, Optical Fiber Communication Conference, Mar. 17–22, *Technical Digest*, postconference edition, Optical Society of America, Washington, DC, 2001, pp. MC1-1–MC1-3.

14. P. Yvernault, D. Durand, D. Mechin, M. Boitel, and D. Pureur, Passive athermal Mach–Zehnder interferometer twin-core optical add/drop multiplexer, *Proc. 27th European Conference on Optical Communication*, Amsterdam, Vol. 6, *Postdeadline Papers*, 2001, pp. 88–89.

15. A. R. Vellekoop and M. K. Smit, Four-channel integrated-optic wavelength demultiplexer with weak polarization dependence, *Journal of Lightwave Technology,* 9:310–314, 1991.

16. H. Takahashi, S. Suzuki, K. Kato, and I. Nishi, Arrayed-waveguide grating for wavelength division multi/demultiplexer with nanometre resolution, *Electronics Letters,* 26:87–88, 1990.

17. C. Dragone, An $N \times N$ optical multiplexer using a planar arrangement of two star couplers, *IEEE Photonics Technology Letters*, 3:812–814, 1991.

18. C. Dragone, C. A. Edwards, and R. C. Kistler, Integrated optics $N \times N$ multiplexer on silicon, *IEEE Photonics Technology Letters*, 3:896–899, 1991.

19. H. Takahashi, S. Suzuki, K. Kato, and I. Nishi, Wavelength multiplexer based on $SiO_2–Ta_2O_5$ arrayed-waveguide grating, *Journal of Lightwave Technology*, 12: 989–995, 1994.

20. B. H. Verbeek and M. K. Smit, PHASAR-based WDM-devices: Principles, design and applications, *IEEE Journal of Selected Topics in Quantum Electronics*, 2:236–250, 1996.

21. S. Valette, State of the art of integrated optics technology at LETI for achieving passive optical components, *Journal of Modern Optics*, 35:993–1005, 1988.

22. C. H. Henry, G. E. Blonder, and R. F. Kazarinov, Glass waveguides on silicon for hybrid optical packaging, *Journal of Lightwave Technology*, 7:1530–1539, 1989.

23. M. Kawachi, Silica waveguides on silicon and their application to integrated-optic components, *Optical Quantum Electronics*, 22:391–416, 1990.

24. Y. Hida, Y. Hibino, T. Kitoh, Y. Inoue, M. Itoh, T. Shibata, A. Sugita, and A. Himeno, 400-channel 25-GHz spacing arrayed-waveguide grating covering a full range of C- and L-bands, *OSA Trends in Optics and Photonics*, Vol. 54, Optical Fiber Communication Conference, Mar. 17–22, *Technical Digest*, postconference edition, Optical Society of America, Washington, DC, 2001, p. WB2.

25. K. Takada, T. Tanaka, M. Abe, T. Yanagisawa, M. Ishii, and K. Okamoto, Beam-adjustment-free crosstalk reduction in 10-GHz-spaced arrayed-waveguide grating via photosensitivity under UV laser irradiation through metal mask, *Electronics Letters*, 36:60–61, 2000.

26. S. Suzuki, S. Sumida, Y. Inoue, M. Ishii, and Y. Ohmori, Polarisation-insensitive arrayed-waveguide gratings using dopant-rich silica-based glass with thermal expansion adjusted to Si substrate, *Electronics Letters*, 33:1173–1174, 1997.

27. Y. Inoue, M. Itoh, Y. Hashizume, Y. Hibino, A. Sugita, and A. Himeno, Birefringence compensating AWG design, in *OSA Trends in Optics and Photonics*, Vol. 54, Optical Fiber Communication Conference, Mar. 17–22, *Technical Digest*, postconference edition, Optical Society of America, Washington, DC, 2001, p. WB4.

28. S. Kobayashi, K. Schmidt, M. Dixon, Y. Hibino, and A. Spector, High operation temperature 1×8 AWG modules for dense WDM applications, *Proc. ECIO'97 8th European Conference on Integrated Optics*, Stockholm, Apr. 2–4, 1997, pp. 80–83.

29. Y. Inoue, A. Kaneko, F. Hanawa, H. Takahashi, K. Hattori, and S. Sumida, Athermal silica-based arrayed-waveguide grating multiplexer, *Electronics Letters*, 33:1945–1947, 1997.

30. C. R. Doerr, R. Pafchek, and L. W. Stulz, Compact and low-loss integrated flat-top passband demux, *Proc. 27th European Conference on Optical Communication*, Amsterdam, Vol. 6, *Postdeadline Papers*, 2001, pp. 24–25.

31. B. Chassagne, K. Aubry, A. Rocher, B. Herbette, V. Dentan, S. Bourzeix, and P. Martin, Passive Athermal Bulk-Optic MUX/DEMUX with Flat-Top Spectral Response, *Proc. 27th European Conference on Optical Communication*, Amsterdam, Vol. 3, 2001, pp. 316–317.

32. T. Chiba, H. Arai, K. Ohira, H. Nonen, H. Okano, and H. Uetsuka, Novel architecture of wavelength interleaving filter with Fourier transform-based MZIs, *OSA Trends in Optics and Photonics*, Vol. 54, Optical Fiber Communication Conference, Mar. 17–22, *Technical*

Digest, postconference edition, Optical Society of America, Washington, DC, 2001, p. WB5.

33. M. Oguma, T. Kitoh, K. Jinguji, T. Shibata, A. Himeno, and Y. Hibino, Flat-top and low-loss WDM filter composed of lattice-form interleave filter and arrayed-waveguide gratings on one chip, *OSA Trends in Optics and Photonics*, Vol. 54, Optical Fiber Communication Conference, Mar. 17–22, *Technical Digest*, postconference edition, Optical Society of America, Washington, DC, 2001, p. WB3.

34. P. C. Becker, N. A. Olsson, and J. R. Simpson, *Erbium-Doped Fiber Amplifiers: Fundamentals and Technology*, Academic Press, San Diego, CA, 1999.

35. R. J. Mears, L. Reekie, I. M. Jauncie, and D. N. Payne, Low-noise erbium-doped fibre amplifier operating at 1.54 mm, *Electronics Letters*, 23:1026–1028, 1987.

36. E. Desurvire, J. R. Simpson, and P. C. Becker, High-gain erbium-doped traveling-wave fiber amplifier, *Optics Letters,* 12:888–890, 1987.

37. M. Nakazawa, Y. Kimura, and K. Suzuki, Efficient Er^{3+}-doped optical fiber amplifier pumped by a 1.48 mm InGaAsP laser diode, *Applied Physics Letters*, 54:295–297, 1989.

38. M. Pessa, J. Nappi, P. Savolainen, M. Toivonen, R. Murison, A. Ovtchinnikov, and H. Asonen, State-of-the-art aluminum-free 980-nm laser diodes, *Journal of Lightwave Technology*, 14:2356–2361, 1996.

39. C. S. Harder, L. Brovelli, H. P. Meier, and A. Oosenbrug, High reliability 980-nm pump lasers for Er amplifiers, *Conference on Optical Fiber Communication, OFC'97*, 1997, p. 350.

40. M. J. Li, Recent progress in fiber dispersion compensators, *Proc. 27th European Conference on Optical Communication*, Amsterdam, Vol. 4, 2001, pp. 486–489.

41. M. Lenzi, L. Cibinetto, G. B. Preve, and S. Brunazzi, Optical waveguide variable attenuators in glass-on-silicon technology, *Proc. ECIO'97 8th European Conference on Integrated Optics*, Stockholm, Apr. 2–4, 1997, pp. 346–349.

42. T. V. Clapp, S. Day, S. Ojiha, and R. G. Peall, Broadband variable optical attenuator in silica waveguide technology, *Proc. 24th European Conference on Optical Communication: ECOC'98*, Vol. 1, *Regular and Invited Papers*, Madrid, Sept. 20–24, Telefonica, Madrid, 1998, pp. 301–302.

43. M. Gustavsson, L. Gillner, and C. P. Larsen, Network requirements on optical switching devices, *Proc. ECIO'97 8th European Conference on Integrated Optics*, Stockholm, Apr. 2–4, 1997, pp. 10–15.

44. F. Dorgeuille, W. Grieshaber, F. Pommereau, C. Porcheron, F. Gaborit, I. Guillemot, J. Y. Emery, and M. Renaud, First array of 8 CG-SOA gates for large-scale WDM space switches, *Proc. 24th European Conference on Optical Communication: ECOC'98*, Madrid, Sept. 20–24, Vol. 1, *Regular and Invited Papers*, Telefonica, Madrid, 1998, pp. 255–256.

45. T. Shibata, M. Okuno, T. Goh, M. Yasu, M. Itoh, M. Ishii, Y. Hibino, A. Sugita, and A. Himeno, Silica-based 16×16 optical matrix switch module with integrated driving circuits, *OSA Trends in Optics and Photonics*, Vol. 54, Optical Fiber Communication Conference, Mar. 17–22, *Technical Digest*, postconference edition, Optical Society of America, Washington, DC, 2001, p. WR1.

46. T. Watanabe, T. Goh, M. Okuno, S. Sohma, T. Shiibata, M. Itoh, M. Kobayashi, M. Ishii, A. Sugita, and Y. Hibino, Silica-based PLC 1×128 thermo-optic switch, *Proc. 27th European Conference on Optical Communication*, Amsterdam, Vol. 2, 2001, pp. 134–135.

47. M. C. Flipse, P. De Dobbelaere, J. Thackara, J. Freeman, B. Hendriksen, A. Ticknor, and G. F. Lipscomb, Reliability assurance of polymer-based solid-state optical switches, *Integrated Optics Devices II: Proc. SPIE*, Vol. 3278, San Jose, CA, Jan. 28–30, G. C. Righini, S. I. Najafi, and B. Jalali, eds., SPIE, Bellingham, WA, 1998, pp. 58–62.

48. A. Borrema, T. Hoekstra, M. B. J. Diemeer, H. Hoekstra, and P. Lambeck, Polymeric 8 × 8 digital optical switch matrix, *Proc. 22nd European Conference on Optical Communication*, Oslo, Sept. 15–19, Telenor, Kjeller, Norway, 1996, pp. 5.59–5.62.

49. N. Keil, H. H. Yao, C. Zawadzki, K. Lösch, K. Satzke, W. Wischmann, J. V. Wirth, J. Schneider, J. Bauer, and M. Bauer, Thermo-optic 1 × 2- and 1 × 8-vertical coupled switches (VCS) using hybrid polymer/silica integration technology, *OSA Trends in Optics and Photonics*, Vol. 54, Optical Fiber Communication Conference, Mar. 17–22, *Technical Digest*, postconference edition, Optical Society of America, Washington, DC, 2001, p. WR2.

50. T. Murphy, S.-Y. Suh, B. Comissiong, A. Chen, R. Irvin, R. Grencavich, and G. Richards, A strictly non-blocking 16 × 16 electrooptic photonic switch module, *Proc. 26th European Conference on Optical Communication, ECOC'2000*, Munich, Sept. 3–7, Vol. 4, VDE Verlag, Berlin, 2000, pp. 93–94.

51. J. E. Fouquet, Compact optical cross-connect switch based on total internal reflection in a fluid-containing planar lightwave circuit, *Proc. Optical Fiber Communication Conference, OFC'2000*, Baltimore, Mar. 5–10, Optical Society of America, Washington, DC, 2000, pp. 204–206.

52. R. Ryf et al., Scalable wavelength-selective crossconnect switch based on MEMS and planar waveguides, *Proc. 27th European Conference on Optical Communication*, Amsterdam, Vol. 6, *Postdeadline Papers*, 2001, pp. 76–77.

53. L. Wosinska and L. Thylen, Large-capacity strictly nonblocking optical cross-connects based on microelectrooptomechanical systems (MEOMS) switch matrices: Reliability performance analysis, *Journal of Lightwave Technology*, 19:1065–1075, 2001.

54. T. Durhuus, B. Mikkelsen, C. Joergensen, S. L. Danielsen, and K. E. Stubkjaer, All-optical wavelength conversion by semiconductor optical amplifiers, *Journal of Lightwave Technology*, 14:942–954, 1996.

55. O. Ait Sab, FEC techniques in submarine transmission systems, *OSA Trends in Optics and Photonics*, Vol. 54, Optical Fiber Communication Conference, Mar. 17–22, *Technical Digest*, postconference edition, Optical Society of America, Washington, DC, 2001, p. TuF1.

56. M. Tomizawa, Y. Miyamoto, T. Kataoka, and Y. Tada, Recent progress and standardization activities on 40 Gbit/s channel technologies, *Proc. 27th European Conference on Optical Communication*, Amsterdam, Vol. 2, pp. 154–157.

57. J.-P. Blondel, Massive WDM systems: Recent developments and future prospects, *Proc. 27th European Conference on Optical Communication*, Amsterdam, Vol. 1, pp. 50–53.

58. T. Ito, K. Fukuchi, and T. Kasamatsu, Enabling technologies for 10 Tb/s transmission capacity and beyond, *Proc. 27th European Conference on Optical Communication*, Amsterdam, Vol. 4, pp. 598–601.

59. S. Bigo, Y. Frignac, G. Charlet, W. Idler, S. Borne, H. Gross, R. Dischler, W. Poehlmann, P. Tran, C. Simonneau, D. Bayart, G. Veith, A. Jourdan, and J.-P. Hamaide, 10.2 Tbit/s (256 × 42.7 Gbit/s PDM/WDM) transmission over 100 km TeraLight fiber with 1.28 bit/s/Hz spectral efficiency, *Proc. Optical Fiber Communication Conference, OFC'2001, Postdeadline Papers*, Optical Society of America, Washington, DC, 2001, p. PD25.

60. K. Fukuchi, T. Kasamatsu, M. Morie, R. Ohhira, T. Ito, K. Sekiya, D. Ogasahara, and T. Ono, 10.92-Tb/s (273 × 40-Gb/s) triple-band/ultra-dense WDM optical-repeatered transmission experiment, *Proc. Optical Fiber Communication Conference, OFC'2001, Postdeadline Papers*, Optical Society of America, Washington, DC, 2001, p. PD24.

61. P. P. Mitra and J. B. Stark, Nonlinear limits to the information capacity of optical fibre communications, *Nature*, 411:1027–1030, 2001.

62. J. M. Kahn and K.-P. Ho, A bottleneck for optical fibres, *Nature*, 411:1007–1010, 2001.

63. T. Miya, Silica-based planar lightwave circuits: Passive and thermally active devices, *IEEE Journal on Selected Topics in Quantum Electronics*, 6:38–45, 2000.

64. K, Kato, and Y. Tohmori, PLC hybrid integration technology and its application to photonic components, *IEEE Journal on Selected Topics in Quantum Electronics*, 6:4–13, 2000.

65. M. K. Smit, Fundamentals and applications of phasar demultiplexers, *Proc. 27th European Conference on Optical Communication*, Amsterdam, Vol. 1, 2001, pp. 2–3.

4 Electro-optic and Wavelength Conversion

BYRAV RAMAMURTHY

University of Nebraska–Lincoln, Lincoln, Nebraska

4.1 INTRODUCTION

Wavelength-division multiplexing (WDM) [7] is a promising technique to utilize the enormous bandwidth of optical fiber. Multiple wavelength-division-multiplexed channels can be operated on a single fiber simultaneously; however, a fundamental requirement in fiber-optic communication is that these channels operate at different wavelengths. These channels can be modulated independently to accommodate dissimilar data formats, including some analog and some digital, if desired. Thus, WDM utilizes the huge bandwidth (around 50 THz) of a single-mode optical fiber while providing channels whose bandwidths (1 to 10 Gbps) are compatible with current electronic processing speeds.

In a WDM network, it is possible to route data to their respective destinations based on their wavelengths. The use of wavelength to route data is referred to as *wavelength routing*, and a network that employs this technique is known as a *wavelength-routed* network [31]. Such a network consists of wavelength-routing switches (or routing nodes) which are interconnected by optical fibers. Some routing nodes (referred to as *cross-connects*) are attached to access stations where data from several end users could be multiplexed onto a single WDM channel. An access station also provides optical-to-electronic (O/E) conversion, and vice versa, to interface the optical network with conventional electronic equipment. A wavelength-routed network that carries data from one access station to another without intermediate O/E conversion is referred to as an *all-optical* wavelength-routed network. Such all-optical wavelength-routed networks have been proposed for building large wide-area networks [8].

To transfer data from one access station to another, a connection needs to be set up at the optical layer similar to the case in a circuit-switched telephone network.

IP over WDM: Building the Next-Generation Optical Internet, Edited by Sudhir Dixit.
ISBN 0-471-21248-2 © 2003 John Wiley & Sons, Inc.

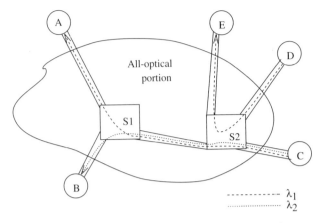

Figure 4.1   All-optical wavelength-routed network.

This operation is performed by determining a path (route) in the network connecting the source station to the destination station and by allocating a common free (or idle) wavelength on all the fiber links in the path. Such an all-optical path is referred to as a *lightpath* or a *clear channel*. The entire bandwidth available on this lightpath is allocated to the connection during its holding time, during which the corresponding wavelength cannot be allocated to any other connection. When a connection is terminated, the associated lightpath is torn down, and the wavelength becomes idle once again on all the links along the route.

Consider the network in Fig. 4.1. It shows a wavelength-routed network containing two WDM cross-connects (S1 and S2) and five access stations (A through E). Three lightpaths have been set up (C to A on wavelength λ_1, C to B on λ_2, and D to E on λ_1). To establish any lightpath, we normally require that the *same* wavelength be allocated on all the links in the path. This requirement is known as the *wavelength continuity constraint*, and wavelength-routed networks with this constraint are referred to as *wavelength-continuous networks*. The wavelength continuity constraint distinguishes the wavelength-continuous network from a circuit-switched network, which blocks calls only when there is no capacity along any of the links in the path assigned to the call. Consider the portion of a network in Fig. 4.2a. Two lightpaths have been established in the network: between nodes 1 and 2 on wavelength λ_1 and between nodes 2 and 3 on wavelength λ_2. Now, suppose that a lightpath needs to be set up between nodes 1 and 3. If only two wavelengths are available in the network, establishing such a lightpath from node 1 to node 3 is now impossible even though there is a free wavelength on each of the links along the path from node 1 to node 3. This is because the available wavelengths on the two links are *different*. Thus, a wavelength-continuous network may suffer from higher blocking than does a circuit-switched network.

It is easy to eliminate the wavelength continuity constraint if we are able to *convert* the data arriving on one wavelength along a link into another wavelength at an intermediate node and forward it along the next link. Such a technique is feasible

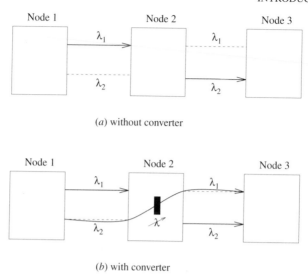

(*a*) without converter

(*b*) with converter

Figure 4.2   Wavelength continuity constraint in a wavelength-routed network.

and is referred to as *wavelength conversion*, and wavelength-routed networks with this capability are referred to as *wavelength-convertible networks* [34]. A wavelength-convertible network that supports complete conversion at all nodes is functionally equivalent to a circuit-switched network (i.e., lightpath requests are blocked only when there is no available capacity on the path). In Fig. 4.2*b*, a wavelength converter at node 2 is employed to convert data from wavelength λ_2 to λ_1. The new lightpath between nodes 1 and 3 can now be established by using wavelength λ_2 on the link from node 1 to node 2 and then by using wavelength λ_1 to reach node 3 from node 2. Notice that a single lightpath in such a wavelength-convertible network can use a different wavelength along each of the links in its path. Thus, wavelength conversion may improve the efficiency in the network by resolving the wavelength conflicts of the lightpaths.

This study examines the role of wavelength converters in a wavelength-routed network. It includes a survey of the enabling device technologies as well as that of the network design and analysis methodologies. This study also attempts to identify important unresolved issues in this field and to uncover challenging research problems. A note on terminology: Wavelength converters have been referred to in the literature as *wavelength shifters, wavelength translators, wavelength changers,* and *frequency converters*. In this study we refer to these devices as *wavelength converters*.

This study is organized as follows (see Fig. 4.3). In Section 4.2, the *technologies* that have made wavelength conversion possible are described. How wavelength converters are built and how switch designs have evolved to incorporate these converters are the focus of this section. In Section 4.3, the *network design, control, and management issues* involved in using the technique of wavelength conversion

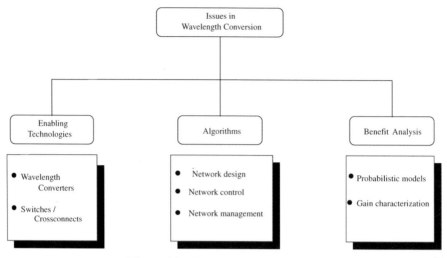

Figure 4.3   Organization of this study.

effectively are highlighted. The approaches adopted to tackle some of these issues are described and new problems in this area are introduced. In Section 4.4, the *benefits* of wavelength conversion are described. Various approaches employed to quantify these benefits are discussed. Section 4.5 concludes this study.

4.2   ENABLING TECHNOLOGIES

Significant advances in optical and electronic device technologies have made wavelength conversion feasible [18]. (For more details the reader is referred to Chapter 3.) Several different techniques have been demonstrated to perform wavelength conversion. A classification and comparison of these techniques is provided in Section 4.2.1. Many novel switch designs have been proposed for utilizing these wavelength converters in a wavelength-convertible network. Some of these techniques are discussed in Section 4.2.2.

4.2.1   Wavelength-Converter Design

The function of a wavelength converter is to convert data on an input wavelength onto a possibly different output wavelength among the N wavelengths in the system (see Fig. 4.4). In this figure and throughout this section, λ_s denotes the input signal wavelength; λ_c, the output (converted) wavelength; λ_p, the pump wavelength; f_s, the input frequency; f_c, the converted frequency; f_p, the pump frequency; and CW, the continuous wave generated as the signal.

An ideal wavelength converter should possess the following characteristics [14]:

- Transparency to bit rates and signal formats
- Fast setup time of output wavelength [38]

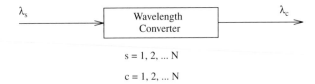

$s = 1, 2, ... N$

$c = 1, 2, ... N$

Figure 4.4   Functionality of a wavelength converter.

- Conversion to both shorter and longer wavelengths
- Moderate input power levels
- Possibility for same input and output wavelengths (no conversion)
- Insensitivity to input signal polarization
- Low-chirp output signal with high extinction ratio* and large signal-to-noise ratio
- Simple implementation

Our classification of the wavelength conversion techniques in this section follows that in [44]. Wavelength conversion techniques can be broadly classified into two types.

1. *Optoelectronic wavelength conversion.* In this method [16], the optical signal to be converted is first translated into the electronic domain using a photodetector (labeled R in Fig. 4.5). The electronic bit stream is stored in the buffer (labeled FIFO for the first in/first out queue mechanism). The electronic signal is then used to drive the input of a tunable laser (labeled T) tuned to the desired wavelength of the output (see Fig. 4.5). This method has been demonstrated for bit rates up to 10 Gbps [47]. However, this method is much more complex and consumes a lot more power than the other methods described below [14]. Moreover, the process of optoelectronic (O/E) conversion adversely affects transparency. All information in the form of phase, frequency and analog amplitude of the optical signal is lost during the conversion process. The highest degree of transparency achievable is *digital transparency*, where digital signals of any bit rates up to a certain limit are accommodated [47].

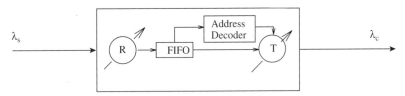

Figure 4.5   Optoelectronic wavelength converter. (From [27].)

---

*The *extinction ratio* is defined as the ratio of the optical power transmitted for a bit 0 to the power transmitted for a bit 1.

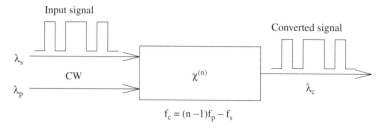

Figure 4.6 Wavelength converter based on nonlinear wave mixing effects.

2. *All-optical wavelength conversion.* In this method, the optical signal is allowed to remain in the optical domain throughout the conversion process. Note that "all-optical," in this context, refers to the fact that there is no optoelectronic conversion involved. Such all-optical methods can be further divided into the following categories and subcategories.

(a) *Wavelength conversion using wave mixing.* Wave mixing arises from a nonlinear optical response of a medium when more than one wave* is present (see Fig. 4.6). It results in the generation of another wave, whose intensity is proportional to the product of the interacting wave intensities. Wave mixing preserves both phase and amplitude information, offering strict transparency. It also allows simultaneous conversion of a set of multiple-input wavelengths to another set of multiple-output wavelengths and could potentially accommodate signals with bit rates exceeding 100 Gbps [47]. In Fig. 4.6, the value $n = 3$ corresponds to four-wave mixing and $n = 2$ corresponds to difference frequency generation. These techniques are described below.

- *Four-wave mixing (FWM).* Four-wave mixing (also referred to as *four-photon mixing*) is a third-order nonlinearity in silica fibers which causes three optical waves of frequencies f_i, f_j, and f_k $(k \neq i,j)$ to interact in a multichannel WDM system [42] to generate a fourth wave of frequency, given by

$$f_{ijk} = f_i + f_j - f_k$$

Four-wave mixing is also achievable in an active medium such as a semiconductor optical amplifier (SOA). This technique provides modulation-format independence [40] and high-bit-rate capabilities [26]. However, the conversion efficiency from pump energy to signal energy of this technique is not very high and it decreases swiftly with increasing conversion span (shift between pump and output signal wavelengths) [48].

---

*A nonlinear response where new waves are generated can also happen with only a single input wave (e.g., Raman scattering).

- *Difference frequency generation (DFG).* DFG is a consequence of a second-order nonlinear interaction of a medium with two optical waves: a pump wave and a signal wave. This technique offers a full range of transparency without adding excess noise to the signal and spectrum inversion capabilities, but it suffers from low efficiency [47]. The main difficulties in implementing this technique lie in the phase matching of interacting waves and in fabricating a low-loss waveguide for high conversion efficiency [47]. In [1], a parametric wavelength-interchanging cross-connect (WIXC) architecture has been proposed that uses DFG-based converters.

(b) *Wavelength conversion using cross modulation.* These techniques utilize active semiconductor optical devices, such as SOAs and lasers. These techniques belong to a class known as optical-gating wavelength conversion [47].

- *Semiconductor optical amplifiers (SOAs) in XGM and XPM mode.* The principle behind using an SOA in the cross-gain modulation (XGM) mode is shown in Fig. 4.7. The intensity-modulated input signal modulates the gain in the SOA due to gain saturation. A continuous-wave (CW) signal at the desired output wavelength (λ_c) is modulated by the gain variation so that it carries the same information as the original input signal. The input signal and the CW signal can be launched either co- or counter-directional into the SOA. The XGM scheme gives a wavelength-converted signal that is inverted compared to the input signal. While the XGM scheme is simple to realize and offers penalty-free conversion at 10 Gbps [14], it suffers from inversion of the converted bit stream and extinction ratio degradation for an input signal "up-converted" to a signal of equal or longer wavelength.

  The operation of a wavelength converter using SOA in cross-phase modulation (XPM) mode is based on the fact that the refractive index of the SOA is dependent on the carrier density in its active region. An incoming signal that depletes the carrier density will modulate the refractive index and thereby result in phase modulation of a CW signal (wavelength λ_c) coupled into the converter [14,24]. The SOA can be integrated into an interferometer so that an intensity-modulated signal format results at the output of the converter. Techniques involving SOAs in XPM mode have been proposed using the nonlinear optical loop mirror (NOLM) [15], Mach–Zender interferometer (MZI) [13], and Michelson

Figure 4.7   Wavelength converter based on XGM in an SOA. (From [14].)

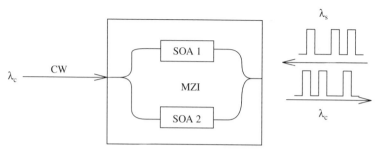

Figure 4.8   Interferometric wavelength converter based on XPM in SOAs. (From [14].)

interferometer (MI) [28]. Figure 4.8 shows an asymmetric MZI wavelength converter based on SOA in XPM mode. With the XPM scheme, the converted output signal can be either inverted or noninverted, unlike in the XGM scheme, where the output is always inverted. The XPM scheme is also very power efficient compared to the XGM scheme [14]. Placing interferometric wavelength converters (IWCs) in the cross-connects is also found to improve the transmission performance of an optical network by reducing the noise in the converted signal [29].

- *Semiconductor lasers.* Using single-mode semiconductor lasers, the lasing-mode intensity of the medium is modulated by input signal light through gain saturation. The output signal obtained is inverted compared to the input signal. This gain suppression mechanism has been employed in a distributed Bragg reflector (DBR) laser to convert signals at 10 Gbps [45].

In this section we reviewed the various techniques used in the design of a wavelength converter. The actual choice of the technology to be employed in wavelength conversion in a network depends on the requirements of the particular system. However, it is clear that optoelectronic converters offer only limited digital transparency. Moreover, deploying multiple optoelectronic converters in a WDM cross-connect requires sophisticated packaging to avoid crosstalk among channels. This leads to increased costs per converter, further making this technology less attractive than all-optical converters [47]. Other disadvantages of optoelectronic converters include complexity and large power consumption [14]. Among all-optical converters, converters based on SOAs using the XGM and XPM conversion schemes presently seem well suited for system use. Converters based on four-wave mixing, although transparent to different modulation formats, perform inefficiently [14]. However, wave mixing converters are the only category of wavelength converters that offer the full range of transparency while allowing simultaneous conversion of a set of input wavelengths to another set of output wavelengths. In this respect, difference frequency generation–based methods offer great promise. Further details on comparison of various wavelength conversion techniques can be found in [14], [30], [39], [44], and [47].

In the next section we examine various switch architectures that have been proposed in the literature for use in a wavelength-convertible network.

4.2.2   Wavelength-Convertible Switch Design

As wavelength converters become readily available, a vital question comes to mind: Where do we place them in the network? An obvious location is in the switches (or cross-connects) in the network. A possible architecture of such a wavelength-convertible switching node is the dedicated wavelength-convertible switch (see Fig. 4.9). In this architecture, each wavelength along each output link in a switch has a *dedicated* wavelength converter (i.e., an $M \times M$ switch in an N-wavelength system requires $M \times N$ converters). The incoming optical signal from a fiber link at the switch is first wavelength demultiplexed into separate wavelengths. Each wavelength is switched to the desired output port by the nonblocking optical switch. The output signal may have its wavelength changed by its wavelength converter. Finally, various wavelengths are multiplexed to form an aggregate signal coupled to an outbound fiber link.

However, the dedicated wavelength-convertible switch is not very cost-efficient since all of the wavelength converters may not be required all the time [19]. An effective method to cut costs is to share the converters. Two architectures have been proposed for switches sharing converters [25]. In the share-per-node structure (see Fig. 4.10a), all the converters at the switching node are collected in a converter bank. (A converter bank is a collection of a few wavelength converters each of which is assumed to have identical characteristics and can convert any input wavelength to any output wavelength.) This bank can be accessed by any of the incoming lightpaths by appropriately configuring the larger optical switch in Fig. 4.10a. In this architecture, only the wavelengths that require conversion are directed to the converter bank. The converted wavelengths are then switched to the

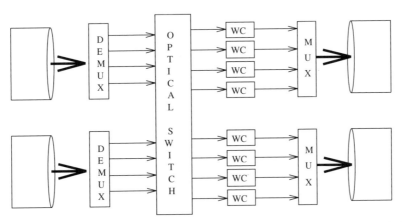

Figure 4.9   Switch with dedicated converters at each output port for each wavelength (WC denotes a wavelength converter). (From [25].)

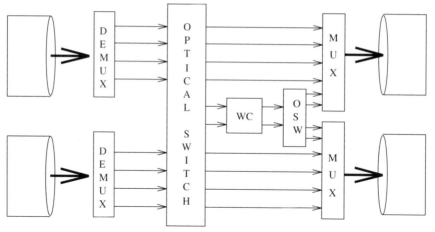

(*a*) Share-per-node wavelength-convertible switch architecture

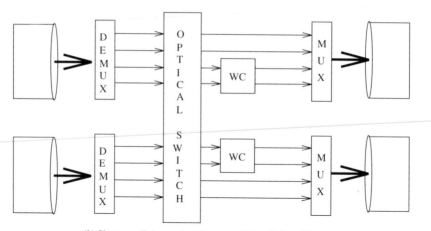

(*b*) Share-per-link wavelength-convertible switch architecture

Figure 4.10   Switches that allow sharing of converters.

appropriate outbound link by the second (small) optical switch. In the share-per-link structure (see Fig. 4.10*b*), each outgoing link is provided with a dedicated converter bank which can be accessed only by those lightpaths traveling on that particular outbound fiber link. The optical switch can be configured appropriately to direct wavelengths toward a particular link, either with or without conversion.

When optoelectronic wavelength conversion is used, the functionality of the wavelength converter can be performed at the access stations instead of at the switches. The share-with-local switch architecture proposed in [25] (see Fig. 4.11) and the simplified network access station architecture proposed in [23] (see Fig. 4.12) fall under this category.

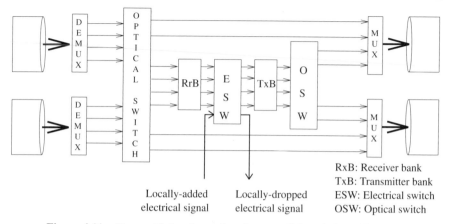

RxB: Receiver bank
TxB: Transmitter bank
ESW: Electrical switch
OSW: Optical switch

Locally-added   Locally-dropped
electrical signal   electrical signal

Figure 4.11   Share-with-local wavelength-convertible switch architecture.

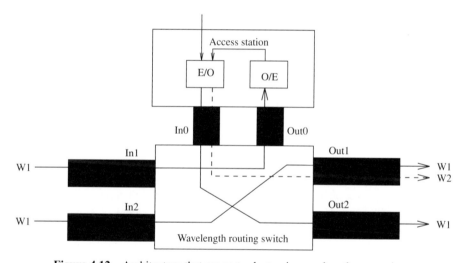

Figure 4.12   Architecture that supports electronic wavelength conversion.

4.3   NETWORK DESIGN, CONTROL, AND MANAGEMENT ISSUES

4.3.1   Network Design

Network designs must evolve to incorporate wavelength conversion effectively. Network designers must choose not only among the various conversion techniques described in Section 4.2.1, but also among the several switch architectures described in Section 4.2.2. An important challenge in the design is to overcome the limitations in using wavelength-conversion technology. These limitations fall into the following three categories:

1. *Sparse location of wavelength converters in the network.* As long as wavelength converters remain expensive [47], it may not be feasible economically to equip all the nodes in a WDM network with these devices. The effects of sparse conversion (i.e., having only a few converting switches in the network) on lightpath blocking have been examined in [41]. An interesting question that has not been answered thoroughly is *where* (optimally) to place these few converters in an arbitrary network and what the likely *upgrade path* is toward full-fledged convert-ibility. A heuristic technique for the placement of these sparse converters in an all-optical network is presented in [19].

2. *Sharing of converters.* Even among the switches capable of wavelength conversion, it may not be cost-effective to equip all the output ports of a switch with this capability. Designs of switch architectures have been proposed (see Section 4.2.2) which allow sharing of converters among the various signals at a switch. It has been shown in [25] that the performance of such a network saturates when the number of converters at a switch increases beyond a certain threshold. An interesting problem is to quantify the dependence of this threshold on the routing algorithm used and the blocking probability desired.

3. *Limited-range wavelength conversion.* Four-wave-mixing-based all-optical wavelength converters provide only a limited-range conversion capability. If the range is limited to k, an input wavelength λ_i can only be converted to wavelengths $\lambda_{\max(i-k,1)}$ through $\lambda_{\min(i+k,N)}$, where N is the number of wavelengths in the system (indexed 1 through N). Analysis shows that networks employing such devices, however, compare favorably with those utilizing converters with full-range capab-ility, under certain conditions [46]. Limited-range wavelength conversion can also be provided at nodes using optoelectronic conversion techniques [36].

Other wavelength-converter techniques have some limitations, too. As seen in Section 4.2.1, a wavelength converter using SOAs in XGM mode suffers greater degradation when the input signal is up-converted to a signal of equal or longer wavelength than when it is down-converted to a shorter wavelength. Moreover, since the signal quality usually worsens after multiple such conversions, the effect of a cascade of these converters can be substantial. The implications of such a device on the design of the network needs to studied further.

Apart from efficient wavelength-convertible switch architectures and their optimal placement, several other design techniques offer promise. Networks equipped with multiple fibers on each link have been considered for potential improvement [20] in wavelength-convertible networks and suggested as a possible alternative to conversion. This work is reviewed in greater detail in Section 4.4. Another important problem is the design of a fault-tolerant wavelength-convertible network [17]. Such a network could reserve capacity on the links to handle disruptions due to link failure caused by a cut in the fiber. Quantitative comparisons need to be developed for the suitability of a wavelength-convertible network in such scenarios.

4.3.2   Network Control

Control algorithms are required in a network to manage its resources effectively. An important task of the control mechanism is to provide routes (i.e., set of fiber links) to the lightpath requests and to assign wavelengths on each of the links along this route while maximizing a desired system parameter (e.g., throughput). Such routing and wavelength assignment (RWA) schemes can be classified into *static* and *dynamic* categories, depending on whether or not the lightpath requests are known a priori. These two categories are described below. We refer the interested reader to Chapters 10, 11, 12, 15, and 16 for more details.

1. *Dynamic routing and wavelength assignment.* In a wavelength-routed optical network, lightpath requests between source–destination pairs arrive at random, and each lightpath has a random holding time, after which it is torn down. These lightpaths need to be set up dynamically by determining a route through the network connecting the source to the destination and assigning a free wavelength along this path. Two lightpaths that have at least a link in common cannot use the same wavelength. Moreover, the same wavelength has to be assigned to a path on all its links. This is the wavelength-continuity constraint described in Section 4.1. This routing and wavelength assignment (RWA) problem, or variants of it, has been studied earlier [2,4,11] for networks *without* wavelength conversion.

Dynamic routing algorithms have been proposed for use in a wavelength-convertible network [10,25]. In [25], the routing algorithm approximates the cost function of routing as the sum of individual costs due to using channels and wavelength converters. For this purpose, an auxiliary graph is created and the shortest-path algorithm is applied on the graph to determine the route. In [10], an algorithm with provably optimal running time has been provided for such a technique employing shortest-path routing. Algorithms have also been studied that use a *fixed path* or *deterministic routing* [37]. In such a scheme there is a fixed path between every source–destination pair in the network. Several RWA heuristics have been designed based on which wavelength to assign to a lightpath along the fixed path and which, if any, lightpaths to block selectively [2]. However, design of efficient routing algorithms that incorporate the limitations in Section 4.3.1 remains an open problem.

2. *Static routing and wavelength assignment.* In contrast to the dynamic routing problem described above, the static RWA problem assumes that all the lightpaths that are to be set up in the network are known initially. The objective is to maximize the total throughput in the network (i.e., the total number of lightpaths that can be established in the network simultaneously). An upper bound on the carried traffic per available wavelength has been obtained (for a network with and without wavelength conversion) by relaxing the corresponding integer linear program (ILP) [37]. Several heuristic-based approaches have been proposed for solving the static RWA problem in a network without wavelength conversion [4,9]. A special case of the static RWA problem, in which all the lightpath requests can be

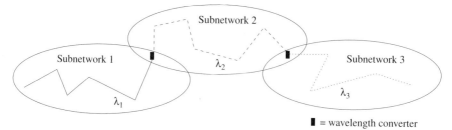

Figure 4.13 Wavelength conversion for distributed network management.

accommodated, is discussed in [36] for networks with limited wavelength conversion . Again, efficient algorithms that incorporate the limitations in Section 4.3.1 for a wavelength-convertible network are still unavailable.

4.3.3 Network Management

Wavelength conversion may be used to promote interoperability across subnetworks that are managed independently. Thus, it supports the distribution of network control and management functionalities among smaller subnetworks by allowing flexible wavelength assignments within each subnetwork [35,47]. As shown in Fig. 4.13, network operators 1, 2, and 3 manage their own subnetworks and may use wavelength conversion for communication *across* subnetworks. In [43], the authors propose to install wavelength converters at the borders between nonoverlapping network partitions in the Cost 239 European Optical Network.

4.4 BENEFIT ANALYSIS

As mentioned above, the availability of full wavelength conversion simplifies management of the network—the wavelength-assignment algorithm in such a network becomes simpler because all the wavelengths can be treated equivalently, and wavelengths used on successive links along a path can be independent of one another. However, the benefits of wavelength conversion in reducing blocking and improving other performance metrics are not nearly as universal or apparent. Although full wavelength conversion eliminates the wavelength continuity constraint (see Section 4.1), the actual performance benefits available in a typical network are found to depend on factors such as connectivity and traffic load. Efforts have been made to quantify these benefits in typical networks using analytical models and simulations. We present a review of several of these studies in the remainder of this section.

4.4.1 Analytical Models

Analytical models have been developed by several researchers to quantify the benefits of wavelength converters. We present below a brief review of the fundamental models available today. All of the models discussed make the following assumptions:

- Each lightpath uses the entire bandwidth of a wavelength channel.
- Each fiber link has the same number (W) of wavelengths.
- Each station has arrays of W transmitters and W receivers.
- The network supports only point-to-point traffic (i.e., there is no multicasting).
- Connection requests are not queued (i.e., if a connection is blocked, it is discarded immediately).

The main differences between the models are highlighted in Table 4.1. The model described in [22] assumes that the loads on different links in the network are independent and that the busy wavelengths on different links are allocated independent of one another. The model presented in [41] incorporates the link-load dependencies using a Markov correlation method and predicts closely the behavior of an arbitrary network. The model also handles the case when only a few nodes in the network have the capability for wavelength conversion. In [6] a Markov chain with state-dependent routing is used to model the wavelength-conversion gain. However, it is assumed that the random variables representing the number of idle wavelengths on different links are independent. The model obtained is computationally intensive and is tractable only for densely connected networks with a few nodes. The model described in [5] makes more simplistic traffic assumptions to avoid the recursive numerical evaluations required in the other models. However, it correctly predicts some unobvious qualitative behavior demonstrated in simulation studies. Below, we describe these approaches in detail.

TABLE 4.1   Comparison of Analytical Models

	Kovačević and Acampora [22]	Subramaniam et al. [41]	Birman [6]	Barry and Humblet [5]
Traffic	Dynamic	Dynamic	Dynamic	Steady-state
Arrival process	Poisson	Poisson	Poisson	Unspecified
Holding time	Exponential	Exponential	Exponential	Unspecified
Routing	Fixed	Fixed	Fixed, least-loaded	Fixed
Wavelength assignment	Random	Random	Random	Random
Link loads	Independent	Correlated (Markovian)	Dependent	Correlated (Markovian)
Wavelengths on adjacent links	Independent	Dependent	Independent	Dependent
Performance metric	Blocking probability p_b	Blocking probability p_b	Blocking probability p_b	Wavelength utilization gain G
Computational complexity	Moderate	Moderate	High	Low

4.4.1.1 *Probabilistic Model with Independent Link-Load Assumption [22]*   An
approximate analytical model is developed for a fixed-path (deterministic) routed
network with an arbitrary topology, both with and without wavelength conversion.
This model is then used along with simulations to study the performance of three
example networks: the nonblocking centralized switch, the two-dimensional-torus
network, and the ring network. The traffic loads and the wavelength-occupancy
probabilities on the links are both assumed to be independent. A wavelength-
assignment strategy is employed in which a lightpath is assigned a wavelength at
random from among the available wavelengths in the path. The blocking probability of the lightpaths is used to study the performance of the network. The
benefits of wavelength conversion are found to be modest in the nonblocking
centralized switch and the ring; however, wavelength conversion is found to
improve significantly the performance of a large two-dimensional torus network.
The analytical model employed in this study cannot be applied to a ring network
because the very high load correlation along the links of a path in a ring network
invalidates the independent link-load assumption.

Details   First, we consider the case when there is no wavelength conversion in the
network. In this case, a connection request is blocked when there is no wavelength
available on every link of the path. The approach in [22] determines the conditional
probability that k wavelengths are available for a connection on a two-hop path and
extends the analysis for an n-hop path.

Let W be the number of wavelengths per fiber, T be the average duration of a
connection, and λ_i be the arrival rate on the ith link of the path. L_i, the average load
offered on the ith link of the path, is then given by $L_i = \lambda_i T$. Let $p_k^{(i)}$ be the
probability that k wavelengths are used on the ith link of the path. Assuming
Poisson arrivals on the link and exponential holding times, we have

$$p_k^{(i)} = \frac{(\lambda_i T)^k}{k!} p_0^{(i)} = \frac{(L_i^k/k!)}{\sum_{l=0}^{W}(L_i^l/l!)} \tag{4.1}$$

For a connection requiring a single hop, the blocking probability is equal to $p_W^{(1)}$, the
probability that all W wavelengths are busy on the link connecting the source and
the destination. Let $q_k^{(n)}$ be the probability that there are k busy wavelengths over the
first n hops of the path. For a one-hop connection, we have $q_k^{(1)} = p_k^{(1)}$, $k \in 1, \ldots, W$.
For a two-hop path, the conditional probability that there are k wavelengths available for a connection given that n_a and n_b wavelengths are free on links a and b
(assuming that the distributions of assigned wavelengths at links a and b are
mutually independent) is

$$R(k|n_a, n_b) = \frac{\binom{n_a}{k}\binom{W - n_a}{n_b - k}}{\binom{W}{n_b}} \tag{4.2}$$

if $\max(0, n_a + n_b - W) \leq k \leq \min(n_a, n_b)$ and is equal to zero otherwise. Using this conditional probability, the distribution of "busy" wavelengths over the two-hop path follows:

$$q_k^{(2)} = \sum_{i=0}^{W} \sum_{j=0}^{W} R(W - k | W - i, W - j) p_i^{(1)} p_j^{(2)} \qquad (4.3)$$

The blocking probability for a two-hop connection is thus $P^{(2)} = q_W^{(2)}$. Hence, for an n-hop path, we have (using recursion)

$$q_k^{(n)} = \sum_{i=0}^{W} \sum_{j=0}^{W} R(W - k | W - i, W - j) q_i^{(n-1)} p_j^{(n)} \qquad (4.4)$$

and

$$P^{(n)} = q_W^{(n)} \qquad (4.5)$$

Next, we consider the case when wavelength conversion is available in the network. Note that a lightpath is blocked only when one or more links on the path have all their wavelengths occupied. Thus, the blocking probability for an n-hop connection is

$$P^{(n)} = 1 - \prod_{i=1}^{n} (1 - p_W^{(i)}) \qquad (4.6)$$

The analysis above for the path blocking probabilities assumes that the link loads along the path are already known. However, in practice, it is the traffic matrix (which represents the offered load between a pair of stations) that is usually known, not the link loads. Hence, the authors estimate the link loads in the network using an approach similar to that in [21]. For a network with wavelength conversion, the arrival process on a link s is independent of the number of connections carried by the link (assuming independent link loads). Thus, the arrivals on the link can be considered to be Poisson arrivals, and the number of occupied wavelengths can be represented by the distribution given in equation (4.1). However, to make the analysis of the network without wavelength conversion tractable, the approach in [22] makes an approximation by assuming Poisson arrivals at the links in this case also. The network blocking probabilities can be obtained by solving the set of coupled nonlinear equations called *Erlang's map*. It is shown in [21] that this set of equations has a unique solution for the network with wavelength conversion. The authors provide an iterative procedure to solve these equations and compute the blocking probability for the network without wavelength conversion.

4.4.1.2 Sparse Wavelength Conversion [41]   Sparse wavelength conversion, in which only a few nodes in the network are equipped with wavelength converters (see Section 4.3.1) is studied in this work.* Two different switching nodes are available in the network: nodes with no wavelength conversion capability and nodes that can convert any incoming wavelength to an arbitrary outgoing wavelength. An analytical model for evaluating the path-blocking performance of such networks is also presented and is shown to be accurate for a variety of network topologies. The model improves on the one in [22] by relaxing the independence assumptions on the loads and wavelength-occupancy probabilities of the links. The authors find that the usefulness of wavelength converters depends on the connectivity of the network. Converters are not very useful in networks with low connectivity, such as the ring, because of the high load correlation between links. Moreover, converters are also found to be of little use in networks with high connectivity, such as the hypercube, because of the small hop lengths. However, converters offer significant benefits in networks with medium connectivity, such as the mesh-torus network, because the link-load correlations are low while the hop lengths are large. The authors show that in most cases, only a small fraction of the nodes have to be equipped with wavelength conversion capability for good performance.

Details   In their model, the authors incorporate the load correlation among the links in the network. In particular, they assume that the load on link i of a path given the loads on links $1, 2, \ldots, i-1$ depends only on the load on link $i-1$. Hence, their analytical model is a Markovian correlation model. First, they derive the conditional free-wavelength distribution on a two-hop path using the following notations:

- $Q(w_f) = \Pr\{w_f$ wavelengths are free on a link$\}$.
- $S(y_f|x_{pf}) = \Pr\{y_f$ wavelengths are free on a link of a path $\mid x_{pf}$ wavelengths are free on the previous link of the path$\}$.
- $U(z_c|y_f, x_{pf}) = \Pr\{z_c$ lightpaths continue to the current link from the previous link of a path $\mid x_{pf}$ wavelengths are free on the previous link, and y_f wavelengths are free on the current link$\}$.
- $R(n_f|x_{ff}, y_f, z_c) = \Pr\{n_f$ wavelengths are free on a two-hop path $\mid x_{ff}$ wavelengths are free on the first hop of the path, y_f wavelengths are free on the second hop, and z_c lightpaths continue from the first to the second hop$\}$.
- $T^{(l)}_{n_f, y_f} = \Pr\{n_f$ wavelengths are free on an l-hop path and y_f wavelengths are free on hop $l\}$.
- $p_l = \Pr\{$an l-hop path is chosen for routing$\}$.

Now consider a two-hop path between nodes 0 and 2 passing through node 1. Let C_l be the number of lightpaths that enter the path at node 0 and leave at node 1, let

---

*Networks without wavelength conversion and those with full wavelength conversion are handled as extreme cases under the same framework of sparse conversion.

C_c be the number of lightpaths that enter the path at node 0 and continue on to the second link, and let C_n be the number of lightpaths that enter the path at node 1 and let λ_l, λ_c, and λ_n be the corresponding lightpath arrival rates. Then the number of lightpaths that use the first link is $C_l + C_c$, and the number of lightpaths that use the second link is $C_c + C_n$. By the assumption of uniform traffic distribution, the arrival rate of lightpaths that enter the path at node 1 is the same as the arrival rate of lightpaths that leave the path at node 1 (i.e., $\lambda_l = \lambda_n$). The quantities C_l, C_c, and C_n can therefore be characterized by a three-dimensional Markov chain, with each state represented by an integer triplet (c_l, c_c, c_n). The probabilities $R(n_f | x_{ff}, y_f, z_c)$, $U(z_c | y_f, x_{pf})$, $S(y_f | x_{pf})$, and $Q(w_f)$ are then derived for the two-hop path. The authors then extend the analysis to determine the blocking probability on a path of arbitrary hop length.

To keep the analysis simple, the authors assume that the effect of lightpath blocking on the load carried along the links is negligible. This assumption, which is valid only for low blocking probabilities, means that the entire load offered to the network is carried along the links. From the lightpath arrival rates at nodes, an approximation for the link arrival rates λ_l and λ_c can be found as follows. Let N be the number of nodes in the network, λ be the lightpath arrival rate at a node, and $\bar{H}$ be the average hop distance. Then the average lightpath arrival rate per link (γ) is given by

$$\gamma = \frac{N\lambda\bar{H}}{L} \tag{4.7}$$

Suppose that there are k exit links per node and that if a lightpath does not leave the network at a node, it chooses one of the k exit links arbitrarily. Then, the arrival rate of lightpaths that continue on to the next link of a path can be estimated as

$$\lambda_c = \gamma \frac{1 - 1/\bar{H}}{k} \tag{4.8}$$

from which we have

$$\lambda_l = \gamma - \lambda_c \tag{4.9}$$

The parameter q, the conversion density of the network, is used to model a network with sparse wavelength conversion. The number of converter nodes in an N-node network is distributed binomially with an average of Nq converters. The blocking probability in a network with sparse wavelength conversion is then computed recursively by conditioning on the event that node i is the last converter on a l-hop path in the network [$1 \leq i \leq (l-1)$].

4.4.1.3 Probabilistic Model for a Class of Networks [6] This study provides an approximate method for calculating the blocking probability in a wavelength-routed network. The model considers Poisson input traffic and uses a Markov chain

model with state-dependent arrival rates. Two different routing schemes are considered: *fixed routing*, where the path from a source to a destination is unique and is known beforehand; and *least-loaded routing* (LLR), an alternate-path scheme where the route from source to destination is taken along the path that has the largest number of idle wavelengths. Analysis and simulations are carried out using fixed routing for networks of arbitrary topology with paths of length at most three hops and using LLR for fully connected networks with paths of one or two hops. The blocking probability is found to be larger without wavelength conversion. This method is, however, computationally intensive and is tractable only for densely connected networks with a few nodes.

Details   We consider a network of arbitrary topology with J links and C wavelengths on each link. A route R is a subset of links from $\{1, 2, \ldots, J\}$. Lightpath requests arrive for route R as a Poisson stream with rate α_R. A lightpath for route R is set up if there is a wavelength w_i such that w_i is idle on all links of route R. The holding times of all lightpaths are exponentially distributed with unit mean.

Let X_R be the random variable denoting the number of idle wavelengths on route R. If $R = \{i, j, k\}$, we may write X_R as $X_{i,j,k}$. Let $\mathbf{X} = (X_1, X_2, \ldots, X_J)$ and let

$$q_j(m) = \Pr[X_j = m], \qquad m = 0, 1, \ldots, C \tag{4.10}$$

be the idle capacity distribution on link j. The author assumes that the random variables X_j are independent, as in [12]. Then

$$q(m) = \prod_{j=1}^{J} q_j(m_j) \tag{4.11}$$

where $\mathbf{m} = (m_1, m_2, \ldots, m_J)$. Further, the author assumes that given m idle wavelengths on link j, the time until the next lightpath is set up on link j is exponentially distributed with parameter $\alpha_j(m)$. It follows that the number of idle wavelengths on link j can be viewed as a birth–death process, so that

$$q_j(m) = \frac{C(C-1)\cdots(C-m+1)}{\alpha_j(1)\alpha_j(2)\cdots\alpha_j(m)} q_j(0) \tag{4.12}$$

where

$$q_j(0) = \left[1 + \sum_{m=1}^{C} \frac{C(C-1)\cdots(C-m+1)}{\alpha_j(1)\alpha_j(2)\cdots\alpha_j(m)} \right]^{-1} \tag{4.13}$$

$\alpha_j(m)$ is obtained by combining the contributions from the request streams to routes that have link j as a member, as follows:

$$\alpha_j(m) = \begin{cases} 0 & \text{if } m = 0 \\ \sum_{R:j\in R} a_R \Pr[X_R > 0 | X_j = m] & \text{if } m = 1, 2, \ldots, C \end{cases} \qquad (4.14)$$

The blocking probability for lightpaths to route R $(L_R = \Pr[X_R = 0])$ is then calculated for routes up to three hops. Similarly, for the case of least-loaded routing (LLR), the author derives the blocking probability (L_R) in a fully connected network.

4.4.1.4 Probabilistic Model Without Independent Link Load Assumption [5]
A model that is more analytically tractable than the ones in [16] and [22] is provided in [5]; however, it uses more simplistic traffic assumptions. The link loads are not assumed to be independent; however, the assumption is retained that a wavelength is used on successive links independent of other wavelengths. The concept of *interference length* (L) (i.e., the expected number of links shared by two lightpaths that share at least one link) is introduced. Analytical expressions for the link utilization and the blocking probability are obtained by considering an average path that spans H (*average hop distance*) links in networks with and without wavelength conversion. The gain (G) due to wavelength conversion is defined as the ratio of the link utilization with wavelength conversion to that without wavelength conversion for the same blocking probability. The gain is found to be directly proportional to the effective path length (H/L). A larger switch size (Δ) tends to increase the blocking probability in networks without wavelength conversion. The model used in [5] is also applicable to ring networks, unlike the work in [22], and it correctly predicts the low gain in utilizing wavelength conversion in ring networks.

Details The simplified model described initially in [5] is based on standard series-independent link assumptions (i.e., a lightpath request sees a network in which a wavelength's use on a fiber link is statistically independent of other fiber links and other wavelengths). However, this model generally tends to overestimate the blocking probability.*

Let there be W wavelengths per fiber link and let ρ be the probability that a wavelength is used on any fiber link. (Since ρW is the expected number of busy wavelengths on any fiber link, ρ is also the fiber utilization of any fiber.) We consider an H-link path for a connection from node A to node B that needs to be set up. First, let us consider a network *with* wavelength converters. The probability P_b' that the connection request from A to B will be blocked equals the probability that along this H-link path, *there exists a fiber link with all of its W wavelengths in use*, so that

$$P_b' = 1 - (1 - \rho^W)^H \qquad (4.15)$$

---

*The link-load independence assumption is relaxed in [5] to provide a more accurate model.

Defining q to be the achievable utilization for a given blocking probability in a wavelength-convertible network, we have

$$q = [1 - (1 - P_b')^{1/H}]^{1/W} \approx \left(\frac{P_b'}{H}\right)^{1/W} \tag{4.16}$$

where the approximation holds for small values of P_b'/H.

Next, let us consider a network *without* wavelength converters. The probability P_b that the connection request from A to B will be blocked equals the probability that along this H-link path, *each wavelength is used on at least one of the H links*, so that

$$P_b = [1 - (1 - \rho)^H]^W \tag{4.17}$$

Defining p to be the achievable utilization for a given blocking probability in a network without wavelength conversion, we have

$$p = 1 - (1 - P_b^{1/W})^{1/H} \approx -\frac{1}{H}\ln(1 - P_b^{1/W}) \tag{4.18}$$

where the approximation holds for large values of H and for $P_b^{1/W}$ not too close to unity. Observe that the achievable utilization is inversely proportional to H, as expected.

We define $G = q/p$ to be a measure of the benefit of wavelength conversion, which is the increase in (fiber or wavelength) utilization for the same blocking probability. From equations (4.16) and (4.18), after setting $P_b = P_b'$, we get

$$G \approx H^{1-(1/W)} \frac{P_b^{1/W}}{-\ln(1 - P_b^{1/W})} \tag{4.19}$$

where the approximation holds for small P_b, large H, and moderate W so that $P_b^{1/W}$ is not too close to unity.

It is also reported in [5] that the gain increases as the blocking probability decreases, but this effect is small for small values of P_b. Also, as W increases, G also increases until it peaks around $W \approx 10$ (for $q \approx 0.5$), and the maximum gain is close to $H/2$. After peaking, G decreases, but very slowly. Generally, it is found that for a moderate to large number of wavelengths, the benefits of wavelength conversion increase with the length of the connection, and decrease (slightly) with an increase in the number of wavelengths. Although this was a simple analysis to study the effects of hop length, a more rigorous treatment incorporating the load dependencies on successive links in a path is also presented in [5].

4.4.2   Related Work on Gain Characterization

In this section we present other significant works that characterize the gain available from networks with wavelength converters.

4.4.2.1   *Bounds on RWA Algorithms With and Without Wavelength Converters [37]*   Upper bounds on the traffic carried (or equivalently, lower bounds on the blocking probability) in a wavelength-routed WDM network are derived in [37]. The bounds are shown to be achievable asymptotically by a fixed RWA algorithm using a large number of wavelengths. The *wavelength reuse factor*, defined as the maximum traffic offered per wavelength for which the blocking probability can be made arbitrarily small by using a sufficiently large number of wavelengths, is found to increase by using wavelength converters in large networks. Simulations show that wavelength converters offer a 10 to 40% increase in the amount of reuse available in the authors' sampling of 14 networks ranging from 16 to 1000 nodes when the number of wavelengths available is small (10 or 32).

4.4.2.2   *Multifiber Networks [20]*   The benefits of wavelength conversion in a network with *multiple* fiber links are studied in [20] by extending the analysis presented in [5] to multifiber networks. Multifiber links are found to reduce the gain obtained due to wavelength conversion, and the number of fibers is found to be more important than the number of wavelengths for a network. It is concluded that a mesh network enjoys a higher utilization gain with wavelength conversion for the same traffic demand than that of a ring or a fully connected network.

4.4.2.3   *Limited-Range Wavelength Conversion [46]*   The effects of limited-range wavelength conversion (see Section 4.3.1) on the performance gains achievable in a network are considered in [46]. The model used in this work captures the functionality of certain all-optical wavelength converters (e.g., those based on four-wave mixing) whose conversion efficiency drops with increasing range. The analytical model follows from [5] but employs both link-load independence and wavelength-independence assumptions. The results obtained indicate that a significant improvement in the blocking performance of the network is obtained when limited-range wavelength converters with as little as one-fourth of the full range are used. Moreover, converters with just half of the full conversion range deliver almost all of the performance improvement offered by an ideal full-range converter.

4.4.2.4   *Minimal Wavelength Conversion in WDM Rings [3]*   In addition to reducing the overall blocking probability, wavelength converters can improve the fairness performance by allowing many long-distance lightpaths, which would have been blocked otherwise, to be established [25]. In [3], the authors define *the unfairness factor* as the ratio of average blocking on the longest path to blocking on the shortest path. The *fairness ratio* is then defined as the ratio of the unfairness

factor without wavelength conversion to that with wavelength conversion. Simulation studies in a 195-node network of 15 interconnected WDM rings with 13 nodes each show significant increase in the fairness ratio, of approximately 10,000, for 32 wavelengths. Similar trends have also been observed in smaller rings. Moreover, for large interconnected rings, this improvement can be achieved with wavelength conversion in just 10 to 20% of the nodes.

4.4.2.5   *Limited Wavelength Conversion in WDM Networks [36]* Limited-range wavelength conversion together with sparse location of converters in the network (see Section 4.3.1) is the focus of this study [36]. The authors consider the *static* case, when all lightpath requests are available at one time and wavelengths are assigned to these lightpaths *off-line*. The wavelength degree of a node is defined as the number of possible wavelengths to which a given input wavelength can be converted at the node. The *load* of a set of lightpath requests is defined as the value $\lambda_{max} = \max_{e \in E} \lambda_e$, where λ_e denotes the number of routes using link e and E denotes the set of links in the network. The authors show that all requests with load $\lambda_{max} \leq W - 1$ have wavelength assignments in a ring network with just one node having *fixed* wavelength conversion (i.e., at this node, each wavelength is converted to a different predetermined output wavelength). They also provide a ring network with two nodes with wavelength degree 2* and the rest of the nodes with no wavelength conversion such that all requests with load $\lambda_{max} \leq W$ have channel assignments. Similarly, a ring network with full wavelength conversion at one node and no conversion at the other nodes is shown to have a channel assignment for all requests with loads at most W. The authors extend their results to star networks, tree networks, and networks with arbitrary topologies.

For information about recent research studies on the benefits of wavelength conversion, we refer the interested reader to [32] and [33].

4.5   SUMMARY

In this study we examined the various facets of the wavelength-conversion technology from its *realization* using current optoelectronic devices to its *incorporation* in a wavelength-routed network design to its *effect* on efficient routing and management algorithms to a *measurement* of its potential benefits under various network conditions. Although understanding of the technology has improved during the past few years, several issues remain unresolved, especially in the context of efficient design mechanisms and routing protocols. Additional efforts in this area are needed to further the performance of such networks using intelligent routing and design methods.

---

*Note that a node without wavelength conversion or with fixed wavelength conversion has wavelength degree 1.

ACKNOWLEDGMENTS

The author thanks Professor Biswanath Mukherjee for his contributions to the material presented in this chapter.

REFERENCES

1. N. Antoniades, K. Bala, S. J. B. Yoo, and G. Ellinas, A parametric wavelength interchanging cross-connect (WIXC) architecture, *IEEE Photonic Technology Letters*, 8(10):1382–1384, Oct. 1996.
2. K. Bala, T. E. Stern, and K. Bala, Algorithms for routing in a linear lightwave network, *Proc. IEEE INFOCOM'91*, pp. 1–9, Bal Harbour, FL, Apr. 1991.
3. K. Bala, E. Bouillet, and G. Ellinas, The benefits of "minimal" wavelength interchange in WDM rings, *Optical Fiber Communication (OFC'97) Technical Digest*, Vol. 6, pp. 120–121, Dallas, TX, Feb. 1997.
4. D. Banerjee and B. Mukherjee, Practical approaches for routing and wavelength assignment in large all-optical wavelength-routed networks, *IEEE Journal on Selected Areas in Communications*, 14(5):903–908, June 1996.
5. R. A. Barry and P. A. Humblet, Models of blocking probability in all-optical networks with and without wavelength changes, *IEEE Journal on Selected Areas in Communications*, 14(5):858–867, June 1996.
6. A. Birman, Computing approximate blocking probabilities for a class of all-optical networks, *IEEE Journal on Selected Areas in Communications*, 14(5):852–857, June 1996.
7. C. A. Brackett, Dense wavelength division multiplexing networks: Principles and applications, *IEEE Journal on Selected Areas in Communications*, 8(6):948–964, Aug. 1990.
8. C. A. Brackett et al., A scalable multiwavelength multihop optical network: A proposal for research on all-optical networks, *IEEE/OSA Journal of Lightwave Technology*, 11(5/6):736–753, May/June 1993.
9. C. Chen and S. Banerjee, A new model for optimal routing and wavelength assignment in wavelength division multiplexed optical networks, *Proc. IEEE INFOCOM'96*, pp. 164–171, San Francisco, 1996.
10. I. Chlamtac, A. Faragó, and T. Zhang, Lightpath (wavelength) routing in large WDM networks, *IEEE Journal on Selected Areas in Communications*, 14(5):909–913, June 1996.
11. I. Chlamtac, A. Ganz, and G. Karmi, Lightpath communications: An approach to high-bandwidth optical WAN's, *IEEE Transactions on Communications*, 40:1171–1182, July 1992.
12. S. Chung, A. Kashper, and K. W. Ross, Computing approximate blocking probabilities for large loss networks with state-dependent routing, *IEEE/ACM Transactions on Networking*, 1(1):105–115, 1993.
13. T. Durhuus et al., All optical wavelength conversion by SOA's in a Mach–Zender configuration, *IEEE Photonic Technology Letters*, 6:53–55, Jan. 1994.

14. T. Durhuus et al., All-optical wavelength conversion by semiconductor optical amplifiers, *IEEE/OSA Journal of Lightwave Technology*, 14(6):942–954, June 1996.

15. M. Eiselt, W. Pieper, and H. G. Weber, Decision gate for all-optical retiming using a semiconductor laser amplifier in a loop mirror configuration, *Electronic Letters*, 29:107–109, Jan. 1993.

16. M. Fujiwara et al., A coherent photonic wavelength-division switching system for broadband networks, *Proc. European Conference on Communication (ECOC'88)*, pp. 139–142, 1988.

17. O. Gerstel, R. Ramaswami, and G. Sasaki, Fault tolerant multiwavelength optical rings with limited wavelength conversion, *Proc. IEEE INFOCOM'97*, pp. 508–516, 1997.

18. B. S. Glance, J. M. Wiesenfeld, U. Koren, and R. W. Wilson, New advances on optical components needed for FDM optical networks, *IEEE/OSA Journal of Lightwave Technology*, 11(5/6):882–890, May/June 1993.

19. J. Iness, Efficient use of optical components in WDM-based optical networks, Ph.D. dissertation, Department of Computer Science, University of California, Davis, CA, Nov. 1997.

20. G. Jeong and E. Ayanoglu, Comparison of wavelength-interchanging and wavelength-selective cross-connects in multiwavelength all-optical networks, in *Proc. IEEE INFOCOM'96*, pp. 156–163, San Francisco, 1996.

21. F. P. Kelly, Blocking probabilities in large circuit switched networks, *Advances in Applied Probability*, 18:473–505, 1986.

22. M. Kovačević and A. S. Acampora, Benefits of wavelength translation in all-optical clear-channel networks, *IEEE Journal on Selected Areas in Communications*, 14(5):868–880, June 1996.

23. M. Kovačević and A. S. Acampora, Electronic wavelength translation in optical networks, *IEEE/OSA Journal of Lightwave Technology*, 14(6):1161–1169, June 1996.

24. J. P. R. Lacey, G. J. Pendock, and R. S. Tucker, Gigabit-per-second all-optical 1300-nm to 1550-nm wavelength conversion using cross-phase modulation in a semiconductor optical amplifier, *Optical Fiber Communication (OFC'96) Technical Digest*, Vol. 2, pp. 125–126, San Jose, CA, 1996.

25. K.-C. Lee and V. O. K. Li, A wavelength convertible optical network, *IEEE/OSA Journal of Lightwave Technology*, 11(5/6):962–970, May/June 1993.

26. R. Ludwig and G. Raybon, BER measurements of frequency converted signals using four-wave mixing in a semiconductor laser amplifier at 1, 2.5, 5, and 10 Gbit/s, *Electronic Letters*, 30:338–339, Jan. 1994.

27. D. J. G. Mestdagh, *Fundamentals of Multiaccess Optical Fiber Networks*, Artech House Optoelectronics Library, Artech House, Norwood, MA, 1995.

28. B. Mikkelsen et al., Polarization insensitive wavelength conversion of 10 Gbit/s signals with SOAs in a Michelson interferometer, *Electronic Letters*, 30(3):260–261, Feb. 1994.

29. B. Mikkelsen et al., All-optical noise reduction capability of interferometric wavelength converters, *Electronic Letters*, 32(6):566–567, Mar. 1996.

30. B. Mikkelsen et al., Wavelength conversion devices, in *Optical Fiber Communication (OFC'96) Technical Digest*, Vol. 2, pp. 121–122, San Jose, CA, 1996.

31. B. Mukherjee, *Optical Communication Networks*, McGraw-Hill, New York, 1997.

32. C. Siva Ram Murthy and M. Gurusamy, *WDM Optical Networks: Concepts, Design and Algorithms*, Prentice Hall, Upper Saddle River, NJ, 2002.

33. B. Ramamurthy, S. Subramaniam, and A. K. Somani, Eds., *Optical Networks*, special issue on wavelength conversion, Vol. 3, Mar./Apr. 2002.

34. B. Ramamurthy and B. Mukherjee, Wavelength conversion in WDM networking, *IEEE Journal on Selected Areas in Communications*, 16(7):1061–1073, Sept. 1998.

35. R. Ramaswami, Optical network architectures, *Optical Fiber Communication (OFC'96) Technical Digest*, San Jose, CA, 1996.

36. R. Ramaswami and G. H. Sasaki, Multiwavelength optical networks with limited wavelength conversion, in *Proc. IEEE INFOCOM'97*, pp. 489–498, San Francisco, 1997.

37. R. Ramaswami and K. N. Sivarajan, Routing and wavelength assignment in all-optical networks, *IEEE/ACM Transactions on Networking*, 3(5):489–500, Oct. 1995.

38. P. Rigole, S. Nilsson, E. Berglind, D. J. Blumenthal, et al., State of the art: Widely tunable lasers (In-plane semiconductor lasers: From ultraviolet to midinfrared), *Proc. SPIE*, Vol. 3001, pp. 382–393, San Jose, CA, Feb. 1997.

39. R. Sabella and E. Iannone, Wavelength conversion in optical transport networks, *Fiber and Integrated Optics*, 15(3):167–191, 1996.

40. R. Schnabel et al., Polarization insensitive frequency conversion of a 10-channel OFDM signal using four-wave mixing in a semiconductor laser amplifier, *IEEE Photonic Technology Letters*, 6:56–58, Jan. 1994.

41. S. Subramaniam, M. Azizoğlu, and A. K. Somani, All-optical networks with sparse wavelength conversion, *IEEE/ACM Transactions on Networking*, 4(4):544–557, Aug. 1996.

42. R. W. Tkach et al., Four-photon mixing and high-speed WDM systems, *IEEE/OSA Journal of Lightwave Technology*, 13(5):841–849, May 1995.

43. N. Wauters, W. Van Parys, B. Van Caenegem, and P. Demeester, Reduction of wavelength blocking through partitioning with wavelength convertors, *Optical Fiber Communication (OFC'97) Technical Digest*, Vol. 6, pp. 122–123, Dallas, TX, Feb. 1997.

44. J. M. Wiesenfeld, Wavelength conversion techniques, *Optical Fiber Communication (OFC'96) Technical Digest*, Vol. Tutorial TuP 1, pp. 71–72, San Jose, CA, 1996.

45. H. Yasaka et al., Finely tunable 10-Gb/s signal wavelength conversion from 1530 to 1560-nm region using a super structure grating distributed Bragg reflector laser, *IEEE Photonic Technology Letters*, 8(6):764–766, June 1996.

46. J. Yates, J. Lacey, D. Everitt, and M. Summerfield, Limited-range wavelength translation in all-optical networks, *Proc. IEEE INFOCOM'96*, pp. 954–961, San Francisco, 1996.

47. S. J. B. Yoo, Wavelength conversion technologies for WDM network applications, *IEEE/OSA Journal of Lightwave Technology*, 14(6):955–966, June 1996.

48. J. Zhou et al., Four-wave mixing wavelength conversion efficiency in semiconductor traveling-wave amplifiers measured to 65 nm of wavelength shift, *IEEE Photonic Technology Letters*, 6(8):984–987, 1994.

5 Contention Resolution in Optical Packet Switching

SHUN YAO and BISWANATH MUKHERJEE

University of California–Davis, Davis, California

SUDHIR DIXIT

Nokia Research Center, Burlington, Massachusetts

5.1 INTRODUCTION

In this chapter we investigate how packet contention can be resolved in the next-generation optical Internet. It is aimed at offering an in-depth discussion of optical contention resolution. Readers unfamiliar with generic architectures of optical packet switching may refer to Section 6.2.

In an optical packet-switched network, contention occurs at a switching node whenever two or more packets try to leave the switch fabric on the same output port, on the same wavelength, at the same time. In electrical packet-switched networks, contention is resolved using the store-and-forward technique, which requires the packets in contention to be stored in a memory bank and to be sent out at a later time when the desired output port becomes available. This is possible because of the availability of electronic random-access memory (RAM). There is no equivalent optical RAM technology; therefore, the optical packet switches need to adopt different approaches for contention resolution. Meanwhile, wavelength-division multiplexing (WDM) networks provide one additional dimension, wavelength, for contention resolution. In this chapter we explore all three dimensions of contention-resolution schemes: wavelength, time, and space, and discuss in detail how to combine these three dimensions into an effective contention-resolution scheme. Readers more interested in a high-level overview of optical packet switching can skip this chapter and refer to Chapter 6.

The basic node architectures for a number of contention-resolution schemes in the wavelength, time, and space domains are described in Section 5.2. We also present simulation experiments and performance results of these schemes and

IP over WDM: Building the Next-Generation Optical Internet, Edited by Sudhir Dixit.
ISBN 0-471-21248-2 © 2003 John Wiley & Sons, Inc.

explain the network behavior observed under these schemes. In Section 5.3 we introduce priority-based routing using preemption. In Section 5.4 we compare slotted and unslotted networks, in Section 5.5 present a hybrid electrical-optical contention-resolution scheme, in Section 5.6 study transmission control protocol (TCP) performance based on the hybrid scheme, and in Section 5.7 summarize the chapter.

5.2   CONTENTION RESOLUTION IN WAVELENGTH, TIME, AND SPACE DOMAINS

The contention-resolution mechanism of the three dimensions are outlined below:

1. *Wavelength conversion* offers effective contention resolution without relying on buffer memory [1–5]. Wavelength converters can convert wavelengths of packets that are contending for the same wavelength of the same output port. It is a powerful and the most preferred contention-resolution scheme (as this study will demonstrate), as it does not cause extra packet latency, jitter, and resequencing problems. The interested reader is referred to Chapter 4 for more details on wavelength conversion.

2. *Optical delay line* (which provides sequential buffering) is a close imitation of the RAM in electrical routers, although it offers a fixed and finite amount of delay. Many previously proposed architectures employ optical delay lines to resolve the contention [6–13]. However, since optical delay lines rely on the propagation delay of the optical signal in silica to buffer the packet in time (i.e., due to their sequential access), they have more limitations than does electrical RAM. To implement large buffer capacity, the switch needs to include a large number of delay lines.

3. The *space deflection approach* [14–18] is a multiple-path routing technique. Packets that lose contention are routed to nodes other than their preferred next-hop nodes, with the expectation that they will eventually be routed to their destinations. The effectiveness of deflection routing depends heavily on the network topology and the traffic pattern offered.

Both wavelength conversion and optical buffering require extra hardware (wavelength converters and lasers for wavelength conversion; fibers and additional switch ports for optical buffering) and control software. Deflection routing can be implemented with extra control software only.

With an orthogonal classification, optical packet-switched networks can be divided into two categories: time-slotted networks with fixed-length packets [19] and unslotted networks with fixed- or variable-size packets [4,8]. In a slotted network, packets of fixed size are placed in time slots. When they arrive at a node, they are aligned before being switched jointly [9]. In an unslotted network, the nodes do not align the packets and switch them one by one "on the fly"; therefore, they do not need synchronization stages. Because of such unslotted operation, they can switch variable-length packets. However, unslotted networks have lower overall throughput than slotted networks, because of the increased packet contention probability [16]. Similar contrast exists between the unslotted and slotted version of the ALOHA network [20]. Due to the lack of workable optical RAM technologies, all-optical networks find it difficult to provide packet-level synchronization, which is required in slotted networks. In addition, it is preferred that a network can accommodate natural IP packets with variable lengths. Therefore, in this chapter we focus primarily on optical contention-resolution schemes in unslotted asynchronous networks.

Figure 5.1 shows the generic node architectures with contention-resolution schemes in time, space, and wavelength domains. Every node has a number of add–drops ports, and this number will vary depending on the nodal degree (i.e., how many input/output fiber pairs the node has). Each add–drop fiber port will correspond to multiple client interfaces, reflecting multiple wavelengths on each fiber. Each client interface input (output) will be connected to a local receiver (transmitter). Different contention-resolution schemes give rise to different architectures.

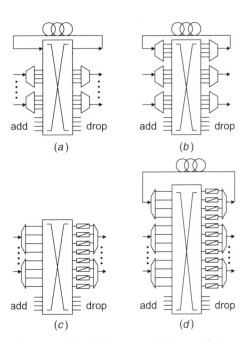

Figure 5.1 Node architectures for different contention-resolution schemes: (*a*) single-wavelength optical buffering; (*b*) multiwavelength optical buffering; (*c*) wavelength conversion; (*d*) wavelength conversion and multiwavelength optical buffering.

5.2.1   Optical Buffering

Optical buffering utilizes one or more optical fiber delay lines looping the signal from the output back to the input of the switch fabric. Figure 5.1a and b illustrate the optical buffering. Time-domain contention resolution requires optical buffering wherein the packet causing the contention enters the optical delay line at the output of the switch fabric, traverses the entire delay line, and reenters the switch fabric. For unslotted operation, the delay line can consist of an arbitrary length of fiber. Simulation experiments find little improvement in network performance when delay lines are made longer than the maximum packet size.

We consider both cases of optical buffering: single wavelength and multiwavelength. Figure 5.1a shows the single-wavelength case, in which the delay line can take only one packet at a time (i.e., there is no multiplexer or demultiplexer at either end). If all the buffers are occupied, the packet in contention needs to seek an alternative contention resolution or must be dropped. Figure 5.1b shows the multiplewavelength case, in which each delay line is terminated by a multiplexer and a demultiplexer. Such a delay line can accommodate multiple packets on different wavelengths. Compared with a single-wavelength delay line, its multiple-wavelength counterpart requires a larger switch fabric and additional hardware, such as a multiplexer and demultiplexer, but achieves larger capacity optical buffering on multiple wavelengths.

5.2.2   Wavelength Conversion

Figure 5.1c and d show contention resolution utilizing wavelength conversion, where the signal on each wavelength from the input fiber is first demultiplexed and sent into the switch, which is capable of recognizing the contention and selecting a suitable wavelength converter leading to the output fiber desired. The wavelength converters can operate with a full degree (i.e., they can convert any incoming wavelength to a fixed desired wavelength) or with a limited range (i.e., they can convert one or several predetermined incoming wavelengths to a fixed desired wavelength).

5.2.3   Space Deflection

Space deflection relies on another neighboring node to route a packet when contention occurs. As a result, the node itself can adopt any node architecture in Fig. 5.1. Space deflection resolves contention at the expense of the inefficient utilization of the network capacity and the switching capacity of another node. Obviously, this is not the best choice among contention-resolution schemes. As discussed in later sections, the node will seek wavelength-domain contention resolution first, time-domain contention resolution second, and space-domain contention resolution third. In practice, contention resolution will often employ a combination of wavelength-, time-, and space-domain contention resolution.

In most deflection networks, certain mechanisms need to be implemented to prevent looping (a packet being sent back to a node it has visited before), such as setting a maximum hop count and discarding all the packets that have passed more hops than this number. This is similar to the time-to-live (TTL) mechanism for routing IP packets.

5.2.4   Combination Schemes

By mixing the basic contention-resolution schemes discussed so far, one can create combination schemes. For example, Fig. 5.1*d* shows the node architecture for wavelength conversion combined with multiwavelength buffering. Note that a packet can be dropped at any node under all of these schemes, due to (1) the nonavailability of a free wavelength at the output port, (2) the nonavailability of a free buffer, and/or (3) the fact that the packet may have reached its maximum hop count.

We define the notations for these schemes as follows:

- *baseline*: no contention resolution. Packet in contention is dropped immediately.
- *N buf, N bufwdm*: buffering. The node has N delay lines, and each delay line can take one or multiple wavelengths at a time.
- *def*: Deflection.
- *wc*: full-range wavelength conversion.
- *wclimC*: limited wavelength conversion. C given wavelengths can be converted to one fixed wavelength.

In the following section, these notations indicate the various approaches and their priorities. For example, *4wav_wc+16buf+def* means a combination of full-range wavelength conversion, single-wavelength buffering with 16 delay lines, and deflection in a four-wavelength system. The order of contention resolution is as follows: *A packet that loses contention will first seek a vacant wavelength on the preferred output port. If no such wavelength exists, it will seek a vacant delay line. If no delay line is available, it will seek a vacant wavelength on the deflection output port. When all of the options above fail, the packet will be dropped.* This contention resolution order provides the best performance in terms of the packet loss rate and packet delay.

5.2.5   Simulation Experiments and Performance Comparison

5.2.5.1   Network Topology and Configuration   For purposes of illustration, the network topology under study is a part of a telco's optical mesh network, as shown in Fig. 5.2. Each link i is L_i kilometers long and consists of two fibers to form a bidirectional link. Every fiber contains W wavelengths, each of which carries a data stream at data rate R. Each node is equipped with an array of W transmitters and an array of W receivers that operate on one of the W wavelengths independently at data rate R. By default, all packets are routed statically via the shortest path.

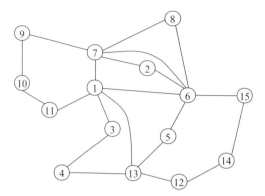

Figure 5.2   Mesh network topology.

It is worth mentioning that because some nodes have many neighbors, the deflection scheme should avoid blind deflections. Deflection is carried out only in hub nodes, which are nodes that have much higher nodal degrees than other nodes to serve as major routing nodes. Nodes 1, 6, 7, and 13 in Fig. 5.2 are examples of hub nodes. Most of the routing is done by the four hub nodes in this example network, and it is possible to have a deflection policy, with which a packet would only be deflected to a node that can eventually lead to the packet's original next-hop node with no more than two extra hops. If no such node exists, the packet will be dropped. Deflection routing requires the node to have a second entry in the routing table that contains preferred deflection ports for each destination.

For example, in Fig. 5.2, if a packet from node 3 is destined for node 9 via node 1, and there is contention at the output port in node 1 leading to node 7, the packet will be deflected to the port leading to node 11; if the port leading to 11 is also busy, the packet will be dropped immediately instead of being deflected to node 13 or node 6. Table 5.1 shows an example of the deflection entries in the routing table at node 7.

With deflection routing, a loop-mitigation mechanism is necessary. A maximum hop count H is set to limit how many hops a packet can travel (each time the packet passes through a delay line or is transmitted from one node to another, it is counted as one hop). Nevertheless, the network topology of Fig. 5.2 and the aforementioned

TABLE 5.1   Deflection Table of Node 7

Next-Hop Node	Deflect to:
1	6
2	6
6	2
8	6
9	Drop

deflection policy can automatically eliminate looping (since the shortest path between any source–destination pair involves no more than two hub nodes, and we can ensure that the deflection table will not cause any looping). The purpose of setting the maximum hop count here is to limit physical impairments (signal-to-noise-ratio degradation, accumulated crosstalk, accumulated insertion loss, etc.) of packets, which is partly introduced by optical buffering.

5.2.5.2 Traffic Generation One of the important requirements for the simulation experiment is to model the network traffic as close to reality as possible. A main characteristic of Internet traffic is its burstiness, or self-similarity. It has been shown in the literature that self-similar traffic can be generated by multiplexing multiple sources of Pareto-distributed on–off periods. In the context of a packet-switched network, on periods correspond to a series of packets sent one after another, and off periods are periods of silence [21].

The probability density function (pdf) and probability distribution function (PDF) of the Pareto distribution are

$$p(x) = \frac{\alpha \cdot b^{\alpha}}{x^{\alpha+1}} \tag{5.1}$$

$$P(x) = \int_{b}^{x} \frac{\alpha \cdot b^{\alpha}}{x^{\alpha+1}} dx = 1 - \frac{b^{\alpha}}{x^{\alpha}} \tag{5.2}$$

where α is the shape parameter (tail index) and b is the minimum value of x. When $\alpha \leq 2$, the variance of the distribution is infinite. When $\alpha \geq 2$, the mean value is infinite as well. For self-similar traffic, α should be between 1 and 2. The Hurst parameter H is given as $H = (3 - \alpha)/2$.

Since $0 < P(x) \leq 1$, the value of x can be generated from a random number $\mathcal{RND}$ with the range $(0, 1]$

$$\frac{b^{\alpha}}{x^{\alpha}} = \mathcal{RND} \tag{5.3}$$

$$x = b \left(\frac{1}{\mathcal{RND}} \right)^{1/\alpha} \tag{5.4}$$

The mean value of Pareto distribution is given by:

$$E(x) = \frac{\alpha \cdot b}{\alpha - 1} \tag{5.5}$$

Once $\alpha_{on}, \alpha_{off}, b_{on}$, and b_{off} are given, the distribution of the on–off periods is determined. b_{on} is the minimum on-period length, equal to the smallest packet size divided by the line rate. The average load of each on–off source, L, is

$$L = \frac{E_{on}}{E_{on} + E_{off}}$$

where E_{on} and E_{off} are the mean value of on and off periods. Therefore,

$$b_{\text{off}} = \left(\frac{1}{L} - 1\right) \frac{\alpha_{\text{on}}(\alpha_{\text{off}} - 1)}{(\alpha_{\text{on}} - 1)\alpha_{\text{off}}} b_{\text{on}} \tag{5.6}$$

During the on period of the on–off source, packets are sent back to back.

In the experiments, network traffic is assumed to consist of IP packets. The nature of IP packets is known to be hard to capture [22]. Statistical data indicate a predominance of small packets, with peaks at the common sizes of 44, 552, 576, and 1500 bytes. Small packets 40 to 44 bytes in length include TCP acknowledgment segments, TCP control segments, and telnet packets carrying single characters. Many TCP implementations that do not implement path maximum transmission unit (MTU) discovery use either 512 or 536 bytes as the default maximum segment size (MSS) for nonlocal IP destinations, yielding a 552- or 576-byte packet size. An MTU size of 1500 bytes is the characteristic of Ethernet-attached hosts. The cumulative distribution of packet sizes in Fig. 5.3 shows that almost 75% of the packets are smaller than the typical TCP MSS of 552 bytes. Nearly 30% of the packets are 40 to 44 bytes in length. On the other hand, over half of the total traffic is carried in packets of size 1500 bytes or larger. This irregular packet-size distribution is difficult to express with a closed-form expression. We adopt a truncated 19-order polynomial, fitted from the statistical data, to faithfully reproduce the IP packet-size distribution (as shown in Fig. 5.3). The number of orders, 19, is the smallest number that can reproduce a visually close match with the steep turns in the statistical data. We set the maximum packet size to be 1500 bytes, since the percentage of packets larger than 1500 bytes is negligibly small.

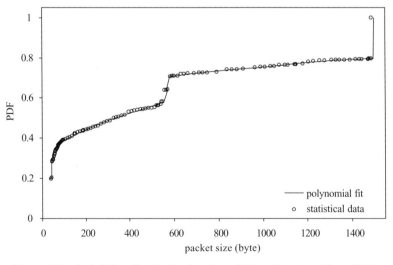

Figure 5.3   Probability distribution function of IP packet sizes. (From [22].)

TABLE 5.2   Network Parameters Used in Simulation

Link Length L_i	Data Rate R	Number of Wavelengths W	Maximum Hop Distance H
20 km	2.5 Gbps	4, 8, 16	8

5.2.5.3 Simulation Metrics   We have chosen four metrics to evaluate network performance, with different contention-resolution schemes: network throughput, packet-loss rate, average end-to-end delay, and average hop distance. They indicate the network utilization, reliability, latency, and physical impairments to the signal. The packet-loss rate is the total number of lost packets divided by the total number of packets generated. The network throughput is defined as

$$\text{network throughput} = \frac{\dfrac{\text{total number of bits delivered successfully}}{\text{network transmission capacity} \times \text{simulation time}}}{\text{ideal average hop distance}}$$

$$\text{network transmission capacity} = (\text{total no. links}) \times (\text{no. wavelengths}) \times (\text{data rate})$$

Network throughput is the fraction of the network resource that delivers data successfully. Because packets can be dropped, a part of the network capacity is wasted in transporting the bits that are dropped. In an ideal situation where no packets are dropped and there is no idle time in any links, the network will be fully utilized and the throughput will reach unity. Average hop distance is the hop distance a packet can travel, averaged over all the possible source–destination pairs in the network. The ideal average hop distance (i.e., no packet dropping) of the network in Fig. 5.2 is 2.42. Table 5.2 shows the values of the parameters used in the simulation experiments.

All the results are plotted against average offered transmitter load (i.e., the total number of bits offered per unit time divided by the line speed). (For example, if the source is generating 0.5 Gbit of data per second and the transmitter/line capacity is 2.5 Gbps, the transmitter load would be 0.2.) With a given average offered transmitter load and a uniform traffic matrix, the average offered link load per wavelength is

ave. link load offered

$$= \frac{\text{ave. offered TX load} \times \text{total no. TX in the network} \times \text{ave. hop distance}}{\text{no. wavelengths} \times \text{total number of unidirectional links in the network}}$$

5.2.5.4 Comparison of the Basic Contention-Resolution Schemes   Figure 5.4*a* compares the network throughput of the schemes incorporating wavelength conversion with different numbers of wavelengths. For reference, it also shows the throughput of *baseline*. We simulate four wavelengths in *baseline*, although the

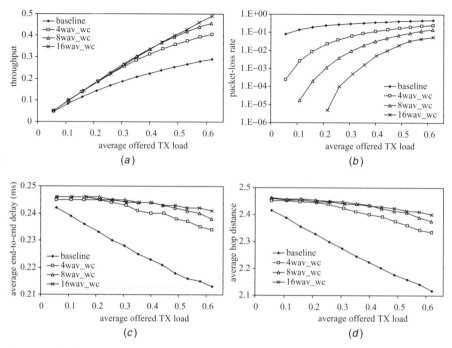

Figure 5.4 Performance comparison of the basic schemes: baseline and wavelength conversion.

number of wavelengths does not affect the results of *baseline* since each wavelength plane operates independently. We find that more wavelengths provide better throughput performance for the wavelength-conversion scheme. Meanwhile, the margin of improvement in throughput decreases when the wavelength number increases; with 16 wavelengths, the network throughput is nearly linear to transmitter load.

Figure 5.4*b* compares the packet-loss rate (represented as a fraction) of these schemes. It is a good complement to Fig. 5.4*a* because the network's throughput has a direct relationship to the packet-loss rate, although the packet-loss rate can reveal more subtle performance differences, especially under light load. To estimate the upper-bound requirement of the packet-loss rate for TCP traffic, we adopt the following criteria: the product of packet-loss rate and the square of the throughput-delay product should be less than unity [23], where the throughput is measured in packets per second on a per-TCP-connection basis. Let us first evaluate the upper bound of the packet-loss rate for a nationwide network whose round-trip delay is approximately 50 ms. For a 1-Mbps TCP connection, the pipe throughput is approximately 100 packets per second. This gives an estimated upper bound of 0.01 for the packet-loss rate. With a smaller optical packet-switched network, the round-trip delay is less; thus, the upper bound of the packet-loss rate can be higher than 0.01 for TCP applications. The transmitter load corresponding to a 0.01

packet-loss rate for *4wav_wc, 8wav_wc,* and *16wav_wc* is approximately 0.16, 0.31, and 0.44, respectively. This clearly indicates the advantage of using more wavelengths.

Figure 5.4*c* compares the average end-to-end packet delays. We can see the general trend of decreasing delay with higher load for all values of wavelengths. This is because when the load increases, the packet-loss rate also increases. Packets with closer destinations are more likely to survive, while packets that need to travel long distances are more likely to be dropped by the network. The overall effect is that when we consider only survived packets in the performance statistics, the delay decreases. This effect is most prominent with *baseline*, because it has the highest packet-loss rate. The same reasoning can be applied to explain the lower delay of *4wav_wc*.

Figure 5.4*d* shows the average hop-distance comparison. Since neither *baseline* nor *wc* involves buffering, the average hop distance is proportional to the average end-to-end delay.

In the schemes incorporating optical buffering, all the optical delay lines are of length 1 km (with propagation delay of 5 µs), enough to hold a packet with maximum length (12,000 bits). We simulate four wavelengths in the network, with different buffering and deflection settings; this is because other number of wavelengths renders similar results, and larger number of wavelengths requires much longer time to simulate. Figure 5.5*a* compares the throughput of different

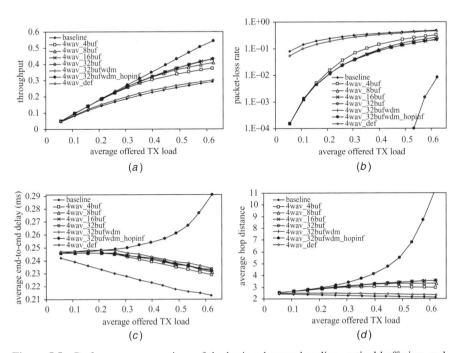

Figure 5.5 Performance comparison of the basic schemes: baseline, optical buffering, and deflection.

optical-buffer schemes with the deflection scheme. For the buffering schemes, we observe the difference in performance between different numbers of optical delay lines. Network throughput first increases with the number of delay lines, but it saturates after the number of delay lines reaches 16: The throughput curves for *4wav_16buf*, *4wav_32buf*, and *4wav_32bufwdm* are almost indistinguishable. This effect is due to the maximum hop count, which is set to 8. To validate this explanation, we also plot another curve, *4wav_32bufwdm_hopinf*, without the maximum-hop-count constraint. It is clear that utilization of optical buffering increases considerably if packets can be buffered an unlimited number of times. For deflection, the throughput of *4wav_def* is only marginally higher than that of *baseline*, indicating that deflection, by itself, is not a very effective contention-resolution technique in this network. For the given topology, only about one-fourth of the nodes can perform deflection. Furthermore, any source node can reach any destination node through at most two hub nodes; therefore, the chance of a packet being deflected is relatively low. Space deflection can be a good approach in a network with high-connectivity topology, such as ShuffleNet [14,16], but is less effective with a low-connectivity topology.

Figure 5.5*b* shows the packet-loss rates. In this network, deflection alone cannot meet the required packet-loss rate of 0.01. The threshold transmitter loads for all the buffering schemes with maximum hop count are approximately 0.2. The packet-loss rate for *4wav_32bufwdm_hopinf* is very low and the upper-bound transmitter load is 0.63.

The plots of end-to-end delay and average hop count in Fig. 5.5*c* and *d* reveal more detail on the effect of setting the maximum hop count. The hop count for all the optical-buffering schemes increases with load, because packets are more likely to be buffered with higher load. However, the end-to-end delay decreases (except for *4wav_32bufwdm_hopinf*) with load. This is because (1) packets destined to closer nodes are more likely to survive, and (2) the buffer delay introduced by optical delay lines is small compared with link propagation delay. Without the maximum-hop-count constraint, both end-to-end delay and average hop count rise quickly with load, indicating that unrestricted buffering can effectively resolve the contention. The amount of extra delay caused by unrestricted buffering is less than 0.5 ms with a transmitter load of 0.63, and this extra delay may be acceptable for most applications. The disadvantage of unrestricted buffering is the physical impairment of the signal (e.g., attenuation, crosstalk), since packets must traverse the switch fabric and delay lines many more times. In this case, optical regeneration may become necessary. We also notice that when the load is light, the end-to-end delay increases for *4wav_def*; this is because under light load, deflection is resolving the contention effectively by deflecting packets to nodes outside the shortest path, thus introducing extra propagation delay. For average end-to-end delay, the effect of the extra propagation delay due to deflections is more prominent than the effect of having more survived packets with closer destinations. This explains the initial increase of delay in the deflection scheme.

Table 5.3 lists the upper bound of average offered transmitter load with acceptable packet-loss rate set at 0.01. *4wav_32bufwdm_hopinf* offers the best

TABLE 5.3 Comparison of Upper-Bound Average Offered Transmitter Load with Packet-Loss Rate of 0.01

Scheme	Maximum TX Load
4wav_wc	0.16
16wav_wc	0.44
4wav_4buf	0.18
4wav_16buf	0.2
4wav_32buf_hopinf	0.63
4wav_def	<0.05

packet-loss performance, but it may be more expensive to implement due to the large number of optical delay lines and switch ports, and it may also require optical amplification/regeneration. Wavelength conversion is also an effective approach to resolve contention, although its effectiveness depends on the number of wavelengths in the system. Deflection is the least effective approach in the example network, but its benefit can be achieved when we combine it with other schemes.

5.2.5.5 Comparison of Combination Schemes In this study we chose four scenarios for different combinations of contention-resolution schemes: *16wav_wc+16buf, 16wav_wc+def, 16wav_16buf+def,* and *16wav_wc+16buf+def.* Figure 5.6*a* shows a throughput comparison of these schemes. The scheme that

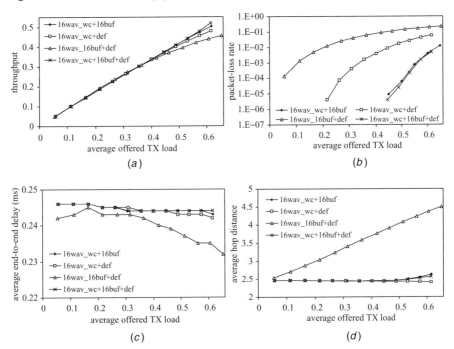

Figure 5.6 Performance comparison of the combination schemes.

incorporates all three dimensions (wavelength, time, and space) for contention resolution offers the best throughput. One can also observe the benefit of using wavelength converters: The schemes involving wavelength conversion perform better under heavy load.

Figure 5.6b compares the packet-loss rates. Although the throughput values for all these schemes are quite close, their packet-loss rates have large differences. The best performer is *16wav_wc+16buf+def*, followed by *16wav_wc+16buf*, *16wav_wc+def*, and *16wav_wc+16buf+def*. The benefit of wavelength conversion and buffering appears to be dominant. The upper-bound transmitter load for a packet-loss rate less than 0.01 is as follows: 0.2 for *16wav_16buf+def*, 0.45 for *16wav_wc+def*, 0.61 for *16wav_wc+16buf*, and 0.65 for *16wav_wc+16buf+def*.

Figure 5.6c and d show the average end-to-end delay and average hop distance of these schemes. The end-to-end delay presents a general trend of decrease for all four schemes, due to the dropping of packets with faraway destinations with increasing traffic load. *16wav_16buf+def* has a more prominent trend of decrease, indicating that buffering and deflection alone can make the network prefer packets with closer destinations. *16wav_16buf+def* also introduces more extra hops, due to the high utilization of optical delay lines.

5.2.5.6   Limited Wavelength Conversion

Both wavelength conversion and optical buffering can effectively resolve contention. However, they require extra hardware, which may increase system cost. Full-range wavelength conversion requires a fast tunable laser as the pump laser for every wavelength converter, and optical buffering requires extra ports on the switch fabric. Here we consider the case of limited wavelength conversion, which can potentially save the amount of required hardware.

Figure 5.7 shows the throughput and packet-loss-rate comparison of the three limited wavelength conversion cases: *16wav_wclim1*, *16wav_wclim2*, and *16wav_wclim4*. For reference, it also shows *4wav_wc*. The performance of these schemes ranks in the following order (from highest to lowest): *16wav_wclim4*, *4wav_wc*, *16wav_wclim2*, *16wav_wclim1*; their respective upper-bound transmitter

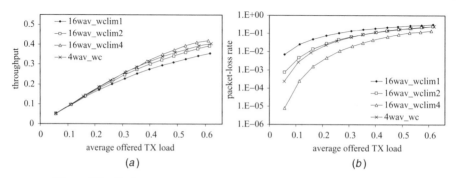

Figure 5.7   Performance comparison of limited wavelength conversion.

load with packet-loss rate of 0.01 are 0.31, 0.17, 0.13, and 0.07. *16wav_wclim4* can accommodate nearly twice the load of *4wav_wc*, indicating that limited wavelength conversion in a system with more wavelengths is better than full-range wavelength conversion in a system with fewer wavelengths.

5.3 PRIORITY-BASED ROUTING

A network built on top of an optical packet-switched network will need not only to provide IP-like connection-less services, but also higher-quality connection-oriented services. This will require the network to have differentiated service qualities on the packet level.

It seems early to delve into the quality of service (QoS) studies of such networks. Since most optical packet switches are still lab prototypes, but it is worthwhile to study network behavior with different packet priorities, which will lead us one step closer to the implementation of class of service (CoS).

5.3.1 Network Architecture and Routing Policies

Figure 5.8 shows the network topology under study. There are six nodes connected by 20-km-long bidirectional fiber links. Every fiber can accommodate four wavelengths, each of which is operating at 2.5 Gbps. The node architectures are shown in Fig. 5.9. Each switch has four input/output ports used for local add–drop, since there are four available wavelengths. Between each source–destination pair, an equal amount of traffic is generated (i.e., uniformly distributed traffic). A packet in contention could experience wavelength conversion, optical buffering, or deflection routing, depending on packet priority and specific routing policy.

The wavelength converters in the study are based on degree 2 wavelength conversion. The four wavelengths are labeled λ_1, λ_2, λ_3, and λ_4, with the wavelength conversion pattern $\lambda_1 \leftrightarrow \lambda_2$, $\lambda_3 \leftrightarrow \lambda_4$. Besides limited wavelength conversion, both optical buffering and deflection are used. Each node is equipped with one multiwavelength delay line. Because of the specific topology of this network, when only shortest-path routing is concerned, it can easily be observed that most packet forwarding happens at nodes 2, 3, 5, and 6; nodes 1 and 4 only generate and receive packets. Therefore, for nodes 1 and 4, the contention takes place primarily at the local drop ports. For this reason, to more effectively resolve contentions, we place wavelength converters only at the output ports of nodes 2, 3, 5,

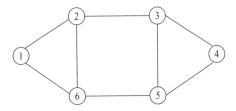

Figure 5.8   Network topology under study.

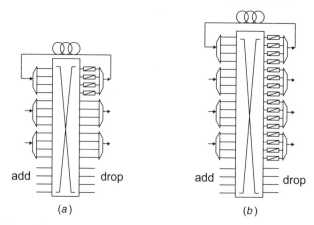

Figure 5.9   (*a*) Architecture for nodes 1 and 4; (*b*) architecture for nodes 2, 3, 5, and 6.

and 6. The characteristics of each contention-resolution scheme are described in much more detail in [24].

There are three packet priority classes: class 3, 2, and 1, class 3 being the highest. In case of contention, a packet will first seek an alternative vacant wavelength on the desired output port; if none is available, it will seek a vacant wavelength in the buffer. If there are no suitable wavelengths in the buffer, it will compare its priority class with the packets currently occupying the port preferred. If the new packet has a higher priority, it can preempt a lower-priority packet and be transmitted successfully (either on its original wavelength or on a converted wavelength). If the two packets in contention are of the same priority, the one arriving later is switched according to the deflection next hop (from the secondary entry in the routing table) and the process above will be repeated. If the secondary routing entry fails as well, the packet will be dropped.

5.3.2   Illustrative Results

The packet class distribution used in the first set of simulation is 10% class 3, 30% class 2, and 60% class 1. Figure 5.10*a* shows the packet-loss-rate comparison. Not surprisingly, class 1 packets have the highest packet-loss rate, which also increases rapidly with increasing load. Class 2 packets have a reasonably good packet-loss rate, which is maintained below 0.01 until the transmitter load reaches 0.2. The increased packet-loss rate for class 3 packets is barely noticeable, and for the entire simulated range (transmitter load < 0.4) the packet-loss rate stays well below 0.01. The simulation results showed that for loads below 0.2, both class 3 and class 2 packets can maintain a packet-loss rate lower than 0.01. The low packet-loss rate achieved by higher-priority packets is at the cost of dropping more lower-priority packets.

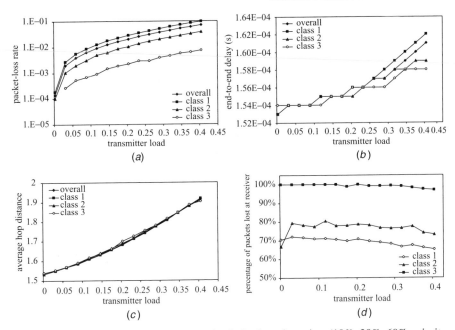

Figure 5.10   Performance comparison of priority-based routing (10%–30%–60% priority distribution).

Figure 5.10*b* shows an end-to-end delay comparison of the three classes. Because the propagation delay between two nodes is much larger than the delay introduced in the fiber delay line, a prominent increase in end-to-end delay indicates that deflection routing is being applied to more packets. Class 1 packets experience the most increase in end-to-end delay as the load increases. This is due to the fact that they are not able to preempt any other class of packets and need to resort to deflection more often. Class 2 and class 3 packets have similar end-to-end delays, which is because both these classes can preempt class 1 packets and can maintain a lower probability of using deflection routing to resolve contentions.

Figure 5.10*c* compares the average hop distance. From the figure it can be observed that all three classes have nearly the same average hop distances, which increase steadily with load. This is due to the effect of optical buffering. During contentions, optical buffering is attempted before preemption or deflection takes place. All three classes are utilizing buffering with similar probability.

The simulation results above indicate that it is possible to achieve desirable network performance in optical packet-switched networks. However, since there is no sophisticated queuing systems as in the electrical domain, the efficiency of bandwidth utilization of such networks will be suboptimal.

To better understand the source of contentions, we also measured the fraction of packets dropped due to contention at the receiver, as shown in Fig. 5.10*d*. All the lost class 3 packets are due to receiver contentions, implying that they can well

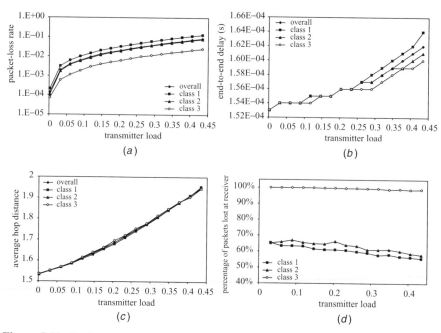

Figure 5.11   Performance comparison of priority-based routing (30%–30%–40% priority distribution).

survive the middle hops but eventually need to compete among themselves for the receivers. This is especially true under the case of a uniform traffic matrix, where each node is receiving packets from the remaining $N - 1$ nodes. For both class 2 and class 1 packets, the fraction of packets lost at the receivers decrease gradually with the load, implying that in-transit contentions intensify faster than the contentions at the receivers.

To further examine the effect of priority distribution on network performance, we have simulated a different priority distribution: 30% class 3, 30% class 2, and 40% class 1. In the packet-loss-rate comparison shown in Fig. 5.11a, class 2 packets experience two times more packet loss than in the previous simulation. This is because there are twice as many class 3 packets, which will preempt both class 2 and class 1 packets, and there are fewer class 1 packets for class 2 packets to preempt.

The end-to-end delay comparison in Fig 5.11b and average hop distance comparison in Fig. 5.11c look very similar to Fig. 5.10b and c, indicating that although packet priority distribution has changed, the probability of packets being buffered or deflected did not change much.

Figure 5.11d shows the fraction of lost packets at the receivers of all the lost packets. Most of the dropped packets in class 3 are still at the receiver side, implying that class 3 packets compete only among themselves and the contentions happen mostly at the receiver. For class 2 and class 1, it appears that a greater fraction of the lost packets are due to contention at the transit switches because of the increased number of class 3 packets. For the class 3 packet-loss rate to be maintained below 0.01, the transmitter load must be less than 0.25; and for class 2 and class 1, the transmitter load thresholds for less than a 0.01 packet-loss rate are 0.15 and 0.1, respectively.

For the three classes of priorities simulated, class 3 has the best performance in terms of latency, signal quality, and packet-loss rate. It will be suitable to carry various mission-critical, real-time traffic, and could be explored further to establish virtual circuitlike connections to accommodate stringent user demands. Class 2, which has higher delay and packet-loss rate, appears to be a good candidate to carry medium-quality connection-oriented traffic (such as TCP or voice) or connection-less data traffic (such as UDP). Class 1 has the highest latency and loss rate. It can be used for applications and data services that do not require real-time connection and can recover from frequent packet losses. It is important to note that the performance of each class is not only related to the specific routing policies and the network topology, but also how the bandwidth is divided between the classes. The priority distribution is crucial to achieving the desired quality of packet delivery for each class. In general, the high quality of the higher-priority service is obtained at the cost of lower-priority services; therefore, the portion of bandwidth assigned to high-priority classes needs to be kept small. In other words, in such optical packet-switched networks, one needs to trade off bandwidth for better flexibility and service quality.

5.4   SLOTTED VERSUS UNSLOTTED NETWORKS

5.4.1   Network Architecture and Routing Policies

We use the same network in Fig. 5.2 for this example. In the slotted network, packets are fixed size (12,000 bits) and placed in a 5-ms time slot. Before the packets enter each switching node, they must be aligned; therefore, switching happens only at the slot boundaries. Synchronization stages are required for the alignment of packets. In the unslotted network, packets are of variable size. The packet-size distribution in this study is set to a negative exponential distribution, with an average of 12,000 bits. No synchronization stages are needed and packets are switched on the fly. The packet arrival in both cases is assumed to be a Poisson process.

The node architecture is a combination of degree 2–limited wavelength conversion, optical buffering, and deflection. The fiber delay line is of one average packet size. In this study the slotted network node is equipped with one fiber delay line (noted as *slotted_1buf*), while the unslotted network node is equipped with 1, 2, 4, 8, or 16 fiber delay lines (noted as *unslotted_1, 2, 4, 8, 16buf*). The traffic generated between any source–destination pair is assumed to be equal. The traffic distribution among these classes is 10% class 3, 30% class 2, and 60% class 1. The priority-based routing policies are as described in Section 5.3.

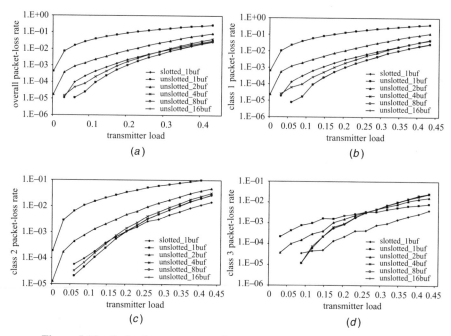

Figure 5.12   Packet-loss-rate comparison of slotted and unslotted networks.

5.4.2   Illustrative Results and Discussion

Figure 5.12*a* shows the overall (including all three classes) packet-loss-rate comparison. There is a large improvement between *unslotted_1buf* and *unslotted_2buf*. The margin of improvement decreases quickly with an increasing number of buffers. There is no significant improvement between *unslotted_8buf* and *unslotted_16buf*. The packet-loss rate of *slotted_1buf* is between *unslotted_4buf* and *unslotted_8buf* and stays well below 0.01 for a transmitter load under 0.3. Although by nature the slotted network performs much better than an unslotted network, it involves complicated synchronization stages at the core switches.

Figure 5.12*b* and *c* show the packet-loss-rate comparison for class 1 and class 2. These figures have trends similar to those in Fig. 5.12*a*. For class 1 packets, *unslotted_4buf* can match the packet-loss rate of *slotted_1buf*, but for class 2 packets, *slotted_1buf* outperforms all the unslotted configurations, regardless of the number of buffers. For class 1 packets to have a packet-loss rate less than 0.01 with a transmitter load below 0.3, there need to be four fiber delay lines, while for class 2, with four fiber delay lines, they can achieve a packet-loss rate less than 0.01 under a transmitter load below 0.35l.

Figure 5.12*d* shows a packet-loss-rate comparison for class 3 packets. Interestingly, the packet-loss rate in the unslotted network increases with the number of buffers. This is against the intuition that more buffering will bring down the packet-loss rate. The reason is that since there is a maximum hop count mechanism and in the routing policies preemption is considered after buffering, when there is sufficient buffer in the switch, a class 3 packet will in most cases be buffered instead of preempting lower-priority packets. Therefore, they are dropped not only because of contention but also because of reaching the maximum hop count. For all the unslotted network configurations, the class 3 packet-loss rate stays well below 0.001, with a transmitter load of less than 0.2, and below 0.01 with a transmitter load of less than 0.35.

5.5 HYBRID CONTENTION RESOLUTION FOR OPTICAL PACKET SWITCHING

Most existing studies on contention resolution have focused on the optical domain of one optical packet switch, or a network interconnected by such switches. An optical packet-switched network should perform packet-based switching optically. However, it still needs to interface with other types of networks to form an end-to-end connectivity. The other networks, especially the client networks, are often electrical. This means that there needs to be an interface at the edge of the optical packet-switched network. In this section we present an architecture that takes advantage of the availability of electrical buffers at the edge, to resolve contentions and lower network cost.

5.5.1 Node Architecture

When packets arrive from client networks, which are mostly electrical, they need to be converted into optical format before being sent to the optical packet-switched network. This conversion is performed at the client interface of the network (Fig. 5.13). The optical packet switch performs two types of packet forwarding: the forwarding of transit packets from other optical packet switches, and the forwarding of local packets received from the client interface. A transit packet will experience possible contention with the local packets as well as other transit

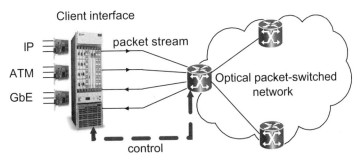

Figure 5.13   Client interface of optical packet-switched networks.

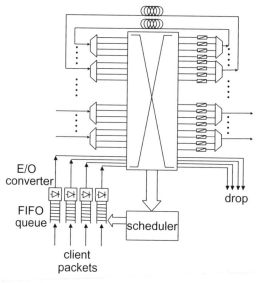

Figure 5.14   Architecture of an optical packet switch incorporating hybrid contention resolution.

packets. In most proposed architectures, contention resolution usually requires a large amount of optical resources, such as wavelength converters and delay lines. In this architecture, the local packets are first queued in the electrical buffers, which can easily be implemented in the electrical part of the client interface. These packets enter the optical switch only when there is no transit packet occupying the preferred wavelength/output port. This buffering mechanism ensures that all the wavelength converters and delay lines are used only for transit packets. Since switching is still carried out by the optical components and there is no O/E/O conversion in the network, the use of electrical buffers at ingress nodes for the client packets does not compromise the all-optical nature of the core network.

Figure 5.14 shows the node architecture that implements the hybrid contention resolution. In the optical portion of the switch, both optical delay lines and wavelength converters are used to resolve contention. In the electrical–optical interface portion of the switch, first in/first out (FIFO) queues are included. All the packets from client networks are queued first. A scheduler observes the state of every wavelength at the output ports of the switch. When the packet's output port clears, the transmitter will convert the packet to optical format and send it to the optical switch fabric.

5.5.2 Simulation Configuration

The network topology is the same as in Fig. 5.2. Every node is equipped with W transmitters (where W is the number of wavelengths), with corresponding FIFO queues. A FIFO queue is fed with a self-similar traffic generator consisting of 12 on–off sources generating IP packets. The traffic is distributed uniformly. Whenever the FIFO queue is not empty and there is a vacant wavelength on the preferred output port, the scheduler will retrieve the packet at the head of the queue and send it to the optical switch fabric. The switch includes a number of optical delay lines, which are of length 12,500 bits.

As shown in Fig. 5.14, there are wavelength converters at each output port of the switch fabric. In the simulation, the number of wavelengths, W, is set to 4, 8, 16, and 32. The number of optical delay lines at each node varies with the nodal degree. We set the number of delay lines equal to (*nodal degree* − 1). Deflection is the third contention resolution.

5.5.3   Illustrative Results

Figure 5.15*a* shows the packet-loss rates plotted against the average offered transmitter load. For the four-wavelength scenario, the packet-loss rate is kept below 0.01 when the transmitter load offered is less than 0.5. For the 32-wavelength case, the acceptable transmitter load is close to 0.65. The light-load portion of the plot is not shown on the figure, because the simulation did not encounter any dropped packets during the simulated time.

Figure 5.15*b* shows the average end-to-end delay. With the average transmitter load lower than 0.6, the end-to-end delay is dominated by propagation delay. The electrical queuing delay in the access FIFO queues are negligibly small. This is because the queued packet does not need to wait for more than a few packets' transmission delay to find a vacant wavelength. With a 2.5-Gbps line rate and native IP packet sizes, the transmission delays of packets are on the order of microseconds. Since the network uses static routing, there will be a most congested link in the network. In this topology it is the link between nodes 1 and 11. There is traffic from a total of 21 source–destination pairs passing this link. It is congested when the average transmitter load reaches 0.66. When the link is approaching congestion, electrical queuing delay and deflection delay begin to dominate. This explains the dramatic increase in the figure.

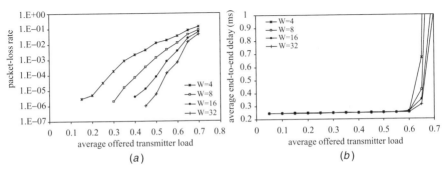

Figure 5.15   (*a*) Packet-loss rate, and (*b*) average end-to-end delay comparison of various simulation scenarios.

5.6   TCP PERFORMANCE WITH OPTICAL PACKET SWITCHING

In Section 5.5 it was proposed that inexpensive electrical buffers at the ingress interfaces be used to reduce packet loss. Since our goal is not only to lower packet-loss rate but also to provide better transport for TCP/IP traffic, in this section we extend the investigation to the TCP performance of optical packet-switched networks. To improve the TCP performance, the work in [25] proposed *optical flow routing*, a type of aggregation that is supposed to reduce packet reordering. In this section we present a packet-aggregation mechanism that allows many packets to be grouped together into a larger entity that can be transported more efficiently through the network. Such a mechanism can help reduce the traffic burstiness and improve system performance.

5.6.1   Node Architecture

Figure 5.16 shows the node architecture. The packet aggregator assembles client packets into larger entities (referred to as *aggregation packets*) in a FIFO manner. It directly interfaces with the client network elements (typically, IP routers), and it consists of a number of FIFO subqueues. Each subqueue buffers packets going to the same destination. A subqueue transmits all the buffered packets in an aggregation packet after a certain period of time, t_a. To avoid unnecessary delay, we set a packet count threshold C such that when the number of buffered packets reaches C, the subqueue will transmit the aggregation packet even if the time from last transmission is less than t_a. This aggregation mechanism can be compared to a bus system (as in public transportation): At any time, there is one bus with one or more

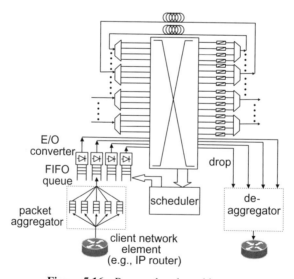

Figure 5.16   Proposed node architecture.

empty seats waiting for passengers for each destination. A bus has a maximum capacity of C passengers and it leaves every t_a seconds. If the bus is full before its scheduled departure time, it will leave early and the next empty bus will pull into the station. The aggregator not only preserves the order of packets but shapes the traffic by injecting more evenly sized aggregation packets at more regular time intervals.

The ingress buffer controls when an aggregation packet can be injected into the optical switch to avoid contention. By using an ingress buffer, we can reduce its contention with transit packets by allowing local aggregation packets to be injected only when there is no transit packet occupying the preferred output port. The local aggregation packets are first stored in the ingress buffer electrically, and they are converted to optical format and then injected into the optical switch. A scheduler is used for constant monitoring of the state of the switch fabric and to control transmission from the ingress buffer.

5.6.2   Simulation Configuration and Numerical Results

Figure 5.17 shows the network topology for the simulation. The simulation experiment uses a file transfer protocol (FTP) session to measure TCP performance. The main performance metric is the transfer time of a large file (assumed to be 1.6 Mbytes in this example). It is reasonable to assume that both hosts have Ethernet interfaces; therefore, the maximum transfer unit (MTU) is 1500 bytes. For the network scenario to be realistic, each link also carries some background IP traffic. Each node is equipped with four transmitters, fed independently by four traffic generators. The intensity of the background traffic is controlled by the average transmitter load offered.

One of the main factors that affect TCP performance is the receiver window size, whose typical values are 8, 32, or 64 kilobytes (kB). The aggregation threshold C can also affect TCP performance. With different C values, the aggregation timer value t_a and the delay-line size should be adjusted accordingly. In the experiments, both t_a and the delay-line size are set equal to the transmission delay of C packets with maximum length (1500 bytes each). The running time for each data point varies between 4 and 75 hours on a 500-MHz Pentium III machine, depending on the TX load. The maximum TX load was 0.5, because larger values made the simulation time prohibitively long.

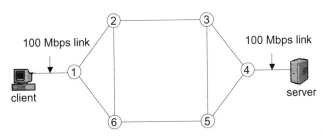

Figure 5.17   Network topology for TCP experiment.

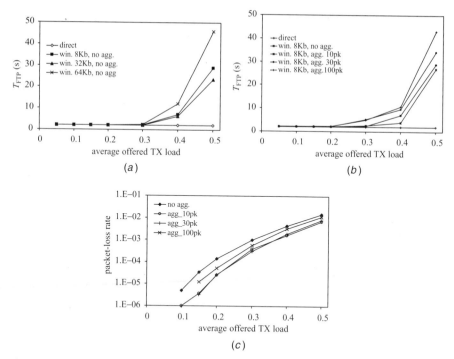

Figure 5.18   Comparison of T_{FTP} for (a) various TCP window sizes, and (b) different aggregation schemes. (c) Comparison of packet-loss rate with various aggregation thresholds.

Figure 5.18a compares the file transfer time T_{FTP} for various TCP window sizes and various values of C. For reference, it also shows the T_{FTP} without background traffic for a client–server pair connected directly through a 100-Mbps link with the same propagation delay (i.e., a link length of 60 km). Without aggregation, a window size of 32 kB provides the best result because the measured TCP round-trip time (RTT) is approximately 3 ms, and the TCP connection's data rate is 100 Mbps. (Note that the optimal window size should be the product of RTT and the data rate.)

Next, in Fig. 5.18b, one can see the effect of the aggregation threshold with different C values of 10, 30, and 100 packets, while window size is equal to 8 kB. For a TX load below 0.2, the aggregation threshold does not have any effect on system performance. As the TX load increases, the 10-packet aggregation scheme has the lowest T_{FTP} value, followed by the 30- and 100-packet schemes. The 10-packet scheme also performs better than the one without aggregation, indicating that aggregation improves TCP performance. However, with more packets aggregated, the performance deteriorates because more queuing delay is introduced in the packet aggregator and ingress buffer. Intuitively, one would think the ideal aggregation packet should contain all the TCP segments sent within one window size. Unfortunately, the aggregator has to hold the first segment for at least the entire transmission delay of all the segments in that window. This defeats the purpose of pipelining in the TCP sliding-window mechanism. Therefore, aggregation does not appear to improve TCP performance directly. However, aggregation reduces the traffic burstiness and lowers the packet-loss rate.

Figure 5.18c shows the packet-loss rates of aggregation schemes with various values of C. The 10- and 30-packet schemes offer the lowest packet-loss rate, followed by the 100-packet and no-aggregation schemes. Large aggregation appears to be less effective because the aggregation threshold C is based on the number of packets instead of bits. Since a large portion of IP traffic consists of very small packets, the size of the aggregation packet can vary dramatically when C is large, and the interdeparture time of the aggregation packets can still be quite bursty. Hence using bit count instead of packet count might be a better approach to aggregation.

5.7 SUMMARY

In this chapter we presented a unified study of the contention-resolution schemes for optical packet switching. Effective contention resolution can be obtained with a combination of wavelength conversion, optical buffering, and space deflection. Priority-based routing with preemption is one possible solution to implementation of class of service. Moreover, it is shown that with the use of electrical buffer at the ingress nodes, an optical–electrical hybrid contention resolution may improve the network performance. The network performance can be further improved by using traffic aggregation at the ingress nodes.

REFERENCES

1. S. L. Danielsen, P. B. Hansen, and K. E. Stubkjear, Wavelength conversion in optical packet switching, *IEEE/OSA Journal of Lightwave Technology*, 16(12):2095–2108, Dec. 1998.

2. V. Eramo and M. Listanti, Packet loss in a bufferless optical WDM switch employing shared tunable wavelegth converters, *IEEE/OSA Journal of Lightwave Technology*, 18(12):1818–1833, Dec. 2000.

3. D. K. Hunter, M. H. M Nizam, M. C. Cia, I. Andonovic, K. M Guild, A. Tzanakaki, M. J. O'Mahony, L. D. Bainbridge, M. F. C Stephens, R. V. Penty, and I. H. White, WASPNET: A wavelength switched packet network, *IEEE Communications*, 37(3), Mar. 99.

4. P. B. Hansen, S. L. Danielsen, and K. E. Studkjaer, Optical packet switching without packet alignment, *Proc. ECOC'98*, p. WdD13, 1998.

5. G. Castanon, L. Tancevski, S. Yagnanarayanan, and L. Tamil, Asymmetric WDM all-optical packet switched routers, *Proc. OFC '00*, Vol. 2, pp. 53–55, 2000.

6. D. K. Hunter, M. C. Chia, and I. Andonovic, Buffering in optical packet switches, *IEEE/OSA Journal of Lightwave Technology*, 16(12):2081–2094, Dec. 1998.

7. L. Tancevski, A. Ge, and G. Castanon, Optical packet switch with partially shared buffers: Design principles, *Proc. OFC'01*, Vol. 2, pp. TuK3–1 to TuK3–3, 2001.

8. L. Tancevski, S. Yegnanarayanan, G. Castanon, L. Tamil, F. Masetti, and T. McDermott, Optical routing of asynchronous, variable length packets, *IEEE Journal on Selected Areas in Communications*, 18:2084–2093, Oct. 2000.

9. C. Guillemot, M. Renaud, P. Gambini, C. Janz, I. Andonovic, R. Bauknecht, B. Bostica, M. Burzio, F. Callegati, M. Casoni, D. Chiaroni, F. Clerot, S. L. Danielsen, F. Dorgeuille, A. Dupas, A. Franzen, P. B. Hansen, D. K. Hunter, and A. K. Kloch, Transparent optical packet switching: The European ACTS KEOPS project approach, *IEEE/OSA Journal of Lightwave Technology*, 16:2117–2133, Dec. 1998.

10. Z. Haas, The "staggering switch": An electronically controlled optical packet switch, *IEEE/OSA Journal of Lightwave Technology*, 11(5/6):925–936, May/June 1993.

11. D. K. Hunter, W. D. Cornwell, T. H. Gilfedder, A. Franzen, and I. Andonovic, SLOB: A switch with large optical buffers for packet switching, *IEEE/OSA Journal of Lightwave Technology*, 16(10):1725–1736, Oct. 1998.

12. I. Chlamtac, A. Fumagalli, L. G. Kazovsky, P. Melman, W. H. Nelson, P. Poggiolini, M. Cerisola, A. N. M. M. Choudhury, T. K. Fong, R. T. Hofmeister, C. Lu, A. Mekkittikul, D. J. M. Sabido IX, C. Suh, and E. W. M. Wong, CORD: Contention resolution by delay lines, *IEEE Journal on Selected Areas in Communications*, 14(5):1014–1029, June 1996.

13. F. Callegati and W. Cerroni, Wavelength allocation algorithms in optical buffers, *Proc. IEEE International Conference on Communications (ICC'01)*, Vol. 2, pp. 499–503, 2001.

14. F. Forghierri, A. Bononi, and P. R. Prucnal, Analysis and comparison of hot-potato and single-buffer deflection routing in very high bit rate optical mesh networks, *IEEE Transactions on Communications*, 43(1):88–98, Jan. 1995.

15. G. Castanon, L. Tancevski, and L. Tamil, Routing in all-optical packet switched irregular mesh networks, *Proc. IEEE Globecom'99*, Vol. b, pp. 1017–1022, Dec. 1999.

16. F. Brogonovo, L. Fratta, and J. Bannister, Unslotted deflection routing in all-optical networks, *Proc. IEEE Globecom'93*, Vol. 1.4, pp. 119–125, 1993.

17. A. S. Acampora and I. A. Shah, Multihop lightwave networks: A comparison of store-and-forward and hot-potato routing, *IEEE Transactions on Communications*, 40:1082–1090, June 1992.

18. J. Fehrer, J. Sauer, and L. Ramfelt, Design and implementation of a prototype optical deflection network, *ACM SIGCOMM*, 24:191–200, 1994.

19. G. Prati, Ed., *Photonic Networks*, Springer-Verlag, London, 1997.

20. A. S. Tanenbaum, *Computer Networks*, Prentice Hall, Upper Saddle River, NJ, 1996.

21. W. Willinger, M. S. Taqqu, R. Sherman, and D. V. Wilson, Self-similarity through high-variability: Statistical analysis of Ethernet LAN traffic at the source level, *IEEE/ACM Transactions on Networking*, 5(1):71–86, Feb. 1997.

22. *http://www.caida.org/outreach/papers/inet98*.

23. T. V. Lakshman and U. Madhow, The performance of TCP/IP for networks with high bandwidth-delay products and random loss, *IEEE/ACM Transactions on Networking*, 5:336–350, June 1997.

24. S. Yao, B. Mukherjee, S. J. B. Yoo, and S. Dixit, All-optical packet-switched networks: a study of contention-resolution schemes in an irregular mesh network with variable-sized packets, *Proc. SPIE OPTICOMM'00*, Vol. 4233, pp. 235–246, Nov. 2000.

25. J. He and D. Simeonidou, A flow-routing approach for optical IP networks, *Proc. OFC'01*, Vol. 1, pp. MN2–1 to MN2–3, 2001.

6 Advances toward Optical Subwavelength Switching

S. YAO and B. MUKHERJEE

University of California–Davis, Davis, California

S. DIXIT

Nokia Research Center, Burlington, Massachusetts

6.1 INTRODUCTION

Over the past few years, Internet traffic has grown rapidly and the optical transport bandwidth has been increasing continuously. These changes are stimulating the evolution of data networks. In such a dynamic environment, a network architecture that accommodates multiple data formats, supports high data rates, and offers flexible bandwidth provisioning is the key feature of the next-generation Internet.

Among all the optical networking paradigms, wavelength-routed networks, in which lightpaths are set up on specific wavelengths, have been the focus of many studies [1,2]. (The reader is referred to Chapter 12 for more details.) Over a short period of a few years, these networks have evolved from textbook subjects to real-life products. Current ongoing efforts to automate and expedite wavelength and bandwidth provisioning in the optical layer indicate the inevitable trends that lead to more intelligent optical networks [3–5]. Migration of certain switching functionality from electronics to optics will remove the incumbent layers that impose unnecessary optical–electrical–optical conversions and unnecessary signal processing. The preoptical reference network layers, defined decades ago, are no longer applicable in the new networking environment (Fig. 6.1). A more flexible wavelength-division multiplexing (WDM) optical network is desired by service providers to meet their versatile traffic demands.

From the demand side, the tremendous increase in the transport networks' bandwidth is stimulating the high volume of gigabit multimedia services. A robust network supporting various kinds of traffic is the cornerstone for the next-generation Internet. However, despite the high throughput that wavelength routers deliver,

IP over WDM: Building the Next-Generation Optical Internet, Edited by Sudhir Dixit.
ISBN 0-471-21248-2 © 2003 John Wiley & Sons, Inc.

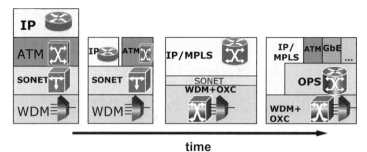

Figure 6.1 Evolution of the data network protocol stack.

they still lack the flexibility in switching granularity. An optical cross-connect can only switch whole wavelengths unless electrical grooming switch fabrics are used. On the other hand, past evolution of the Internet has proved that a future-proof network not only needs high-capacity circuit switching but also needs high-performance switching with finer granularities. The key to the successful accommodation of heterogeneous traffic lies in the deployment of cost-effective switching schemes that provide easy access to the large bandwidth that WDM offers. Therefore, a number of research groups have proposed various optical subwavelength switching techniques, such as optical packet switching, photonic slot routing, and optical burst switching. In the rest of this chapter we introduce these optical subwavelength switching techniques.

6.2 OPTICAL PACKET SWITCHING

Optical packet switching is optical switching with the finest granularity. Incoming packets are switched all-optically without being converted to electrical signal. It is the most flexible and also the most demanding switching scheme.

6.2.1 Slotted Networks

In general, there are two categories of optical packet-switched networks: slotted (synchronous) and unslotted (asynchronous) networks. When individual photonic switches form a network, at the input ports of each node, packets can arrive at different times. Since the switch fabric can change its state incrementally (set up one input/output connection at an arbitrary time) or jointly (set up multiple input/output connections together at the same time), it is possible to switch multiple aligned packets together or to switch each packet individually on the fly. In both cases, bit-level synchronization and fast clock recovery are necessary for packet header recognition and packet delineation.

In a slotted network, all the packets have the same size. A fixed-size time slot contains both the payload and the header. The time slot has a longer duration than

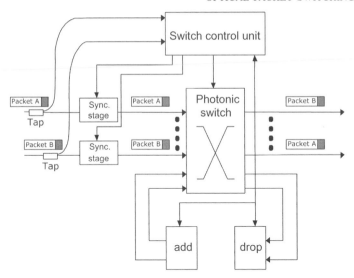

Figure 6.2   Generic node architecture of the slotted network.

the whole packet to provide guard time. All the input packets arriving at the input ports need to be aligned in phase with one another before entering the switch fabric (Fig. 6.2). To synchronize all the incoming packets successfully, it is important to study what types of delay variation a packet could experience, which we discuss below.

The time for a packet to travel through a certain distance of the fiber depends on the fiber length, chromatic dispersion, and temperature variation. When WDM is used, the effect of chromatic dispersion has to be considered. Chromatic dispersion results in different propagation speed for packets transmitted on different wavelengths; therefore, different propagation delays occur. For example, with a typical fiber dispersion of 20 ps/nm per kilometer (where ps is the time unit for delay variation, nm the unit for wavelength difference, and km the unit for propagation distance), a wavelength variation of 30 nm (consistent with the typical erbium-doped fiber amplifier's 1530- to 1560-nm window) and a propagation distance of 100 km, the propagation delay variation would be about 60 ns. If dispersion compensation fibers are used, the delay variation above can be reduced by one order of magnitude. The packet propagation speed also varies with temperature, with a typical figure of 40 ps/°C per kilometer. Under a temperature-variation range of 0 to 25°C, 100 km of fiber introduces 100 ns of delay variation.

The delay variations mentioned above are relatively slow with respect to time; they can be compensated statically instead of dynamically (on a packet-by-packet basis). How much delay each packet experiences inside a node depends on the switch fabric and contention resolution scheme. Depending on the implementation of the switch fabric, a packet can take different paths with unequal lengths within the switch fabric. All the considerations given to delay variations in the internode

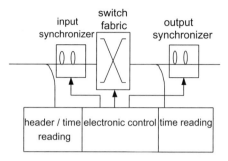

Figure 6.3 Function block diagram of synchronization of the packets.

links apply here. It is worth noting that the fast time jitter (compared with the slow delay variation) induced by dispersion between different wavelengths and unequal optical paths varies from packet to packet at the output of the switch; therefore, a fast output synchronization interface might be required. Thermal effects within a node are smaller here because it varies more slowly and can be controlled easily.

Figure 6.3 is a function block diagram showing the node architecture for slotted networks. A tap splits a small amount of power from the incoming packets for the header reading. The header processing circuits recognize a preamble at the beginning of the packet and then read the header information. It also passes the timing information of the incoming packet to the control unit to configure the synchronization stages and the switch fabric. The input synchronization stage aligns packets before they enter the switch fabric. The output synchronization stage, which is not shown in Fig. 6.2, is to further compensate the fast time jitter that occurs inside the node. It may or may not be necessary, depending on the actual packet format and node architecture.

Packet delineation is essential for both header reading and switch configuration. During packet delineation the incoming bits are locked in phase with the clock in order for the node to read the header information. The traditional phase-locked-loop approach is not applicable because it requires too many bits. Burst-mode receivers have been shown to achieve bit-level synchronization within the nanosecond range. References [6] to [8] describe another burst-mode receiver setup in which the transmitter frequency multiplexes its bit clock with the baseband data and modulates the optical carrier with the composite signal. The data and clock travel along the fiber with negligible dispersion. At the receiver end, a photodiode detects the optical signal. Its radio-frequency output is amplified and split. The data and clock are first separated by a low-pass filter that cuts off the baseband and then by a narrow bandpass filter centered at the clock frequency. The retrieved clock is fed into the analog receiver. If the delay from the output of the photodiode to the input of the receiver is matched, the clock and data will be in bit synchronization for all incoming packets.

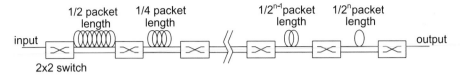

Figure 6.4   Scheme for input synchronization stage in a node. (From [9].)

Since packets are entering a node from different links, for all the previously stated reasons they could arrive totally out of phase with one another. Figure 6.4 shows a typical synchronization stage consisting of a series of switches and delay lines, as they appear in the input synchronization stage of a node.

Once the header processor recognizes the bit pattern and performs packet delineation, it identifies the packet start time, and the control unit will calculate the necessary delay and configure the correct path through these switched delay lines. The length of the delay lines are in an exponential sequence between the 2×2 switches (i.e., the first delay line is equal to $\frac{1}{2}$ time slot duration, the second delay is equal to $\frac{1}{4}$ time slot duration, etc.). The resolution of this scheme is $1/2^n$ of the time slot duration, where n is the number of delay lines. This type of synchronization scheme is suitable for both static (slow) and dynamic (fast) synchronization. At the system initialization the synchronization is set up to compensate for delay variations between different inputs and to keep this configuration throughout the system operation time (static). For packet-based (dynamic) synchronization, much faster switches are necessary to operate during guard time.

From a physical point of view, such a packet-synchronization scheme introduces insertion loss and crosstalk due to the switches used. Cascading the switches will inevitably require optical amplification, which will result in degraded signal-to-noise ratio. Meanwhile, the crosstalk accumulated through the switches will also increase the bit-error rate. In a multinode network the power penalty brought by all the synchronization stages may significantly impair system performance.

6.2.2 Unslotted Networks

In an unslotted network, packets may or may not have the same size. Packets arrive and enter the switch without being aligned. Therefore, the packet-by-packet switch action could take place at any point in time. Obviously, in unslotted networks, the chance for contention is larger because the behavior of the packets is more unpredictable (similar to the contention in the slotted and unslotted ALOHA networks [10]). On the other hand, unslotted networks are more flexible than slotted networks, since they are better at accommodating packets of variable sizes.

Figure 6.5 shows the general node architecture and packet behavior for unslotted networks. (Note the absence of synchronization stages and packet alignment.) The fixed-length fiber delay lines hold the packet when header processing and switch reconfiguration are taking place. There is no packet alignment stage, and all the

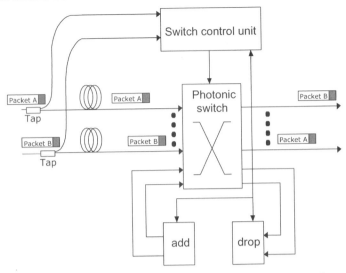

Figure 6.5   Generic node architecture of the unslotted network.

packets experience the same amount of delay with the same relative position in which they arrived, provided that there is no contention. The unslotted networks circumvent the requirement of synchronization stages. However, given the same traffic load, the network throughput is lower than that of the slotted networks because contention is more likely to occur.

6.2.3   Contention Resolution

In a packet-switched network each packet has to go through a number of switches to reach its destination. When the packets are being switched, contention occurs whenever two or more packets are trying to leave the switch from the same output port, on the same wavelength, at the same time. How the contention is resolved has a great effect on network performance. Here we look at three types of contention resolutions: optical buffering, deflection routing, and wavelength conversion. For a more detailed, quantitative discussion of contention resolution, the reader is referred to Chapter 5.

6.2.3.1   Optical Buffering   In electronic switches, contention is usually resolved by a store-and-forward technique, which means that the switch will first store packets in contention in a queue and send them out later when the output port becomes available. This is possible because of the available random-access memory (RAM). In an optical switch there is no ready-to-use random-access optical memory. An optical buffer can only consist of optical delay lines. The main difference between electronic RAM and an optical buffer is that optical buffers are fixed-length fibers. Once a packet has entered the fiber, it must emerge from the other end after a fixed amount of time; there is no way to retrieve the packet any time earlier (except for recirculation fiber loops, which are discussed later.)

There are various designs of node architecture applying optical buffer and there are various ways to categorize them. One way is to compare them to the buffering in electronic switches (input, output, shared, and recirculating buffering). There is also a simpler, more direct way: single-stage buffering or multistage buffering, forward buffering, or feedback buffering. (A stage is a single continuous piece of delay line.)

The first example, proposed in the European ACTS KEOPS (keys to optical packet switching), is a broadcast-and-select space switch using a single-stage forward buffering scheme as contention resolution (Fig. 6.6). The wavelength converters encode the packet streams entering each input; therefore, the packets on each input are distinguished by a separate wavelength. The streams are then combined by a multiplexer and distributed to k groups of delay lines of different lengths, which give the packets necessary delays to resolve contention. By means of semiconductor optical amplifier (SOA) gates and passive couplers, each output port is able to select the packets with proper delays. At the final stage, the demultiplexer, SOA gates, and multiplexer can select one packet from a specific input port. In this architecture there is only one stage of buffer, and each delay line feeds forward to the next part of the switch. Since each packet is broadcast to all delay lines and every output port, it is possible both to offer multicast operation and implement packet priorities. The drawback is the use of a great number of components and controls, which increase the cost considerably. For example, it needs n wavelength converters, $n \cdot k + n^2$ SOA gates, and $2n + 1$ multiplexers/demultiplexers.

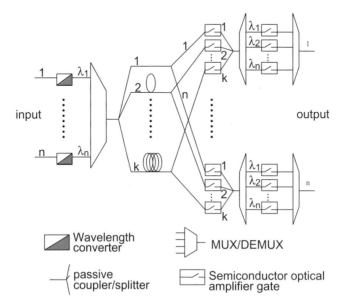

Figure 6.6   Broadcast and select switch proposed in the KEOPS project.

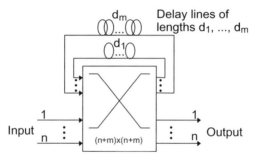

Figure 6.7   Shared-memory optical packet switch.

The second example is the shared-memory optical packet (SMOP) switch [11], which belongs in the single-stage feedback category. Figure 6.7 shows its principle of operation. The lengths of the delay lines could be $1, 2, 3, \ldots, m$ packet duration. The $(n + m) \times (n + m)$ space switch can switch a packet either directly to an output port or to one of the delay lines, according to how much delay the packet needs. Delay lines of length greater than one packet duration greatly reduce the number of recirculation loops needed, resulting in a reduced need for amplifiers and less noise. This scheme also allows prioritized packet switching since a lower-priority packet may be preempted by being sent to another circulation. Since the number of recirculations a packet is to take is unpredictable, some packets could suffer more power loss than others, making optical amplification necessary. This will inevitably introduce additional signal-to-noise ratio degradation to the recirculating packets.

Another case of single-stage feedback optical buffering is the fiber loop memory switch concept introduced in research and development for the advanced communications in Europe (RACE) asynchronous transfer mode optical switching (ATMOS) project, shown in Fig. 6.8. The buffer is based on a fiber loop delay line containing multiple wavelength channels. When contention occurs, the input packet is converted to one of the available wavelengths in the loop and kept circulating by activating the corresponding passive fixed filter (i.e., by turning on the related SOA

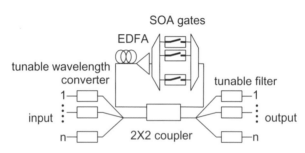

Figure 6.8   Fiber loop memory switch from ATMOS.

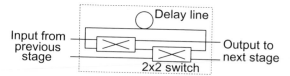

Figure 6.9 A 2 × 2 switching element containing optical buffer.

gate). At the input of the loop, half of the power enters the loop and the other half goes toward the outputs through the passive coupler. When the contention is resolved, the packet is switched to the destination link by proper tuning of the corresponding output tunable filter. At the same time, the passive filter in the loop is turned off to erase the packet in the buffer. It is possible for incoming packets to preempt those that are already waiting; hence this type of switch can also implement packet priorities.

For multistage feedforward buffering examples, several node architectures applying cascaded 2 × 2 switching elements containing optical buffer [12] have been proposed (Fig. 6.9). Each of these switching elements provides buffering of one or more packet duration delays in case of contention. A larger switch fabric can be constructed by cascading a number of these 2 × 2 elements.

There are various designs of optical buffering, such as the staggering switch [13] and the switch with large optical buffers (SLOBs) [14]. Packet-loss rate, network latency, hardware cost, control circuit complexity, and packet reordering are among the many important issues to be considered in the design, which depends on the network specification (i.e., network dimension, topology, traffic load and pattern, etc.).

6.2.3.2 *Deflection Routing*   Optical buffering was to a great extent inspired by their conventional electronic network counterparts. In the electronic networks, the link bandwidth is much less than today's optical fiber's capacity, and many efforts were put into increasing link utilization. In a network deploying optical buffers, each packet is guaranteed to arrive at its destination along the shortest possible path, and for a given connectivity the expected number of hops is minimized. Implementing optical buffers involves a great amount of hardware and complex electronic controls. Another issue that arises with optical buffer is that the optical signal suffers from power loss in the delay lines, and optical amplifiers are often necessary. The accumulated noise from the cascaded amplifiers can severely limit the network size at very high bit rates unless signal regeneration is applied. In deflection routing, as the name implies, contention is resolved as follows: If two or more packets need to use the same output link to achieve minimum distance routing, only one will be routed along the desired link while others are forwarded on paths that may lead to a non-shortest-path route. Hence, for each source–destination pair, the number of hops taken by a packet is no longer fixed. Deflection routing does not necessarily exclude the use of optical buffers. The simplest form of

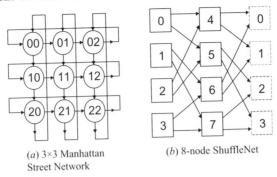

(a) 3×3 Manhattan
Street Network

(b) 8-node ShuffleNet

Figure 6.10 Manhattan Street Network and ShuffleNet.

deflection routing is hot-potato routing [15], which is a special case where buffers are not provided at all.

Most studies on deflection have focused on regular network topologies with a uniform traffic load. These logical topologies can be built on different physical topologies, such as ring, star, or mesh. Figure 6.10 shows the two most typical logical topologies used for network performance simulation: the Manhattan Street Network (MSN) and ShuffleNet. Each node in these two topologies has two input ports and two output ports. Figure 6.11 shows example node architectures for MSN or ShuffleNet.

Studies have been conducted to determine the impact of different routing strategies on network performance, such as delays, average number of hops (i.e., the number of switches a packet has to traverse between the source and destination node) for each packet, network aggregate capacity (the number of packets a network can process within a certain period of time), and so on. A comparison between store-and-forward and hot-potato routing with the ShuffleNet topology shows that the average number of hops for each packet is larger for hot-potato routing, because not all the packets take the shortest route toward their destinations [15]. As the number of users (or number of nodes) increases, both the average number of hops and aggregate capacities increase for both routing strategies. In multihop networks, where information from a source node to a destination node

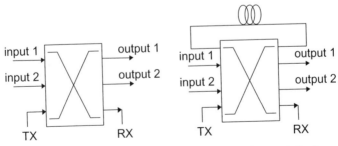

Figure 6.11 Example of node architecture for hot-potato routing, or deflection with limited buffer routing in MSN or ShuffleNet.

may be routed through intermediate nodes [16], only a portion of the network capacity is used for newly generated traffic. A certain amount of network capacity is taken up by bypassing traffic as packets hop from one node to another to reach the destinations. The overall capacity of the network is inversely proportional to the average number of hops and proportional to the number of nodes and the capacity of each link between two nodes. Store-and-forward routing can maximize the network capacity as the number of nodes increases. It has also been shown that even for networks containing several thousand nodes, the aggregate capacity of hot-potato routing is not worse than 25% of that for store-and forward routing [17]. Another more intuitive explanation is that in hot-potato routing the nodes use the entire network as a big buffer and route the packet in contention to the rest of the network. It trades off network throughput for simpler hardware implementation.

In MSN and ShuffleNet, there are three characteristics that determine the performance of the network with deflection routing:

1. *Diameter.* It is the maximum distance in number of hops between any node pair in the network. The diameter is a good indicator of how compact a network is.
2. *Deflection cost.* This is a maximum increase in path length in the number of hops due to a single deflection.
3. *Don't care nodes.* For a given destination, any node that has both its output links as part of the shortest path is a don't care node.

A high percentage of don't care nodes helps keep the number of deflections to a low level even at high loads. The performance of ShuffleNet is better because the initial advantage in diameter of ShuffleNet over Manhattan Street Network is preserved under heavy traffic by the high percentage of don't care nodes. The expected number of hops noticeable depends on the routing algorithm. For store-and-forward with infinite buffers, the average number of hops is a minimum, since the packets always take the shortest path to destination. However, the queuing delay could diverge to infinity when the network approaches saturation (i.e., when probability of packet generation in each time slot approaches 1). For deflection routing the average number of hops becomes an increasing function of link load and the throughput is therefore lower than that with store-and-forward.

So far, the routing strategies described above have assumed slotted (synchronous) network operation, which, as shown earlier, involves complex and expensive packet alignment stages. Since we are examining deflection routing here, which means none or little optical buffer is used, what will happen if we have an asynchronous network operation? What is the network performance if we take away the packet alignment stages and use deflection routing at the same time? Asynchronous network suffers from more severe congestion as the offered load increases, and its throughput would collapse completely when the load exceeds a certain threshold. The reason is that with the increasing congestion there are more and more packets starting to "wander" around in the network (due to deflection routing) and they further lower the network capacity for processing newly generated packets; meanwhile, more packets are being generated. The entire scenario forms a vicious cycle, and as a result, the network throughput will eventually collapse. To avoid such total collapse the number of hops a packet traverses has to be monitored and kept under a maximum value.

One way to improve the network throughput and reduce congestion is to provide limited optical buffering to such asynchronous deflection networks. The corresponding performance improvement is very encouraging in the high-load region. Also, the congestion is greatly reduced with more than one recirculating loop. One practical concern of the asynchronous deflection routing with limited optical buffer is the number of times a packet is allowed to recirculate in the loop. Optical amplification imposes noise to the signal. Network latency also increases with the number of circulations. Therefore, it is necessary to establish an optimal maximum number of recirculations for the packets.

Deflection routing plays a prominent role in many optical network architectures, since it can be implemented with no or modest optical buffering. Asynchronous (unslotted) deflection routing combined with limited buffering can help to avoid complex synchronization schemes and provide a decent performance with careful design. In general, deflection routing presents more choices to the network designer, although many problems, such as packet reordering, remain to be studied more thoroughly.

6.2.3.3 *Wavelength Conversion*   Optical buffering and deflection routing could be regarded as deflection in general, one in the time domain and the other in the space domain. With today's enabling technology in WDM, the wavelength domain presents us with one more dimension of solution. Both buffering and deflection have their advantages and disadvantages: buffering offers better network throughput but involves more hardware and controls; deflection is easier to implement but cannot offer an ideal network performance. When combined with wavelength conversion, their disadvantages could be overcome or minimized. In this section we examine some interesting combinations.

In a switch node applying wavelength conversion and buffering, the input stage demultiplexes wavelength channels, and the wavelength converters locate available wavelengths for certain output ports. The space switch selects the appropriate output port.

Wavelength conversion has been shown to reduce the number of optical buffers and to reduce packet-loss probability. When the nodes are provided with a number of optical receivers/transmitters equal to the number of wavelengths, hot-potato routing in conjunction with wavelength conversion becomes an interesting option for mesh topologies such as Manhattan Street Network and ShuffleNet [18].

One advantage of wavelength conversion is that it provides noise suppression and signal reshaping. In a network with only a small number of wavelengths, buffering might be more desirable. In a network with a large number of wavelengths and full wavelength conversion, buffers may not be necessary.

There are several possible combinations of optical buffering, wavelength conversion, and deflection routing. It presents an open research problem to decide which scheme offers a low implementation cost, low packet delay, low packet-loss rate, high network throughput, and so on, depending on the network specifications.

6.2.4 Header and Packet Format

In electronic networks, the packet header is transmitted serially with payload data at the same data rate, such as IP packets and ATM cells. Electronic routers or switches will process the header information at the same data rate as the payload. In an optical network, the bandwidth is much higher than their electronic counterparts. A typical wavelength channel has a line speed of 2.5 Gbps (OC-48). Although there are various techniques to detect and recognize packet headers at Gbps speed, either electronically or optically, it is costly to implement electronic header processors operating at such high speed.

Among several different proposed solutions, packet switching with a subcarrier multiplexed (SCM) header is attracting increasing interest. In this approach the header and payload data are multiplexed on the same wavelength (optical carrier). In the current that modulates the laser transmitter, payload data are encoded at the baseband while header bits are encoded on a properly chosen subcarrier frequency at a lower bit rate, as shown in Fig. 6.12. The header information on different wavelengths can be retrieved by detecting a small fraction of the light in the fiber with just a conventional photodetector, without any type of optical filtering. In the output current of the photodetector, various data streams from different wavelengths jam at baseband, but the subcarrier remains distinct and the header can be retrieved by an electrical filter (Fig. 6.13).

A nice feature of subcarrier multiplexed header is that the header can be transmitted in parallel with the payload data and it can take up the entire payload transmission time. Of course, the header can also be transmitted serially with the payload if desired. One potential pitfall of subcarrier multiplexed header is its possible limit on the payload data rate. If the payload data rate is increased, the baseband will expand and might eventually overlap with the subcarrier frequency, which is limited by microwave electronics.

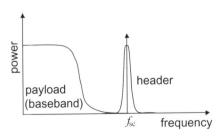

Figure 6.12 Power spectrum of the laser modulation current.

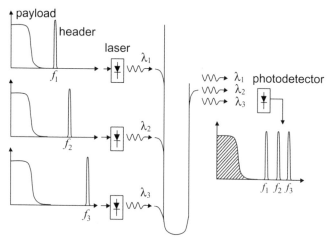

Figure 6.13   Header retrieval in SCM.

In many routing and switching protocols, packet headers have to be updated at each node. There have been several approaches proposed on optical header replacement for headers transmitted serially with the payload data stream. Optical header replacement could be done by blocking the old header with a fast optical switch and inserting the new header, generated locally by another laser, at proper time. One important issue here is that in WDM networks the new header should be precisely at the same wavelength as the payload data; otherwise, serious problems could arise because of dispersion, nonlinearity, or wavelength-sensitive devices in the network.

It has been proposed that the header updating be done by transmitting the payload and header on separate wavelengths, demultiplexing the header for optoelectronic conversion, electronic processing, and retransmission on the header wavelength. This approach suffers from fiber dispersion, which separates the header and payload as the packet propagates through the network. Subcarrier multiplexed headers have far fewer dispersion problems since they are very close to the baseband frequency. SCM header could be removed by narrowband optical filters, but it would be very sensitive to wavelength drift. Previous practical SCM header replacement schemes are limited to full optoelectronic conversion of the entire packet followed by electronic filtering, remodulation, and retransmission on a new laser. Vaughn and Blumenthal [19] have proposed a technique to update the SCM header with simultaneous wavelength conversion of baseband payload using SOAs. It involves a two-stage process: First, simultaneous SCM header suppression and wavelength conversion of the baseband payload is achieved due to the low-pass frequency response of cross-gain modulation in the SOAs; then header replacement is achieved by optically remodulating the wavelength-converted signal with a new header at the original subcarrier frequency.

Packet length is another issue of concern for network designers. A short packet might not give good throughput because a greater percentage of the bandwidth is given to the header or guard time between time slots. On the other hand, a long packet would need longer optical buffers and not provide a granularity that is fine enough. From a physical point of view, balancing the packet-error rate (PER) between payload and header is very important. PER is different from bit-error rate (BER); it is the probability of the entire packet received in error. PER increases with BER and the number of bits contained in the packet. For efficient network operation, the PER for payload and header should be about the same, in order to deliver the packets as successfully as possible [20]. Payload usually contains many more bits than the header. If the header is updated at every node traversed, the bits in payload will have to suffer more physical impairment than the bits in the header. Another fact is that if SCM is used, the header is usually transmitted at a lower bit rate than the payload data. All these facts lead to a big advantage of lower BER for header bits over the payload bits. Therefore, it is imperative to optimize the amount of power to be tapped from the packet at each node and the packet length in order to achieve a balanced PER for payload and header at the destination node.

6.2.5   Optical Label Switching

Meagher et al. [21] documented the first demonstration of computer-to-computer communication using an optical packet switching mechanism (i.e., optical label switching) implemented in a testbed at Telcordia. This approach adopts the unslotted operation and uses SCM for optical header (optical label) encoding. The network demonstrated utilizes deflection routing, wavelength conversion, and a preemption scheme based on packet priority. It also proposes an optical single-sideband (OSSB) subcarrier header erasing and replacement technique [22].

6.3   PHOTONIC SLOT ROUTING

One recently proposed approach to optical packet switching is photonic slot routing (PSR) [23,24]. In PSR, time is divided into photonic slots. All the wavelengths are slotted synchronously. Each photonic slot, across all the wavelengths, has a pre-assigned destination; therefore, all the packets transmitted in one photonic slot on different wavelengths are routed together through the same path. By requiring the packets to have the same destination, the photonic slot is routed as a single entity, without the need for demultiplexing individual wavelengths at intermediate nodes. Thus, wavelength-insensitive components may be used at each node, resulting in less complexity, faster routing, and lower network cost [23,25]. PSR reduces the optical hardware complexity by shifting the burden to the electrical-buffer management; therefore, most of the intelligence of PSR resides in the design of the access-control protocols.

PSR has been studied in bus and ring networks [23,25] as well as in mesh networks [24]. Next, we introduce the main PSR architecture and some of the proposed PSR protocols.

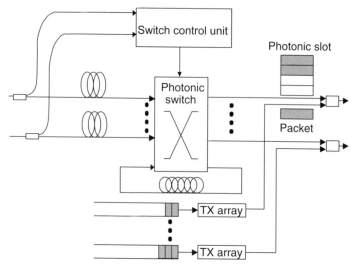

Figure 6.14 PSR node architecture.

6.3.1 Architecture for PSR Mesh Networks

In a PSR mesh network, each node consists of a wavelength-insensitive optical packet switch, optical delay lines, and electronic buffers (for ingress packet buffering). Each node (while acting as a source) maintains a separate queue for each destination node. A diagram of the node architecture is shown in Fig. 6.14.

A photonic slot (Fig. 6.15) spans across all the WDM channels wavelengths in the network, and each wavelength can carry a single packet. At the input fiber, a wavelength-selective drop device (such as a filter) extracts the header of each photonic slot, which may be carried on a separate wavelength (as shown in Fig. 6.15). The header contains information such as the destination of the slot and which

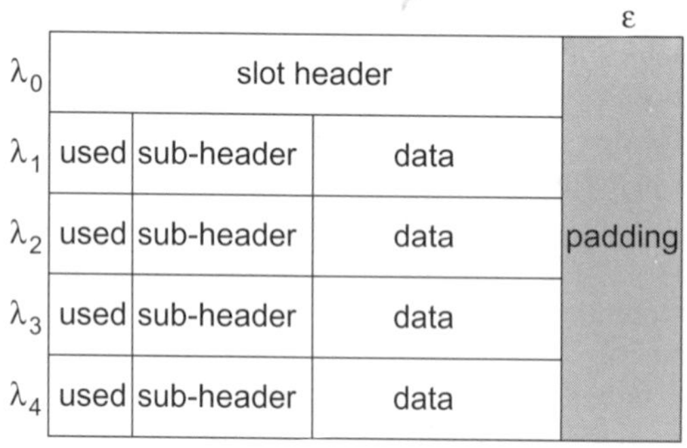

Figure 6.15 Format of a photonic slot.

wavelengths in the slot are occupied by packets. The destination information is used to determine/configure the switch setting so that the slot can be routed appropriately toward its destination according to a standard routing algorithm. The slot-occupancy information determines which wavelengths are still available in the current slot. Because it takes some time to process the slot header and to configure the optical packet switch, delay lines are necessary for each input fiber.

On a given output fiber link, the node may insert packets into existing photonic slots which are headed for the same destination, or the node may transmit newly created photonic slots if no other slots are contending for the link. The packet insertion may be performed by an optical coupler. For example, in Fig. 6.14, a photonic slot departing on the top output fiber link contains packets on wavelengths λ_1 and λ_2. The node may then insert a packet into the photonic slot on wavelength λ_3.

Contention can occur between two or more photonic slots at each of the output fibers. One approach for resolving contention is to buffer the photonic slots optically when they lose a contention. In this case, optical delay lines are used at each node [26]. Alternative approaches for dealing with contention include deflection routing and dropping photonic slots.

One of the challenges in implementing PSR in a mesh environment is maintaining synchronization of photonic slots at each node. Each of the incoming photonic slots must arrive simultaneously on all input fibers in order to be routed through the optical packet switch at the same time. Synchronization schemes described previously can be applied here.

Another physical issue in PSR is dispersion, which causes each wavelength to travel at a different speed along a fiber link, thus resulting in the spreading of different wavelengths inside a photonic slot. To compensate for slot dilation, padding can be added to extend the photonic slot length by some amount ϵ (see Fig. 6.15). A limitation of this approach is that it will result in reduced network utilization. The value of ϵ will increase with the diameter of the network; thus, to maintain reasonable utilization, we may need to limit the size of the network.

6.3.2 PSR Access Control Protocols

The basic PSR access control protocols control how a node inserts packets to a photonic slot. When a node has packets to send, it seeks for either an empty slot (not assigned to any destination yet) or an available slot that is already assigned to the same destination as the packet to be sent. How empty slots are assigned and how packets are inserted in the nonempty slots can largely determine the network performance and fairness. When a node receives an empty slot, it has to decide whether or not to transmit packets into this slot and which destination to be assigned to the slot.

Before discussing slot-assignment algorithms, it is helpful for us to look at how packets can be inserted into slots (when transmission is allowed). Policies are needed to determine how many packets should be inserted into the slot. There are several variations of policies that have different effects on fairness, packet delay, network throughput, and contention probability.

- *Greedy insertion.* A transmitter fills a slot whenever it has packets to transmit, and it fills the slot with as many packets as the slot can hold.
- *Lower-bound insertion.* A transmitter cannot fill an empty slot with packets unless it has at least a certain number of packets for the same destination. The transmitter follows the greedy-insertion approach for placing packets in nonempty slots.
- *Upper-bound insertion.* A transmitter can only fill a slot with up to a certain number of packets. The upper bound for the number of packets may differ depending on whether or not the slot is empty and on the position of the node in the path to the destination.

Greedy insertion performs better under low load by fully utilizing the network resources. Under higher load it will result in increased contention at downstream nodes and in fewer transmission opportunities for downstream nodes.

Lower-bound insertion will avoid contention up to some load limit. Every node waits for a certain number of packets to arrive before it starts to transmit. Hence, the number of nonempty slots generated from a source node is less than the case in which there is no lower bound. When the load limit is exceeded, a node will always receive enough packets to exceed the lower bound in the time between slot arrivals. Therefore, this policy becomes greedy insertion under high load.

Upper-bound insertion, which limits the number of packets that a node may transmit in a slot, provides a greater number of transmission opportunities for downstream nodes, thereby improving fairness. To determine the degree to which a node can fill a slot, it may be necessary to determine the position of a node in the path to the destination. However, this approach would require too much information for a node to collect and remember. Also, even with full knowledge of a node's location, network resources may be wasted under low load.

Now let us return to the slot assignment problem. The simplest algorithm is based on packet arrivals. Upon receiving an empty slot, a node randomly chooses one of its nonempty queues and inserts a number of packets into that slot according to the packet insertion policy. Whenever there is a nonempty queue for a given output fiber, the departing slots on this fiber will always have destinations assigned to them and the slots will not be empty.

This algorithm works well under light load; however, under heavy load, it causes more contention and unfairness in resource allocation. Nodes located toward the network edge receive more empty slots than nodes located in the internal region of the network.

To ensure that the upstream nodes will not overwhelm the downstream nodes, a slot assignment algorithm should be based on capacity allocation. The main goal for such an algorithm is to maximize network throughput, minimize contention, and provide fair bandwidth allocation. Chlamtac et al. [25] propose a transmission control protocol based on slot preassignment for PSR in a ring topology. In this approach, a TDM frame consists of a number of slots, and the source and destination of each slot in the TDM frame is determined by a network-wide TDM schedule. The number of slots in the TDM frame assigned to each source–

destination pair is determined in such a way that fair bandwidth distribution among the source–destination pairs, as well as contention-free slot routing at intermediate nodes, are achieved. A similar protocol for mesh topology is proposed in [24]. However, this TDM frame construction problem is shown to be NP-hard. Zang et al. [27] propose to assign destinations probabilistically to arriving slots. This approach performs a capacity-allocation algorithm, which takes the traffic matrix and routing matrix as inputs and determines the fraction of slots on each link that a source node could use for a certain destination.

6.4 OPTICAL BURST SWITCHING

Like PSR, optical burst switching (OBS) is an approach that attempts to shift the computation and control complexity from the optical domain to the electrical domain, from the core to the edge of the network. OBS has the switching granularity between a circuit and a packet. Burst switching was first proposed for voice and data communications in the early 1980s [28]. It was introduced to optical networks in the mid-1990s [29–31]. In an IP over WDM context, a burst formed at the network edge contains a number of IP packets and can be of tens of kilobytes to a few megabytes long. In OBS, a control packet, separate from the data burst that carries the payload, is sent first. The switches along the path will be configured (certain input/output ports will be connected and reserved) according to the information carried in the control packet as it propagates along. Shortly after the control packet is sent, the optical burst leaves the source node (without any acknowledgment about the reservation along the switches on the path) and will go through the configured switches. Compared with optical packet switching, OBS eliminates the need for optical buffering of payload, while the header is processed by pushing the buffering function to the edge nodes, where electrical buffer is available.

Figure 6.16 shows a typical OBS node architecture. The control packet that precedes the data burst is carried on a separate control wavelength. The switch control unit configures the switch and establishes an internal all-optical path inside the switch according to the control packet. Once the switch is configured, the data burst will arrive and go through the switch without any extra latency.

Similar to PSR, the intelligence of OBS is implemented primarily in the burst assembly and transmission algorithm and protocol rather than the switch architecture itself. Two important aspects in the design of OBS protocol are (1) the duration or size of the burst and how the configured switch state is released, and (2) the relation between the control packet and the payload data burst. When bandwidth is reserved in the switch, the control packet may or may not indicate the duration of the burst. If the burst duration is not known to the switch, the reservation will not be released until the switch is explicitly told to do so. This can be done by sending another release control packet, or by detecting the end of a burst by the switch itself. The importance of the offset between control packet and data burst is best demonstrated in a proposed protocol called *just enough time* (JET) [30]. In JET, the source node is assumed to have knowledge of the explicit route for the data burst

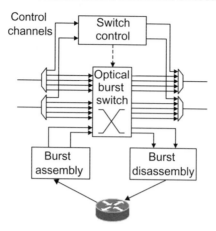

Figure 6.16   OBS node architecture.

and the control packet processing time δ at each switch (Fig. 6.17). The offset between data burst and control packet should be no less than the total processing delay of the control packet for all the switches along the routing path. This ensures that any switch along the path will have been configured when the burst arrives. Also, since the burst does not arrive immediately after the control packet is processed, the switch does not need to reserve the bandwidth (establish the required input/output connection) until the burst actually arrives. This delayed reservation makes more efficient use of the bandwidth. The reader is referred to Chapter 13 for more details on OBS.

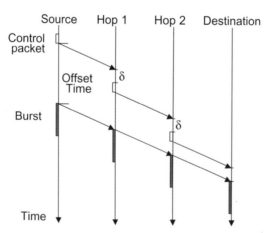

Figure 6.17   Offset time and delay reservation in the JET protocol.

6.5 SUMMARY

The scope of optical subwavelength switching is extensive. This topic involves routing, synchronization, contention resolution, header format/updating, switch fabrics, physical impairment of the transmission, network control, protocol, and so on. Only a handful of the important aspects were covered here. Optical packet switching is promising because it offers a large capacity and data transparency. However, after many years of research, this technology has not yet been applied in actual products, because of (1) the lack of deep and fast optical memories and (2) the poor level of integration. It is our belief that these issues will be overcome not only through technical breakthroughs but also through clever network design, making optimal use of optics and electronics wherever they fit best.

REFERENCES

1. I. Chlamtac, A. Farago, and T. Zang, Lightpath routing in large WDM networks, *IEEE Journal on Selected Areas in Communications*, 14(5):909–913, June 1994.
2. B. Mukherjee, D. Banerjee, S. Ramamurthy, and A. Mukherjee, Some principles for designing a wide-area WDM optical network, *IEEE/ACM Transactions on Networking*, 4(5):124–129, Oct. 1996.
3. D. O. Awduche, Y. Rekhter, J. Drake, and R. Coltun, Multi-protocol lambda switching: Combining MPLS traffic engineering control with optical crossconnects, IETF draft: *draft-ietf-awduche-mpls-te-optical-03.txt*, Oct. 2001.
4. A. Banerjee, J. Drake, J. P. Lang, B. Turner, K. Kompella, and Y. Rekhter, Generalized multiprotocol label switching: An overview of routing and management enhancements, *IEEE Communications*, 39:144–150, Jan. 2001.
5. A. Banerjee, J. Drake, J. P. Lang, B. Turner, D. Awduche, L. Berger, K. Kompella, and Y. Rekhter, Generalized multiprotocol label switching: An overview of signaling enhancements and recovery techniques, *IEEE Communications*, 39:144–151, July 2001.
6. A. Tajima, A 10-Gbit/s optical asynchronous cell/packet receiver with a fast bit synchronization circuit, *Proc. OFC'99*, pp. 111–113, 1999.
7. Y. Yamada, S. Mino, and K. Habara, Ultrafast clock recovery for burst-mode optical packet communication, *Proc. OFC'99*, pp. 114–116, 1999.
8. K. V. Shrikahande et al., HORNET: A packet-over-WDM multiple access metropolitan area ring network, *IEEE/OSA Journal of Lightwave Technology*, 18(10), Oct. 2000.
9. P. R. Prucnal, Optically processed self-routing synchronization, and contention resolution for 1D and 2D photonic switching architectures, *IEEE Journal on Quantum Electronics*, 29(2):600–612, Feb. 1993.
10. A. S. Tanenbaum, *Computer Networks*, Prentice Hall, Upper Saddle River, NJ, 1996.
11. M. J. Karol, A shared-memory optical packet (ATM) switch, *Proc. 6th IEEE Workshop on Local and Metropolitan Area Networks*, pp. 205–211, 1993.
12. R. L. Cruz and J. T. Tsai, COD: Alternative architectures for high-speed packet switching, *IEEE/ACM Transactions on Networking*, 4:11–21, Feb, 1996.

13. Z. Haas, The "staggering switch": An electronically controlled optical packet switch, *IEEE/OSA Journal of Lightwave Technology*, 11(5/6):925–936, May/June 1993.

14. D. K. Hunter, W. D. Cornwell, T. H. Gilfedder, A. Franzen, and I. Andonovic, SLOB: A switch with large optical buffers for packet switching, *IEEE/OSA Journal of Lightwave Technology*, 16(10):1725–1736, Oct. 1998.

15. A. S. Acampora and I. A. Shah, Multihop lightwave networks: A comparison of store-and-forward and hot-potato routing, *IEEE Transactions on Communications*, 40:1082–1090, June 1992.

16. B. Mukherjee, *Optical Communication Networks*, McGraw-Hill, New York, 1997.

17. F. Forghierri, A. Bononi, and P. R. Prucnal, Analysis and comparison of hot-potato and single-buffer deflection routing in very high bit rate optical mesh networks, *IEEE Transactions on Communications*, 43(1):88–98, Jan. 1995.

18. A. Bononi, G. A. Castanon, and O. K. Tonguz, Analysis of hot-potato optical networks with wavelength conversion, *IEEE/OSA Journal of Lightwave Technology*, 17:525–534, Apr. 1999.

19. M. D. Vaughn and D. J. Blumenthal, All-optical undating of subcarrier encoded packet header with simultaneous wavelength conversion of baseband payload in semiconductor optical amplifiers, *IEEE Photonics Technology Letters*, 9:827–829, 1997.

20. B. Ramamurthy, D. Datta, H. Feng, J. P. Heritage, and B. Mukherjee, Impact of transmission impairments on the teletraffic performance of wavelength-routed optical networks, *IEEE/OSA Journal of Lightwave Technology*, 17(10):1713–1723, 1999.

21. B. Meagher et al., Design and implementation of ultra-low latency optical label switching for packet-switched WDM networks, *IEEE/OSA Journal of Lightwave Technology*, 18(12):1978–1987, Dec. 2000.

22. W. Way, Y. Lin, and G. K. Chang, A novel optical label swapping technique using erasable optical single-sideband subcarrier label, *Proc. OFC'00*, Vol. WD4, pp. 1–4, Mar. 2000.

23. I. Chlamtac, V. Elek, A. Fumagalli, and C. Szabo, Scalable WDM access network architecture based on photonic slot routing, *IEEE/ACM Transactions on Networking*, 7(1):1–9, Feb. 1999.

24. I. Chlamtac, A. Fumagalli, and G. Wedzinga, Slot routing as a solution for optically transparent scalable WDM wide area networks, *Photonic Network Communications*, 1(1):9–21, June 1999.

25. I. Chlamtac, V. Elek, and A. Fumagalli, A fair slot routing solution for scalability in all-optical packet-switched networks, *Journal of High Speed Networks*, 6(3):181–196, 1997.

26. I. Chlamtac, A. Fumagalli, L. G. Kazovsky, P. Melman, W. H. Nelson, P. Poggiolini, M. Cerisola, A. N. M. M. Choudhury, T. K. Fong, R. T. Hofmeister, C. Lu, A. Mekkittikul, D. J. M. Sabido IX, C. Suh, and E. W. M. Wong, CORD: Contention resolution by delay lines, *IEEE Journal on Selected Areas in Communications*, 14(5):1014–1029, June 1996.

27. H. Zang, J. P. Jue, and B. Mukherjee, Capacity allocation and contention resolution in a photonic slot routing all-optical WDM mesh network, *IEEE/OSA Journal of Lightwave Technology*, 18(12):1728–1741, Dec. 2000.

28. S. R. Amstutz, Burst switching: an introduction, *IEEE Communications*, 21(8):36–42, Nov. 1983.

29. G. C. Hudek and D. J. Muder, Signaling analysis for a multi-switch all-optical network, *Proc. IEEE International Conference on Communications (ICC'95)*, pp. 1206–1210, June 1995.

30. M. Yoo and C. Qiao, Just-enough-time(JET): A high speed protocol for bursty traffic in optical networks, *Digest of IEEE/LEOS Summer Topical Mettings on Technologies for a Global Information Infrastructure*, pp. 26–27, Aug. 1997.

31. J. S. Turner, *Trabit Burst Switching*, Technical Report WUCS-97-49, Washington University, St. Louis, MD, Dec. 1997.

7 Multiprotocol over WDM in the Access and Metropolitan Networks

MAURICE GAGNAIRE

École Nationale Supérieure des Télécommunications, Paris, France

7.1 INTRODUCTION

The principal advances carried out in the field of optical devices and systems have been outlined in earlier chapters. Dense wavelength-division multiplexing (DWDM), optical amplification, or optical add–drop multiplexing are some of these major advances. In this chapter we aim at analyzing carriers' network evolution in the context of optical access networks and optical metropolitan area networks. In Section 7.2 we outline with a global perspective the successive steps that have marked the past or that will mark the near future of optical carriers' networks. The geographical coverage of carriers' networks may be divided into three subdomains: access networks, metropolitan area networks (MANs), and core networks. Core networks are based on either circuit switching or packet switching. The former corresponds to public switched telephone networks (PSTNs). The latter refers either to connection-oriented packet-switching networks [asynchronous transfer mode (ATM) or frame relay] or to connectionless-oriented packet-switching networks (IP). Section 7.3 aims at presenting the current characteristics of these various portions of public networks. We then focus in Section 7.4 on the main challenges that have to be satisfied in the access and the metro facing the evolution of users' requirements. Several broadcast and select network architectures have been considered to satisfy these requirements. In Section 7.5 we discuss the main performance issues for the metro and the access. Section 7.6 is dedicated to the description of metropolitan optical networks based either on passive optical stars or active optical rings. Two variants of optical metropolitan rings, called dynamic packet transport (DPT) and resilient packet ring (RPR), have been proposed by two major vendors. The RPR network should be the basis of a coming IEEE standard. Section 7.7 focuses on optical access networks, also known as passive optical

IP over WDM: Building the Next-Generation Optical Internet, Edited by Sudhir Dixit.
ISBN 0-471-21248-2 © 2003 John Wiley & Sons, Inc.

networks (PONs). Two variants of PONs are presented successively: ATM PONs (APONs) and broadband PONs (BPONs). Section 7.8 is a summary of this chapter.

7.2   TRANSPORT OPTICAL NETWORKS: A GLOBAL PERSPECTIVE

During the 1980s, most carriers deployed unidirectional optical fibers in their long-haul infrastructure. For this first generation of optical carrier networks, the G.652 single-mode optical fiber enables spans between two successive electrical nodes around 40 km for an online data rate of about 100 Mbps. Point-to-point optical links are connected according to an irregular meshed topology. The plesiochronous digital hierarchy (PDH) was widely deployed during the first half of the same decade to supervise optical transmission. While PDH was being developed, PSTN evolved from analog to digital at the beginning of the 1970s. Let us briefly recall the basic principles of PDH. Once voice signals are digitized at 64 kbps, a first level of PDH multiplexers merge these individual data channels into periodic frames thanks to time-division multiplexing (TDM). There is no clock dispersion between voice coders and these first-level PDH multiplexers, both types of equipment being located in the same building. This is not the case between higher-order multiplexers, which are in general distant from each other. The concept of *plesiochronism* refers to the relative independence between the clocks used in higher-order multiplexers. Two different PDH hierarchies have been adopted in North-America (T1 at 1.544 Mbps, T2 at 6.132 Mbps, T3 at 44.736 Mbps and T4 at 274.176 Mbps) and in Europe (E1 at 2.048 Mbps, E2 at 8.448 Mbps, E3 at 34.368 Mbps, and E4 at 139.264 Mbps). The plesiochronous digital hierarchy has a few limitations:

- It lacks multiplexing flexibility.
- In each node, compensation of clock dispersion between the incoming signal and the output signal is carried out by means of bit-oriented stuffing/destuffing. Such a technique is applicable to bit rates limited to a few hundreds of Mbps.
- It enables supervision of the quality of transmission limited to point-to-point links. It does not allow sharing of information concerning the end-to-end quality of transmission.
- Different PDH hierarchies exist in Europe, Japan, and North America. This disparity induces a relative incompatibility between equipment from different vendors. The highest achievable bit rate with PDH transmission is either 140 Mbps or 565 Mbps, depending on the countries.

SONET (Synchronous Optical Network)/SDH (Synchronous Digital Hierarchy) was developed in the mid 1980s by Bellcore[*] to eliminate these drawbacks. In comparison to PDH, several benefits characterize SONET/SDH:

---

[*]Bellcore is now Telcordia.

TABLE 7.1 SONET/SDH Hierarchy

SONET (Bellcore)	Data Rate	SDH (ITU)
STS-1/OC-1	51.84 Mbps	
STS-3/OC-3	155.52 Mbps	STM-1
STS-9/OC-9	466.56 Mbps	
STS-12/OC-12	622.08 Mbps	STM-4
STS-18/OC-18	933.12 Mbps	
STS-24/OC-24	1,244.16 Mbps	
STS-36/OC-36	1,866.24 Mbps	
STS-48/OC-48	2,488.32 Mbps	STM-16
STS-192/OC-192	9,953.3 Gbps	STM-64
STS-768/OC-768	39,813 Gbps	STM-256

- SONET/SDH increases network survivability considerably, thanks to automatic protection switching (APS) and to its end-to-end supervision capability.

- Instead of using a heavy bit stuffing/destuffing technique to compensate clock dispersion, SONET/SDH manages plesiochronism by means of a pointer mechanism. This pointer coded in each frame header makes it possible to localize very rapidly a tributary unit (also called a *virtual container*) transported in the payload of a frame. The pointer mechanism coupled with a positive or negative justification at the byte level enables on-line bit rates up to the Gbps range.

- Unlike PDH, SONET/SDH is adopted by all the vendors and operators around the world.

- Unlike PDH, which operates on point-to-point links, SONET/SDH is applicable to a mix of meshed linear and ring topologies.

When SONET/SDH was being developed in 1984, the great majority of traffic in public networks was made of circuit-oriented phone calls[*] (POTS traffic). The proliferation of local area networks (LANs) in the mid-1970s and later of Internet has required the development of successive generations of packet-oriented networks (X.25, frame relay, ATM, etc.). In most instances, these packet-oriented networks run over a SONET/SDH transport. As a reminder, we show in Table 7.1 the standardized data rates of the SONET/SDH hierarchy.

At the beginning of the 1990s, manufacturers and carriers adopted SONET/SDH protection and restoration principles to optical ring topologies. Compared to the existing irregular meshed topologies, ring topologies enable a fast and automatic restoration of the network in case of fiber cut or node failure. This is why several carriers began from this date to redesign their global infrastructure progressively on the basis of adjacent SONET/SDH self-healing rings. Figure 7.1 illustrates a carrier network based on hierarchical SONET/SDH rings.

---

[*]The term *POTS* (plain old telephone service) *traffic* is used to refer to circuit-oriented phone cells.

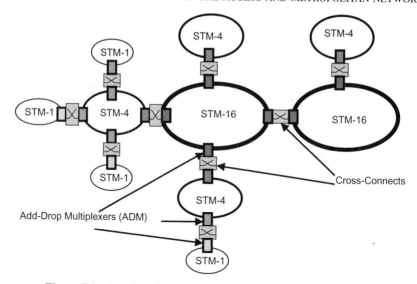

Figure 7.1   Ring-based topology for a long-distance carrier network.

After the year 2000, the ratio of POTS traffic over data traffic has been inverted in several public networks. This evolution is known as the *convergence toward data-centric networks*. The ATM technique was considered in the 1980s and 1990s as the best approach to merge voice and data traffic on the same packet-switched network. Today, carriers are in fact still obliged to manage two types of network infrastructures, one for the circuit-switched traffic and the other for packet-switched traffic. Figures 7.2 and 7.3 illustrate these two parallel versions of carrier networks, including the three geographical subnets we mentioned in our introduction (access network, metropolitan area networks, and core networks). Figure 7.2 illustrates a global carrier's network configuration when the core network is circuit-switched oriented (PSTN). PSTN networks are based on a set of hierarchical switches: end offices,[*] toll offices (or central office), and intermediate switching offices. We have shown in this figure the presence of network access servers (NASs). An NAS is used to enable point-to-point connection between an end user's voice-band modem and the point-of-presence (POP) of an ISP (Internet service provider). The server NAS has the twin role of identifying end users and metering the duration and data volume associated with the connections of these end users to Internet servers.

In the case of packet-switched networks (see Fig. 7.3), the core network is based on packet-switching technology, such as X.25, frame relay, or ATM. We have shown in this figure the presence of broadband access servers (BASs). The role of a BAS is quite similar to the role of a NAS. A BAS is used to enable communication between an ADSL user and the POP of an ISP. A BAS identifies endusers and meters the duration and the data volume associated with their connections to

[*]End offices are used as traffic concentrators.

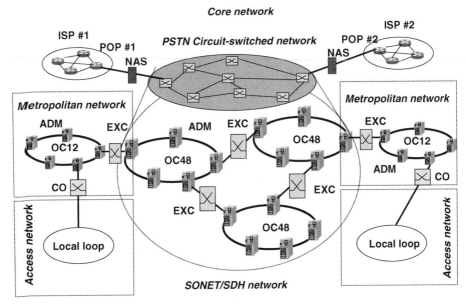

Figure 7.2   Global network configuration based on PSTN networks.

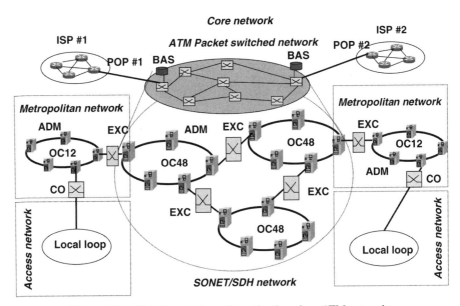

Figure 7.3   Global network configuration based on ATM networks.

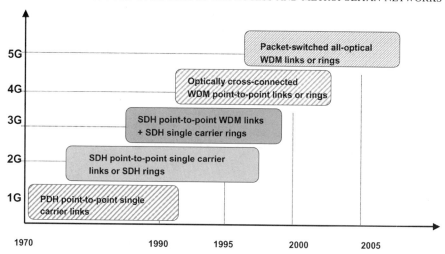

Figure 7.4   Historical perspective of optical networking.

Internet servers. Both the NAS and the BAS dynamically allocate an IP address to each end user by means of the PPP (point-to-point) protocol.

As shown in Figs. 7.2 and 7.3, metropolitan area networks and core networks run in most cases on a SONET/SDH infrastructure at the physical layer. A core network's topology is in general a mix of linear topologies and ring topologies, whereas metropolitan area networks are based on rings topologies. Access networks do not use SONET/SDH.* Although it is not represented on Figs. 7.2 and 7.3, ISPs' routers benefit in most cases from the guaranteed quality of service offered by PSTN or ATM. This is why, in practice, ISPs' networks also use SONET/SDH equipment at the physical layer.

To conclude this overview, it is possible to classify the evolution of optical networks into five generations that are illustrated by Fig. 7.4. The first generation dating from the 1970s was characterized by the introduction of PDH point-to-point optical fibers in long-haul telecommunications networks. At the end of the 1980s, SONET/SDH replaced PDH in the second-generation optical networks. At the middle of the 1990s, WDM began to be introduced to increase the capacity of point-to-point optical fibers by several orders of magnitude. The first optical cross-connects were introduced at the very beginning of the year 2000 in order to interconnect point-to-point WDM links or rings. Today, only static lightpath establishment is considered for this fourth generation of optical networks. Let us recall that a lightpath (also called a *clear channel*) corresponds to a point-to-point all-optical virtual connection between two distant electrical nodes of the network. A few research programs have demonstrated the feasibility of optical packet switches. The fifth generation that could appear at the end of this decade should

---

*We see in the following that in rare cases, some carries may use SONET/SDH technology in the upper part of their access network, called a *feeder network*.

see the arrival of all-optical packet switching in core networks. In that case, the lifetime of a lightpath should be comparable to the round-trip time through all-optical clouds. According to the state of the technology and the complexity of the routing and wavelength allocation problem (RWA), dynamic all-optical packet switching remains rather prospective.

7.3 CURRENT PUBLIC NETWORK ARCHITECTURE

7.3.1   Access Networks

An access network represents the last portion of a carrier's network, from a central office (CO) to a set of end users. According to Erlang's law, a telephone user is on average active at most 10% of the day. This is why telephone lines are concentrated at an end office (EO) before reaching the CO. Voice calls are transmitted as analog signals over twisted copper pairs from the users to the EOs. Voice digitization is carried out at the input of the EO, which is itself connected to the CO by means of a toll connecting trunk.

As it is illustrated by Fig. 7.5, several end offices are point-to-point connected to a same central office. This star configuration presents a serious drawback in case of cable cut between one of the end offices and the central office. When such an event occurs, about 5000 users are disconnected from PSTN and it takes in general

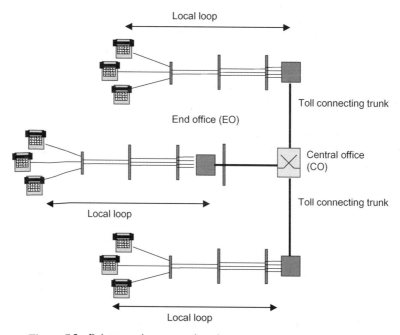

Figure 7.5   Point-to-point connections between concentrators and a CO.

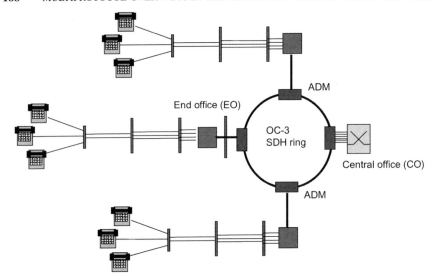

Figure 7.6   SONET/SDH self-healing rings in the access network.

several weeks to repair the cable trunk. To accelerate network recovery in such circumstances, several carriers have decided to replace the star infrastructure at the top of the access network by a SONET/SDH self-healing ring (see Fig. 7.6). In this case, each EO is connected to an add-drop multiplexer (ADM). Thus, digital voice signals generated by the end offices are concentrated into T1/E1 PDH frames. These frames are then inserted into SONET/SDH virtual containers circulating on the ring. In reference to Figs. 7.2 and 7.3, SONET/SDH rings in the feeder network typically run at the OC-3 (optical channel 3) or STM-1 (synchronous time multiplex1) data rate equal to 155 Mbps.

Analog modems have been in use for about 15 years by residential users to connect to ISPs' servers via twisted pairs in the local loop. The data rate offered by such modems is limited to 34 kbps in most cases. Progressively, ADSL modems are introduced to upgrade the local loop capacity up to a few Mbps in the downstream direction and to a few hundred kbps in the upstream direction.

7.3.2   Metropolitan Area Networks

As described in Figs. 7.2 and 7.3, metropolitan area networks (MANs) are used to concentrate upstream traffic coming from different local loops before forwarding this traffic toward core networks. In third-generation public networks, MANs are based on SONET/SDH rings. Such rings are still widely exploited in current public networks. Today, a MAN typically consists of an OC-12 SONET/SDH self-healing ring operating at 622 Mbps. Some of the ADMs inserted along the rings are used to offer enterprises high-speed access to core networks. Other ADMs of the same rings are used to collect residential traffic coming from various central offices. Between

1990 and 2000, the emergence of DWDM has motivated the design and development of all-optical MANs based on various topologies (star, folded bus, or dual rings). Facing single-wavelength SONET/SDH rings, these new-generation MANs have not yet been deployed in carrier' networks.

7.3.3 Core Networks

Because of data-centric convergence, in this section we refer only to broadband core networks based on the ATM technique (Fig. 7.3). ATM switches which are now widely deployed in carriers' networks enable different end-to-end quality-of-service guarantees. The ATM technique has been designed to facilitate efficient multiplexing of variable-bit-rate sources within the same network. Due to its complexity, the Q.2931 signaling inherent in ATM is in most cases unable to provide switched virtual circuits in real time. ATM virtual connections are then preestablished for the long-term. Today, a large fraction of the traffic transported by ATM networks is IP traffic. Semipermanent ATM virtual connections are used as tunnels between distant IP routers in order to improve the transit delay of IP packets. As an example, such ATM connections are used between a CO and the various BASs of a packet-switched core network. Typically, SONET/SDH rings in the core network operate at the OC-48 (equivalent to the ITU-T STM-16) or OC-192 (equivalent to the ITU-T STM-64) data rates, corresponding to 2.5 and 10 Gb/s, respectively. As shown in Fig. 7.3, core SONET/SDH rings are today interconnected by means of electrical cross-connects (EXCs).

During the last 10 years, the capacity of several optical fibers installed in core networks has been upgraded thanks to WDM. As mentioned at the end of Section 7.2, the next step for the carriers is to link these point-to-point WDM optical fibers by means of optical cross-connects (OXCs). A lot of research is ongoing to determine the algorithms that would facilitate all-optical network planning and dimensioning. For the short term, all-optical core networks made up of OXCs and optical fibers are seen at a layer rather independent from the electrical layer. This optical layer should offer in the first phase a set of semipermanent lightpaths that could be used by electrical nodes (IP routers, ATM switches, SONET/SDH ADMs, etc.). This approach, known as the *overlay model,* is currently discussed within the ITU-T under the name *automatic switched optical network* (ASON) [1]. The ASON standard aims at specifying a control plane and an optical user network interface (O-UNI) for the fourth- and fifth-generation optical networks (see Fig. 7.4) [2]. Figure 7.7 illustrates the principle of an ASON network. In this figure we illustrate an example of a lightpath linking two optical cross-connects (OXC 1 and OXC 2). Two electrical nodes (two IP routers) connect with this lightpath via O-UNI signaling with OXC 1 and OXC 2, respectively. This lightpath is represented on the figure by means of a bold dashed line.

Three kinds of optical connections should be supported in an ASON network:

1. *Semipermanent lightpaths.* These lightpaths are established for the long term (network planning) under the control of a network management system.

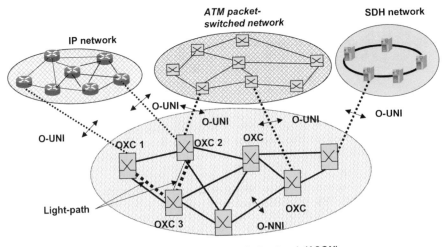

Figure 7.7 Principle of an ASON network.

2. *Soft permanent lightpaths.* These lightpaths are characterized by a certain lifetime related to the traffic intensity supported. Soft permanent lightpaths are set up between ASON nodes (OXCs) by means of a signaling protocol specified as O-NNI (optical network-to-network interface). Like semipermanent lightpaths, soft permanent lightpaths are established for the medium term, that is, at least for a few hours or longer.

3. *Switched optical lightpaths.* Such lightpaths are set up in real time at the request of a customer node (see Fig. 7.7): for instance, an IP router, an ATM switch, or a SONET/SDH ADM. Switched lightpaths require specification of the O-UNI.

7.3.4 ISP Networks

The Internet is made of several autonomous systems (AS), each of them managed by different ISPs. An autonomous domain is itself a set of IP routers linked to each other by means of optical fibers. Two kinds of routers are used in the Internet, either internal or external. Internal routers are used within an AS. External routers enable interconnection of autonomous domains with each other. Within an autonomous domain, the same routing algorithm is adopted: for instance, RIP (routing internet protocol), OSPF (open shortest path first), or IS-IS (intermediate system to intermediate system). These three algorithms route IP packets on a shortest-path basis. More powerful routers called BGP (border gateway protocol) routers are used as external routers. Web servers, voice servers, or video servers are connected to some of the routers of an autonomous domain. Several factors penalize interactivity between end users and Internet servers. Local loops based on dial-up modems are a well-known bottleneck. ADSL modems reduce this penalty by preventing IP packets to transit through a PSTN network. Nevertheless, the performance of ADSL access is also affected by transit delays through ISP networks.

7.4 CHALLENGES IN ACCESS AND METROPOLITAN NETWORKS

In Section 7.2 we reviewed a global perspective of the evolution of carriers' networks. In this section, we discuss some of the challenges that will characterize for the next few years this evolution in the access and metro networks. In the context of deregulation of the telecommunications' market, carriers consider access networks and metropolitan area networks as strategic. Indeed, the last mile of a carrier's network enables direct contact with customers. A strong increase in core network capacity has been observed these last 10 years, passing from 2.5 Gbps (second generation) to several tens of Gbps (third generation) per optical fiber. With such a capacity, the cost of the bit transported has collapsed. In other words, the source of income for the carriers is today less in the volume of transported data than in the number of innovative services they can offer to customers. The offer of new services, such as interactive Internet access or high-speed data transfer, requires a new generation of access and metropolitan area networks. Access networks aim at transporting the traffic generated by residential users and SMEs (small and medium enterprises). Metropolitan area networks, which federate several access networks, aim at transporting the traffic generated by large enterprises in addition to the traffic received from the access networks.

7.4.1 Access Networks

The ILECs (incumbent local exchange carriers) that own copper wires in the local loop benefit from their existing infrastructure. About 800 millions of such pairs are installed today all over the world. The CLECs (competitive local exchange carriers) that do not benefit from such a heritage have two options for offering interactive services to their customers. The first option is leasing copper wires from an ILEC, and the second option is to install their own local loop infrastructure. The first option requires unbundling and an obligation from the ILEC to lease a fraction of its copper infrastructure to competitors. In many countries, regulatory bodies have today adopted the principle of unbundling. Nevertheless, CLECs estimate the leasing cost of copper pairs so high that it puts the CLECs at a great disadvantage. This is why numerous CLECs have opted to focus on radio access systems. Several radio access technologies, also known as wireless in the loop (WITL), have been developed in the past 10 years. WITL systems are easy and cost-effective to install and maintain. Other technologies, based on hybrid fiber coaxial (HFC), power line communication (PLC), and satellite networks, are considered as alternative solutions.

Although optical fiber access technologies date from the end of the 1980s, many carriers consider them too expensive to be competitive with WITL and ADSL. We list below some of the limitations and constraints that characterize optical access systems in comparison to other technologies:

- Compared to a copper infrastructure or to a WITL system, fiber in the loop (FITL) imposes a big investment in terms of civil engineering to install optical fibers. The setting of dedicated point-to-point optical fibers from each end user to dedicated subscriber line interface circuits at the central office probably remains too expensive. Meanwhile, this drawback is minimized if the optical infrastructure shares common sections of optical fiber between several end users. Passive optical networks (PONs) facilitate such an economical sharing of optical resource. Very recently, several carriers such as Verizon[*] and BellSouth in the United States and NTT in Japan have investigated the economic feasibility of fiber to the home (FTTH). The results of these investigations suggest that recent progress in PONs makes the cost of deployment of FTTH only slightly more expensive than the cost of deployment of copper wires in the last mile [3]. We describe this evolution in more detail in Section 7.7.

- The great advantage of WITL over FITL is its incremental cost. The WITL equipment is based on additive modules that can be implemented progressively as user demand increases. Typically, each module corresponds to a given pair of transmitter and receiver radio frequencies. The base stations located at the center of a WITL cell may support several of these pairs in a given frequency band.

- Concerning POTS traffic, traffic concentration is preferable in the access network to reduce the number of line interface cards at the central office. In Section 7.3.1 we discussed the fact that this concentration is carried out in existing networks by means of end offices. The same kind of constraint exists for packet-oriented traffic, the cost of a packet switch also being proportional to the number of line interface cards.

- At the beginning of the 1990s, the lifetime of optoelectronic devices (transmitters and receivers) was rather short (a few years). This characteristic had a strong impact on the maintenance cost of optical networks. In the last 10 years, this lifetime has been lengthened significantly. Additionally, the cost of optical modems that have to be installed at each end of the access network depends on their complexity. The choice of the topology may have a strong impact on this complexity in terms of power ranging, synchronization, and multiple access.

7.4.2   Metropolitan Networks

We underlined in Section 7.3.2 that both professional users and residential users were connected to metropolitan area networks. Traditionally, professional users are

---

[*]Verizon was formerly Bell Atlantic.

connected directly to the ADMs of metro SONET/SDH rings, whereas residential users are connected to these same rings indirectly via the central offices and the copper access networks. For some years now, two specific services are offered to the professional: virtual private networks (VPNs) and remote data storage. Both of these services require large bandwidth capacities, up to the Gbps range in the most stringent cases. We list below some of the challenges that WDM faces in the metro networks.

- Until now, DWDM equipment has been designed for the long-haul market. One of the challenges for DWDM manufacturers today is to adapt the cost of long-haul WDM devices to the MANs. Indeed, a MAN covers a much reduced number of potential users compared to a long-haul network.

- For the same reason, there is also a challenge for carriers who have to manage optical MANs at a reduced cost in comparison to their long-haul network [4]. For the past few years, municipal and regional authorities have considered metropolitan area networks as key to economic expansion. In several European countries, the local administrations invest heavily in the deployment of dark fibers.

- We underlined in Section 7.2 the fact that existing STM-4 SONET/SDH metro rings have been designed essentially for the transport of voice traffic. The main drawback of such rings is their unsuitability for the transport of data traffic, and more specifically of Internet traffic. In a data-centric world, IP traffic is very variable in terms of throughput and directionality (statistical behavior). Unlike Internet traffic, voice connections are symmetrical and predictable. We know that unlike voice traffic, statistical multiplexing of IP flows is self-similar with long-range dependence. Buffer dimensioning in electrical nodes (cross-connects, ADMs, switches, etc.) is much more complex in the context of long-range dependence traffic than in the case of multiplexed voice traffic.

- The concept of automatic switched optical network (ASON) introduced in Section 7.3.3 has been initially considered within the ITU-T for transport networks. Today, network operators estimate that metropolitan area networks should also benefit from this concept. The control plane required by ASON could facilitate traffic engineering. The virtual private network could benefit Optical VPN within ASON-compatible MANs. The rapid provisioning of switched optical lightpaths should also be very useful for remote data storage services.

Between 1990 and 2000, several investigations were carried out to determine the unsuitability of SONET/SDH rings for the new metro context. For that purpose, various types of WDM networks based on broadcast and select topologies have been investigated. The term broadcast and select means that a packet sent by a source node naturally propagates to any destination node without explicit routing or switching along the path. Each inactive node only has to check permanently the destination address of all the packets it detects on the medium in order to determine those which are dedicated to it. These networks rely essentially on three possible topologies: bus, star, and ring. By extending the broadcast and select LAN topology to the metro environment, it is possible to benefit from a connectionless service well adapted to data traffic. In such networks, lightpath establishment is much simpler than in irregular meshed topologies. Such a simplification relies on the design of new medium-access Control (MAC) protocols.

7.5   PERFORMANCE AND QoS EVALUATION

Two main criteria characterize the performance of broadcast and select networks: medium utilization ratio and mean access delay. In Sections 7.5.1 and 7.5.2, we describe the meaning of these two standards in the context of LANs and MANs which authorize a single successful transmission at the same time. We show in Section 7.5.3 how these two criterions may be merged in a single expression by means of the concept of power of MAC protocol.[*] In Section 7.5.4 we describe the main quality-of-service (QoS) parameters of data networks and their application to access and metro networks.

7.5.1   Medium Utilization Ratio

In the context of a broadcast and select network, protocol efficiency Φ_{max} corresponds to the maximum achievable throughput Φ on the medium. The expression for the value of Φ_{max} is determined when the packet arrival rate and the packet size follow, respectively, the same distributions in different nodes of the network. These two statistical processes are independent and identically distributed on the various nodes of the network. One also assumes that the destination of each packet generated by a node is distributed equally over all the possible destinations. Let $\lambda(i)$ be the packet generation rate in a node i. Let ρ be the global offered load to the network, that is, the summation of $\lambda(i)$ for any i. If C stands for the medium capacity, the offered load, in most cases, is normalized to this capacity to evaluate the intrinsic performance of the MAC protocol independently from the technology (the speed of electronics). Typically, the effective throughput on the medium Φ is evaluated for different values of ρ. Like ρ, Φ is normalized to C. Over a long period of observation, it is possible to evaluate, either by means of computer simulations or by means of analytical formulations, the evolution of Φ versus ρ (see Fig. 7.8). The MAC protocol efficiency corresponds to the maximum achievable medium utilization ratio Φ_{max}. It is expressed is a percentage. Let us recall that we consider in this section only the case of LANs or MANs enabling a single successful access at the same time. Such networks can be modeled by single-server queueing systems.

Fig. 7.8 shows three curves C_1, C_2 and C_3. Curve, C_1 represents the case of an ideal MAC protocol for which, starting from very low loads, Φ increases linearly

[*]The Concept of power of MAC protocol has been introduced by Leonard Kleinrock (University of California–Los Angeles).

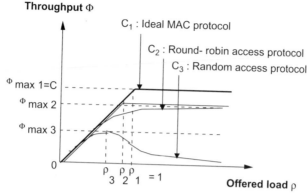

Figure 7.8   MAC protocol efficiency (single-server case).

with ρ up to $\Phi_{max1} = 1$. Realistic MAC protocols only approach this ideal behavior. For instance, curves C_2 and C_3 refer to MAC protocols based on round-robin access or on random access, respectively.

- In the case of round-robin protocols (e.g., token ring protocols), the upper bound of Φ_{max2} is strictly under 100%, due to the fraction of time during which no information is sent on the medium.
- In the case of random access protocols (e.g., CSMA/CD protocols), the upper bound Φ_{max3} may be quite bad because of collisions. The value of Φ reaches zero when ρ tends to infinity. The reason why Φ_{max} decreases for offered loads greater than a certain value ρ_3 is due to the fact that from this value, the MAC protocol spends more time trying to solve collisions than in sending packets successfully on the medium.

7.5.2   Mean Access Delay

Mean access delay W is defined as the expectation of the time elapsed for each packet between its instant of generation and its instant of transmission (with success) on the medium. Mean access delay W is determined under the same assumptions as for the evaluation of Φ_{max}. Fig. 7.9 depicts the fluctuation of W versus ρ for an ideal MAC protocol (curve C_1), for a round-robin access protocol (curve C_2) or for a random access protocol (curve C_3). In the case of the ideal protocol, W is characterized by a minimum W_{min} under low load ($W_{min} = 0$) until $\rho < 1$ and which becomes infinite when $\rho = 1$ (overload situation with the absence of steady state). Again, we may conclude from Fig. 7.9 that ideal MAC protocols do not exist. Let us consider the case of token ring protocols. Let T_{prop} be the time required by a frame to propagate all around the ring. The minimum access delay W_{min2} in that case corresponds to $T_{prop}/2$. This minimum is $W_{min3} = 0$ in the case of CSMA/CD. As in Section 7.5.1, let us recall that we consider here only the case of LANs, or MANs, enabling a single successful access at the same time.

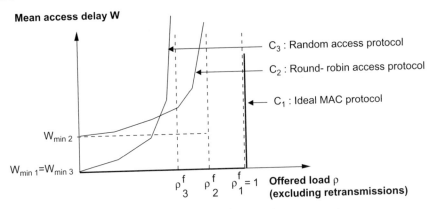

Figure 7.9   Mean access delay (single-server case).

7.5.3 Power of a MAC Protocol

One notices that maximizing Φ_{max} and minimizing W_{min} for the same MAC protocol is a difficult objective. Many investigations have shown that improving Φ_{max} implies degrading W_{min}, and vice versa. Thus, in the case of token ring protocols, we see through Figs. 7.8 and 7.9 that Φ_{max2} is high under high load conditions, but W_{min2} is not optimized under low load. Conversely, in the case of Ethernet (CSMA/CD protocol), W_{min3} is null at low load, but Φ_{max3} is not optimum under high load. Leonard Kleinrock summarizes this double objective by means of the concept of power P of a MAC protocol. The parameter P can be defined as the ratio Φ_{max}/W_{min}. A way to optimize P is to replace the single-server approach to the first-generation LANs by a multiple-server approach.[*] The expression *single-server approach* refers to the queueing model of a MAC protocol that enables successful transmission of a single station at a time. Token ring and Ethernet, which are typical examples of single-server MAC protocols, may be modeled by means of an M/G/1 queueing system. Fig. 7.10a describes an M/G/1 queueing system with multiple queues. The N queues of the model represent the N nodes of the network. The global generation rate $\lambda_1 + \lambda_2 \cdots + \lambda_N$, representing the set of new generated packets and of retransmitted packets is modeled by a Poisson process.[†] The fact that at most a single packet is served at a time is modeled by a single server in the system.[‡] The running time of the MAC protocol for each packet is modeled by a general service distribution.[§]

---

[*]The reader not familiar with queueing theory is referred to [5].
[†]The letter "M" in M/G/1 refers to this property.
[‡]The number "1" in M/G/1 refers to this second property.
[§]A general service distribution is characterized in general by its average duration and its standard deviation. The letter "G" in M/G/1 refers to this property.

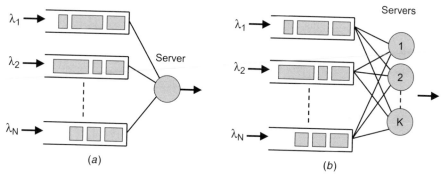

Figure 7.10   (*a*) M/G/1 and (*b*) M/G/K queueing systems.

The expression *multiple-server approach* refers to MAC protocols that enable simultaneous multiple transmissions without collisions on the medium. *Multitoken rings*, also known as *slotted rings*, are a typical example of such systems. They are created by dividing the length of the ring into an integer number of slots (let's say K) of fixed duration. A busy bit (corresponding to a form of token) is inserted at the beginning of each of these slots. An active node on the ring detecting a slot with a busy bit equal to zero may insert a packet in the payload of this slot. When the slot has been filled in, the busy bit is set to "1." The destination node extracts the packet from the slot payload and resets the busy bit to zero. Such a network can be modeled by means of an M/G/K queueing system (see Fig. 7.10*b*). Again, the global generation rate $\lambda_1 + \lambda_2 + \cdots \lambda_N$ representing the set of generated packets by the N nodes of the network is modeled by a Poisson process. The fact that K slots are accessible simultaneously is modeled by means of K parallel servers with general service distribution.

Compared to the single-server approach, the multiple-server approach enables an increase of power P of a MAC protocol. Indeed, in the case of slotted rings, for example, the value of W_{min} is equal to T_{prop} /2.K and the value of Φ_{max} is closer to 100% than in the case of token rings, due to the reduced fraction of time during which no information is sent on the medium. The multiple-server approach has been adopted in most of the gigabit LANs and MANs developed during the 1990s (DQDB, Metaring, FDDI-2, Hangman, CRMA, etc.) [6]. The DPT and RPR MAC protocols described in Sections 7.6.2.3 and 7.6.2.4, respectively, are of these types.

7.5.4   QoS Provisioning

In terms of volume, the data traffic, and more specifically, the Internet traffic, is becoming more predominant in carrier networks. This evolution is perceptible in the metro and access networks. The current format of IP packets is Version 4 (also called IPv4). Fig. 7.11 shows the format of an IPv4 packet. Without getting into the meaning of all the fields of this format, let us only consider that the source and destination addresses are coded in 4 bytes and that the variable size of the packet's

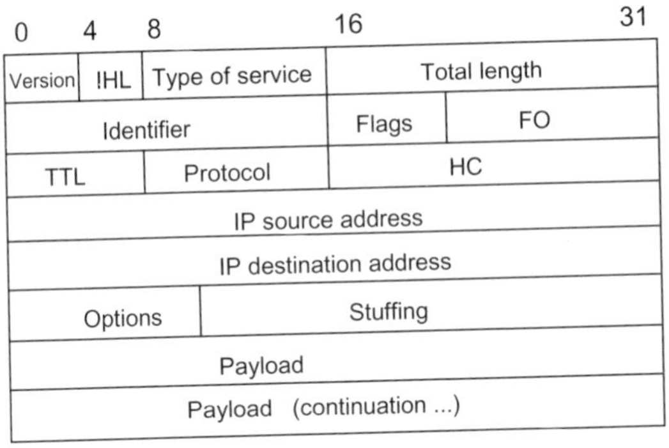

0 4 8			16		31
Version	IHL	Type of service	Total length		
Identifier			Flags	FO	
TTL		Protocol	HC		
IP source address					
IP destination address					
Options			Stuffing		
Payload					
Payload (continuation ...)					

Figure 7.11   IPv4 packet format.

payload enables transport of TCP segments.[*] In the IPv4 packet header, a specific field of 8 bits is called TOS (type of service), which enables to differentiate the service discipline and the buffer management strategy to be applied to the packet in an IP router.

7.5.4.1 QoS Parameters in the Metro and in the Access   During these last 10 years, several investigations have aimed at providing quality of service (QoS) in IP networks to support real-time traffic (voice and video) in addition to asynchronous data traffic. For that purpose, IP network carriers must be able to propose to service providers and to the end users different grades of QoS by means of service level agreements (SLAs). In packet networks, SLA specification relies on four QoS parameters:

1. *Availability.* This parameter corresponds to the capability of the network to return to the operational state after the cable cut or a node's failure. In existing optical networks, availability is guaranteed due to automatic protection switching built into the SONET/SDH equipment. Availability can be quantified by means of the mean time to repair (MTTR).

2. *Reliability.* A network may induce several errors due to the noise introduced by optical links or by the various equipment that is used along the path. Errors may also occur in case of buffer overflow. In IP networks, reliability may be offered by the TCP end-to-end protocol. The dynamics of the TCP protocol in case of error recovery is incompatible with the constraints of real-time services such as interactive voice. In that case, TCP is replaced by the UDP (user datagram protocol) or RTP (real-time protocol) transport protocols.

3. *End-to-end delay.* Real-time applications such as interactive voice or interactive video require stringent end-to-end delays. The average end-to-end

[*]In most case, IP packet size distribution is trimodal, with peaks around 40 bytes, 600 bytes, and 1500 bytes.

delay of the packets must remain under a certain threshold. The variance of end-to-end delay, also know as the *jitter*, must also remain under a certain threshold. These two constraints must be satisfied to facilitate voice or video decoding at the receiver node.

4. *Network capacity.* Each application requires a specific guaranteed minimum bandwidth or a specific mean bandwidth. These bandwidth requirements depend directly on the available capacity of the network on a continuous basis.

Availability, reliability, and end-to-end delay are partially correlated parameters. Their provisioning in IP networks may be carried out by means of four types of mechanisms:

1. *Policing.* Unlike best-effort IP networks, new-generation IP networks must be able to control the traffic generated by the end users. This control can be carried out by means of admission control, rate control (policing), or traffic shaping.

2. *Marking.* According to a given SLA, the packets generated by an end user may be considered compliant or noncompliant by the access router. In case of nonconformity, the packets are marked. A marked packet may suffer from a degraded service within the network in reference to the negotiated SLA.

3. *Congestion avoidance.* To prevent any buffer overflow within IP routers, nonconforming marked packets may be discarded.

4. *Queueing.* The internal architecture of IP routers may be designed to assign different queues to each SLA. IP packets may also be seen as flows (e.g., in the case of voice over IP). A per flow queueing is then implemented in the routers.

Three successive enhancements have been proposed within IETF to introduce QoS in IP networks: integrated services (IntServ), differentiated services (DiffServ) and multiprotocol label switching (MPLS). The IntServ approach is abandoned today due to its lack of scalability. MPLS that aims at guaranteeing QoS requires the implementation of a control plane within IP networks. In the context of MANs, both the DiffServ and MPLS approaches have been considered.

7.5.4.2 DiffServ Approach   The DiffServ approach aims at classifying IP packets according to different service level agreements (SLAs). According to its SLA, a packet is stored and served on the basis of specific queueing and scheduling strategies within a DiffServ-compatible router. A *DiffServ cloud* is a set of DiffServ routers in which we distinguish edge routers and internal routers. The role of an edge router is to classify the IP packets it receives from various non-DiffServ routers into a small number of aggregated flows. Packet classification is based on the interpretation of the type-of-service (TOS) byte associated to each IP packet header. For instance, an enterprise may negotiate with its ISP a given SLA.

According to the value of this SLA, the first 6 bits of the TOS field of the packets generated by the enterprise's router are coded with a specific value. This value, called the *DiffServ code point* (DSCP), is interpreted in edge routers in such a way that the corresponding packet is served with a specific forwarding by each internal router. This forwarding behavior is called per-hop behavior (PHB). Two PHBs have been specified by the IETF:

1. *Expedited forwarding (EF-PHB)*. The EF-PHB guarantees a minimum rate for the aggregate EF flows sharing the various output ports of a router. Inside a DiffServ cloud, EF-PHB also enables low loss, low latency, and low jitter. For that purpose, an absolute priority is provided to the EF packets compared to AF packets for access to an output port.
2. *Assured forwarding PHB (AF-PHB)*. The AF-PHB considers four levels of best-effort services called AF1 to AF4. Unlike EF-PHB, AF-PHB offers different levels of forwarding assurances to IP packets. A certain amount of buffer space and bandwidth is allocated to each AF class in the internal routers by means of the random early discard (RED) mechanism.

In IP networks, traffic classification may be carried out at various levels of the OSI model: at the application level (by means of the user's identity, the URL, etc.), at layer 4 (by means of TCP or UDP ports), at layer 3 (by means of IP packet addresses, the TOS fields, etc.), and at layer 2 (by means of the IEEE 802.1p priority protocol or the IEEE 802.1q VLAN (virtual LAN) identifier). Updating an IP network for DiffServ service provisioning is a relatively easy and inexpensive operation. Indeed, DiffServ only requires a modification of the internal architecture and functionalities of IP routers, especially in terms of service disciplines and buffer management. DiffServ does not mandate an update of all the routers of an autonomous system. Indeed, classical IP routers may interoperate with DiffServ routers. In Section 7.6.1.3 we describe a possible implementation of the DiffServ service over a WDM all-optical MAN.

7.5.4.3 MPLS Approach The basic objective of MPLS is to provide QoS guarantees in IP networks. In comparison to DiffServ, updating an IP network with MPLS is more complex and more costly. Indeed, MPLS requires the introduction of a control plane in all the routers of an autonomous system (AS) or in all the routers of a subset of an AS to define an MPLS-compatible cloud. This control plane relies on a signaling protocol, the RSVP (resource reservation protocol) protocol, which enables the setup and release of virtual circuits called LSPs (label-switched paths). The main objective of MPLS is to facilitate traffic engineering in IP networks. Traffic engineering refers to *intelligent routing*, which consists of replacing, if necessary, shortest-path routing by constraint-based routing. On the basis of the dynamic traffic matrix and the fluctuating load of the IP-MPLS cloud, MPLS facilitates the satisfaction of some of the QoS criteria mentioned in Section 7.5.4.1. Two types of routers are distinguished in MPLS clouds, *internal routers* (called LSRs for label-switched routers) and *edge routers*.

Most of the intelligence of MPLS is implemented in the edge routers. Edge routers that receive traffic from classical IP routers proceed to traffic classification and aggregation and to forwarding equivalence Class (FEC) assignment. Thus, using RSVP, an edge router at the border of an MPLS cloud is able to decide on the setup and release of LSPs with QoS adapted to different IP flows. The FEC assignment to incoming IP packets is carried out by means of a label assignment mechanism. A label that may be compared to a virtual channel identifier is made up of 20 bits and is associated with each IP packet within the MPLS cloud. Like ATM virtual channel identifiers, labels have a local meaning. MPLS enables association of several levels of labels to a packet using a label stacking mechanism. Within the MPLS cloud, only the most recent label located at the top of a list of labels is analyzed by an LSR. Each LSR replaces layer 3 IP routing by layer 2 forwarding. Currently, IP routers are coupled with ATM switches to implement this forwarding mechanism. When an IP packet leaves an MPLS cloud, its labels are progressively destacked according to a LIFO (last in first out) discipline. When a packet reaches an egress edge router, all its labels have been removed and it is again under the control of classical IP routing.

7.6 OPTICAL METROPOLITAN NETWORKS

As mentioned in Section 7.4.2, broadcast and select topologies have been widely investigated for metro networks. Essentially, two types of networks have been considered successively: WDM passive optical stars and WDM rings. In a passive optical star, nodes are linked together physically by means of a passive optical tree. Modulated optical streams generated by the various nodes of the network are first combined by means of passive couplers. The set of these optical streams is then duplicated by means of passive splitters in order to be received by all the nodes of the network. Finally, each receiver selects among the various optical streams those that are designated for it. WDM rings also correspond to a broadcast and select medium, a transmitted optical stream being detected automatically by all the other nodes along the ring. Due to the state and cost of the technology, WDM rings available on the market are based on active nodes that use an O/E/O (optical-to-electric and electric-to-optical) conversion. Between WDM star and WDM ring, only the former currently enables all-optical transparency. Several investigations and testbeds have recently shown the feasibility of all-optical WDM rings using optical add–drop multiplexers (OADMs). Such all-optical rings should be the best solution for multiservice MANs. However, they remain economically unattractive, at least for the short term.

7.6.1 Passive Optical Stars

Figure 7.12 is an example of passive optical star network. Optical star networks are based on 1×2 or 2×1 passive optical couplers. Optical transmitters and optical receivers are represented on the left-hand side and on the right-hand side of the

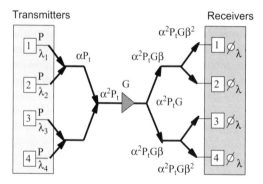

Figure 7.12   Passive all-optical star single-hop MAN.

figure, respectively. The parameters α and β stand for the attenuation of a 2×1 and a 1×2 coupler, respectively. Four transceiver combinations are possible depending on whether the transmitters are fixed (FT) or tunable (TT) and the receivers are fixed (FR) or tunable (TR). The option FT + FR is estimated to be not economically feasible because of its cost and lack of scalability. Several types of technologies have been considered for the tuning of optical transmitters and receivers (electro-optic, acousto-optic, mechanical, etc.). Depending on the technology adopted, tuning times of between a few nanoseconds and a few milliseconds are achievable. The proposed configurations assume either tunable transmitters and tunable receivers (TT + TR) or fixed transmitters and tunable receivers (FT + TR). The FT + TR option is motivated primarily by the good agility of tunable receivers and the low cost of fixed transmitters. Let N stand for the number of nodes in the network. In the early versions of WDM passive optical stars, N was quite limited (a few tens at the maximum) due to power budget constraints. After 1995, this lack of scalability of the optical stars was solved by the introduction of erbium-doped fiber amplifiers (EDFA) at the center of the star (see Fig. 7.12).

Passive optical star topology enables a fair power budget which is identical between any source i and any receiver j. In our example, this power budget is equal to the ratio of the received power P_r with the transmitted power P_t. One has $P_r/P_t = \alpha^2 G \beta^2$, where G stands for the gain of the optical amplifier. If one precludes the FT + FR combination, any other transceiver combination is subject to collision. Indeed, in the case of FT + TR, two transmitters sending a packet simultaneously towards the same receiver cannot be received simultaneously by this receiver. In the case of TT + FR, two transmitters sending a packet simultaneously to the same receiver will collide in the first half of the optical star. Finally, the TT + TR combination is also subject to collision, where several nodes want to send a packet simultaneously to the same receiver. Ideally, according to an average traffic matrix, the best combination consists of the optimum number of tunable transmitters and tunable receivers to implement in the nodes of the network. In the example of Fig. 7.12, each node i is equipped with a fixed transmitter on λ_i and with a tunable receiver covering $(N-1)$ of these wavelengths. In most of the testbeds, a data rate of 2.5 Gbps on each optical channel has been adopted.

7.6.1.1 MAC Protocols for Passive Optical Stars

In this section we assume that FT + TR (case of the example in Fig. 7.12). Let us call $\Lambda = \{\lambda_1, \lambda_2, \ldots, \lambda_N\}$ the set of all the wavelengths used in the network. Several medium access control protocols have been proposed during the last 10 years to prevent collisions in passive optical star topologies [7–10]. These proposals are either connection-oriented or connectionless. In both cases, either an in-band or out-of-band signaling channel is required.

1. *In-band signaling.* A node i that wants to communicate with a destination node j sends a request for transmission periodically to this destination node on wavelength λ_i. At this instant, node j may be either inactive or active. If it is active, node j has its receiver tuned on to a given wavelength λ_j different from λ_i and ignores the request. If it is inactive, node j scans cyclically $(N-1)$ wavelengths of the network. Once it detects a signal on λ_i, node j fixes its tuning on λ_i and sends back an acknowledgment to node i, which has already tuned its own receiver on λ_j. Once node i receives this acknowledgment, a symmetrical data transfer without collision may be initiated between nodes i and j. The main drawback of the in-band signaling approach is the inherent delay in connection establishment (at least a round-trip time). For highly bursty traffic sources such as Internet traffic, the MAC protocol efficiency Φ_{max} (see Section 7.5.1) may be degraded if the duration of a burst is lower than the round-trip time through the optical star. TCP segment size is in most cases less than 2000 bytes. Thus, the duration of an IP packet transmission is at most 6.4 μs. Assuming an optical signal velocity of 5 μs/km, the network size must be at most of 640 m. The typical range of a MAN being at least 100 km, this in-band signaling channel approach is unsuitable. One advantage of in-band signaling is that it enables variable-size packet transmission.

2. *Out-of-band signaling.* Each node i is equipped with a fixed transmitter and a fixed receiver on λ_s corresponding to the out-of-band signaling channel. Whenever a node has a packet pending for transmission, it sends a reservation request on λ_s to all the other nodes of the network. On the basis of a distributed wavelength allocation algorithm, each station is then able to determine if it can send a packet, and on which wavelength. To prevent any contention on the signaling channel, access to the wavelength λ_s is based either on a TDMA (time division multiple access) scheme or on a CDMA (code division multiple access) scheme. In most of the proposals published in the literature, the signaling channel and the data channels are synchronized and based on fixed time slots of the same duration. The dynamic allocation scheme (DAS) based on a random scheduling algorithm (RSA) proposed in [11] and [12] is one of the typical examples of MAC protocols that have been developed at the beginning of the 1990s for passive optical stars with out-of-band signaling. We describe below the basic principles and the limitations of this protocol.

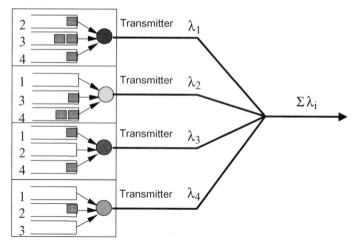

Figure 7.13   Node configuration in the RSA MAC protocol.

Fig. 7.13 illustrates a configuration of the nodes when the RSA protocol is implemented in a network with $N = 4$ nodes. Each node owns two laser diodes, one on its characteristic wavelength λ_i for data transmission and the other one on the signaling channel λ_s to request transmission. Each node implements per-destination queueing (in our example, each has three distinct buffers).

The principle of the RSA protocol is described in Fig. 7.14. The duration Δ of a data slot is equal to the duration of a signaling slot. A signaling slot is divided into

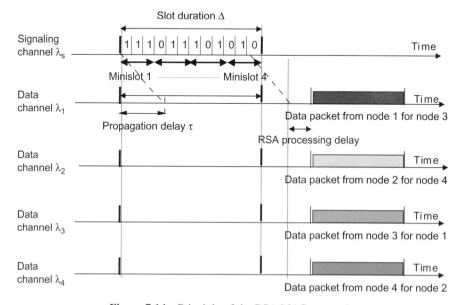

Figure 7.14   Principle of the RSA MAC protocol.

N minislots, each of these minislots being made up of $N-1$ bits. Minislots are indexed from 1 to N. By convention, the $N-1$ bits of the nth minislot correspond to the $N-1$ queues of node n. For instance, let us consider the second minislot ($n=2$) in our example of Fig. 7.14. The three bits of this minislot indicate the state of each transmission queue in node $i=2$. The value of the three successive bits 0, 1, 1 of the minislot aims at informing all the nodes of the network that node 2 has no pending packet for node 1 and at least one pending packet for nodes 3 and 4. Fig. 7.14 describes the state of the signaling channel when the state of the network is the one given in Fig. 7.13. All the nodes of the optical star are synchronized by equalizing (artificially) propagation delays from any sender to any receiver on the basis of the most distant nodes. For that purpose, a network initialization is carried out to evaluate this upper end-to-end delay bound τ. On the basis of the measured value of τ, one determines the additional delay $T_{\text{prop}}(j)$ to implement in each receiver j by means of electronic buffering. Thus, slot by slot, each node i is regularly informed of the pending packets in the other nodes of the network. Once all the nodes have received the current signaling slot, they run separately the same random generator seed, of which the input, is the value of the received signaling slot. Let T be the set of the transmitters: $T=\{t_1, t_2, \ldots, t_N\}$ and R be the set of the receivers: $R=\{r_1, r_2, \ldots, r_N\}$. The algorithm associated with this random generator seed is as follows:

1. $k := 1$.
2. Choose randomly a transmitter t_k among those that are active ($t_k \in T$).
3. Choose randomly in the node with transmitter t_k a nonempty queue corresponding to receiver r_k ($r_k \in R$).
4. $k := k + 1$.
5. Repeat until $k := N$:
   (a) Choose randomly a transmitter t_k among those that are active, such as
   $t_k =\in T - \{t_1, t_2, \ldots, t_{k-1}\}$.
   (b) Choose randomly in the node with transmitter t_k a nonempty queue corresponding to receiver r_k, such as $r_k \in R - \{r_1, r_2, \ldots, r_{k-1}\}$.
   (c) $k := k + 1$.

At the end of this algorithm, each node i knows which of its pending packets are allowed to be sent on its proper wavelength λ_i. As an example, let us consider the possible results of the RSA corresponding to the signaling slot shown in Fig. 7.14. It is possible to describe by means of a bipartite graph the state of the transmission queues of the different nodes. In Fig. 7.15, each vertex t_i or r_j corresponds a transmitter or a receiver, respectively. An edge exists from t_i to r_j if there is at least one pending packet for receiver r_j in the node with transmitter t_i. To prevent any contention, the MAC protocol has to select a fraction of the required edges in such a way that at most only a single edge arrives at each receiver.

Fig. 7.16 describes a possible result of the RSA. We see that in the next data slot, nodes 1, 2, 3, and 4 are allowed to send a packet to nodes 3, 4, 1, and 2,

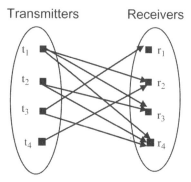

Figure 7.15   Bipartite graph associated with a signaling slot.

respectively. Meanwhile, the RSA does not ensure an optimum matching between transmitters and receivers. Fig. 7.17 illustrates another possible result of the RSA associated with the signaling slot of Fig. 7.14. We see that due to the random nature of the RSA, edges may be selected in an order such that some transmission requests

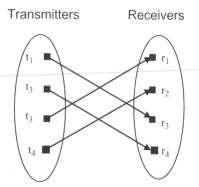

Figure 7.16   Possible result of the RSA algorithm.

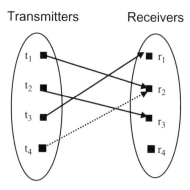

Figure 7.17   Nonoptimum matching of the RSA algorithm.

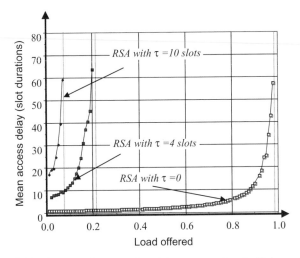

Figure 7.18   Impact of propagation delay τ on protocol efficiency $\Phi_{\max}$.

cannot be satisfied. In the example of Fig. 7.17, transmitter t_4, the last one to be selected by the algorithm, cannot send a packet to receiver r_2.

7.6.1.2   Network Performance   The RSA suffers from two main drawbacks: the cost of its signaling channel and its strong dependency on propagation delays. Thus, the maximum number of nodes $N_{\max}$ that may be connected on a network is = $[L^{+0.5}]$, where L stands for the size in bits of the signaling slot. Simulation results (see Fig. 7.18) show the disruptive effect of propagation delay τ on protocol efficiency $\Phi_{\max}$. If τ is neglected, $\Phi_{\max} = 98\%$, which means that each optical channel is used at almost 100% of its capacity. With $\tau = 4$ slots or $\tau = 10$ slots, this efficiency falls down to 20% and 8%, respectively. The assumptions considered for these simulations are those specified in Section 7.5.1.

7.6.1.3   Passive Optical Stars Instead of DiffServ Cloud   Several improvements have been proposed to reduce these drawbacks [13–16]. These new protocols help overcome the scalability limitations of the network by using a more sophisticated signaling slot configuration. By pipelining the transmission requests, it is possible to remove the need for a node to resend its request bit slot by slot until a successful transmission occurs. At last, some MAC protocols have been extended to enable variable-size packet transmission, where contiguous slot reservation is carried out. The most recent versions of the protocols proposed for WDM optical stars aim at integrating real-time services with data services [17–19]. In [20], such an integration is obtained in strict conformance with the DiffServ specifications of the IETF for Internet traffic (cf. Section 7.5.4.2). The basic idea of this proposal

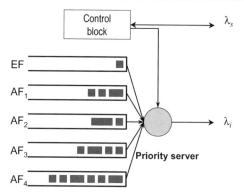

Figure 7.19 Node configuration for a DiffServ-compatible passive optical star.

assumes that each node connected to the optical star is an edge router of a DiffServ cloud. Figure 7.19 describes a node configuration in which several queues are considered: one queue for the aggregated EF traffic, and distinct queues for the four AF traffic classes.

In each node, the EF aggregated traffic benefits from a minimum specified bit rate. Within the same node, this guaranteed minimum rate is independent of concurrent AF traffic sources. This means that an EF packet is always served before an AF packet. Within each AF class, packets are marked with one of three possible dropping probability levels. In case of congestion, this dropping probability determines the relative importance of the packet within the class. The forwarding of an AF packet of class k on the optical star network depends on three criteria:

1. The amount of resources allocated to class k
2. The current load of class k
3. In case of congestion within class k, the dropping probability of the packet

In internal or edge routers supporting DiffServ, the random early discard (RED) mechanism is implemented to offer a certain amount of buffer space and bandwidth to each AF class. Fig. 7.20 describes the principle of the RED policy for a given AF queue. For each AF queue of class k, one associates two thresholds, *min-k* and *max-k*. The RED algorithm calculates the average queue size q_a at each packet arrival time using a low-pass filter with an exponential moving average. If q_a is lower than *min-k*, the packet is accepted in the queue. If q_a is larger than *max-k*, the packet is dropped. If $min\text{-}k < q_a < max\text{-}k$, the packet is dropped with probability p, p increasing linearly from 0 to *Pmax*. In [20], these rules have been mapped on the MAC protocol of a passive optical star to satisfy the IETF DiffServ specifications.

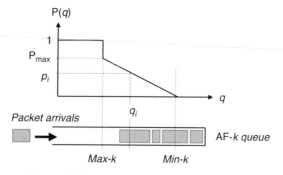

Figure 7.20   Principle of the RED algorithm.

7.6.2   Dual Counterrotating Rings

7.6.2.1   MANs Based on WDM SONET/SDH Rings   In Section 7.2 we discussed that most of the metropolitan area networks used in carrier networks are based on SONET/SDH rings with electric add–drop multiplexers (ADMs). Fig. 7.21 illustrates the typical configuration of a MAN of that type working at 622 Mbps. In Section 7.4.2 we noted that present-day SONET/SDH equipment has been designed essentially for the transport of TDM traffic. Today, enterprises are connected to MANs by means of TDM multiplexers. A metropolitan SONET/SDH ring is between 100 and 200 km long. It is itself connected to a higher-capacity ring (called a *regional ring* in the Fig. 7.22) by means of an electrical cross-connect (EXC). This regional ring is also a SONET/SDH ring, operating, for instance, at the

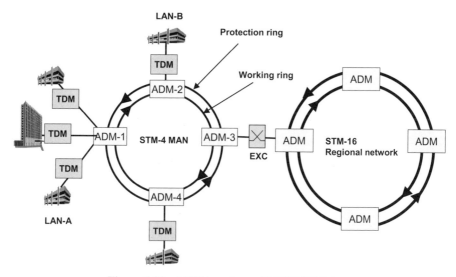

Figure 7.21   MAN based on a SONET/SDH ring.

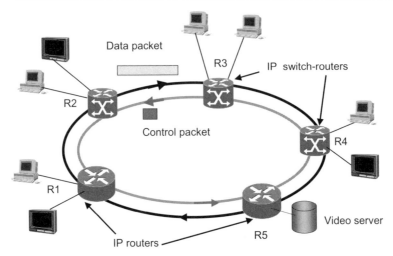

Figure 7.22   Configuration of a DPT ring.

STM-16 rate (2.5 Gbps). Currently, dual counterrotating rings is the most cost-effective resilient physical topology compared to linear meshed topologies.

In SONET/SDH networks, point-to-point links between multiplexers may be protected against cable cuts (and also again the failure of an optical interface) by means of two types of automatic protection switching (APS) techniques called "1 + 1" and "1:1". The 1 + 1 and 1:1 protection techniques and their generalization 1 + N and 1:N have been standardized by the ITU-T in the G.783 recommendation. A restoration time under 50 ms is normally required so that PSTN calls are not lost during protection switching. The various techniques that have been developed for protection and restoration of SONET/SDH are described in detail in Chapter 15.

Semipermanent point-to-point connections are established on SONET/SDH rings. For instance, a virtual private network (VPN) service can be offered by a carrier to interconnect the distant LANs of an enterprise (LAN-A and LAN-B in Fig. 7.21). The MAC frames generated by LAN-A are inserted in TDM frames (e.g., E1 or T1 frames) to reach the nearest add–drop multiplexer (ADM-1). Within this ADM, E1/T1 frames are encapsulated into virtual containers (VC-12 or VC11), themselves encapsulated in higher-order virtual containers (VC-4 or VC-3). These virtual containers are finally inserted in STM-4 frames, which circulate periodically on both rings. The receiving ADM drops these higher-order VCs from the SDH frames received. The original PDH frames (E1/T1) are then sent to the destination LAN-B.

Currently, a few SONET/SDH MANs have been upgraded to support multiple wavelengths. Globally, the protection/restoration principles we have described in this section remain unchanged. Since the inception of WDM, it is as though several independent SONET/SDH rings of a different wavelength were superimposed on the same optical fiber infrastructure. Today, several prospective investigations are carried out to investigate the feasibility of all-optical WDM SONET/SDH rings in which electrical ADMs are replaced by OADMs [21,22].

7.6.2.2   *Limitations of SONET/SDH Rings in the Metro*   Carrier networks today transport a large part of data traffic for distant LAN interconnection, Internet access, client server applications, or data storage. In this context, SONET/SDH rings as they are used in the metro pose several limitations.

1. *Dynamic access and bandwidth allocation*. The TDM nature of SONET/ SDH is unsuited to the variable bit rate of data applications. TDM assumes fixed bandwidth reservation for point-to-point connections. Because SONET/SDH is circuit oriented, the bandwidth reserved for a point-to-point link is dimensioned either on the basis of the peak bit rate of the connection (which corresponds to over provisioning) or on the basis of an intermediate value between the peak bit rate and the mean bit rate of the connection. In the first case, waste in capacity increases with traffic burstiness. In the second case, data may be lost within ADMs due to buffer overflow.

2. *Circuit provisioning delay*. The complexity and cost involved in point-to-point connection management on a SONET/SDH ring increases with the number of these connections. This is why PDH service provisioning through a SONET/SDH ring involves long lead times (several weeks to several months). The most recent SONET/SDH equipment is now able to provision automatically but partially point-to-point circuits. Indeed, carriers have to manually manage their proper engineering rules as soon as the number of connections exceeds a certain threshold if they want to optimize their network resources. In the case of Internet traffic, the duration of a circuit connection through the SDH infrastructure may be too long, compared to the duration of the transmission of a few IP packets. Traffic grooming in SONET/SDH networks, that is, traffic concatenation in virtual containers, may be a solution to this problem.

3. *Wasted bandwidth*. Ideally, at each instant, the unused bandwidth on the ring should be available to serve instantaneous traffic peaks. The bandwidth granularity offered by a carrier to a customer via such MANs corresponds to the granularity of the PDH hierarchy. This granularity is too coarse compared to the large variety of data rates that may be required by the end user. For instance, let us consider the interconnection of two distant 10 BaseT Ethernets at 10 Mbps. In Europe, this interconnection can be implemented, for instance, by means of six E1 tributaries, which correspond, in fact, to an 11.52 Mbps capacity. This interconnection can also be carried out by means of a single E3 tributary at 34 Mbps. The transport of IP packets or Ethernet frames over a MAN based on a SONET/SDH ring implies additional protocol conversions overhead. These overheads represent a useless waste for the corporate customers who have to pay the carrier in proportion to the volume of data transported. SONET/SDH protection may result in wasting up to 50% of the network capacity, such as with dedicated protection rings (DPRINGs) or subnetwork connection protection rings (SNC-P rings).

4. *Group communication*. Group communication is a type of service that did not exist when SONET/SDH was designed. Video or audio conferencing requires multicast communications that SONET/SDH rings are unable to support.

5. *Equipment cost*. One last issue that cannot be neglected is the high cost of SONET/SDH cross-connects or ADMs. As we discuss in Section 7.7.1, new-generation high-speed Ethernet switches have been considered recently as a promising alternative for the metro. At comparable data rates, such equipment currently offers a cost reduction of an order of magnitude compared to SONET/SDH line cards.

7.6.2.3   *Dynamic Packet Transport*

The dynamic packet transport (DPT) network proposed by Cisco [23] aims at reducing the limitations of SONET/SDH rings in the metro that have been outlined in Section 7.6.2.2. Like classical SONET/SDH rings, a DPT network is based on a dual counterrotating optical ring. The two rings of the network called *outer ring* and *inner ring* are both used for data transmission. The choice of the ring to send a packet is based on the shortest path in terms of the number of hops. Each node on the ring is either a first-generation IP router or a new IP switch router. Each input/output port of these routers is equipped with a DPT line interface card. The main difference between a SONET/SDH ring and a DPT ring is in the utilization of a MAC protocol implemented in the DPT line interface cards. This MAC protocol called spatial reuse protocol (SRP) enables dynamic access and dynamic bandwidth allocation. As specified in the IETF white paper titled "Dynamic Packet Transport Technology and Applications Overview" dating from 1999 (RFC 2892), the SRP protocol can be applied on top of various physical layers. In its current implementation, DPT relies on SONET/SDH framing applied to a mono-wavelength optical channel along each of the two rings. The typical line bit rate of a DPT ring corresponds to the STM-4 speed at 622 Mbps.[*] Extensions of the DPT principles to a multiwavelength dual ring are considered for further extensions. Fig. 7.22 describes the configuration of a DPT network.

The protocol stack adopted on each line interface card is proper to the DPT node configuration. The SRP protocol assumes a layer 3 switching capability at each node to insert (or add) or to extract (or drop) IP packets from the network. Thus, IP packets are encapsulated into SRP frames at the MAC layer. These SRP frames are themselves inserted as tributaries within SONET/SDH-equivalent frames. Figure 7.23 illustrates the format of an SRP frame.

The header of the SRP frame is made up of 32 bits that correspond to the following fields:

- *TTL (time-to-live)*. This 11-bit field carries the hop count. It is decremented at each hop. When TTL reaches 0, the packet is stripped from the ring.
- *RI (ring identifier)*. This bit indicates the ring used (inner or outer).

---

[*]Extensions of these rates to STM-16 at 2.5 Gbps or to STM-64 at 10 Gbps are announced by the vendor.

TTL (11 bits)		RI (1 bit)	D (1 bit)	Priority (3 bits)
Mode (3 bits)	Usage (12 bits)			P (1 bit)
Destination MAC address (3 × 2 bytes = 48 bits)				
Source MAC address (3 × 2 bytes = 48 bits)				
Protocol type (2 bytes)				
Payload (up to 9194 bytes)				
FCS (4 bytes)				

Figure 7.23   SRP frame format.

- *D.* This single-bit field indicates if the destination must strip the frame from the ring or not.
- *Prio (priority).* This 3-bit-long field enables eight priority levels for a data packet.
- *Mode.* This 3-bit field identifies the type of packet (data, or one of the possible control packets).
- *Usage.* This 12-bits field indicates the currently usable bandwidth for the node considered.
- *P (parity).* This field controls the validity of the previous bits.

The subsequent fields of the SRP frame header are the destination and the source addresses of the nodes along the ring. The standard 48-bit IEEE MAC address format is respected by these two fields. The "protocol type" field of 2 bytes specifies if the data packet contained in the payload is an IPv4 packet, an address resolution protocol (ARP), packet or SRP control packet. The data payload enables the transport of IP packets with a maximum transfer units (MTU) of 9194 bytes. An SRP frame ends with a 4-byte frame check sequence (FCS) that enables a receiver node to control the validity of the destination and source addresses, protocol type, and payload.[*] In unicast communication, the destination node is responsible for packet stripping from the ring. For instance, if router *R*1 has a packet for router *R*3, it uses the outer ring for its transmission. Unlike SONET/SDH rings, DPT rings enable multicast communications. Let us assume that a video server is connected to router *R*5; users connected to routers *R*1, *R*2, and *R*4 may receive the same video sequence from *R*5. In that case, frame stripping is carried out either by the sender *R*5 or by the last destination node on the ring (*R*4 in our example). This last option is adopted if the last destination node along the ring within the multicast group is very distant from the source node. In this situation, a specific time-to-live (TTL) stamp corresponding to the number of hops between *R*5 and *R*4 (i.e., 4) is assigned to the packet containing the video sequence. The DPT node architecture which is comparable to a slotted ring with buffer insertion [24], is illustrated in Fig. 7.24.

---

[*]Let us recall that the header of the SRP frame is itself protected by means of the P bit.

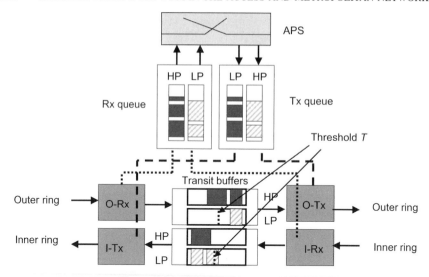

Figure 7.24   DPT node's configuration.

Two traffic classes are considered on the network: high-priority (HP) traffic and low-priority (LP) traffic. As shown in Fig. 7.24, an upstream packet arriving, for instance, from the outer ring is first recognized by the node on the basis of its header's value. Such an operation may occur only after an optical-to-electrical conversion has been carried out by the O-Rx. To decide if this packet has to be regenerated to the next node, or if it has arrived at its destination, intermediate buffering is necessary (transit buffers). If the packet needs to be retransmitted to the next node, an electrical-to-optical conversion is carried out by the O-Tx. Locally, two sets of buffers are used to store either received packets (Rx queue) or pending packets for transmission (Tx queue). A starvation phenomenon may occur if, for instance, router $R2$ becomes active with a packet for $R4$, whereas $R1$ is sending a very long file to $R3$ on the outer ring (see Fig. 7.22). In that case the SRP protocol must be able to allocate access to the outer ring fairly for both $R1$ and $R2$. For that purpose, the SRP protocol includes a fairness mechanism based on the use of control packets sent in the reverse direction (the inner ring). As mentioned above, both rings can be used simultaneously for data transmission. This means that control packets are also used on the outer ring in order to manage bandwidth balancing on the inner ring. A control packet is itself inserted in a SRP frame. Fig. 7.25 illustrates the format of an SRP frame containing a control packet. We recognize on this figure the 18-byte SRP frame header and the 4-byte FCS mentioned in Fig. 7.23. A specific TTL is assigned to control packets.

Control packets are considered high-priority packets and are transmitted only to adjacent nodes. The semantic associated with control packets corresponds to either protection switching messages, to topology discovery packets, or to the SRP fairness algorithm. The SRP protocol uses a fairness algorithm to share equally the available capacity for low-priority traffic among the various active nodes. This

TTL (11 bits)		RI (1 bit)	D (1 bit)	Priority (3 bits)
Mode (3 bits)	Usage (12 bits)			P (1 bit)
Destination MAC address (3 × 2 bytes = 48 bits)				
Source MAC address (3 × 2 bytes = 48 bits)				
Protocol Type (2 byte)				
Control version (1 byte)	Type of control message (1 byte)			
CRC (similar to TCP CRC; 2 bytes)				
TTL control (different from TTL of data packets; 2 bytes)				
Control packet payload (up to 9188 bytes)				
FCS (4 bytes)				

Figure 7.25   SRP control packet format.

available capacity is equal to the network capacity minus the bandwidth used by high-priority traffic. The basic principle of this fairness algorithm consists of a given node i sending a control packet to its upstream neighbor $(i-1)$ as soon as the low-priority transit buffer in this node i is full (congestion state). This control packet informs node $(i-1)$ of the low-priority use by node i. On the basis of this information, node $(i-1)$ adjusts its low-priority transmission rate to facilitate decongestion in node i. In practice, each node owns a counter that dynamically determines its low-priority traffic transmission credit.

Protection and restoration techniques called IPS (intelligent protection system) developed for the DPT network are quite similar to automatic protection switching (APS) adopted in SONET/SDH rings. These techniques do not use the SONET/SDH overhead bytes (MSOH, RSOH, POH) and do not need a reserved bandwidth for protection [25]. They are supposed to restore the network in case of fiber cut or of node's failure in less than 50 ms. Let us now describe the basic principles of the SRP protocol:

- A high-priority forwarded packet in the HP transit buffer is always regenerated first even in the presence of pending high-priority packets in the HP Tx queue.
- A high-priority packet pending for transmission in the HP Tx queue is sent prior to a forwarded low-priority packet in the LP transit buffer as long as this transit buffer is not full.
- A low-priority forwarded packet in the LP transit buffer is served even in the presence of HP packets in HP Tx queue if the LP transit buffer is full.
- When the LP transit buffer occupancy is not full and lower than a given threshold T, a low-priority packet in the LP transit buffer is served prior to pending low-priority packets in LP Tx buffer.
- When the LP transit buffer is not full, whatever the filling of this buffer relative to the threshold T, a low-priority packet in the LP transit buffer may

be transmitted if the low-priority credit of transmission of the station has not expired.

- If the transit buffer occupancy is not full and higher than T, and if the local node still has low-priority credits for transmission, a low-priority packet from the LP Tx buffer may be transmitted.

The SRP fairness algorithm aims at a fair share of the available capacity between the various low-priority active sources along the ring. This fair share is based on two counters (usage counters) implemented in each DPT equipment. These two counters dynamically control the number of transmitted low-priority forwarded packets and pending low-priority packets, respectively. A strong similarity exists between the SRP fairness algorithm and the global fairness algorithm (GFA) developed for the metaring Gigabit network [26,27]. In conclusion, the DPT technology is a proprietary solution for cost-effective resilient data-centric MANs.

7.6.2.4 Resilient Packet Ring and IEEE 802.17 Like DPT, the resilient packet ring (RPR) network is another approach for cumulating the benefits of a dual-counter rotating ring in terms of survivability with those of a MAC protocol in terms of dynamic bandwidth allocation. Unlike DPT, which appears as a mono-vendor technology, RPR is supported by various service providers and vendors. The IEEE (Institute of Electrical and Electronic Engineers), traditionally in charge of LAN and MAN standardization at the global level, is promoting the 802.17 standard inspired by the RPR technology [28]. In its basic principles, RPR networks are quite similar to DPT networks. Both networks use dynamic packet-switched oriented access to the medium. Both of them optimize bandwidth utilization by means of destination release and spatial reuse. Like the intelligent protection switching (IPS) of DPT networks, a specific service protection has been developed for RPR. This service aims at offering the same level of resiliency as the automatic protection switching (APS) of SONET/SDH (50 ms restoration duration) but with a much lighter protocol overhead. Like DPT, RPR uses both the rings to transport data. Similar to DPT, RPR enables multicast communication. The IEEE 802.17 working group has targeted the completion of the standard for March 2003.

7.7   OPTICAL ACCESS NETWORKS

For the last 10 years, various technologies have been competing for the market of high-speed access networks. During this period, most of the carriers have estimated that the cost of optical access systems was prohibitive compared to other access technologies, such as radio, copper wire, or coaxial cable. Indeed, two major drawbacks characterize optical access networks: the cost of civil engineering required to install the optical fiber and the cost of optoelectronic devices. Today, one can estimate that only the first of these two arguments remains really valid. With the massive deployment of DWDM point-to-point links in transport networks since 1990, the unit cost of optical devices and systems has considerably decined. Despite these difficulties, a group of seven telcos (France Telecom, Deutsche Telekom, British Telecom, NTT, KPN, Telefonica, and Telecom Italia) decided in 1995 to analyze the technical and economical feasibility of optical access networks. This association of major operators is known as the FSAN (full-service access network) initiative. Other members joined the FSAN initiative in 1996 (Bell South, GTE, Swiss PTT, and Telstra). The fact that several other major carriers joined the FSAN may be justified by two evolutions:

1. Toward the end of the 1990s, important advances were made in the field of optical devices and systems. For instance, the life duration of laser diodes and photodetectors increased considerably. Optical amplifiers with 40-dB gains, such as EDFAs, enabled longer-range use without O/E/O conversion. Optical tributaries could be inserted directly into optical fiber rings without any fiber cut, due to the availability of optical add–drop multiplexers (OADMs). The optical cross-connects became feasible due to wavelength converters minimizing the use of optical buffers. Finally, fiber Bragg gratings or dispersion shifted fibers became very useful to compensate for chromatic dispersion.

2. In terms of the market, the last mile of the carrier networks concerns essentially the residential users and small and medium enterprises[*] (SMEs). During the last decade, numerous SMEs have subscribed to ADSL services to get high-speed access to public data networks. Today, several of them estimate that in the near future (two or three years), they should need higher-performing solutions requiring access rates of several tens of Mbps in the downstream direction and a few Mbps in the upstream direction. In fact, full-rate ADSL performance is very promising, with 8 Mbps in the downstream direction and 800 Kbps in the upstream direction, respectively. However, such rates are almost unreachable, for two reasons. First, each subscriber line is unique and is more or less adapted to large spectrum signals.[†] Second, POTS splitters located at each end of the subscriber line for separating analog and digital signals imply strong limitations on the bit rate. This is the reason why the ADSL lite technology has been developed. ADSL lite enables around 1 Mbps and 300 kbps in the downstream and upstream directions, respectively. The VDSL (very high speed digital subscriber line) technology aims at very high bit rates (up to 50 Mbps for downstream traffic) but over a very short range (at most 1500 m). Many doubt today that such bit rates could be achievable on twisted pairs in the local loop, due to the serious problems of electromagnetic compatibility between VDSL and external radio systems. For all these reasons, one may observe that a lot of interest in optical access has emerged in the past few years.

[*]Let us recall here that metropolitan area networks provide high-speed access to data networks essentially for medium-sized or large enterprises.
[†]An ADSL modem needs a 1-MHz spectrum on copper wires.

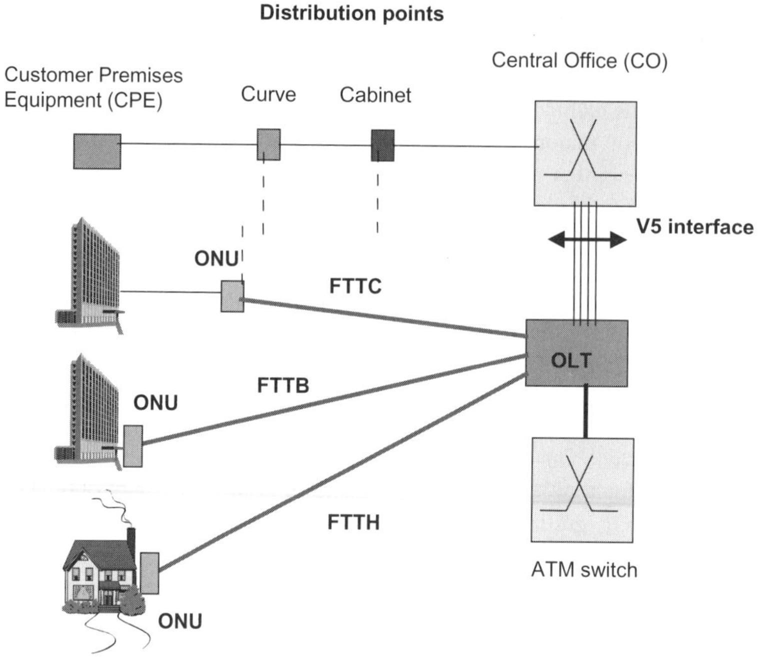

Figure 7.26   Various FTTx alternatives.

Bell Canada, Verizon, Qwest, and Singtel are some of the most recent new-comers to the FSAN initiative.[*] The first task of the FSAN members was to determine the optical localization of the two ends of an optical access network. Because optical access networks are designed to transport digital voice commu-nication, the head of the optical access network is located close to the central office. The localization of the other end of the optical access network has been the subject of many discussions. Should this termination be located at customer premises (in CPE), at the foot of the building, or at intermediate points along the existing copper infrastructure, such as the cabinet or the curves? Fig. 7.26 illustrates these three possibilities, known as FTTH (fiber to the home), FTTB (fiber to the building), and FTTC (fiber to the curb). The two terminations of an optical access network are the OLT (optical line termination) at the central-office side and the ONU (optical network unit) at the customer side.

Synchronous services (voice and video) as well as data services (TCP/IP) are assumed to be transported in a packet-switched mode. Depending on its topology, an optical access network may use upstream traffic concentration. In the case of dedicated point-to-point optical links, an OLT and an ONU must be dedicated to each end user. Such a configuration is too expensive, in terms of both civil

---

[*]The FSAN consortium group had 21 telecommunications operators at the time of publication of this book. It cooperates with various standarization bodies, such as ITU-T, ETSI, or the ATM Forum.

engineering and terminal equipment. As shown in Fig. 7.26, depending on the various options investigated, voice traffic is either forwarded to a central office via the V.5 signaling interface or to a packet switch. Although several options have been investigated, the FSAN members finally recommended an all-optical infrastructure based on a passive optical tree, also known as a passive optical network (PON). The three main benefits of PONs may be underlined:

1. A passive optical tree does not require any electrical equipment in the network itself. This helps minimize the maintenance cost of the entire system.

2. The tree topology enables sharing of a large fraction of the optical medium between several end users.

3. Despite the initial cost of civil engineering, once the fiber cable has been installed, the infrastructure remains usable for at least the next 10 years. It is extremely easy to upgrade the capacity of this infrastructure using WDM.

Within a short time, a consensus emerged among FSAN members to consider only the case of PONs connected directly to ATM switches. Historically, two versions of PONs have been developed successively. The first, A-PON for ATM over PON, assumes that all end-users share dynamically the same wavelength in the upstream direction. The most recent version is a simple evolution of the A-PON approach renamed B-PON (broadband PON). This new approach aims at underlining the fact that optical access systems are not limited to end-to-end ATM. They are also able to provide frame relay access, Ethernet access, video distribution (analog or digital), or high-speed leased-line services. We describe A-PON and B-PON in the following sections.

7.7.1 A-PON Access Systems

The FSAN initiative proposed a hybrid alternative combining a PON with several copper wire terminations [29]. At the time, ATM was the only standardized and implemented technology enabling multiplexing of different types of services with various quality-of-service (QoS) requirements. For this reason, the FSAN consortium decided to base PON access systems on ATM, explaining the origin of the A-PON acronym.

7.7.1.1 A-PON Physical Layer Figure 7.27 illustrates the configuration of an A-PON access system. The OLT is a modem located at the head of the optical tree. An ONU corresponds to a modem located at the low end of the optical tree. In the case of FTTB and FTTH, the optical infrastructure is deployed from the ATM switch to the customer premises. Fig. 7.27 refers more specifically to the case of FTTC access, which should concern the majority of residential users. In that case, the optical infrastructure typically ends at the level of a distribution point along the copper infrastructure (a cabinet or a curb). VDSL technology is seen as the ideal complement of an A-PON in the context of FTTC. The last mile between each ONU and customer premises reuses the installed copper wires. High-speed

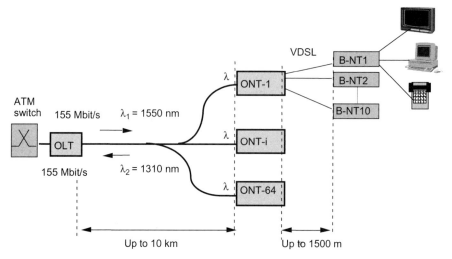

Figure 7.27   Passive optical network configuration.

transmission is achieved by means of VDSL modems installed at each end of the twisted pairs. Due to the attenuation constraints of passive splitters along the PON and the attenuation of the optical fiber, the range of a PON is either 10 or 20 km, according to the number of ONTs: 64 or 32, respectively. Several splitting points may be used between the OLT and an ONU, but the global attenuation must remain under 30 dB, as specified in the standard.

Full-duplex transmission is carried out by using two distinct wavelengths at 1310 nm and 1550 nm for upstream and downstream traffic, respectively (see Fig. 7.28). Two possible data rates are considered for downstream traffic, either 622 or 155 Mbps, whereas a single speed is considered for upstream traffic (155 Mbps). Once the optical fiber has been installed, a carrier uses two ranging operations:

1. *Distance ranging*. Distance ranging consists of evaluation of the round-trip time (RTT) between each ONU and the OLT. This distance can be deduced from the

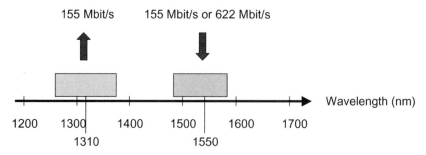

Figure 7.28   Spectrum utilization in an A-PON system.

received signal-to-noise ratio (SNR) of a pulse sent with a fixed power by the OLT. On the basis of the highest value of these measured RTTs, each ONU sets an artificial delay corresponding to a fixed queueing delay applied to both the upstream and downstream traffic. By this method, propagation delays between the OLT and each ONU are artificially equalized. We shall see in the next section the importance of distance ranging for the MAC protocol.

2. *Power ranging.* Power ranging aims at equalizing the SNR for each propagation path between the OLT and the various ONUs. If such a ranging procedure was not used, the transmission power of the ONUs being fixed and based on the worst case of the longest distance between the OLT and an ONU, the receivers of the nearest ONUs could be saturated. Power ranging thus helps in adapting the receivers' sensitivity. This adaptation is static in the downstream direction, SNR being fixed. In the upstream direction, the receiver of the OLT, which knows in advance which ONU is allowed to transmit, is able to adjust its sensitivity dynamically packet by packet.

Once the network has been initialized and becomes operational, another type of ranging must be carried out permanently at the OLT. Indeed, clock distortion being proportional to distance, the phase-locked-loop (PLL) of the OLT has to be updated for each burst received from the various ONUs. Downstream traffic from the central office to end users is naturally broadcast toward all the ONUs using TDM (time-division multiplexing). An encryption function such as DES (data encryption standard) is considered at the physical layer of an A-PON to provide confidentiality.

7.7.1.2   A-PON MAC Protocol   Upstream traffic generated by the various ONUs share the common capacity of the tree by means of a MAC protocol. Many investigations have been carried out in the design of such a MAC protocol, to satisfy three objectives:

1. Preventing collisions
2. Maximizing the medium utilization ratio (cf. Section 7.5.1)
3. Minimizing the perturbation in the time structure of the various ATM connections

For that purpose, a request-permit mechanism is considered (see Fig. 7.29). For each ATM cell pending for transmission within an ONU, a bandwidth request is sent to the OLT. On the basis of the instantaneous unused bandwidth and on the characteristics of the considered ATM connection, the OLT assigns to the considered ONT a permit. This permit specifies at which instant this ONT will be authorized to send its ATM cell. Such a scheduling of ATM cells in the upstream direction justifies the distance equalization mentioned in Section, 7.7.1.1.

On the one hand, the downstream traffic is made up of a continuous succession of 53-byte ATM cells. Permits are transmitted in bursts from the OLT to the ONUs in dedicated PLOAM (physical layer operation and maintenance) cells. On the other hand, upstream ATM cells are encapsulated in A-PON packets. An A-PON

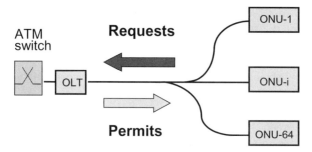

Figure 7.29   Request/permit of the A-PON MAC protocol.

packet format has been specified. The entire packet is 56 bytes long, including a 3-byte overhead followed by a payload containing an ATM cell. The 3-byte header of an A-PON packet is used first to resynchronize, burst by burst, the phase-locked loop of the receiver (the OLT). The same header is also used as a guard time between upstream slots.[*] One notices that downstream ATM cells do not require such an encapsulation and are sent directly to the ONUs. To facilitate the request-permit mechanism, a frame format has been specified by ITU-T for both upstream and downstream traffic. Fig. 7.30 illustrates the format of the G.983 frames. We see from the figure that assuming a symmetrical data rate of 155 Mbps, an upstream frame of size 56 ATM cells coincides with a downstream frame of size 53 APON packets, also called *slots*. Among the 53 upstream slots, one slot, called the MBS (multi-burst slot), is used for the transport of batches of requests.

The specification of dynamic bandwidth assignment on A-PON systems is a complex task if one aims at keeping the ATM connections conformant at the output

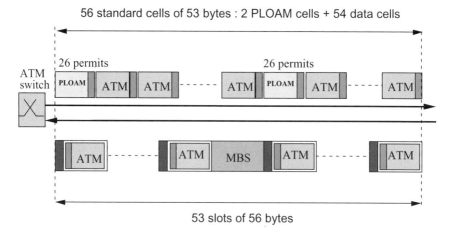

Figure 7.30   Upstream frame and downstream frame G.983 format.

[*] A slot corresponds to a 56-byte APON packet.

of the PON for upstream traffic. Indeed, an ATM cell may be detected in strict conformance with its traffic contract within an ONU and as nonconforming at the output of the OLT. Let us recall that such a conformance for both CBR and VBR connections is evaluated by means of the generic cell rate algorithm (GCRA). Nonconformance may be introduced on upstream traffic because of the disruptive effect of the MAC protocol on ATM connections. Let us consider, for instance, the case of a CBR voice connection. In principle, the cells belonging to this connection should leave the ONU with a strict periodicity corresponding to the peak cell rate of the CBR flow. These CBR requests cannot be sent with the same periodicity to the OLT but have to wait for the next MBS slot. Similarly, the permits generated at the OLT are also transmitted by bursts on the basis of the PLOAM cells periodicity. Both the request and permit mechanisms added to the multiplexing of requests and permits within the ONUs and the OLT, respectively, are at the origin of the perturbation introduced by the A-PON MAC protocol. Generally, it is well known from the design of ATM switches that buffer management strategies and queueing disciplines have a strong impact in case of tandem and parallel mulitplexing on the QoS of real-time connections. It has been shown in [30] and [31] that the use of time stamps associated with pending real-time cells within the ONUs enables considerable improvement in the conformity of these connections.

Network survivability is apparently an important aspect that has been neglected in the A-PON specification. Indeed, the entire PON system fails if a fiber cut occurs between the OLT and the first splitting point, and more generally, if a fiber cut occurs in any branch of the tree. This gap is today judiciously filled in the B-PON specifications.

7.7.2   B-PON Access System

As mentioned at beginning of this section, the broadband PON (B-PON) concept is an extension of the A-PON specifications. The various aspects of the broadband passive optical network (B-PON) have been standardized by the ITU-T [32–34] on the basis of the numerous contributions of the members of the FSAN initiative. The G.983.1 recommendation [32] specifies the physical layer characteristics of a B-PON system. The G.983.2 recommendation [33] specifies the management interface between the head and leaves of the optical tree. The most recent recommendation, G.983.3 [34], dating from 2001, specifies that an additional wavelength band may be used on the system: for instance, to broadcast a video channel or even for using several optical carriers to increase the capacity of the system. Even if an ATM-based transmission is considered at the physical layer, a B-PON system aims at providing various types of high-speed services, such as Ethernet access, video distribution or the equivalent of a leased line. According to the FSAN objectives, a B-PON aims at delivering multimedia services to both business and residential users up to 10 Mbps per user. Two major aspects differentiate a B-PON from an A-PON: a more sophisticated use of WDM and a survivable physical layer.

7.7.2.1 B-PON Survivability   Four possible topologies have been investigated within the FSAN to protect and to restore the optical tree in case of a fiber cut. Only

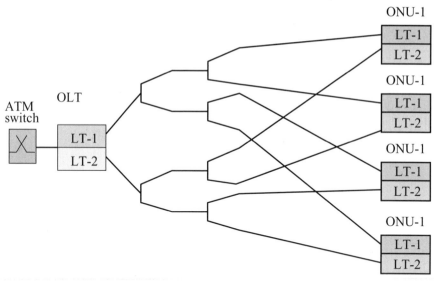

Figure 7.31 Possible strategy for B-PON survivability.

two of these topologies have finally been selected in the ITU-T recommendation. Fig. 7.31 illustrates one of the two alternatives. Protection is obtained by means of a full duplication of the optical tree. In each optical modem, either the OLT or the ONU is duplicated to provide a primary interface and a secondary interface. The protection scheme adopted is of the 1:1 type. This means that both the primary (working) and the secondary (standby) interfaces are used in a normal situation. By convention, only low-priority best-effort traffic is transported on the secondary infrastructure. A B-PON may accommodate simultaneously a mixed situation with protected and nonprotected ONUs with 50 ms of restoration time [35].

A few details referring to the physical layer have been added to the A-PON specification. For instance, each passive splitter induces a parasite reflectance. This reflectance must be such that for any physical configuration, the optical path loss remains under 30 dB. We have seen that in the case of A-PON systems, encryption at the physical layer on the downstream flow was considered to provide confidentiality. For economic reasons, B-PON specifications replace this encryption by a churning function, which permutes at the OLT and reorders the downstream ATM cells at the ONUs. A churning key specifying the rules of these permutations is exchanged, at most, every second between the OLT and each ONU. The churning function, which is activated and released at the setup and release of each ATM virtual path, thus avoids complex implementation at the physical layer.

7.7.2.2 Extended WDM The G.983.3 recommendation specifies that by extension of A-PON bandwidth utilization (cf. Fig. 7.28), a B-PON offers increased service capability, including broadcast/multicast video delivery services or high-speed leased line services. The OLT and the ONUs are replaced by extended versions of this equipment (E-OLT and E-ONUs). Whereas a single bandwidth, around 1550 nm is specified for downstream traffic in A-PON systems, a B-PON system considers a narrower version of this bandwidth, limited to 20 nm (1480 nm, 1500 nm), for transporting downstream traffic. Such a reduced bandwidth, called the *basic band* in the recommendation, is compatible with the use of conventional distributed feedback laser diodes. An additional bandwidth, referred to as the *enhancement band*, is defined either for the provision of DWDM services or for the delivery of video services.

7.8 SUMMARY

Due to the recent advances in optical communication and the thrust for packet-oriented carrier networks, metropolitan area networks and access networks should evolve considerably in the near future. Two main types of topologies are considered for the MANs: WDM passive optical stars and SONET/SDH active rings. In the very short term, solutions based on single-wavelength rings, such as the pending IEEE 802.17 standard, should enable SDH survivability with the multiplexing flexibility of MAC protocols. In the future, WDM rings using optical add–drop multiplexers (OADMs) should be developed to provide all-optical transparency. It should then be possible to assign dynamically various logical topologies on top of the physical topology. Instantaneous traffic requests could be satisfied by this flexible logical topology. Research has been ongoing to investigate this innovative approach for the metropolitan domain [36]. In the access, B-PON systems should be the basic solution for broadband access. Finally, because of the reduced cost of Ethernet interfaces compared to SDH interfaces, Ethernet-oriented services such as E-PON (Ethernet PON) should be of great commercial interest in coming years. Chapter 8 is fully dedicated to this topic.

REFERENCES

1. ITU-T Recommendation G.ason, *Architecture for the Automatic Switched Optical Network*, Draft V.0.51, Paris, Mar. 2001.
2. Optical Internet Forum, *User Network Interface (O-UNI) 1.0 Specification*, OIF2000.125.3, Dec. 2000.
3. D. Wrede, Current prospects for FTTH deployments: A business case summary, Lightwave, pp. 308–314, Mar. 2001.
4. A. Jourdan, L. Tancevski, and T. Pfeiffer, Which proportion of optics in future metropolitan area networks, *Alcatel Telecommunications Review*, pp. 219–221, third quarter 2001.
5. M. Schwartz, *Telecommunication Networks: Protocols, Modeling and Analysis*, Addison-Wesley, Reading, MA, 1987.
6. M. Gagnaire and G. Macario, Bandwidth allocation on the metaring gigabit network, *Proc. IFIP-TC6, International Conference on Information Networks and Data Communications*, Funchal, Madeira Island, Portugal, Apr. 18–21, North-Holland Amsterdam, 1994.

7. M. Chen, N. Dono, and R. Ramaswami, A media access protocol for packet-switched wavelength division multi-access metropolitan area networks, *IEEE Journal on Selected Areas in Telecommunications*, 8:1048–1057, 1990.

8. M. Guizani, High speed protocol for an all-optical packet switched metropolitan area network, *International Journal on Network Management*, 7:9–17, 1997.

9. P. Humblet, R. Ramaswami, and K. Sivarajan, An efficient communication protocol for high-speed packet-switched multi-channel networks, *IEEE Journal on Selected Areas in Communications*, 11:568–578, 1993.

10. D. Levine and I. Akyildiz, PROTON: A media access control protocol for optical networks with star topology, *IEEE/ACM Transactions on Networking*, 3:158–168, 1995.

11. R. Chipalkatti and Z. Zhang, A. Acampora, High speed communication protocols for optical star coupler using WDM, *Proc. IEEE INFOCOM'92*, pp. 2124–2133, 1992.

12. R. Chipalkatti, Z. Zhang, and A. Acampora, Protocols for optical star-coupled networks using WDM: Performance and complexity study, *IEEE Journal on Selected Areas in Communications*, 11:579–589, 1993.

13. R. Chipalkatti and A. Acampora, Protocols for optical star couplers networks using WDM: Performance and complexity study, *IEEE Journal on Selected Areas in Communications*, 11(4), 1994.

14. M. Gagnaire and A. Berrada, An improved random scheduling algorithm for access to passive optical star networks, *European Conference on Networks and Optical Communications (NOC'97)*, 1997.

15. E. H. Dinan and M. Gagnaire, An Efficient MAC protocol for packet-switched WDM photonic network, *Telecommunications Systems Journal*, 16(1/2):135–146, 2001.

16. T. Stroesslin and M. Gagnaire, A flexible MAC protocol for all-optical WDM metropolitan area network, *IEEE IPCCC*, Phoenix, AZ, Feb. 2000.

17. A. Yan, A. Ganz, and C. M. Krishna, A distributive adaptive protocol providing real-time services on WDM-based LANs, *IEEE Journal of Lightwave Technology*, 14(6):1245–1253, 1996.

18. B. Li and Y. Qin, Traffic scheduling in a photonic packet switching system with QoS guarantee, *IEEE Journal of Lightwave Technology*, 16(12):2281–2295, 1998.

19. K. Bengi, Medium access control protocols for passive optical star WDM single-hop networks efficiently integrating real-time and data services, *Optical Networks*, 2(4): 88–100, 2001.

20. J. Kuri and M. Gagnaire, A MAC protocol for IP differentiated services in a WDM metropolitan area network, *Photonic Network Communications Journal*, 3(1/2): 63–74, 2001.

21. A. Lardies, R. Jagannathan, A. Fumagalli, I. Cerutti, and M. Tacca, A flexible WDM ring network design and dimensioning methodology, *Proc. ONDM'2000*, Athens, Kluwer Academic, Amsterdam, 2001.

22. P. Arijs and P. Demeester, Dimensioning of non-hierarchical interconnected WDM ring network, *Proc. ONDM'2000*, Athens, Kluwer Academic, Amesterdam, 2001.

23. D. A. Schupke, Reliability models of SRP rings, *Americas Conference on Information Systems (AMCIS)*, Long Beach, CA, Aug. 10–13, 2000.

24. H. T. Kung, Gigabit local area networks: A system perspective, *IEEE Communications*, pp. 79–89, Apr. 1992.

25. H.-M. Foisel, M. Jaeger, F.-J. Wesphal, K. Ovsthus, and J.-C Bischoff, Evaluation of IP over WDM network architectures, *Photonic Communications Network*, 3:41–48, 2001.

26. I. Cidon and Y. Ofek, Metaring: A full duplex ring with fairness and spatial reuse, *IEEE Transactions on Communications*, 41(1):110–120, Jan. 1993.

27. M. Gagnaire, Behavior of the metaring gigabit network under multipriority traffic, *Proc. IFIP-WG6.4, High Performance Networking '94*, Grenoble, France, June 27–July 1, North-Holland, Amsterdam, 1994.

28. IEEE 802.17 Resilient Packet Ring (RPR) Working Group, *http://grouper.ieee.org/groups/802/17/index.htm*.

29. J. R. Stern et al., The full service access network requirements Specifications, 8th International Workshop on Optical Access Networks, Atlanta, GA, Mar. 2–5, 1997.

30. M. Gagnaire and S. Stojanovski, Stream traffic management over an ATM passive optical network*, Computer Networks Journal*, 32(5):571–586, Apr. 2000.

31. S. Stojanovski and M. Gagnaire, Traffic shaping and MAC-PON protocols: To integrate or not? Mini-Conference on Access Networks, IEEE Globecom, Sydney, Australia, Nov. 1998.

32. ITU-T Recommendation G.983.1, *Broadband Optical Access Systems Based on Passive Optical Networks (PON)*, 1998.

33. ITU-T Recommendation G.983.2, *ONT Management and Control Interface Specification for ATM-PON*, 1999.

34. ITU-T Recommendation G.983.3, *Broadband Optical Access Systems with Increased Service Capability by Wavelength Allocation*, 2001.

35. F. J. Effeneberger, H. Ichibangase, and H. Yamashita, Advances in broadband passive optical networking technologies, *IEEE Communications*, pp. 118–124, Dec. 2001.

36. M. Ajmone Marsan, E. Leonardi, M. Lo, and F. Neri, Modeling slotted WDM rings with discrete-time Markovian models, *Computer Communications Journal*, 32(5):617–631, May 15–2000.

8 Ethernet Passive Optical Networks

GLEN KRAMER and BISWANATH MUKHERJEE

University of California–Davis, Davis, California

ARIEL MAISLOS

Passave Networks, Tel- Aviv, Israel

8.1 INTRODUCTION

In recent years the telecommunications backbone has experienced substantial growth; however, little has changed in the access network. The tremendous growth of Internet traffic has accentuated the aggravating lag of access network capacity. The "last mile" still remains the bottleneck between high-capacity local area networks (LANs) and the backbone network. The most widely deployed broadband solutions today are digital subscriber line (DSL) and cable modem (CM) networks. Although they are an improvement over 56-kbps dial-up lines, they are unable to provide enough bandwidth for emerging services such as video on demand (VoD), interactive gaming, or two-way video conferencing. A new technology is required; one that is inexpensive, simple, scalable, and capable of delivering bundled voice, data, and video services to an end user over a single network. Ethernet passive optical networks (EPONs), which represent the convergence of low-cost Ethernet equipment and low-cost fiber infrastructure, appear to be the best candidate for the next-generation access network.

8.1.1 Traffic Growth

Data traffic is increasing at an unprecedented rate. Sustainable data traffic growth rate of over 100% per year has been observed since 1990. There were periods when a combination of economic and technological factors resulted in even larger growth rates (e.g., 1000% increase per year in 1995 and 1996) [1]. This trend is likely to continue in the future. Simply put, more and more users are going online, and those who are already online are spending more time there and are using more bandwidth-intensive applications. Market research shows that after upgrading to

IP over WDM: Building the Next-Generation Optical Internet, Edited by Sudhir Dixit.
ISBN 0-471-21248-2 © 2003 John Wiley & Sons, Inc.

a broadband connection, users spend about 35% more time online than before [2]. Voice traffic is also growing, but at a much slower rate of 8% annually. According to most analysts, data traffic has already surpassed voice traffic. More and more subscribers telecommute and require the same network performance as they see on corporate LANs. More services and new applications will become available as bandwidth per user increases.

Neither DSL nor cable modems can keep up with such demand. Both technologies are built on top of existing communication infrastructure not optimized for data traffic. In cable modem networks, only a few radio-frequency (RF) channels are dedicated for data, while the majority of bandwidth is tied up servicing legacy analog video. DSL copper networks do not allow sufficient data rates at required distances due to signal distortion and crosstalk. Most network operators have come to the realization that a new, data-centric solution is necessary. Such a technology would be optimized for Internet protocol (IP) data traffic. The remaining services, such a voice or video, will converge into a digital format and a true full-service network will emerge.

8.1.2   Evolution of the First Mile

The *first mile*? Once called the last mile, the networking community has renamed this network segment the first mile, to symbolize its priority and importance.[*] The first mile connects the service provider central offices to business and residential subscribers. Also referred to as the *subscriber access network* or *local loop*, it is the network infrastructure at the neighborhood level. Residential subscribers demand first-mile access solutions that are broadband, offer Internet media-rich services, and are comparable in price with existing networks.

Incumbent telephone companies responded to Internet access demand by deploying DSL technology. DSL uses the same twisted pair as that used by telephone lines and requires a DSL modem at the customer premises and a DSL access multiplexer (DSLAM) in the central office (CO). The data rate provided by DSL is typically offered in a range from 128 kbps to 1.5 Mbps. Although this is significantly faster than an analog modem, it is well shy of being considered broadband, in that it cannot support emerging voice, data, and video applications. In addition, the physical area that one central office can cover with DSL is limited to distances of less than 18,000 ft (5.5 km), which covers approximately 60% of potential subscribers. And even though to increase DSL coverage remote DSLAMs (R-DSLAMs) may be deployed closer to subscribers, network operators, in general, do not provide DSL services to subscribers located more than 12,000 ft from a CO, due to increased costs [3].

Cable television companies responded to Internet service demand by integrating data services over their coaxial cable networks, which were originally designed for

---

[*]Ethernet in the First Mile Alliance was formed in December 2001 by Alloptic, Cisco Systems, Elastic Networks, Ericsson, Extreme Networks, Finisar, Intel, NTT, and World Wide Packets. For more information, see [22].

analog video broadcast. Typically, these hybrid fiber coax (HFC) networks have fiber running between a video head end or a hub to a curbside optical node, with the final drop to the subscriber being coaxial cable, repeaters, and tap couplers. The drawback of this architecture is that each shared optical node has less than 36 Mbps effective data throughput, which is typically divided among 2000 homes, resulting in frustrating slow speed during peak hours. To alleviate bandwidth bottlenecks, optical fibers, and thus optical nodes, are penetrating deeper into the first mile.

The next wave of local access deployment promises to bring fiber to the building (FTTB) and fiber to the home (FTTH). Unlike previous architectures, where fiber is used as a feeder to shorten the lengths of copper and coaxial networks, these new deployments use optical fiber throughout the access network. New optical fiber network architectures are emerging that are capable of supporting gigabit per second speeds, at costs comparable to those of DSL and HFC networks.

8.1.3 Next-Generation Access Network

Optical fiber is capable of delivering bandwidth-intensive, integrated voice, data, and video services at distances beyond 20 km in the subscriber access network. A logical way to deploy optical fiber in the local access network is to use a point-to-point (PtP) topology, where dedicated fiber runs from the CO to each end-user subscriber (Fig. 8.1a). Although this is a simple architecture, in most cases it is cost prohibitive, due to the fact that it requires significant outside plant fiber deployment as well as connector termination space in the local exchange. Considering N subscribers at an average distance L kilometers from the central office, a PtP design

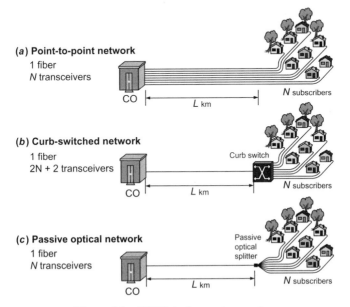

Figure 8.1 FTTH deployment scenarios.

requires $2N$ transceivers and NL total fiber length (assuming that a single fiber is used for bidirectional transmission).

To reduce fiber deployment, it is possible to deploy a remote switch (concentrator) close to the neighborhood. That will reduce the fiber consumption to only L kilometers (assuming negligible distance between the switch and customers), but will actually increase the number of transceivers to $2N + 2$, as there is one more link added to the network (Fig. 8.1b). In addition, curb-switched network architecture requires electrical power as well as backup power at the curb switch. Currently, one of the highest costs for local exchange carriers (LECs) is providing and maintaining electrical power in the local loop.

Therefore, it is logical to replace the hardened (environmentally protected) active curb-side switch with an inexpensive passive optical splitter. The passive optical network (PON) is a technology viewed by many as an attractive solution to the first mile problem [4,5]; a PON minimizes the number of optical transceivers, central office terminations, and fiber deployment. A PON is a point-to-multipoint (PtMP) optical network with no active elements in the signal path from source to destination. The only interior elements used in PON are passive optical components such as optical fiber, splices, and splitters. An access network based on a single-fiber PON requires only $N + 1$ transceivers and L kilometers of fiber (Fig. 8.1c).

8.2   OVERVIEW OF PON TECHNOLOGIES

8.2.1   Optical Splitters/Combiners

A passive optical network employs a passive (not requiring any power) device to split optical signal (power) from one fiber into several fibers and, reciprocally, to combine optical signals from multiple fibers into one. This device is an optical coupler. In its simplest form, an optical coupler consists of two fibers fused together. Signal power received on any input port is split between both output ports. The splitting ratio of a splitter can be controlled by the length of the fused region and therefore is a constant parameter.

$N \times N$ couplers are manufactured by staggering multiple 2×2 couplers (Fig. 8.2) or by using planar waveguide technology. Couplers are characterized by the following parameters:

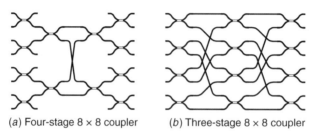

(a) Four-stage 8 × 8 coupler   (b) Three-stage 8 × 8 coupler

Figure 8.2   Two 8×8 couplers created from multiple 2×2 couplers.

- *Splitting loss:* power level at the coupler's output versus power level at its input, measured in decibels. For an ideal 2×2 coupler, this value is 3 dB. Figure 8.2 illustrates two topologies for 8×8 couplers based on 2×2 couplers. In a four-stage topology (Fig. 8.2*a*), only $\frac{1}{16}$ of the input power is delivered to each output. Figure 8.2*b* shows a more efficient design called a *multistage interconnection network* [6]. In this arrangement, each output receives $\frac{1}{8}$ of the input power.

- *Insertion loss:* power loss resulting from imperfections of the manufacturing process. Typically, this value ranges from 0.1 to 1 dB.

- *Directivity:* amount of input power leaked from one input port to another input port. Couplers are highly directional devices, with the directivity parameter reaching 40 to 50 dB.

Very often, couplers are manufactured to have only one input or one output. A coupler with only one input is referred to as a *splitter*. A coupler with only one output is called a *combiner*. Sometimes, 2×2 couplers are made highly asymmetric (with splitting ratios 5/95 or 10/90). This kind of coupler is used to branch off a small portion of signal power, for example, for monitoring purposes. Such devices are called *tap couplers*.

8.2.2 PON Topologies

Logically, the first mile is a PtMP network, with a CO servicing multiple subscribers. There are several multipoint topologies suitable for the access network, including tree, tree and branch, ring, or bus (Fig. 8.3). Using $1:2$ optical tap

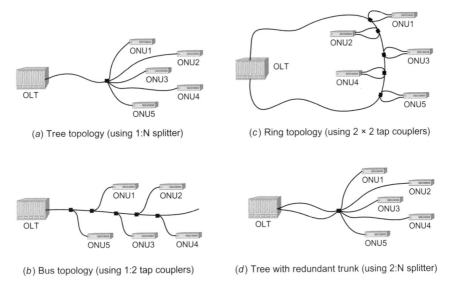

(*a*) Tree topology (using 1:N splitter)

(*c*) Ring topology (using 2 × 2 tap couplers)

(*b*) Bus topology (using 1:2 tap couplers)

(*d*) Tree with redundant trunk (using 2:N splitter)

Figure 8.3 PON topologies.

couplers and $1:N$ optical splitters, PONs can be deployed flexibly in any of these topologies. In addition, PONs can be deployed in redundant configurations such as double rings or double trees; or redundancy may be added to only a part of the PON, say the trunk of the tree (Fig. 8.3d) (also refer to [7] for more redundant topologies).

All transmissions in a PON are performed between an optical line terminal (OLT) and optical network units (ONUs) (Fig. 8.3). The OLT resides in the CO and connects the optical access network to the metropolitan area network (MAN) or wide area network (WAN), also known as a *backbone* or *long-haul network*. The ONU is located either at the end-user location (FTTH and FTTB) or at the curb, resulting in fiber to the curb (FTTC) architecture.

The advantages of using PONs in subscriber access networks are numerous:

- PONs allow for a long reach between the CO and customer premises, operating at distances over 20 km.
- PONs minimize fiber deployment in both the CO and the local loop.
- PONs provide higher bandwidth due to deeper fiber penetration, offering Gbps solutions.
- Operating in the downstream as a broadcast network, PONs allow for video broadcasting either as IP video or analog video.
- PONs eliminate the necessity of installing active multiplexers at the splitting locations, thus relieving network operators from the gruesome task of maintaining active curb-side units and providing power to them. Instead of active devices in these locations, PONs use small passive optical splitters, located in splice trays, and deployed as part of the optical fiber cable plant.
- Being optically transparent end to end, PONs allow upgrades to higher bit rates or additional wavelengths.

8.2.3 WDM versus TDM PONs

In the downstream direction (from OLT to ONUs), a PON is a point-to-multipoint network. The OLT typically has the entire downstream bandwidth available to it at all times. In the upstream direction, a PON is a multipoint-to-point network: Multiple ONUs transmit all toward one OLT. Directional properties of a passive splitter/combiner are such that an ONU's transmission cannot be detected by other ONUs. However, data streams from different ONUs transmitted simultaneously may still collide. Thus, in the upstream direction (from user to network), PON should employ some channel separation mechanism to avoid data collisions and fairly share the trunk fiber channel capacity and resources.

One possible way of separating the ONU's upstream channels is to use wavelength-division multiplexing (WDM), in which each ONU operates on a different wavelength. Although it is a simple solution (from a theoretical perspective), it remains cost-prohibitive for an access network. A WDM solution would require either a tunable receiver or a receiver array at the OLT to receive multiple channels.

An even more serious problem for network operators would be wavelength-specific ONU inventory: Instead of having just one type of ONU, there would be multiple types of ONUs based on their laser wavelength. Each ONU will have to use a laser with narrow and controlled spectral width and thus will become more expensive. It would also be more problematic for an unqualified user to replace a defective ONU because a unit with the wrong wavelength may interfere with some other ONU in the PON. Using tunable lasers in ONUs may solve the inventory problem but is too expensive at the current state of technology. For these reasons, a WDM PON network is not an attractive solution in today's environment.

Several alternative solutions based on WDM have been proposed: specifically, wavelength-routed PON (WRPON). A WRPON uses an arrayed waveguide grating (AWG) instead of a wavelength-independent optical splitter/combiner. We refer the reader to [8] for a detailed overview of these approaches. In one variation, ONUs use external modulators to modulate the signal received from the OLT and send it back upstream. This solution, however, is not cheap either; it requires additional amplifiers at or close to the ONUs to compensate for signal attenuation after round-trip propagation, and it requires more expensive optics to limit the reflections, since both downstream and upstream channels used the same wavelength. Also to allow independent (nonarbitrated) transmission from each of N ONUs, the OLT must have N receivers, one for each ONU.

In another variation, ONUs contain cheap laser-emitting diodes (LEDs) whose wide spectral band was sliced by the AWG on the upstream path. This approach still requires multiple receivers at the OLT. If, however, a single tunable receiver is used at the OLT, a data stream from only one ONU can be received at a time, which in effect makes it a time-division-multiplexed (TDM) PON.

In a TDM PON, simultaneous transmissions from several ONUs will collide when reaching the combiner. To prevent data collisions, each ONU must transmit in its own transmission window (time slot). One of the major advantages of a TDM PON is that all ONUs can operate on the same wavelength with absolutely identical components. The OLT will also need a single receiver. A transceiver in an ONU must operate at the full line rate, even though the bandwidth available to the ONU is lower. However, this property also allows the TDM PON to efficiently change the bandwidth allocated to each ONU by changing the assigned time-slot size, or even to employ statistical multiplexing to fully utilize the bandwidth available in the PON.

In a subscriber access network, most of the traffic flows downstream (from network to users) and upstream (from users to the network), but not peer to peer (user to user). Thus, it seems reasonable to separate the downstream and upstream channels. A simple channel separation can be based on space-division multiplexing (SDM), where separate PONs provided for downstream and for upstream transmissions. To save optical fiber and reduce cost of repair and maintenance, a single fiber may be used for bidirectional transmission. In this case, two wavelengths are used: typically, 1310 nm ($\lambda 1$) for upstream transmission and 1550 nm ($\lambda 2$) for downstream transmission (Fig. 8.4). The channel capacity on each wavelength can be flexibly divided between the ONUs.

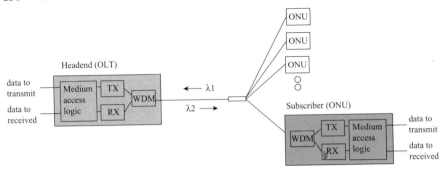

Figure 8.4   PON using a single fiber.

Time sharing appears to be the preferred method today for optical channel sharing in an access network, as it allows for a single upstream wavelength, such as 1310 nm, and a single transceiver in the OLT, resulting in a cost-effective solution.

8.2.4   Burst-Mode Transceivers

Due to unequal distances between CO and ONUs, optical signal attenuation in the PON is not the same for each ONU. The power level received at the OLT will be different for each time slot (called the *near–far problem*). Figure 8.5 depicts power

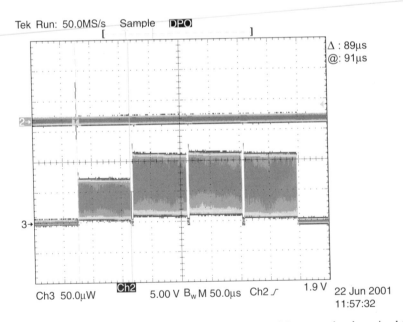

Figure 8.5   Near–far problem in a TDM PON: a snapshot of the power level received from four time slots.

levels of four time slots received by the OLT from four different ONUs in a TDM PON. As shown, one ONU's signal strength is lower at the OLT, probably due to its longer distance. If the receiver in OLT is adjusted to receive high-power signal properly from a close ONU, it may mistakenly read 1's as zeros when receiving a weak signal from a distant ONU. In the opposite case, if the receiver is trained on a weak signal, it may read zeros as 1's when receiving a strong signal.

To detect the incoming bit stream properly, the OLT receiver must quickly be able to adjust its zero–one threshold at the beginning of each time slot received (i.e., it should operate in *burst mode*). A burst mode receiver is necessary only in the OLT. The ONUs read a continuous bit stream (data or idles) sent by the OLT and do not need to readjust quickly.

An alternative approach is to allow ONUs to adjust their transmitter powers such that power levels received by OLT from all ONUs become the same. This method is not particularly favored by transceiver designers, as it makes the ONU hardware more complicated, requires special signaling protocol for feedback from the OLT to each ONU, and most important, may degrade the performance of all ONUs to that of a most distant unit.

Another issue is that it is not enough just to disallow ONUs to send any data. The problem is that even in the absence of data, lasers generate spontaneous emission noise. Spontaneous emission noise from several ONUs located close to the OLT can easily obscure the signal from a distant ONU (*capture effect*). Thus, an ONU must shut down its laser between the time slots. Because a laser cools down when it is turned off, and warms up when it is turned on, the power it emits may fluctuate at the beginning of a transmission. It is important that the laser be able to stabilize quickly after being turned on.

8.3 ETHERNET PON ACCESS NETWORK

Ethernet PON (EPON) is a PON-based network that carries data traffic encapsulated in Ethernet frames (defined in the IEEE 802.3 standard). It uses a standard 8b/10b line coding (8 user bits encoded as 10 line bits) and operates at standard Ethernet speed.

8.3.1 Why Ethernet?

Passive optical networking has been considered for the access network for quite some time, even well before the Internet spurred bandwidth demand. The full-service access network (FSAN) recommendation (ITU G.983) defines a PON-based optical access network that uses ATM as its layer 2 protocol. In 1995, when the FSAN initiative was started, ATM had high hopes of becoming the prevalent technology in the LAN, MAN, and backbone. However, since that time, Ethernet technology has leapfrogged ATM. Ethernet has become a universally accepted standard, with over 320 million port deployments worldwide, offering staggering economies of scale [9]. High-speed gigabit Ethernet deployment is widely accelerating and 10-gigabit Ethernet products are becoming available. Ethernet, which is easy to scale and manage, is winning new grounds in MAN and WAN. Considering the fact that 95% of LANs use Ethernet, it becomes clear that ATM PON may not be the best choice to interconnect two Ethernet networks.

One of ATM's shortcomings is the fact that a dropped or corrupted ATM cell will invalidate the entire IP datagram. However, the remaining cells carrying the portions of the same IP datagram will propagate further, thus consuming network resources unnecessarily. Also, ATM imposes a cell tax on variable-length IP packets. For example, for the trimodal packet-size distribution reported in [10], the cell tax is approximately 13% (i.e., to send the same amount of user's data, an ATM network must transmit 13% more bytes than an Ethernet network (counting 64-bit preamble and 96-bit minimum interframe gap (IFG) in Ethernet and 12 bytes of overhead associated with ATM adaptation layer 5 (AAL-5)). Finally, and perhaps most important, ATM did not live up to its promise of becoming an inexpensive technology—vendors are in decline and manufacturing volumes are relatively low. ATM switches and network cards are significantly more expensive (roughly eightfold) than Ethernet switches and network cards [9].

On the other hand, Ethernet looks like a logical choice for an IP data-optimized access network. Newly adopted quality-of-service (QoS) techniques have made Ethernet networks capable of supporting voice, data, and video. These techniques include full-duplex transmission mode, prioritization (P802.1p), and virtual LAN (VLAN) tagging (P802.1Q). Ethernet is an inexpensive technology that is ubiquitous and interoperable with a variety of legacy equipment. In the remainder of the chapter we focus on EPONs.

8.3.2 Principle of Operation

The IEEE 802.3 standard defines two basic configurations for an Ethernet network. In one configuration it can be deployed over a shared medium using the carrier sense multiple access with collision detection (CSMA/CD) protocol. In another configuration, stations may be connected through a switch using full-duplex point-to-point links. Properties of EPON are such that it cannot be considered either a shared medium or a point-to-point network; rather, it is a combination of both.

In the downstream direction, Ethernet frames transmitted by the OLT pass through a $1:N$ passive splitter and reach each ONU. N is typically between 4 and 64. This behavior is similar to that of a shared-medium network. Because Ethernet is broadcast by nature in the downstream direction (from network to user), it fits perfectly with Ethernet PON architecture: Packets are broadcast by the OLT and extracted by their destination ONU based on the media-access control (MAC) address (Fig. 8.6).

In the upstream direction, due to the directional properties of a passive optical combiner, data frames from any ONU will reach only the OLT, and not other ONUs. In that sense, in the upstream direction, the behavior of EPON is similar to that of a point-to-point architecture. However, unlike in a true point-to-point network, in EPON data frames from different ONUs transmitted simultaneously

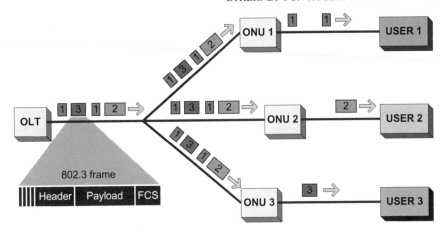

Figure 8.6 Downstream traffic in EPON.

still may collide. Thus, in the upstream direction (from users to network) the ONUs need to employ some arbitration mechanism to avoid data collisions and share the fiber-channel capacity fairly.

A contention-based media access mechanism (something similar to CSMA/CD) is difficult to implement because ONUs cannot detect a collision at the OLT (due to the directional properties of optical splitter/combiner). An OLT could detect a collision and inform ONUs by sending a jam signal; however, propagation delays in PON, which can exceed 20 km in length, can greatly reduce the efficiency of such a scheme. Contention-based schemes also have the drawback of providing a non-deterministic service (i.e., node throughput and channel utilization may be described as statistical averages). There is no guarantee of a node getting access to the media in any small interval of time. It is not a problem for CSMA/CD-based enterprise networks where links are short, typically overprovisioned, and traffic consists predominantly of data. Subscriber access networks, however, in addition to data, must support voice and video services and thus must provide some guarantees on timely delivery of these traffic types.

To introduce determinism in frame delivery, various noncontention schemes have been proposed. Figure 8.7 illustrates an upstream time-shared data flow in an EPON. All ONUs are synchronized to a common time reference and each ONU is allocated a time slot. Each time slot is capable of carrying several Ethernet frames. An ONU should buffer frames received from a subscriber until its timeslot arrives. When its time slot arrives, the ONU would "burst" all stored frames at full channel speed, which must correspond to one of standard Ethernet rates (10/100/1000/ 10,000 Mbps). If there are no frames in the buffer to fill the entire time slot, idle 10-bit characters are transmitted. The possible time slot allocation schemes could range from a static allocation [fixed time-division multiple access (TDMA)] to a dynamically adapting scheme based on instantaneous queue size in every ONU (statistical multiplexing scheme). More allocation schemes are possible, including

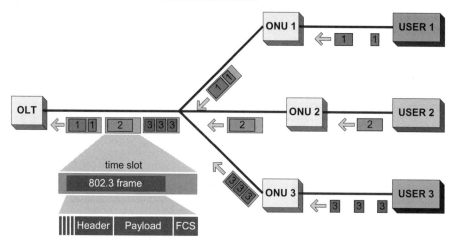

Figure 8.7 Upstream traffic in EPON.

schemes utilizing notions of traffic priority, quality of service (QoS), service-level agreements (SLAs), oversubscription ratios, and so on.

Decentralized approaches to implement a dynamic slot assignment scheme are also possible, in which ONUs decide when to send data and for how long. These schemes are somewhat similar to a token-passing approach, except that in this case it is a passive ring. In such a scheme, every ONU, before sending its data, will send a special message announcing how many bytes it is about to send. The ONU that is scheduled next (say, in round-robin fashion) will monitor transmission of the previous ONU and will time its transmission such that it arrives at the OLT right after transmission from the previous ONU. Thus, there will be no collision and no bandwidth will be wasted. This scheme is similar to hub polling [11]. However, this scheme has a major limitation: It requires connectivity (communicability) between ONUs. That imposes some constraints on PON topology; namely, the network should be deployed as a ring or as a broadcasting star. This requirement is not desirable as it may require more fiber to be deployed, or fiber plant with different topology might already be predeployed. In general, a preferred algorithm will support any point-to-multipoint PON topology.

In an optical access network, we can count only on connectivity from the OLT to every ONU (downstream traffic) and every ONU to the OLT (upstream traffic). That is true for all PON topologies. Therefore, the OLT remains the only device that can arbitrate time-division access to the shared channel. The challenge of implementing an OLT-based dynamic arbitration scheme is in the fact that the OLT does not know how many bytes of data each ONU has buffered. The burstiness of data traffic precludes a queue occupancy prediction with any reasonable accuracy. If the OLT is to make an accurate time slot assignment, it should know the state of a given ONU exactly. One solution may be to use a polling scheme based on grant and request messages. Requests are sent from an ONU to report changes in an ONU's state (e.g., the amount of buffered data). The OLT processes all requests and allocates different transmission windows (time slots) to ONUs. Slot-assignment information is delivered to ONUs using grant messages.

The advantage of having centralized intelligence for the slot-allocation algorithm is that the OLT knows the state of the entire network and can switch to another allocation scheme based on that information; the ONUs don't need to monitor the network state or negotiate and acknowledge new parameters. That will make ONUs simpler and cheaper and the entire network more robust.

8.3.3 Multipoint Control Protocol

To support a time slot allocation by the OLT, the multipoint control protocol (MPCP) is being developed by the IEEE 802.3ah task force. This protocol relies on two Ethernet messages: A GATE and REPORT. A GATE message is sent from OLT to an ONU and used to assign a transmission time slot. A REPORT message is used by an ONU to convey its local conditions (such as buffer occupancy, etc.) to the OLT to help it make intelligent allocation decisions. Both GATE and REPORT messages are MAC control frames (type 88-08) and are processed by the MAC control sublayer.

There are two modes of operation of MPCP: auto-discovery (initialization) and normal operation. Auto-discovery mode is used to detect newly connected ONUs and learn the round-trip delay and MAC address of that ONU, and, in addition, perhaps some additional parameters yet to be defined. Normal mode is used to assign transmission opportunities to all initialized ONUs.

Since more than one ONU can require initialization at one time, auto-discovery is a contention-based procedure. At a high level, it works as follows:

1. OLT allocates an initialization slot, an interval of time when no previously initialized ONUs are allowed to transmit. The length of this initialization slot must be at least ⟨*transmission size*⟩ + ⟨*maximum round-trip time*⟩ − ⟨*minimum round-trip time*⟩, where ⟨*transmission size*⟩ is the length of the transmission window that an uninitialized ONU can use.

2. OLT sends an initialization GATE message advertising the start time of the initialization slot and its length. While relaying this message from a higher layer to the MAC layer, MPCP will time-stamp it with its local time.

3. Only uninitialized ONUs will respond to the initialization GATE message. Upon receiving the initialization GATE message, an ONU will set its local time to the arriving time stamp in the initialization GATE message.

4. When the local clock located in the ONU reaches the start time of the initialization slot (also delivered in the GATE message), the ONU will transmit its own message (initialization REPORT). The REPORT message will contain the ONU's source address and a time stamp representing a local ONU's time when the REPORT message was sent.

5. When the OLT receives the REPORT from an uninitialized ONU, it learns its MAC address and round-trip time. As illustrated in Fig. 8.8, the round-trip

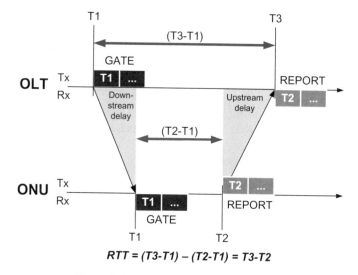

Figure 8.8   Round-trip-time measurement.

time of an ONU is exactly the time difference between the time the REPORT is received at the OLT and the time stamp contained in the REPORT.

Since multiple uninitialized ONUs may respond to the same initialization GATE message, the REPORT messages may collide. In that case, the ONUs whose REPORTs have collided will not get any slot assignments for their normal operation. If an ONU does not receive a slot assignment within some timeout interval, it will infer that a collision has occurred and will attempt to initialize again after skipping some random number of initialization GATE messages. The number of messages to skip is chosen randomly from an interval that doubles after each inferred collision (i.e., using exponential backoff).

Below we illustrate the normal operation of MPCP. It is important to notice that MPCP is not concerned with particular bandwidth-allocation schemes; rather, it is a supporting protocol necessary to deliver these decisions from the OLT to the ONUs.

1. From its higher layer (MAC control client), MPCP gets a request to transmit a GATE message to a particular ONU with the following information: time when that ONU should start transmission and length of the transmission (Fig. 8.9).

2. MPCP layer (in OLT and each ONU) maintains a clock. Upon passing a GATE message from its higher layer to MAC, MPCP time-stamps it with its local time.

3. Upon receiving a GATE message matching that ONU's MAC address (GATE messages are unicast), the ONU will program its local registers with

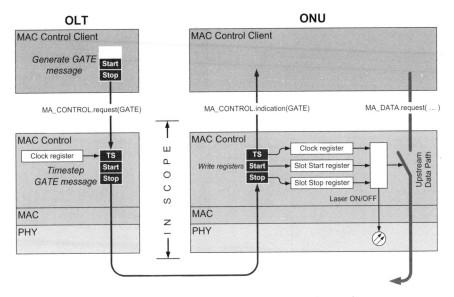

Figure 8.9   Multipoint control protocol: GATE operation.

transmission start and transmission length times. The ONU will also verify the time when the GATE message arrived is close to the time-stamp value contained within the message. If the difference in values exceeds some predefined threshold, the ONU will assume that it has lost its synchronization and will switch itself into an uninitialized mode. In that mode, the ONU is not allowed to transmit. It will monitor its incoming traffic, waiting for the next initialization GATE message to perform initialization.

4. If the time the GATE message is received is close to the time-stamp value in the GATE message, the ONU will update its local clock to that of the time stamp. When the local time reaches the start transmission value, the ONU will start transmitting. That transmission may include multiple Ethernet frames. The ONU will ensure that no frames are fragmented. If the next frame does not fit in the remainder of the time slot, it will be deferred until the next time slot.

REPORT messages are sent by ONUs in the assigned transmission windows together with data frames. REPORT messages can be sent automatically or on-demand. A REPORT message is generated in the MAC control client layer and is time-stamped in the MAC control (Fig. 8.10). Typically, REPORT would contain the desired size of next time slot based on ONU's queue size. When requesting a time slot, an ONU should account for additional overhead: namely, 64-bit frame preamble and 96-bit IFG associated with every frame.

When a time-stamped REPORT message arrives at the OLT, it is passed to the MAC control client layer responsible for making the bandwidth-allocation decision. Additionally, the OLT will recalculate the round-trip time to the source ONU as

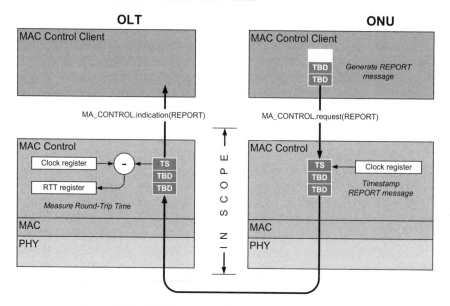

Figure 8.10   Multipoint control protocol: REPORT operation.

shown in Fig. 8.8. Some small deviation of the new RTT from the RTT measured previously may be caused by changes in fiber refractive index resulting from temperature drift. A large deviation should alarm the OLT about the ONU's potential missynchronization and should prevent the OLT from further granting any transmissions to that ONU until it is reinitialized.

The description above represents a framework of the protocol being developed for the EPON. There are many more details that remain to be discussed and agreed upon. This work is currently being conducted in the IEEE 802.3ah task force, a standards group charged with the development of the Ethernet solution for the subscriber access network.

8.3.4   EPON Compliance with 802 Architecture

The IEEE 802 architecture defines two types of media: a shared medium and full duplex. In a shared medium, all stations are connected to a single access domain where at most one station can transmit at a time and all stations can receive all the time. The full-duplex segment is a point-to-point link connecting two stations (or a station and a bridge) such that both stations can transmit and receive simultaneously. Relying on the definitions above, bridges never forward a frame back to its ingress port. In other words, it is assumed that all the stations connected to the same port on the bridge can communicate with one another without the bridge's help. This bridge behavior has led to an interesting problem: users connected to different ONUs on the same PON are unable to communicate with one another without data

Figure 8.11   Link ID field embedded in frame preamble.

being processed at layer 3 (network layer) or above. This raises a question of compliance with IEEE 802 architecture, particularly with P802.1D bridging.

To resolve this issue and to ensure seamless integration with other Ethernet networks, devices attached to the EPON medium will have an additional sublayer that, based on its configuration, will emulate either a shared medium or a point-to-point medium. This sublayer is referred to as a shared-medium emulation (SME) or point-to-point emulation (PtPE) sublayer. This sublayer must reside below the MAC layer to preserve the existing Ethernet MAC operation defined in IEEE standard P802.3. Operation of the emulation layer relies on tagging of Ethernet frames with tags unique for each ONU (Fig. 8.11). These tags, called *link ID*, are placed in the preamble before each frame. To guarantee uniqueness of link IDs, each ONU is assigned one or more tags by the OLT during initial registration phase.

8.3.4.1   *Point-to-Point Emulation*

In point-to-point emulation (PtPE) mode, the OLT must have N MAC ports (interfaces), one for each ONU (Fig. 8.12). When sending a frame downstream (from the OLT to an ONU), the PtPE sublayer in the OLT will insert the link ID associated with a particular MAC port from which the frame arrived (Fig. 8.12*a*). Even though the frame will be delivered to each ONU, only one PtPE sublayer will match that frame's link ID with the value assigned to the ONU and will accept the frame and pass it to its MAC layer for further verification. MAC layers in all other ONUs will never see that frame. In this sense, it appears as if the frame was sent on a point-to-point link to only one ONU.

In the upstream direction, the ONU will insert its assigned link ID in the preamble of each frame transmitted. The PtPE sublayer in the OLT will demultiplex the frame to the proper MAC port based on the unique link ID (Fig. 8.12*b*). The PtPE configuration is clearly compatible with bridging, as each ONU is connected to an independent bridge port. The bridge placed in the OLT (Fig. 8.13) will relay inter-ONU traffic between its ports.

8.3.4.2   *Shared-Medium Emulation*

In shared-medium emulation (SME), frames transmitted by *any* node (OLT or any ONU) should be received by *every* node (OLT and every ONU). In the downstream direction, the OLT will insert a "broadcast" link ID which will be accepted by every ONU (Fig. 8.14*a*). To ensure shared-medium operation for upstream data (frames sent by ONUs), the SME sublayer in OLT must mirror all frames back downstream to be received by all other ONUs (Fig. 8.14*b*). To avoid frame duplication when an ONU receives its own frame, the

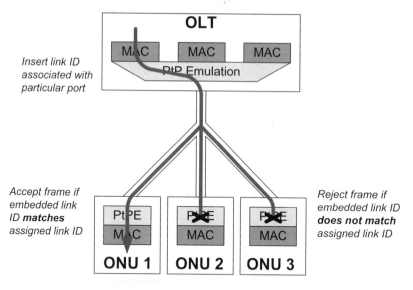

Insert link ID associated with particular port

*Accept frame if embedded link ID **matches** assigned link ID*

*Reject frame if embedded link ID **does not match** assigned link ID*

(*a*) **Downstream Transmission**

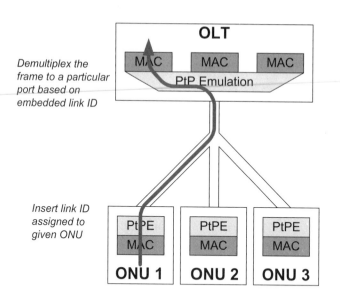

Demultiplex the frame to a particular port based on embedded link ID

Insert link ID assigned to given ONU

(*b*) **Upstream Transmission**

Figure 8.12 Point-to-point emulation.

SME sublayer in an ONU accepts a frame only if the frame's link ID is different from the link ID assigned to that ONU. The shared-medium emulation requires only one MAC port in the OLT. Physical-layer functionality (SME sublayer) provides ONU-to-ONU communicability, eliminating the need for a bridge.

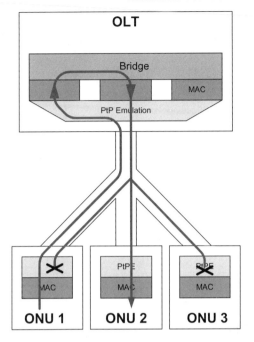

Figure 8.13   Bridging between ONUs with point-to-point emulation.

8.3.4.3   Combined PtPE and SME Mode   While both PtPE and SME options provide solutions for P802.1 standards compliance issues, both of them also have drawbacks, specifically when considered for an application in a subscriber access networks. The PtPE mode precludes the possibility of having a single-copy multicast/broadcast when the OLT sends one frame received by several ONUs. This feature is very important for services such as video broadcast or any real-time broadcast services. To support such services, the OLT operating in the PtPE mode must duplicate broadcast packets, each time with a different link ID. Shared-medium emulation, on the other hand, provides multicast/broadcast capabilities. However, because *every* upstream frame is reflected downstream, it wastes a lot of downstream bandwidth.

To achieve optimal operation, it is feasible to deploy a PON with point-to-point and shared-medium emulation simultaneously. In such a configuration in an EPON with N ONUs, the OLT will contain $N + 1$ MACs: one for each ONU (PtPE) and one for broadcasting (Fig. 8.15). Each ONU must have two MACs: one for a shared medium and one for point-to-point emulated link. To separate the traffic optimally, higher layers (above MAC) will decide to which port to send the data (e.g., by using VLANs). Only data that should be broadcast will be sent to the port connected to the emulated shared-medium segment.

8.3.4.4   Open Issues   The work on emulation sublayer design is still in progress. A serious challenge that needs to be solved is that the emulation sublayer must be

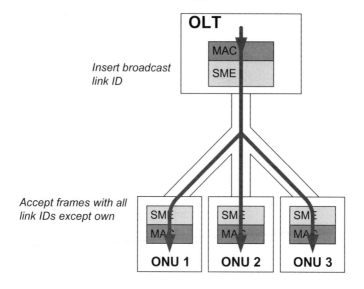

(*a*) **Downstream Transmission**

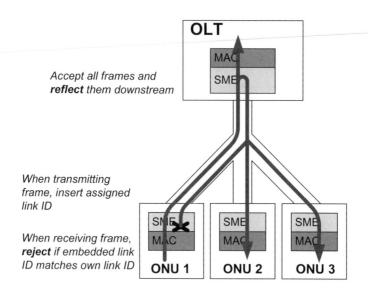

(*b*) **Upstream Transmission**

Figure 8.14 Shared-medium emulation.

able to multiplex several data flows into one flow. In PtPE mode, the emulation layer may receive data frames from multiple MAC ports simultaneously. In SME mode this happens when an ONU-to-ONU data frame competes with a network-to-ONU data frame for the downstream channel. The apparent drawback of this competition

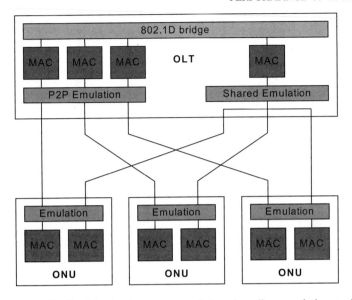

Figure 8.15   Combined point-to-point and shared-medium emulation mode.

is that, now, some frames may have to be discarded below the MAC sublayer, which may make the BER dependent on the traffic load. To drop frames intelligently, the emulation sublayer should be aware of the sender's or the recipient's SLA and frame priority. All these features are strictly out of scope of the IEEE 802 standard and do not belong in the PHY layer. Additionally, even if frames are not dropped in the emulation sublayer, MAC-to-MAC delay may not be constant, due to head-of-line blocking, which may have a detrimental effect on QoS.

An alternative proposal suggests putting the emulation sublayer in the MAC control layer. In this case, the link ID information should propagate transparently through the MAC, and this will require MAC modifications. Another, more subtle problem is that since frame filtering is now performed above the MAC layer, in PtPE mode, every MAC will see all frames before they are filtered out based on link ID. This means that a corrupted and invalid frame will increment error counters in all MACs as opposed to only one MAC at the other end of its virtual PtP link. This, of course, will invalidate layer-management facilities provisioned by the standard. Finding solutions for the above-mentioned issues, as well as converging on a best place for the emulation sublayer, remains on the list of open issues for the IEEE 802.3ah task force.

8.4   PERFORMANCE OF EPON

The performance of an EPON depends on the particular bandwidth-allocation scheme. Choosing the best allocation scheme, however, is not a trivial task. If all users belong to the same administrative domain (say, a corporate or campus network), full statistical multiplexing would make sense—network administrators would like to get most out of the available bandwidth. However, subscriber access networks are not private LANs and the objective is to ensure service-level agreement (SLA) compliance for each user. Using statistical multiplexing mechanisms to give each user best-effort bandwidth may complicate billing and offset the user's drive to upgrade to a higher bandwidth. Also, subscribers may get used to and expect the performance that they get during low-activity hours when lots of best-effort bandwidth is available. Then, at peak hours, the same users would perceive the service as unsatisfactory, even though they get what is guaranteed by their SLA. An optimized bandwidth-allocation algorithm will ultimately depend on the future SLA and the billing model used by the service provider.

This notion has led to a fixed-pipe model for an access network. *Fixed pipe* assumes that each user will agree to and pay for a fixed bandwidth regardless of the network conditions or applications using it. Because the contracted bandwidth must be available at any time, this model does not support oversubscription. Correspondingly, network operators are not eager to give users an additional best-effort bandwidth. It is not easy to charge for, and users are not willing to pay for what is difficult to measure. In a sense, this model operates like a fixed circuit given to each customer.

Recently, however, there has been a shift to a new paradigm. Since bandwidth is getting cheaper, the revenues the service providers get from data traffic are decreasing. Correspondingly, many carriers complain that to accommodate the increased traffic on their networks, they have to upgrade their networks often, and thus their capital expenses increase, but the revenue remains flat or even decreases. In recent years it has become apparent that raw bandwidth cannot generate enough revenue. The new thinking among telecommunication operators calls for service-based billing in which users pay for the services they get and not for the guaranteed bandwidth they are provisioned. In this model, the network operators are willing to employ statistical multiplexing to be able to support more services over the network.

Below we compare EPON performance operated in fixed TDMA (fixed pipe) and statistical multiplexed modes.

8.4.1 Model Description

In this study we consider an access network consisting of an OLT and N ONUs connected using a passive optical network (Fig. 8.16). Every ONU is assigned a downstream propagation delay (from the OLT to the ONU) and an upstream propagation delay (from the ONU to the OLT). Whereas with a tree topology both downstream and upstream delays are the same, with a ring topology delays will be different. To keep the model general, we assume independent delays and select them randomly (uniformly) over the interval $(50\,\mu s,\ 100\,\mu s)$. These values correspond to distances between the OLT and ONUs ranging from 10 to 20 km.

From the access side, traffic may arrive at an ONU from a single user or from a gateway of a LAN (i.e., traffic may be aggregated from a number of users). Ethernet

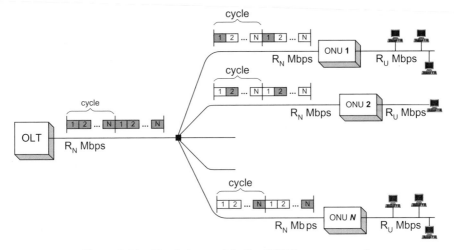

Figure 8.16 Simulation model of an EPON access network.

frames should be buffered in the ONU until the ONU is allowed to transmit the packets. The transmission speed of the PON and the user access link may not necessarily be the same. In our model we consider R_U Mbps to be the user data rate (rate of access link from a user to an ONU) and R_N Mbps to be the network data rate (upstream slotted link from an ONU to the OLT) (see Fig. 8.16). We should mention here that if $R_N \geq N \times R_U$, the bandwidth utilization problem does not exist, as the system throughput is higher than the peak aggregated load from all ONUs. In this study we consider a system with $N = 16$ and R_U and R_N being 100 and 1000 Mbps, respectively.

A set of N time slots together with their associated guard intervals is called a *cycle*. In other words, a cycle is a time interval between two successive time slots assigned to one ONU (Fig. 8.16). We denote cycle time by T. Making T too large will result in increased delay for all the packets, including high-priority (real-time) packets. Making T too small will result in more bandwidth being wasted by guard intervals.

To obtain an accurate and realistic performance analysis, it is important to simulate system behavior with appropriate traffic injected into the system. There is an extensive study showing that most network traffic flows [i.e., generated by http, ftp, variable-bit-rate (VBR) video applications, etc.] can be characterized by self-similarity and long-range dependence (LRD) (see [12] for an extensive reference list). To generate self-similar traffic, we used the method described in [13], where the resulting traffic is an aggregation of multiple streams, each consisting of alternating Pareto-distributed on–off periods.

Figure 8.17 illustrates the way the traffic was generated in an individual ONU. Within the on period, every source generates packets back to back (with a 96-bit interframe gap and 64-bit preamble in between). Every source assigns a specific priority value to all its packets. Packets generated by n sources are aggregated

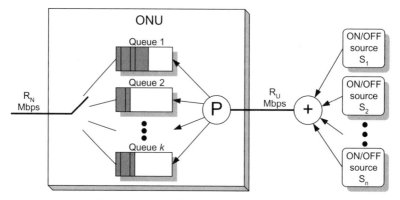

Figure 8.17   Traffic generation in the ONU.

(multiplexed) on a single line such that packets from different sources do not overlap. After that the packets are forwarded to the respective queues based on their priority assignments, and the queues are served in order of their priorities.

Each on–off source generates load

$$\tilde{\phi}_i = \frac{E[\mathrm{ON}_i]}{E[\mathrm{ON}_i] + E[\mathrm{OFF}_i]} \qquad (8.1)$$

where $E[\mathrm{ON}_i]$ and $E[\mathrm{OFF}_i]$ are expected lengths (durations) of on and off periods of source i. Load aggregated from all n sources in an ONU is called *offered ONU load* (OOL) and denoted by ϕ:

$$\phi = \sum_{i=1}^{n} \tilde{\phi}_i \qquad (8.2)$$

Offered network load (ONL) Φ is the sum of the loads offered by each ONU and scaled based on R_D and R_U rates. Clearly, since the network throughput is less than the aggregated peak bandwidth from all ONUs, the ONL can exceed 1:

$$\Phi = \frac{R_U}{R_N} \sum_{j=1}^{N} \phi^{[j]} \qquad (8.3)$$

It is important to differentiate between offered load and *effective load*. The effective ONU load (EOL) is denoted ϖ and results from the data (packets) that have been sent out by the ONUs. Thus, the EOL is equal to the OOL only if the packet-loss rate is zero. In general, $\varpi \le \phi$.

The EOL generated by the ONU j is denoted $\varpi^{[j]}$. *Effective network load* (ENL) Ω is just a sum of the EOLs generated by all ONUs, with a corresponding scaling coefficient based on the PON and user link bit rates:

$$\Omega = \frac{R_U}{R_N} \sum_{j=1}^{N} \varpi^{[j]} \qquad (8.4)$$

Every ONU may have k queues that are served in order of their priority (priority queuing is discussed in Section 8.5.3.1). Every ONU has a finite buffer of size Q. The memory is allocated to different queues based on demand and priority; that is, if the entire buffer is occupied and a frame with a higher priority arrives, the lowest-priority nonempty queue will drop one or more frames, so that the higher-priority queue can store the new packet. In our simulations, buffer size Q was set to 10 Mbytes.

8.4.2   Bandwidth-Allocation Schemes

The essence of the MPCP protocol is in assigning a variable-sized slot (transmission window) to each ONU based on decisions made by some bandwidth-allocation scheme. To prevent the upstream channel from being monopolized by one ONU with high data volume, there should be a maximum transmission window size limit assigned to every ONU. We denote an ONU-specific maximum transmission window size by $W_{MAX}^{[i]}$ (in bytes). The choice of specific values of $W_{MAX}^{[i]}$ determines the maximum granting cycle time T_{MAX} under heavy load conditions:

$$T_{MAX} = \sum_{i=1}^{N}\left(G + \frac{8 \times W_{MAX}^{[i]}}{R}\right) \tag{8.5}$$

where $W_{MAX}^{[i]}$ is the maximum window size for the ith ONU (bytes), G the guard interval (seconds), N the number of ONUs, and R the line rate (bps).

The guard intervals provide protection for fluctuations of round-trip time of different ONUs. Additionally, the OLT receiver needs some time to readjust its sensitivity due to the fact that signals from different ONUs may have different power levels (near–far problem). Making T_{MAX} too large will result in increased delay for all Ethernet frames, including those carrying high-priority (real-time) IP packets. Making T_{MAX} too small will result in more bandwidth being wasted by guard intervals.

It is the ONU's responsibility to ensure that the frame it is about to send fits in the remainder of the time slot. If the frame does not fit, it should be deferred until the next time slot, leaving the current timeslot underutilized (not filled completely with Ethernet frames). In Section 8.5.1 we investigate the time slot utilization issues in more detail.

In addition to the maximum cycle time, the $W_{MAX}^{[i]}$ value also determines the guaranteed bandwidth available to ONU i. Let $\Lambda_{MIN}^{[i]}$ denote the (minimum) guaranteed bandwidth of ONU i (bps). Obviously,

$$\Lambda_{MIN}^{[i]} = \frac{8 \times W_{MAX}^{[i]}}{T_{MAX}} \tag{8.6}$$

that is, the ONU is guaranteed to be able to send at least $W_{MAX}^{[i]}$ bytes (or $8 \times W_{MAX}^{[i]}$ bits) in at most T_{MAX} time. Of course, an ONUs bandwidth will be limited to its guaranteed bandwidth only if all other ONUs in the system also use all of their available bandwidth. If at least one ONU has less data, it will be granted a shorter transmission window, thus making the granting cycle time shorter, and therefore the available bandwidth to all other ONUs will increase proportionally to their $W_{MAX}^{[i]}$. This is the mechanism behind dynamic bandwidth distribution described in [15]: by adapting the cycle time to the instantaneous network load (i.e., queue occupancy), the bandwidth is distributed automatically to ONUs based on their loads. In the extreme case, when only one ONU has data to send, the bandwidth available to that ONU will be

$$\Lambda_{MAX}^{[i]} = \frac{8 \times W_{MAX}^{[i]}}{N \times G + (8 \times W_{MAX}^{[i]})/R} \tag{8.7}$$

In our simulations, we assume that all ONUs have the same guaranteed bandwidth (i.e., $W_{MAX}^{[i]} = W_{MAX}, \forall i$). This results in

$$T_{MAX} = N \left(G + \frac{8 \times W_{MAX}}{R} \right) \tag{8.8}$$

We believe that $T_{MAX} = 2$ ms and $G = 5$ μs are reasonable choices. They make $W_{MAX} = 15,000$ bytes. With these choices of parameters, every ONU will get a guaranteed bandwidth of 60 Mbps and a maximum (best-effort) bandwidth of 600 Mbps [see equations (8.6) and (8.7)].

The following algorithm was considered in our simulation study.

RTT[i]   table containing round-trip times for each ONU (originally
                populated during auto-discovery phase; updated constantly
                during normal operation)
V   size of transmission window (in bytes) requested by ONU
W   size of transmission window (in bytes) granted to ONU
ch_avail   time when channel becomes available for next transmission
local_time   read-only register containing OLT's local clock values
guard   guard interval (constant)
delta   time interval to process GATE message in the ONU (minimum
                time between GATE arrival and beginning of time slot)

```
ch_avail = 0;
repeat forever
{
    FOR i from 1 to N
    {
        /* wait for REPORT from ONU i */
        until(REPORT from ONU i arrived)
            /* do nothing */;

        /* get round-trip time */
        RTT[i] = local_time - REPORT.timestamp
        /*get requested slot size */
        V = REPORT.slot_size
```

```
    /* update channel availability time
       to make sure we don't schedule slot
       for the past time */
    if(ch_avail < local_time + delta + RTT[i])
       ch_avail = local_time + delta + RTT[i]

    /* make timeslot allocation decision
       (specific allocation schemes are
       considered below)*/
    W = f(V)

    /* create GATE message */
    GATE.slot_start = ch_avail−RTT[i]
    GATE.slot_size = W

    /* update channel availability time for next ONU*/
    ch_avail = ch_avail + guard + time(W)

    /* send GATE message to ONU i */
    send(i, GATE)
  }
}
```

The remaining question is how the OLT should determine the window size granted if the window size requested is less than the predefined maximum ($W^{[i]} < W_{MAX}$). Table 8.1 defines a few approaches (services) the OLT may take in making its decision.

TABLE 8.1   Grant Scheduling Services Used in Simulation

Service	Formula	Description
Fixed	$W^{[i]} = W_{MAX}$	This scheduling discipline ignores the window size requested and always grants the maximum window. As a result, it has a constant cycle time T_{MAX}. Essentially, this is a fixed-pipe model and corresponds to the fixed TDMA PON system described in [14].
Limited	$W^{[i]} = \min \begin{cases} V^{[i]} \\ W_{MAX} \end{cases}$ $V^{[i]}$ = window size requested	This discipline grants the number of bytes requested but no more than W_{MAX}. It is the most conservative scheme and has the shortest cycle of all the schemes.
Gated	$W^{[i]} = V^{[i]}$	This service discipline does not impose the W_{MAX} limit on the window size granted (i.e., it will always authorize an ONU to send as much data as it has requested). Of course, without a limiting parameter, the cycle time may increase unboundedly if the load offered exceeds the network throughput. In this discipline, such a limiting factor is the buffer size Q (i.e., an ONU cannot store more than Q bytes and thus will never request more than Q bytes). *(Continued)*

TABLE 8.1   *(Continued)*

Service	Formula	Description
Constant credit	$W^{[i]} = \min \begin{cases} V^{[i]} + \text{const} \\ W_{\text{MAX}} \end{cases}$	This scheme adds a constant credit to the window size requested. The idea behind adding the credit is the following: Assume that x bytes arrived between the time when an ONU sent a REPORT message and the beginning of the time slot assigned to the ONU as the result of processing the REPORT. If the window size granted equals the window requested $+x$ (i.e., it has a credit of size x), these x bytes will experience a shorter delay, and thus the average delay will reduce.
Linear credit	$W^{[i]} = \min \begin{cases} V^{[i]} \times \text{const} \\ W_{\text{MAX}} \end{cases}$	This scheme uses an approach similar to that of the constant credit scheme. However, the size of the credit is proportional to the window requested. The reasoning here is the following: LRD traffic possesses a certain degree of predictability (see [16]), so if we observe a long burst of data, this burst is likely to continue for some time into the future. Correspondingly, the arrival of more data during the last cycle time may signal that we are observing a burst of packets.
Elastic	$W^{[i]} = \min \begin{cases} V^{[i]} \\ NW_{\text{MAX}} - \sum_{j=i-N}^{i-1} W^{[j]} \end{cases}$	Elastic service is an attempt to get rid of a fixed maximum window limit. The only limiting factor is the maximum cycle time T_{MAX}. The maximum window is granted in such a way that the accumulated size of the last N grants (including the one being granted) does not exceed NW_{max} bytes ($N =$ number of ONUs). Thus, if only one ONU has data to send, it may get a grant of size up to NW_{max}.

8.4.3   Simulation Results

First, let us take a look at the components of the packet delay (Fig. 8.18). The packet delay d is equal to

$$d = d_{\text{poll}} + d_{\text{cycle}} + d_{\text{queue}} \tag{8.9}$$

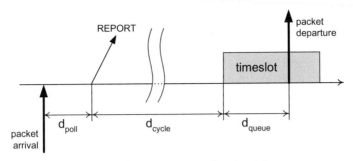

Figure 8.18 Components of packet delay.

where

$d_{poll},$ = time between packet arrival and next REPORT sent by that ONU. On average this delay equals half the cycle time.

d_{cycle} = time interval from ONU's request for a transmission window until the beginning of the time slot in which this frame is to be transmitted. This delay may span multiple cycles (i.e., a frame may have to skip several time slots before it reaches the head of the queue), depending on how many frames there were in the queue at the time of the new arrival.

d_{queue} = delay from the beginning of the time slot until the beginning of frame transmission. On average, this delay is equal to half the slot time and is insignificant compared to the previous two.

Figure 8.19 illustrates the mean packet delay for different time-slot allocation services as a function of an ONU's offered load ϕ. In this simulation, all ONUs have identical load; therefore, the offered network load Φ is equal to $N\phi$.

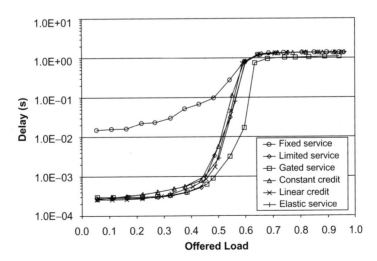

Figure 8.19 Mean packet delay.

As can be seen in Fig. 8.19, all granting services except fixed and gated have almost coinciding plots. We discuss fixed and gated service results below. As for the rest of the schemes, no other method gives a detectable improvement in packet delay. The explanation to this lies in the fact that all these methods are trying to send more data by way of increasing the granted window size. Although that may result in a decrease or elimination of the d_{cycle} delay component for some packets, it will increase the cycle time and thus result in an increase in the d_{poll} component for all the packets.

The fixed service plot is interesting as an illustration of the LRD traffic. Even at the very light load of 5%, the average packet delay is already quite high (about 15 ms). This is because most packets arrive in very large packet trains. In fact, the packet trains were so large that the 10-Mbyte buffers overflowed and about 0.14% of packets were dropped. Why do we observe this anomalous behavior only with fixed service? The reason is that the other services have a much shorter cycle time; there is just not enough time in a cycle to receive more bytes than W_{MAX}; thus the queue never builds up. In fixed service, on the other hand, the cycle is large (fixed) regardless of the network load and several bursts that arrive close to each other can easily overflow the buffer.

Before we continue with our discussion of gated service, we would like to present the simulation results for the average queue size (Fig. 8.20). This picture is similar to the mean delay plot: Fixed service has a larger queue due to larger cycle time.

Let us now turn our attention to the delay and queue size plots for gated service. It can be noticed that gated service provides a considerable improvement in the midrange load between 45 and 65%. At 60% load, for example, the delay and average queue size are approximately 40 times less than that with other services. This happens because gated service has higher channel utilization due to the fact

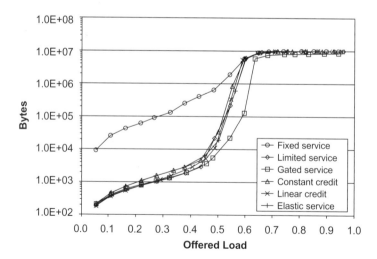

Figure 8.20   Average queue size.

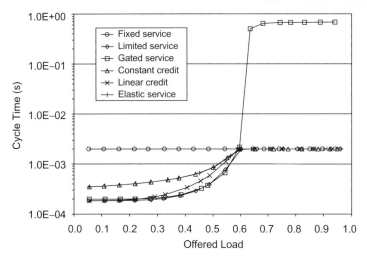

Figure 8.21   Mean cycle times for various service disciplines.

that the cycle time is much larger, and as a result, fewer guard intervals are used per unit of time. For the same reason, its saturation delay is a little bit lower than in other services (see Fig. 8.19)—the entire buffer contents are being transferred in one jumbo transmission rather then in batches of W_{MAX} bytes with a guard time in front of each batch.

Next we will show that even though gated service has lower delay and average queue size, it is not a suitable service for an access network under consideration. The problem lies in the much longer cycle time (see Fig. 8.21). As a result, the d_{poll} delay is much larger and therefore the packet latency is also much higher. Clearly, large d_{cycle} and d_{queue} delay components can be avoided for high-priority packets by using priority queuing. But d_{poll} is a fundamental delay, which cannot be avoided in general. This makes gated service not feasible for an access network.

Thus, we conclude that neither of the service disciplines discussed is better than limited service, which is most conservative of all. As such, for the remainder of this study, we focus our attention on the limited service discipline. In the next section we analyze the fairness and QoS characteristics of limited service.

8.4.4   Performance of Limited Service

In this section we analyze the performance of one ONU (called *tagged ONU*) while varying its offered load (ϕ_i) independent of its ambient load (effective load Ω generated by the rest of the ONUs). In Fig. 8.22 we present the average packet delay. All system parameters remained the same as in the previous simulation.

When the effective network load is low, all packets in a tagged source experience very little delay, irrespective of the ONU's offered load. This is a manifestation of dynamic bandwidth allocation—when the network load is low, the tagged source gets more bandwidth. The opposite situation—low offered load at the ONU and

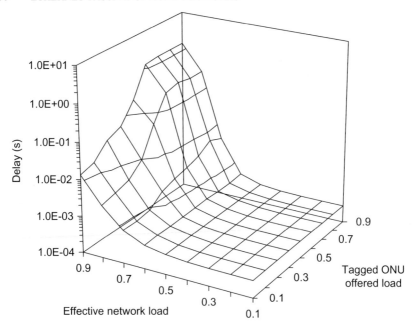

Figure 8.22 Average packet delay as a function of effective network load and ONU offered load.

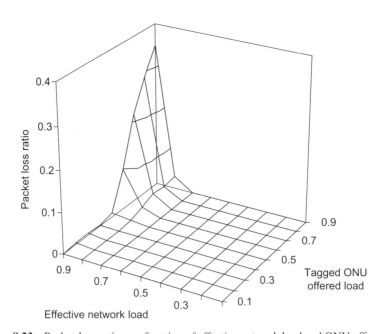

Figure 8.23 Packet-loss ratio as a function of effective network load and ONU offered load.

high effective network load—results in a higher delay. The only reason for this is the burstiness (i.e., long-range dependence) of the traffic. This is the same phenomenon as that observed with fixed service: High network load results in increased cycle time. This cycle time is large enough to receive more than W_{MAX} bytes of data during a burst. Hence, the d_{cycle} delay for some packets will increase beyond one cycle time. We will discuss a way to combat this phenomenon by using the priority queuing scheme described in Section 8.5.3.

Figure 8.23 shows the probability of a packet loss (due to buffer overflow) in a tagged ONU i as a function of its offered load (ϕ_i) and the effective load of the entire network (Ω). The buffer size Q was set to 10 Mbytes as in previous simulations. Once again we observe that the packet-loss ratio is zero or negligible if the effective network load is less than 80%. When the network load is above 80% and the tagged ONU offered load is above 50%, we observe considerable packet loss due to buffer overflow.

In the simulations above, all traffic was treated as belonging to only one class (i.e., all frames have the same priority). In Section 8.5.3.1 we discuss EPON performance for multiple classes of service.

8.5   CONSIDERATIONS FOR IP-BASED SERVICES OVER EPON

The driving force behind extending Ethernet into the subscriber access area is Ethernet's efficiency for delivering IP packets. Data and telecom network convergence will lead to more and more telecommunication services migrating to a variable-length packet-based data network. To ensure successful convergence, this migration should be accompanied by implementation of specific mechanisms traditionally available in telecom networks only.

Being designed with IP layer in mind, EPON is expected to operate seamlessly with IP-based traffic flows, similar to any switched Ethernet network. One distinction with the typical switched architecture is that in an EPON, the user's throughput is slotted (gated; i.e., packets cannot be transmitted by an ONU at any time). This feature results in two issues unique to EPONs: slot utilization by variable-length packets and slot scheduling to support real-time and controlled-load traffic classes.

8.5.1   Slot Utilization Problem

The slot utilization problem is related to the fact that Ethernet frames cannot be fragmented, and as a result, variable-length packets don't fill the given slot completely. When a next packet to be transmitted by an ONU does not fit in a remaining slot, it will have to wait for the next slot, thus leaving an unused remainder in the current slot. This problem manifests itself in several different situations. One example of this behavior can be observed when OLT grants to an ONU a slot smaller than what the ONU requested (i.e., when $W^{[i]} > W_{\text{MAX}}$). Because the OLT has no knowledge of a packet's delineation in ONU's buffer, it

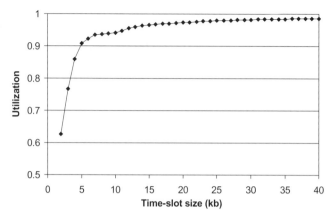

Figure 8.24 Slot utilization for various slot sizes.

cannot grant a slot exactly matching the cumulative size of the packets such that this size does not exceed W_{MAX}. Another example of slots not being filled completely can be found when a priority queue is implemented in an ONU. In this situation, even if the granted slot size exactly matches the ONU's requested slot size, the order of packets in the ONU's queue may change due to high-priority packets arriving after the last request. That, of course, will change packet delineation and will not allow the packets to fill the slot completely.

The fact that there is an unused remainder at the end of the slot means that the user's throughput is less than the bandwidth given to the user by a network operator in accordance with a particular SLA. Figure 8.24 presents slot utilization for packet traces obtained on an Ethernet LAN in Bellcore [17]. Increasing the slot size improves the utilization (i.e., the user's throughput approaches the bandwidth assigned by the operator but has detrimental effects on the data latency, as larger slots increase the overall cycle time).

Slot utilization can also be improved by employing smarter packet scheduling (e.g., the bin-packing problem). Rather than stopping the transmission when the head-of-the-queue frame exceeds the remainder of the slot, the algorithm may look ahead in the buffer and pick a smaller packet for immediate transmission (first-fit scheduling). However, as it turns out, first-fit scheduling is not such a good approach. To understand the problem, we need to look at the effects of packets reordering from the perspective of TCP/IP payload carried by Ethernet frames. Even though TCP will restore the proper sequence of packets, an excessive reordering may have the following consequences:

1. According to the fast retransmission protocol, the TCP receiver will send an immediate ACK for any out-of-order packet, whereas for in-order packets, it may generate a cumulative acknowledgment (typically, for every other packet) [18]. This will lead to more unnecessary packets being placed in the network.

2. Second, and more important, packet reordering in ONUs may result in a situation where n later packets are being transmitted before an earlier packet. This would generate n ACKs ($n - 1$ duplicate ACKs) for the earlier packet. If n exceeds a predefined threshold, it will trigger packet retransmission and reduction of the TCP's congestion window size (the *cwnd* parameter). Currently, the threshold value in most TCP/IP protocol stacks is set to 3 (refer to the fast retransmission protocol in [18] or other references).

Even if special care is taken at the ONU to limit out-of-order packets to one or two, the rest of the end-to-end path may contribute additional reordering. Although true reordering typically generates fewer than three duplicate ACKs and is ignored by the TCP sender, together with reordering introduced by the ONU, the number of duplicate ACKs may exceed 3, thus forcing the sender to retransmit a packet. As a result, the overall throughput of the user's data may decrease.

So, what is the solution? It is reasonable to assume that the traffic entering the ONU is an aggregate of multiple flows. In the case of business users, it would be the aggregated flows from multiple workstations. In the case of a residential user, we still may expect multiple connections at the same time. This is because, as a converged access network, PON will carry not only data, but also voice-over-IP (VoIP) and video traffic. In addition, home appliances are becoming network plug-and-play devices. The conclusion is that if we have multiple connections, we can reorder packets that belong to different connections and never reorder them if they belong to the same connection. Connections can be distinguished by examining the source–destination address pairs and source–destination port numbers. This will require an ONU to look up layer 3 and layer 4 information in the packets. Thus, the important trade-off decision that EPON designers have to make is whether it makes sense to increase the required processing power considerably in an ONU to improve the bandwidth utilization.

8.5.2 Circuit Emulation (TDM over IP)

The migration of TDM circuit-switched networks to IP packet-switched networks is progressing at a rapid pace. But even though the next-generation access network will be optimized for IP data traffic, legacy equipment [e.g., RF set-top boxes, analog TV sets, TDM private branch exchanges (PBXs)] and legacy services [T1/E1, Integrated Services Digital Network (ISDN), Plain Old Telephone Service (POTS), etc.] will remain in use in the foreseeable future. Therefore, it is critical for next-generation access networks, such as Ethernet PONs, to be able to provide both IP-based services and jitter-sensitive and time-critical legacy services that have traditionally not been the focus of Ethernet.

The issue in implementing a circuit-over-packet emulation scheme is mostly related to clock distribution. In one scheme, users provide a clock to their respective ONUs, which in turn is delivered to the OLT. But since the ONUs cannot transmit all the time, the clock information must be delivered in packets. The OLT will regenerate the clock using this information. It is somewhat trivial to impose a constraint that the OLT should be a clock master for all downstream ONU devices. In this scenario, an ONU will recover the clock from its receive channel, use it in its transmit channel, and distribute it to all legacy devices connected to it.

8.5.3 Real-Time Video and Voice over IP

Performance of a packet-based network can conveniently be characterized by several parameters: bandwidth, packet delay (latency) and delay variation (jitter), and packet-loss ratio. Quality of service (QoS) refers to a network's ability to provide bounds on some or all of these parameters. It is useful to further differentiate statistical QoS from guaranteed QoS. Statistical QoS refers to a case when the specified bounds can be exceeded with some small probability. Correspondingly, guaranteed QoS refers to a network architecture where parameters are guaranteed to stay within the bounds for the entire duration of a connection. A network is required to provide QoS (i.e., bounds on performance parameters) to ensure proper operation of real-time services such as video over packets (digital video conferencing, VoD), voice over IP (VoIP), real-time transactions, and so on. To be able to guarantee QoS for higher-layer services, the QoS must be maintained in all the traversed network segments, including the access network portion of the end-to-end path. In this section we focus on QoS in the EPON access network.

The original Ethernet standard based on the CSMA/CD MAC protocol was developed without any concern for QoS. All connections (traffic flows) were treated equally and were given best-effort service from the network. The first major step in allowing QoS in the Ethernet was the introduction of the full-duplex mode. *Full-duplex MAC* (otherwise, called *null-MAC*) can transmit data frames at any time; this eliminated nondeterministic delay in accessing the medium. In a full-duplex link (segment), once a packet is given to a transmitting MAC layer, its delay, jitter, and loss probability are known or predictable all the way to the receiving MAC layer. Delay and jitter may be affected by head-of-line blocking when the MAC port is busy transmitting the preceding frame at the time when the next one arrives. However, with a 1-Gbps channel, this delay variation becomes negligible since the maximum-sized Ethernet frame is transmitted in only about 12 μs. It is important to note that the full-duplex MAC does not make the Ethernet a QoS-capable network: Switches located in junction points still may provide nondeterministic, best-effort services.

The next step in enabling QoS in the Ethernet was the introduction of two new standards extensions: P802.1p, "Supplement to MAC Bridges: Traffic Class Expediting and Dynamic Multicast Filtering" (later merged with P802.1D) and P802.1Q, "Virtual Bridged Local Area Networks." P802.1Q defines a frame format extension allowing Ethernet frames to carry priority information. P802.1p specifies the default bridge (switch) behavior for various priority classes; specifically, it allows a queue in a bridge to be serviced only when all higher-priority queues are empty. The standard distinguishes the following traffic classes:

1. *Network control:* characterized by a "must get there" requirement to maintain and support the network infrastructure
2. *Voice:* characterized by less than 10 ms delay, and hence maximum jitter (one-way transmission through the LAN infrastructure of a single campus)
3. *Video:* characterized by less than 100 ms delay
4. *Controlled load:* important business applications subject to some form of admission control, be that preplanning of the network requirement at one extreme or bandwidth reservation per flow at the time the flow is started at the other end
5. *Excellent effort:* or "CEO's best effort," the best-effort services that an information services organization would deliver to its most important customers
6. *Best effort:* LAN traffic as we know it today
7. *Background:* bulk transfers and other activities that are permitted on the network but that should not affect the use of the network by other users and applications

If a bridge or a switch has fewer than seven queues, some of the traffic classes are grouped together. Table 8.2 illustrates the standards-recommended grouping of traffic classes.

Both full-duplex and P802.1p/P802.1Q standards extensions are important but not sufficient QoS enablers. The remaining part is admission control. Without it, each priority class may intermittently degrade to best-effort performance. Here, EPON can provide a simple and robust method for performing admission control. In Section 8.3.3 we mentioned that multipoint control protocol (MPCP) relies on GATE messages sent from the OLT to ONUs to allocate a transmission window. A very simple protocol modification may allow a single GATE message to grant multiple windows, one for each priority class. REPORT message is also to be extended to report queue states for each priority class. Alternatively, admission control can be left to higher-layer intelligence in ONUs. In this case, the higher layer will know when the next transmission window will arrive and how large it will be and will schedule packets for transmission accordingly.

8.5.3.1 Performance of COS-Aware EPON

In this section we investigate how priority queuing will allow us to provide a delay bound for some services. Below we describe a simulation setup in which the data arriving from a user are classified in three priority classes and directed to different queues in the ONU. The queues then are serviced in order of their priority; a lower-priority queue is serviced only when all higher-priority queues are empty. In this experiment, the tagged ONU has a constant load. We investigate the performance of each class as the ambient network load varies.

The *best-effort* (BE) class has the lowest priority. This priority level is used for nonreal-time data transfer. There is no delivery or delay guarantees in this service. The BE queue in the ONU is served only if higher-priority queues are empty. Since

TABLE 8.2 Mapping of Traffic Classes into Priority Queues (P802.1p)

Number of Queues	Traffic Types Queue Assignments						
1	Network control	Voice	Video	Controlled load	Excellent effort	Best effort	Background
2	Network control	Voice	Video	Controlled load	Excellent effort	Best effort	Background
3	Network control	Voice	Video	Controlled load	Excellent effort	Best effort	Background
4	Network control	Voice	Video	Controlled load	Excellent effort	Best effort	Background
5	Network control	Voice	Video	Controlled load	Excellent effort	Best effort	Background
6	Network control	Voice	Video	Controlled load	Excellent effort	Best effort	Background
7	Network control	Voice	Video	Controlled load	Excellent effort	Best effort	Background

all queues in our system share the same buffer, the packets arriving at higher-priority queues may displace the BE packets that are already in the BE queue. In our experiment, the tagged source has the BE traffic with an average load of 0.4 (40 Mbps).

The *assured forwarding* (AF) class has a higher priority than the BE class. The AF queue is served before the BE queue. In our experiment, the AF traffic consisted of a VBR stream with an average bit rate of 16 Mbps. This corresponds to three simultaneous MPEG-2-coded video streams [19]. Since the AF traffic is also highly bursty (LRD), it is possible that some packets in long bursts will be lost. This will happen if the entire buffer is occupied by AF or higher-priority packets.

The *guaranteed forwarding* (GF) priority class was used to emulate a T1 line in the packet-based access network. The GF class has the highest priority and can displace the BE and AF data from their queues if there is not enough buffer space to store the GF packet. A new GF packet will be lost only if the entire buffer is occupied by GF packets. The GF queue is served before the AF and BE queues. The T1 data arriving from the user is packetized in the ONU by placing 24 bytes of data in a packet. Including Ethernet and UDP/IP headers results in a 70-byte frame generated every 125 μs. Hence, the T1 data consume the bandwidth equal to 4.48 Mbps. Of course, we could put 48 bytes of T1 data in one packet and send one 94-byte packet every 250 μs. This would consume only 3.008 Mbps but will increase the packetization delay.

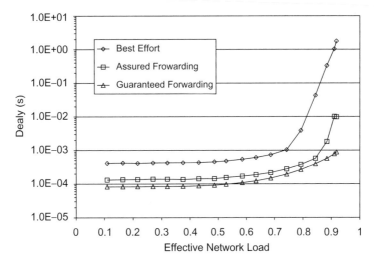

Figure 8.25   Average packet delay for various classes of traffic as a function of the effective network load.

Figures 8.25 and 8.26 show the average and maximum packet delay for each type of traffic. The average load of the tagged ONU was set to 40 Mbps of BE data, 16 Mbps of AF data, and 4.48 Mbps of GF data, a total of about 60 Mbps. The horizontal axis shows the effective network load. These figures show how the traffic parameters depend on the overall network load.

We can see that the BE traffic suffers the most delay when the ambient load increases. Its delay increase and the simulation results show that some packets are discarded when the network load exceeds 80%. The AF data also experience an increased delay, but no packet losses are observed. The increased delay in the AF traffic can be attributed to long bursts of data. Clearly, applying some kind of traffic shaping/policing limiting the burst size at the source would improve the situation. The GF data experience a very slight increase in both average and maximum delays. This is due to the fact that the packets were generated with a constant rate (i.e., no data bursts). The average delay in this case exactly follows the average cycle time, being one-half of that. The maximum delay is equal to the maximum observed cycle time, and for any effective network load it is bounded by T_{MAX} (2 ms with our chosen set of parameters).

Of course, to restore the proper T1 rate, a shaping buffer (queue with constant departure rate) should be employed at the receiving end (in the OLT). After receiving a packet (or a group of packets) from an ONU, the shaping buffer should have at least 2 ms worth of data (i.e., 384 bytes of T1 data). This is because the next packet from the same ONU may arrive after the maximum delay of 2 ms. When such a delayed packet arrives, it should still find the nonempty buffer. Let us say that the minimum buffer occupancy is 24 bytes, which is 125 μs of T1 transmission

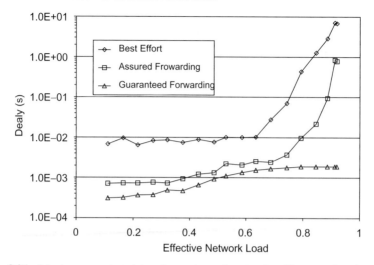

Figure 8.26 Maximum packet delay for various classes of traffic as a function of the effective network load.

time (we call it a *buffer underrun protection time*). In this case, the overall latency experienced by T1 data will consist of:

1. 125 μs of packetization delay in ONU
2. Polling delay in ONU (up to T_{MAX})
3. Up to 100 μs of propagation delay (assuming a maximum distance of 20 km)
4. Wait time in the shaping buffer

Items 2 and 4 together are equal to T_{MAX} plus buffer underrun protection time (i.e., 2.125 ms). Thus, the total latency is about 2.35 ms. If this latency is too high for a given specification, it can be reduced by decreasing T_{MAX} [see equation (8.5)].

8.6 SECURITY ISSUES

Security has never been a strong part of Ethernet networks. In point-to-point full-duplex Ethernet, security is not a critical issue because there are only two communicating stations using a private channel. In shared half-duplex Ethernet, security concerns are minimized because users belong to a single administrative domain and are subject to the same set of policies.

EPON, however, has a different set of requirements, due primarily to its intended use in subscriber access environment. EPON serves noncooperative, private users, but on the other hand, has a broadcasting downstream channel, potentially available to any interested party capable of operating an end station in promiscuous mode. In general, to ensure EPON security, network operators must be able to guarantee subscriber privacy and must be provided mechanisms to control subscriber's access to the infrastructure.

In a residential access environment, individual users expect their data to remain private. For the business access application, this requirement is fundamental. The two main problems associated with lack of privacy are subscriber's susceptibility to eavesdropping by neighbors (a subscriber issue), and susceptibility to theft of service (a service provider issue). Let us explore these two problems:

8.6.1   Eavesdropping

In EPON, eavesdropping is possible by operating an ONU in promiscuous mode. Being exposed to all downstream traffic, such an ONU can listen to traffic intended for other ONUs. Point-to-point emulation adds link IDs (see Section 8.3.4) that allow an ONU to recognize frames intended for it, and filter out the rest. However, this mechanism does not offer the required security, as an ONU might disable this filtering and monitor all traffic.

The upstream transmission in an EPON is relatively more secure. All upstream traffic is multiplexed and is visible only to the OLT (due to directivity of a passive combiner). Although reflections might occur in the passive combiner, sending some upstream signal downstream again. But the downstream transmission is in a different wavelength than the upstream transmissions. Thus, the ONU is "blind" to reflected traffic that is not processed in the receive circuitry. The upstream can also be intercepted at the PON splitter/combiner, as splitters and combiners are most often manufactured as symmetrical devices (i.e., even though only one coupler port is connected to the trunk fiber, more ports are available). A special device sensitive to the upstream wavelength can be connected facing downstream to one such unused port. This device will then be able to intercept all upstream communications.

8.6.2   Theft of Service

Theft of service occurs when a subscriber impersonates a neighbor and transmits frames that are not billed under the impersonator's account. OLT obtains the identity of the subscriber through link ID inserted by each ONU in the frame preambles. This link ID can be faked by the malicious ONU when transmitting in the upstream direction. Of course, to be able to transmit in the hijacked time slot, the impersonating ONU should also be able to eavesdrop to receive GATE messages addressed to a victim.

8.6.3   Solving the Problem Using Encryption

Encryption of downstream transmission prevents eavesdropping when the encryption key is not shared. Thus, a point-to-point tunnel is created that allows private communication between the OLT and the different ONUs. Encryption of the upstream transmission prevents interception of the upstream traffic when a tap is added at the PON splitter. Upstream encryption also prevents ONU impersonation: data arriving from an ONU must be encrypted with a key available to that ONU only. There exist secure methods of key distribution, but they are outside the scope of this book.

Encryption and decryption may be implemented at the physical layer, data link layer, or in higher layers. Implementing encryption above the MAC sublayer will encrypt the MAC frame payload only and leave the headers in plain text. In that scenario, the transmitting MAC will calculate frame check sequence (FCS) for the encrypted payload, and the receiving MAC will verify the received frame integrity before passing the payload to a higher sublayer for decryption. This scheme prevents malicious ONUs from reading the payload, but they may still learn other ONUs MAC addresses.

Alternatively, encryption can be implemented below the MAC. In that scheme, the encryption engine will encrypt the entire bit stream, including the frame headers and FCS. At the receiving end, decryptor will decrypt the data before passing it to MAC for verification. Since encryption keys are different for different ONUs, frames not destined to a given ONU will not decrypt into a properly formed frame and will be rejected. In this scheme no information may be learned by a malicious ONU. Implementing an encryption layer below MAC appears to be a more secure and reliable method.

8.6.3.1 *Encryption Method* The downstream transmission in an EPON is a frame-based communication channel in which each frame is addressed to a different destination. As each frame is an independent piece of information, the encryption method cannot be stream-based. The most appropriate solution is a block-based cipher that encrypts each frame separately.

The link ID field located in each frame's preamble is used to identify a tunnel between the OLT and an ONU (PtP emulation). This header may also be used to support an encryption mechanism in the EPON. For this purpose one of the reserved bytes in the header will be used as a key index (key identifier) (Fig. 8.27). Based on the value of this field, it will be possible to determine whether the frame is encrypted, and what key was used.

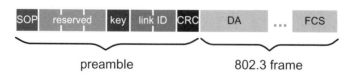

Figure 8.27   Frame preamble with embedded link ID and encryption key index.

Each ONU holds a key that is valid for the current session. The key identifier references the key in the ONU. This behavior allows a smooth transition from one valid session to the next when rekeying the session. A default key identifier is used for frames sent unencrypted. This mechanism has a built-in expansion path, as different key indexes can lead to different cipher algorithms and allow for eventual implementation of conditional access systems at layer 2.

Periodic rekeying allows maintaining the security of the established tunnels indefinitely. As block ciphers use fixed-size blocks, and the Ethernet frames are of variable length, block boundary may be dissimilar to packet boundary, and the last block will be stuffed to reach the required block size. As zero stuffing is a potential weakness in the encryption, an alternative method is used in which the last n bits ($n < 128$) will be XOR-ed, with the result of a second cipher iteration of the next-to-last block.

The advanced encryption standard (AES) algorithm, originally designed to replace the aging data encryption standard (DES), is considered for Ethernet PONs. This algorithm allows the use of 128-, 192-, or 256-bit keys.

8.7   EPON UPGRADE SCENARIOS

By the amount of bandwidth it makes available to subscribers, EPON is a giant step forward compared to DSL and cable modem technologies. With a line rate of 1 Gbps (current target for IEEE 802.3ah efforts) and 16 to 64 subscribers per EPON, each user will get between 15 and 60 Mbps bandwidth. Still, unavoidably, as more bandwidth-intensive services become available to users, this capacity will get exhausted. It is therefore crucial for the success of EPON technology to provide a smooth path for future upgrades. It is hard to envision what upgrade scenario will be most favorable at the time when the EPON capacity will become a limiting factor. The best scenario will not require a forklift upgrade (i.e., will allow incremental expenses) and will capitalize on the most matured technology. In this section we consider three possible directions: wavelength upgrade, rate upgrade, and spatial upgrade.

8.7.1   Wavelength Upgrade

If WDM technology matured enough to provide high-volume, low-cost components, it may become economically feasible to migrate EPON to a multiwavelength configuration. In this scheme, some of the EPON ONUs will migrate to new wavelengths for both upstream and downstream traffic. While the data rate on each wavelength will remain the same, there will be fewer ONUs to share that bandwidth capacity. This procedure can be repeated again in the future and eventually will lead to a WDM PON system where each subscriber is allocated its individual wavelength(s). The major cost factor for such an upgrade is the necessity of tunable transmitters and receivers in ONUs. Also, the OLT must have multiple transceivers (tunable or fixed)—one for each wavelength. To allow incremental upgrade, a new spectral region should be allocated for premium ONUs. This will allow nonpremium ONUs to continue using cheap 1310-nm lasers with high spectral width. Only the premium ONUs will have to be replaced by new ones operating at different wavelength or having tunable transceivers.

8.7.2 Rate Upgrade

With finalizing of the 10-Gbps Ethernet standard by the IEEE, rate upgrade appears as an attractive solution. To allow incremental upgrade cost, only a subset of ONUs may be upgraded to operate at higher rates. Thus, the rate-upgrade scenario will call for a mixed-rate EPON where some ONUs operate at 1 Gbps and some at 10 Gbps. This upgrade would require high-speed electronics in the OLT, which is able to operate at both rates. The main challenge with this kind of upgrade is that dispersion penalties become much more severe. The narrow margin of available power budget may not allow the desired distances or concentration ratios.

8.7.3 Spatial Upgrade

It is reasonable to expect the fiber prices to reduce enough to consider a spatial-upgrade scenario in which a subset of users is migrated to a separate EPON. In this scenario, a new trunk fiber is deployed from the central office to the splitter, and some branches are reattached to a new trunk fiber (and a new splitter). To avoid the cost of additional fiber deployment, this upgrade fiber can be predeployed at the time of the original deployment. Alternatively, some network operators consider EPON deployment with a splitter located in the central office. This EPON configuration will require as much fiber to be deployed as in the point-to-point configuration, but it will still require only one transceiver in the OLT. This will allow having much higher-density equipment, which is very important in a limited CO space available to competitive local exchange carriers (CLECs), who have to rent CO space from incumbent LECs (ILECs). In a scenario where a splitter is located in the central office, to upgrade to higher bandwidth, some users will be reconnected to another EPON in the patch panel located in the same central office. Given the current state of the technology, this seems to be the most cost-efficient type of upgrade. Eventually, spatial upgrade may lead to a point-to-point architecture with an independent fiber running to each subscriber.

8.8 IEEE P802.3ah STATUS

The standards work for Ethernet in the local subscriber access network is being done in the IEEE P802.3ah Ethernet in the First Mile Task Force. This group received approval to operate as a task force from the IEEE Standards Association (IEEE-SA) Standards Board in September 2001. The P802.3ah Ethernet in the First Mile Task Force is bringing Ethernet to the local subscriber loop, focusing on both residential and business access networks. Although at first glance this appears to be a simple task, in reality the requirements of local exchange carriers are vastly different from those of enterprise managers for which Ethernet has been designed. To "evolve" Ethernet for local subscriber networks, P802.3ah is focused on four primary standards definitions:

1. Ethernet over copper
2. Ethernet over point-to-point (PtP) fiber
3. Ethernet over point-to-multipoint (PtMP) fiber
4. Operation, administration, and maintenance (OAM)

Thus, the Ethernet in the First Mile (EFM) Task Force is focused on both copper and fiber standards, optimized for the first mile, and glued together by a common operation, administration, and maintenance system. This is a particularly strong vision, as it allows a local network operator a choice of Ethernet flavors using a common hardware and management platform. In each of these subject areas, new physical layer specifications are being discussed to meet the requirements of service providers while preserving the integrity of Ethernet. Standards for Ethernet in the First Mile are anticipated by September 2003, with baseline proposals emerging as early as March 2002.

The Ethernet-over-point-to-multipoint (PtMP) track is focusing on the lower layers of an EPON network. This involves a physical (PHY)-layer specification, with possibly minimal modifications to the 802.3 MAC. The standards work for P2MP fiber-based Ethernet is in progress, while the multipoint control protocol framework is emerging. This emerging protocol uses MAC control messaging (similar to the Ethernet PAUSE message) to coordinate multipoint-to-point up-stream Ethernet frame traffic. Materials concerning the P802.3ah standards effort can be found at [20] and presentations materials at [21].

8.9 SUMMARY

Unlike the backbone network, which received an abundance of investment in long-haul fiber routes during the Internet boom, optical technology has not been deployed widely in the access network. It is possible that EPONs and point-to-point optical Ethernet offer the best possibility for a turnaround in the telecom sector. As service providers invest in optical access technologies, this will enable new applications, stimulating revenue growth and driving more traffic onto the backbone routes. The large increase in access network bandwidth provided by EPONs and point-to-point optical Ethernet will eventually stimulate renewed investment in metro and long-haul fiber routes.

The subscriber access network is constrained by equipment and infrastructure not originally designed for high-bandwidth IP data. Whether riding on shorter copper drops or optical fiber, Ethernet is emerging as the future broadband protocol of choice, offering plug-and-play simplicity, IP efficiency, and low cost. Of particular interest are Ethernet passive optical networks, which combine low-cost point-to-multipoint optical infrastructure with low-cost, high-bandwidth Ethernet. The future broadband access network is likely to be a combination of point-to-point and point-to-multipoint Ethernet, optimized for transporting IP data as well as time-critical voice and video.

REFERENCES

1. K. G. Coffman and A. M. Odlyzko, Internet growth: Is there a "Moore's law" for data traffic? in *Handbook of Massive Data Sets*, J. Abello, P. M. Pardalos, and M. G. C. Resende, Eds., Kluwer Academic, Dordrecht, The Netherlands, 2001.

2. *Broadband 2001:A Comprehensive Analysis of Demand, Supply, Economics, and Industry Dynamics in the U.S. Broadband Market*, J.P. Morgan Securities, New York, Apr. 2001.

3. *Access Network Systems: North America—Optical Access, DLC and PON Technology and Market Report*, Report RHK-RPT-0548, RHK Telecommunication Industry Analysis, San Francisco, June 2001.

4. G. Pesavento and M. Kelsey, PONs for the broadband local loop, *Lightwave*, 16(10):68–74, Sept, 1999.

5. B. Lung, PON architecture "futureproofs" FTTH, *Lightwave*, 16(10):104–107, Sept. 1999.

6. B. Mukherjee, *Optical Communication Networks*, McGraw-Hill, New York, 1997.

7. F. J. Effenberger, H. Ichibangase, and H. Yamashita, Advances in broadband passive optical networking technologies, *IEEE Communications*, 39(12):118–124, Dec. 2001.

8. R. Ramaswami and K. N. Sivarajan, *Optical Networks: A Practical Perspective*, Morgan Kaufmann, San Francisco, 1998.

9. S. Clavenna, Metro Optical Ethernet, *Lightreading* (*www.lightreading.com*), Nov. 2000.

10. K. Claffy, G. Miller, and K. Thompson, The nature of the beast: Recent traffic measurements from an Internet backbone, *Proc. INET'98*, Geneva, July 1998. Available at *http://www.isoc.org/inet98/proceedings/6g/6g_3.htm*.

11. J. L. Hammond and P. J. P. O'Reilly, *Performance Analysis of Local Computer Networks*, Addison-Wesley, Reading, MA, 1987.

12. W. Willinger, M. S. Taqqu, and A. Erramilli, A bibliographical guide to self-similar traffic and performance modeling for modern high-speed networks, in *Stochastic Networks*, F. P. Kelly, S. Zachary, and I. Ziedins, Eds., Oxford University Press, Oxford, 1996, pp. 339–366.

13. W. Willinger, M. Taqqu, R. Sherman, and D. Wilson, Self-similarity through high-variability: Statistical analysis of Ethernet LAN traffic at the source level, *Proc. ACM SIGCOMM'95*, Cambridge, MA, Aug. 1995, pp. 100–113.

14. G. Kramer, B. Mukherjee, and G. Pesavento, Ethernet PON (ePON): Design and analysis of an optical access network, *Photonic Network Communications*, 3(3):307–319, July 2001.

15. G. Kramer, B. Mukherjee, and G. Pesavento, Interleaved polling with adaptive cycle time (IPACT): A dynamic bandwidth distribution scheme in an optical access network, *Photonic Network Communications*, 4(1):89–107, Jan. 2002.

16. K. Park and W. Willinger, Self-similar network traffic: An overview, in *Self-Similar Network Traffic and Performance Evaluation*, K. Park and W. Willinger, Eds., Wiley-Interscience, New York, 2000.

17. W. Leland, M. Taqqu, W. Willinger, and D. Wilson, On the self-similar nature of ethernet traffic (extended version), *IEEE/ACM Transactions on Networking*, 2(1):1–15, Feb. 1994.

18. W. R. Stevens, *TCP/IP Illustrated*, Vol. 1, Addison-Wesley, Reading, MA, 1994.

19. M. W. Garrett and W. Willinger, Analysis, modeling and generation of self-similar VBR video traffic, *Proc. ACM Sigcomm'94*, London, Sept. 1994, pp. 269–280.

20. IEEE P802.3ah Task Force, *http://www.ieee802.org/3/efm*.

21. IEEE P802.3ah Task Force, *http://www.ieee802.org/3/efm/public*.

22. Ethernet in the First Mile Alliance, *http://www.efmalliance.org*.

9 Terabit Switching and Routing Network Elements

TI-SHIANG WANG and SUDHIR DIXIT

Nokia Research Center, Burlington, Massachusetts

9.1 INTRODUCTION

Bandwidth-hungry applications and incessant consumer demand have provided carriers with the impetus to expand network capacity at a phenomenal pace. As a result, some carriers are relying on the approach of adding more switches and routers to keep up with the demand for more bandwidth. Others are installing fiber systems using dense wavelength-division multiplexing (DWDM) systems able to exploit high bandwidth of the fiber, which can carry up to 80 OC-48 (2.5-Gbps) channels per fiber. However, the central-office space necessary for this additional equipment is often very difficult to attain. Even when space is made available, the sheer mechanical and thermal requirements of more racks and boxes impose an entirely new set of problems that must be overcome before services can even be provisioned. In addition, data-centric traffic on the Internet has increased, although circuit-switching requirements remain and will persist well into the future. This proliferation of data traffic is being driven by a number of factors, including the fact that the number of Internet users worldwide is increasing dramatically. To keep pace with the growing demand, transmission speeds have increased to the level of gigabits or even terabits per second.

To use this huge bandwidth efficiently, as well as to satisfy the increasing Internet bandwidth demand, carriers are working to define scalable system architectures able to provide multigigabit capacity now and multiterabit capacity in the near future. Both constituencies also realize that these terabit-class switches and routers will need to do much more than pass cells and packets at significantly higher aggregate line rates. From the marketing analysis and prediction point of view, Ryan Hankin and Kent (RHK) predicts that the core gigabit/terabit switch and router market will jump to $5.53 billion in 2003. In addition, IDC points out that the

IP over WDM: Building the Next-Generation Optical Internet, Edited by Sudhir Dixit.
ISBN 0-471-21248-2 © 2003 John Wiley & Sons, Inc.

number of points of presence (POPs) and mega-POPs that will require terabit-class routers between now and 2003 is between 2000 and 3000.

There are several limitations on the existing network infrastructure. The existing public networks are based on circuit switch technology to provide only voice services. As data traffic carried over the existing network infrastructure increases, carriers increase the capacity of their networks by overlaying devices designed to increase data transmission rates based on network standards such as Synchronous Optical Network/Synchronous Digital Hierarchy (SONET/SDH), which is not scalable and not efficient for packet/data applications. In the existing network infrastructure, the efficiency of data transmission is being increased by adopting packet-switching technologies such as asynchronous transfer mode (ATM) and Internet protocol (IP), which divide data traffic into individual packets and transmit them independently over the transport network, such as SONET/SDH.

As far as the existing routers/switches are concerned, there is a chasm between the capabilities of existing routers and the optical transmission network, which produces bottlenecks in the existing network infrastructure. This chasm results in large part because of the limited ability of existing router architectures to adapt to the evolving and increasing bandwidth demands of carriers. That is, huge bandwidth in the backbone cannot be exploited fully or utilized sufficiently via existing routers or switches. Current router offerings employ a centralized architecture with a limited number of interfaces available and limited scalability. This limitation requires that carriers cluster multiple routers to emulate the functionality of a single large router with greater capacity in optical transmission speeds, as shown in Fig. 9.1.

Clustering routers requires a significant number of interfaces, most of which are then dedicated solely to interconnecting multiple routers and switches. In other words, clustering has experienced inefficiency because most of these interfaces are used for interconnection between routers and/or switches instead of transmission. At the same time, existing routers are unable to dynamically enable the provisioning of new services without disrupting the entire network. This limitation increases

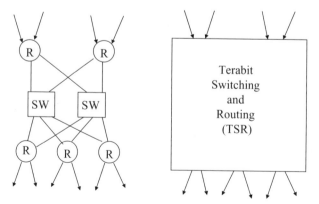

Figure 9.1   Cluster multiple routers and terabit switch/router.

the time and effort necessary to deliver new services or to reconfigure the network based on the bandwidth demand by the traffic. On the other hand, next-generation transport platforms must also be able to handle efficiently all the prevailing logical layer protocols, such ATM, IP, multiprotocol label switching (MPLS), gigabit Ethernet, and circuit-switched time-division multiplexing (TDM). Providing this level of service elasticity at terabit-rate aggregate bandwidths will require an entirely new network element. This new network element will need a high-density switch core, which is able to manage the switching function of the massive traffic volume. In addition, an intelligent controller and potentially hundreds of line cards able to aggregate and process the varied services are required.

9.2   RELATED ACTIVITIES [1–19]

Because of its huge bandwidth and flexibility, optical fibers and wavelength-division multiplexing (WDM) technology enable the carriers to expand the band-width with existing fiber infrastructure while introducing the interconnecting switch modules to alleviate the drawbacks of electronic switches such as dense inter-connection requirements and electromagnetic interference. How to make use of huge bandwidth (ca. Tbps) of optical fiber to build terabit packet switches is also a challenging issue that remains to be answered. In the past few years many studies have been performed on the design of terabit switching and routing (TSR) node.

In TSR, to perform the packet-switching function, information is encapsulated in packets consisting of a header and payload. The header contains information pertaining to packet destination, while payload carries the actual information. That is, a TSR has three major functions: switching, buffering, and header translation. Switching ensures that each packet emerges at the correct output, depending on the information contained in the packet header. This information might be related to the output destination or service requirements. Buffering is indispensable in packet-switching systems since one or more packets may arrive during the same time slot on different inputs wishing to go to the same output. An important factor when designing TSR is contention resolution, since multiple packets may arrive asyn-chronously at the same time to go to the same output port. Depending on the processing format of the packet (i.e., optically or electronically), the terabit switching and routing can be categorized as transparent or opaque.

9.2.1   Transparent Terabit Switching and Routing [2–9]

With the help of advanced technology in optical devices (particularly WDM technology), several research activities have been in progress to implement the required TSR all-optically to bring the transparency performance onto the network. The implementation of transparent TSR is to process packets optically. That is, each payload of a packet is kept in the optical domain until it arrives at its destination. The header of a packet is processed optically as well. In addition, transparent TSR provides major advances in bit rate/signal format independence, network integration capacity, and operational simplicity. That is, packets are allowed to bypass extensive electronic signal processing at intermediate switching nodes. Furthermore, both logical control and contention resolution are assumed to be handled by an electronic controller. In addition, the payload of a packet is carried and stored in the optical buffers.

Optical buffers can be implemented only by delaying the transmission time of optical packets by fiber lines. For transparent TSR, two kinds of optical buffering schemes have been proposed: fiber delay line and random access. A comprehensive survey of optical buffering schemes and their comparison is presented in [1] and in Chapter 5. Optical buffering schemes based on fiber delay lines are used to stagger the contended packets into different fiber delay lines, which represent different times to send out the packets to the same output port. To provide some sharing property for these fiber delay lines, wavelength conversion and some scheduling algorithms are also implemented in the design of TSR [8]. In optical buffering based on a random access scheme, all the packets with different wavelengths share the same fiber delay line and packets (or wavelengths), which can be stored in or read out by controlling optical switches [7].

Before performing the switching function in the TSR, the front-end processor is required to perform some key functions. These functions include packet delineation, header overwrite, and packet synchronization. The packet delineation unit identifies packet boundaries. Once the packet boundaries are identified, the header (e.g., VCI/VPI fields in ATM switching) is replaced optically with the new values. The packet synchronization unit is used to align the phase of incoming packets to make sure that they are switched in the optical switching fabric without any corruption. In other words, the range of the front-end processor capacity should be scaled easily from gigabit per second to multiple terabits per second.

9.2.2   Opaque Terabit Switching and Routing [1,10–19]

Compared to the transparent TSR, in opaque TSR the payload of each incoming packet and its header is processed electrically at each TSR. That is, for the electronic-buffered TSR, all the logical control, contention resolution, and packet storage are handled electronically. The contended packets are stored in electronic buffers. Either optical switches or electrical switches can be used for interconnection and transmission between electronic input and output ports. In particular, due to limited resources of available wavelength and optical device technology, interconnection networks using optical switches should be built up by several small modular WDM subsystems. These can be scaled up in space, time, and wavelength domains to meet the requirement of terabit per second capacity in the future. Optical interconnection networks should be designed with a modular structure to allow scaling up in order to provide large switching size (in space domain), to handle a high data rate (in time domain), and to be state of the art (next-generation TSRs in wavelength domain). The best combination of these three domains will be a key to the success of next-generation TSRs.

TABLE 9.1   Summary of Work in Terabit Switching and Routing (TSR)

TSR Type	Authors	Buffering Placement				Arbitration Scheme	Terabit Capability	Throughput/ Delay (%)	Multicast
		Input	Input/Output	Output	Shared				
Transparent	Hass [2]					×			
	Chiaroni et al. [3]			×	×			100	×
	Guillemotet et al. [4]			×	×			100	×
	Duan et al. [5]	×							
	Sasayama et al. [6]			×					
	Choa and Chao [7]			×				100	×
	Hunter et al. [8]				×				
	Raffaelli [9]			×		×			
Opaque	Wang and Dixit [1]		×				×		×
	Arthurs et al. [10]	×				×		31.8	
	Lee et al. [11]	×				×			×
	Cisneros and Brackett [12]	×			×	×	×	58.6	
	Munter et al. [13]		×			×			
	Nakahira et al. [14]		×						
	Yamanaka et al. [15]		×				×		
	Chao et al. [16]		×			×	×	100	
	Maeno et al. [17]						×		×
	Araki et al. [18]						×		×
	Chao et al. [19]		×			×	×	100	×

As mentioned before, buffering is indispensable in the packet-switching systems, and several demonstrations of optical loop buffer have been made to store the contended packets optically. However, the storage time (or the number of circulation) of packets in an optical loop buffer is limited by the amplified spontaneous emissions (ASEs), which are generated by the optical amplifiers that degrade the signal-to-noise performance. Thus, WDM technology seems unattractive for TSR based on optical buffering in the near future. On the other hand, the role of WDM technology will come in the form of optical interconnection between high-performance electronic switches. Several opaque TSRs using optical interconnection networks (OINs) have been proposed relying on advanced optical devices. The optical interconnection network (OIN) is designed to interconnect electronic switches/routers at both input and output sides. In this approach, electronic buffers are placed inherently at the input and output ports of the interconnection networks. In other words, the front-end processor mentioned above will, inherently, be implemented electronically.

To achieve the best throughput/delay performance, fast arbitration schemes are required to overcome head-of-line (HOL) blocking if an input-buffered scheme is applied with the help of powerful logical control using electronic technologies and devices. The OIN-based TSR derives strength from electronic logical operation and avoids the issues of optical buffering as mentioned above. However, somewhat complicated and centralized electronic controls are required so that the control complexity increases. Furthermore, OIN-based TSRs do not have the capability of providing transparent services in terms of data rates and protocols. In Table 9.1 we summarize previous work in TSR in terms of buffering placement, arbitration schemes supporting terabit capacity, throughput/delay performance, and multicast function.

9.3 KEY ISSUES AND REQUIREMENTS FOR TSR NETWORKS

In an effort to respond to the challenges posed by the increasing volume of data traffic on the existing public networks, carriers focus on optimizing their next-generation optical networks for more efficient data transmission in addition to providing new data communications services. As a result, carriers are demanding terabit switching and routing with the following requirements:

9.3.1 Traffic Aggregation

Today's typical metro network supports a number of end-user services. These services are usually transported over the regional and/or long-haul networks, which are built around DWDM equipment. Because the cost per wavelength is quite high, end-user services, especially low-speed services, need to be groomed to maximize utilization of each wavelength. Thus, a router/switch with terabit capacity such as TSR should work with fiber and support traffic aggregation, including OC-3/12/48/192 and SDH-1/4/16/64 add–drop multiplexing (ADM). TSR will have the capability to provide digital cross-connecting DWDM terminals, ATM and Ethernet switches on a single platform.

9.3.2   Modular Structure and Greater Granularity

To provide flexibility, it is necessary to be able to offer granularity from T1 (1.5 Mbps) through OC-192 (10 Gbps) in the same box. Scalability must be achieved across a wide range without disrupting the network as it grows over an extended period of time. Thus, TSR requires a modular approach rather than a single-chassis approach. In other words, TSR should be modular and scalable, adding packet routing/forwarding in the interface modules. Furthermore, its switching fabric can be built within a range terabits or above with small switching modules.

9.3.3   Scalability

With data growing two- to fourfold annually, gigabit routers would have to be replaced every two to three years. Thus terabit-class routers/switches are required to allow growth by small and cost-effective increments of switch capacity, without changing the existing investment and infrastructure to accept the current traffic load while allowing a smooth migration to heavier loads of both voice and data in the future.

9.3.4   Multiple Protocol Interfaces

In addition to providing a smooth migration to an IP-centric core, or vice versa, TSRs have to preserve the investments in legacy equipment, such as SONET, ATM, and GbE. Furthermore, TSRs should provide interfaces directly to DWDM transponders. This allows optical internetworking that effectively harnesses the bandwidth created by fiber and DWDM advancements.

9.3.5   Quality of Service

QoS is essentially the ability to assign different priorities to different types of traffic, which is crucial to the delivery of time-sensitive data streams such as voice and video. Queuing delays on certain network elements, intermittent latency, and dropped packets from congested links make it difficult to provide predictable performance, even with today's best-effort Internet. Some customers are willing to pay a premium for more predictable service tied to service-level agreements (SLAs). Next-generation switches or routers must maintain control mechanisms to enforce preferred access to network resources for high-priority flows. As new applications such as video streaming and real-time packetized voice migrate to IP, the network must meet their requirements as well as those of legacy services. Terabit switching and routing must incorporate packet prioritization, network engineering, traffic congestion management and control, and ultimately, bandwidth management to enable carriers to deliver QoS guarantees to their customers. This is in addition to providing capacity in the Tbps range.

Furthermore, the multitude of layers produces bandwidth inefficiencies and adds to the latency of connections and QoS assurances. These layers are unaware of each other, causing duplication of network services and generating unnecessary overhead at each layer. Thus, it is necessary to reduce the operations cost by integrating functionality from layers 1 and 3 to simplify protocol stack, where IP traffic is either carried by SONET/SDH over the WDM or fiber layer, or is mapped directly to the WDM layer. TSRs should be transparent to packet data rate and format in addition to extending the success of point-to-point links in an all-optical switched network.

9.4   ARCHITECTURES AND FUNCTIONALITIES

9.4.1   Generic Architecture

In Fig. 9.2 a generic terabit switching and routing node architecture is shown. Basically, each TSR node consists of four subsystems: input port interface (IPI), switching fabric, output port interface (OPI), and electronic controller. Depending on the architecture, both IPI and OPI perform one or more of the following functions:

- Buffering
- Packet filtering and classification

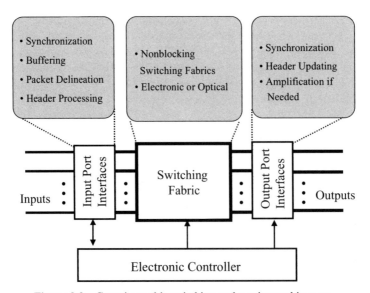

Figure 9.2   Generic terabit switching and routing architecture.

- Synchronization
- Packet delineation
- Header processing and updating
- Optical or optoelectronic conversion if needed

Switching fabric, implemented either optically or electronically, is controlled by the electronic controller. It performs the switching function and routes or switches packets to their destinations. The electronic controller is responsible for controlling, monitoring, and providing control protocols for the TSR. Each subsystem of a TSR is described in detail in the following sections.

9.4.2 Input Port Interface and Output Port Interface

In the IPI/OPI, synchronization function is performed to align packets at multiple input ports in order to correlate the packet positions with actual switching events. Packet delineation circuits are used to find the beginning of each packet. These two functions can be implemented in both transparent and opaque TSR using different approaches. In the opaque optical switch, electronic buffers or memory stacks inherently provide these functions. On the other hand, extra circuits with some fiber delay lines and optical switching components are necessary to perform these functions in the transparent optical switches. In addition, the header information of each packet is tapped at the IPI for processing and updated by new header information at the OPI. To perform the IP routing function, packet classification and filtering are also performed here.

In the OPI, in addition to performing header updating function, amplifiers are used to ensure that these packets are routed through multiple switches in terms of routing, timing, and proper signal level. For transparent optical switches, optical amplifiers such as semiconductor optical amplifier (SOAs) and erbium-doped fiber amplifier (EDFA) can be integrated with other subsystems in the switch node to perform amplification optically without regenerating the packets in their electrical formats. In the opaque optical switches, the amplification and header update can be done in the electronic domain if necessary.

Because of packet switching, every router must be able to store packets to deal with multiple packets destined for the same output. Basically, packet buffers (or queues) can be placed at the inputs, outputs, or both ends of the router. Theoretically, all these buffering schemes can be implemented either optically or electronically. However, high complexity in terms of components, interface, and control will be experienced for transparent TSR.

9.4.3 Buffering Schemes

9.4.3.1 Output Buffering   Placing buffers at the output, as shown in Fig. 9.3, enables shaping the traffic transmission. This allows throughput/delay performance to reach 100%. However, implementing output buffers requires the switching fabric

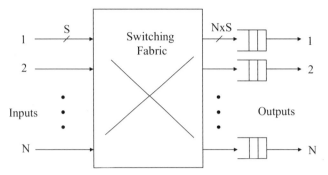

Figure 9.3 Output-buffering TSR architecture.

to be capable of forwarding traffic to a single output at the aggregate speed of all inputs in the case of all input packets intended for the same output simultaneously. For example, if the link speed of input is S, the line speed of output will be $N \times S$ if all the inputs have the same destination. This fast access operation limits the switching capacity and makes the implementation difficult. Clearly, for high-speed ports (in the Gbps range), this kind of architecture is not feasible and only a small switching fabric size can be implemented.

9.4.3.2 *Input Buffering*   On the other hand, placing the buffers at the inputs creates difficulties in shaping the traffic and can result in head-of-line (HOL) blocking, as shown in Fig. 9.4. Each input link has its dedicated buffer. HOL blocking occurs when the packet at the head of a buffer cannot be transmitted to an output due to a contending packet from another input, while a packet farther back in the buffer is blocked, although its destination port is open to receive the packet. The HOL blocking problem limits the throughput of input buffered TSR to approximately 58% of the maximum aggregate input rate.

To overcome HOL blocking, multiple queues must be used at each input as shown in Fig. 9.5. This type of queuing, called *virtual output queuing*, (VOQ), can

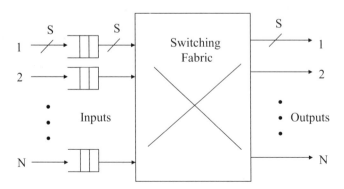

Figure 9.4 Input-buffering TSR architecture.

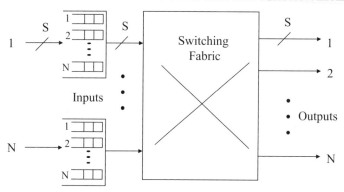

Figure 9.5   Virtual output queuing TSR architecture.

perform as well as output queuing, depending on the arbitration algorithm used to choose which queue to send a packet from at every given packet time slot. One should notice that there is no high-speed operation for input buffering scheme. That is, the line speed is the same (i.e., S) in the entire switch.

9.4.3.3 Input/Output Buffering   To compensate for the drawbacks of input buffering and output buffering mentioned above, input/output buffering with a moderate speedup has been widely considered as the most feasible solution for terabit routers and switches. One of the challenges to building terabit routers/ switches based on the input/output buffering lies in the stringent arbitration time to solve output contention. Traditional arbiters handle all inputs together and the arbitration time is proportional to the number of inputs. As a result, the switch size or capacity is limited given a fixed amount of arbitration time. A distributive arbitration scheme to provide the terabit capacity has been proposed [20].

9.4.4   Switching Fabric

Although placement of buffering schemes affects the performance and complexity while designing TSR, the switching fabric plays another important role in routing the packets and resolving the contention in real time. The design of a switching fabric depends on the switch requirements, the maturity in technology of devices, and the contention approach used in the switch (i.e., input, output, shared, and input/output buffering scheme). Basically, switching fabric can be categorized as optical switching or electrical switching. Optical switching involves switching signals optically. Optical switches are generally defined as switches that use either mirrors or refractive effects to redirect the stream of light. The signals carried on the fiber are optical and would stay optical through the switch, with no need to translate to electrical and back to optical. There is another benefit to an optical switch. It operates independent of the bit rate of the optical signal, especially for the signal on the fiber with a high data rate, such as OC-48, OC-192, and OC-768. Alternatively, electrical switches will act on the electrical bit stream derived from the optical signal using transponders. Transponders perform optical-to-electrical (O/E) and electrical-to-optical (E/O) conversion. They also clean up the signal and provide wavelength interchange if needed. From the implementation point of view, the switching fabric can be built in either single-stage or multistage configuration.

9.4.4.1 Single-Stage Approach   Single-stage switching fabric implementations limit this scaling limitation. One example is the shared-memory switch [21]. Traditionally, $N \times N$ shared memory architecture consists of a multiplexer, a demultiplexer, and a shared-memory stack. The multiplexed data rate is N times the interface rate using TDM operation at input and output to write in and read out the memory stack. It requires high-speed memories even if the serial-to-parallel conversion circuits are introduced. Also, a small size of switching function can be achieved.

To overcome the constraint of accessing the memory stack and to provide an alternative to using TDM operation at the input and output of the memory, a switch architecture with shared multiple memories is introduced. For this architecture, multiple memories are shared among all inputs and outputs through two crossbar switches, as shown in Fig. 9.6. Although the high-speed operation to access these memory stacks can be relaxed, maintaining the sequence of packets and efficiently using these memories remains an issue. In addition, the size of the switch is still small and it is difficult to achieve the capacity at the terabit level. The scalability of bandwidth is limited as well.

Crossbar switching fabric is another example of a single-stage approach. A crossbar switch, shown in Fig. 9.7, enables higher capacity than a shared-memory design. Each switching element in the switching fabric is regulated by the electronic controller as bar or cross state. For example, packets in input ports 4 and 5 are destined to output ports 2 and 5, respectively. Then the associated switching elements are controlled by the bar states, and the rest of the elements are maintained in cross states. However, crossbar switching architecture suffers from other limitations. In a crossbar, in addition to the capacity, the size of the switch

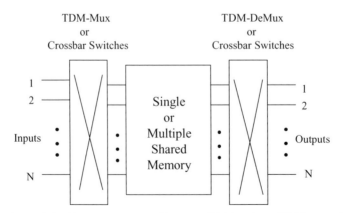

Figure 9.6   Shared-memory architecture (TDM and non-TDM cases).

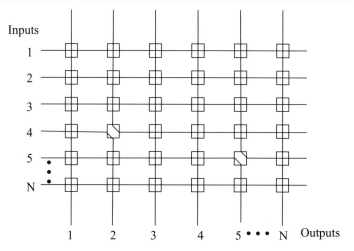

Inputs

Outputs

Figure 9.7 Crossbar architecture.

grows proportionally to the square of the number of switch ports. In addition, multi-cast function cannot be achieved without adding extra switching fabric (i.e., copy network or switching fabric), which increases the complexity.

9.4.4.2 Multistage Approach Multistage designs, such as the well-known Banyan family of fabrics along with others, such as Delta and Omega fabrics, can be modularly implemented and scaled more efficiently (e.g., three-stage Clos and broadcast and select [1,22]), as shown in Fig. 9.8. Modular multistage switching fabrics do not experience memory limitations and provide low complexity in the switching size. Each modular switch in the multistage switching fabric can be implemented as crossbar architecture. The crossbar architecture enables multiple inputs to connect to multiple outputs simultaneously, as long as different inputs are connected to different outputs. In Fig. 9.8, an $mk \times nl$ three-stage architecture is

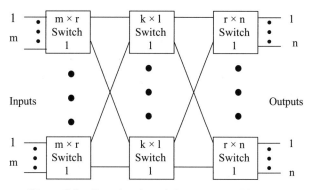

Figure 9.8 Generic $mk \times nl$ three-stage architecture.

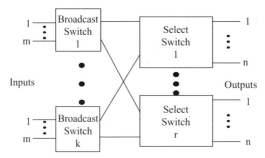

Figure 9.9   An $mk \times nr$ broadcast-and-select architecture.

shown. To assure strict nonblocking, $m + n \geq 2r - 1$ [22]. The multistage architecture is very suitable for building high-performance switching fabrics. It is possible to connect several crossbars together to achieve even greater bandwidth and more inputs.

Alternatively, compared to multistage architecture, a broadcast-and-select architecture can be easy to implement optically and to provide multicast function. Basically, the function of each broadcast switch is twofold. It includes the multiplexing function of incoming packets and the distribution of these packets to all select switching modules, as shown in Fig. 9.9. Each switching module performs the selection/switching function to select the specific packets to their destined output ports. The broadcast function can be implemented using passive optical couplers, such as the star-coupler and power splitter/combiner. Basically the select switches of a broadcast-and-select architecture consist of crossbar-like switching fabric and some optical filters [1]. That is, one or many optical packets or signals can be selected, depending on the architecture, for the packet-switching function. Some optical switching devices with fast switching speeds, such as SOA gates, lithium niobates, and optical external modulators, have been considered for use in the future optical packet-switching applications. Thanks to current advanced optoelectronic integrated chip (OEIC) and planar lightwave circuit (PLC) technologies [23,24], optical active components (e.g., laser diodes, SOA gates, optical modulators) and passive devices (e.g., optical couplers, optical multiplexers/demultiplexers) can be integrated monolithically in the same package or substrate. These technologies enable a wide variety of functions that will be required in building the optical switching matrix and providing cost minimization. Furthermore, the reliability in optical switching matrix can be enhanced as well.

9.5   IMPLICATION OF IP ON OPTICAL NETWORK NODES

9.5.1   IP-Based IPI and OPI

Basically the IPIs and OPIs are interconnected by the switching fabric (using either optical or electronic technology). To meet the stringent time constraint imposed on IP forwarding, especially IP table lookup, this operation is separated from the construction of routing tables and the execution of routing protocols, called a *route controller* (RC). IP forwarding is implemented in hardware at each IPI/OPI, whereas the construction of routing tables and the execution of routing protocols are handled and implemented in the electronic controller. With the separation, the routing table lookup and QoS control will move out of IPI/OPI; hence each IPI/OPI becomes more lightweight. A lightweight IPI/OPI can achieve higher forwarding speed and therefore accommodate higher-speed link interface. In addition, with the separation, better load-balancing performance can be achieved without adding the complexity of the IP forwarding operations. This is because the RC is shared by all the forwarding operations in the IPI/OPI.

An IPI provides several functions. A simple packet scheduler (e.g., round-robin operation) is used to arrange the packet arrivals from different interfaces. This packet scheduler can be implemented using a first-in, first-out (FIFO) buffer per interface to store incoming packets to prevent packet loss. The output packets of the scheduler enter the input switch interface, in which packet segmentation takes place. While a packet is segmented, its IP header is first checked by the input packet filter for network security and flow classification. Afterward, the header is sent to the forwarding engine for IP table lookup, deciding which OPI(s) this packet is destined for.

One should note that data segments are stored in a FIFO waiting for arbitration before being forwarded through the switching fabric if input buffering is applied. The forwarding sequence is packet by packet for each input switch interface in order to simplify the reassembly. The input port number is added to each segment before it enters the switching fabric for ensuring correct packet reassembly at the OPIs.

At each OPI, segments of a packet arriving at an output port may be interleaved with those from other input ports. While a packet is being reassembled, its IP header can be sent to the output packet filter for outbound filtering and then to the output forwarding engine for another IP route lookup. Here it will decide for which outgoing interface(s) this packet should be destined. The packets are then broadcast at the output line interface to all the desirable interfaces. For the outgoing path, outgoing packets are buffered in the OPI before being scheduled. There might be multiple buffers at each OPI with different priorities. QoS-aware scheduling schemes are used to enforce the differentiation between different service classes. Currently, an Internet backbone router may have to deal with more than 40,000 routes in its routing table. As the number of routes keeps increasing, computing and maintaining the routing tables by the RC becomes even more challenging. One solution could be to use multiple RCs and interconnecting them to achieve both higher processing speed and load balancing. The detailed operation and design of IPI/OPI are described in [19].

9.5.2 IP-Based Electronic Controller

Basically, the electronic controller is responsible for the arbitration if input buffering is applied in the switching node design and route controller function.

9.5.2.1 Route Controller The RC performs three main tasks: (1) executing routing protocols (such as RIP and DVMRP) and other control protocols (such as ICMP, IGMP, and SNMP), exchanging control messages via the IPIs with neighbor routers and network management servers, and maintaining a routing information base (RIB) in the system; (2) based on the RIB computing, updating, and distributing the IP unicast/multicast forwarding table (also called the forwarding information base) for every IPI; and (3) performing packet filter management. All the communications are through a control plane, which can be implemented either by a specific bus, such as a PCI bus, or by the core switch, such as the OIN.

9.5.2.2 Arbitration Scheme Pure output buffering has the best delay/throughput performance but is limited to a certain size, due to the constraint of memory speed. Two design techniques overcome this bottleneck. The first is to build very wide memories that can load an entire packet into a single memory cycle by deploying memory elements in parallel and feeding them into a packetwide data bus. The second approach to building fast output buffers is to integrate a port controller and the associated buffers on a single chip. This approach allows the read/write control logic to access buffers in memory, where an entire row is read/written at a time. As the input line speed is high (i.e., above 2.5 Gb/s), both of these approaches are costly and not easily implementable. In contrast, pure input buffering suffers from HOL blocking and thus has low throughput and a longer delay. Input/output buffering and doubling of the switch's internal speed is used to yield almost 100% throughput and low average input buffer delay. The output contention among the input packets is resolved by the arbitration scheme.

9.6 WHY MULTIPROTOCOL LABEL SWITCHING?

9.6.1 Without MPLS

Although scalable hardware that supports more IP traffic capacity at terabit level is a key element, the IP platform needs scalable software to complement the hardware capabilities. These would support information exchange at all levels within the network. Support for these services requires protocols that enable efficient allocation of bandwidth among users in a flexible and dynamic manner. Currently, the routing software provides support for the routing protocols, such as IS-IS, open shortest path first (OSPF), and BGP. IS-IS is an interior gateway protocol (IGP) that provides dynamic routing between routers using the shortest-path first algorithm (SPF) or Dijkstra algorithm. OSPF uses the SPF to route packets within a single autonomous system (AS) or domain. BGP is an exterior gateway protocol (EGP) that performs routing between multiple domains or ASs in IP networks. Currently, the choice of a routing path is not coupled with any knowledge of the necessary path characteristics required at the application level nor the ability to find better-suited routes based on the awareness of the overall topology and available resources of the network. These protocols would simply find the shortest path from source to destination, regardless of the utilization along the path. In other words, the traditional routing and packet forwarding cannot provide the resources, such as scalability, QoS, and traffic engineering, to meet future needs. MPLS addresses the key problems faced by today's networks—speed, scalability, QoS, management, and traffic engineering—by providing a mechanism for efficient routing and resource reservation.

9.6.2   With MPLS [25–27]

MPLS enhanced the performance and scalability of IP networks by replacing traditional IP packet forwarding (i.e., using complicated address-matching algorithms) with fast, simple, ATM-like switching based on labels. This functionality enabled basic traffic engineering capabilities and brought virtual private networking (VPN) services to IP networks. With the help of MPLS, terabit switching and routing will be more intelligent and scalable. MPLS is composed of a number of different protocol specifications that work together to give the network the omniscient knowledge required to balance traffic patterns. It extends IS-IS and OSPF protocols to make them bandwidth aware so that they understand the long-term expected utilization of each link. At each TSR, these routing protocols are used to advertise information about each of its link's reserved value to all other TSRs in the network. Each TSR learns about the current reservation of status of links in the network and understands how much unreserved bandwidth is available for new traffic flows. Additionally, MPLS uses a signaling protocol to reserve bandwidth along the path selected. If all routers still have sufficient capacity, an additional bandwidth is reserved along this path. Otherwise, this path is rejected, and the routing protocol attempts to find another path.

Another advantage of using MPLS is to collapse network layers as shown in Fig. 9.10. As we know, many networks today are stuck with a multilayer model: DWDM, SONET, ATM, and IP. It moves from layer 0 to layer 3. Generalized MPLS (GMPLS) is supposed to allow IP to be engaged directly with DWDM, eliminating SONET and ATM, which are the two most expensive and hard-to-scale layers. Through constraint-based routing, MPLS will provide ATM-style QoS and traffic engineering at the IP layer. Similar routing techniques can provide fast reroute capability for protection and restoration.

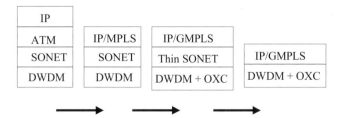

Figure 9.10   Collapsing network layers in tomorrow's optical network.

Supporting traffic engineering in TSR is very important in today's core networks because it provides the ability to move traffic to alternative paths, avoiding bottlenecks. The traffic engineering allows service providers to maximize network efficiency and ensure smooth operational service to their customers. In addition, traffic engineering is essential to increase throughput across a network for various traffic types while decreasing congestion. When a more desirable route becomes available, high-priority classes may be distributed seamlessly to the new path without loss of service quality. Basic traffic engineering is implicit in MPLS because label switch paths are created independently based on user-defined policies and current network information. The ingress node, which is the first MPLS-enabled node at the network edge, can calculate the optimum path based on traffic requirements and network conditions. Updated information about network conditions is regularly broadcast, automatically or on demand, among MPLS nodes using OSPF and IS-IS enhanced with traffic engineering extensions for MPLS (now known as OSPF-TE and IS-IS-TE).

MPLS enhances standard routing protocols by providing fine control over how traffic moves through the network, including routing traffic around link failures and congestion. The resources the traffic flow requires and the resources actually available in the network determine the route for the traffic flow. MPLS sets a specific route for a data packet. This route is defined by assigning a label to the packet. These labels contain information based on the routing table entry as well as specific values for prioritizing packets, thus enhancing network efficiency. For dynamic traffic engineering and QoS, the ingress node would use either of the following enhanced MPLS control protocols to identify and reserve the best label-switched path. CR-LDP (constraint-based routing label distribution protocol) is an extension to the basic MPLS label distribution protocol (LDP). Unlike basic LDP, CR-LDP supports explicit routing parameters. Constraint-based routing extends this concept by advertising additional information. Resource information (e.g., link capacity and utilization) and link characteristics (i.e., delay) are included in the routing protocol. This communication allows routers to take network resource information and link capabilities into account when selecting paths. Alternatively, RSVP-TE (resource reservation protocol with traffic engineering) is an extension to the decade-old RSVP protocol and supports capabilities somewhat similar to CR-LDP. The resource reservation protocol with traffic engineering extensions is a label distribution protocol. The RSVP-TE implementation enables service providers to optimize network performance.

MPLS provides a way to optimize packet flows within IP networks standards-based signaling, routing, and path computation intelligence. These capabilities enable the quality-of-service mechanisms required for delivering profitable services, ensuring efficient network resource utilization, and unifying the network infrastructure even as it carries multiple protocols. MPLS works with multiple protocols by putting its own label onto their data cells/packets inserted between layer 2 and layer 3. Packets from the data plane [i.e., ATM, frame relay, IP, Ethernet, and packet over SONET (PoS)] of all these protocols can be carried on a common MPLS core. Regardless of the underlying protocol, packets are forwarded by the MPLS label attached to them. There is no need to read all that detail in the originally assigned layer 2 or layer 3 headers, or to be subject to the limitations of the underlying protocol. That is, MPLS is transparent to the data packets. The reader is encouraged to refer to Chapter 14 for details on MPLS and GMPLS.

9.7 SUMMARY

As the demands on the network core continue to increase, terabit switch routers can be expected to adopt a data-centric approach, handling circuit and multiservice in the core that consists of IP/MPLS-based switch routers interfaced directly to DWDM transponders. While optical switches and terabit switches/routers provide an intermediate step toward higher-performance and higher-throughput TDM-based public network, terabit switch routing allows one to create an entirely new type of packetized network. Terabit switch/routers will simplify the core network, reduce the operational costs, improve network performance, and provide quality of service in the future.

Generalized MPLS extends the reach of the MPLS control plane beyond routers and ATM switches all the way to physical layer devices such as optical cross-connects, photonic switches, and legacy TDM devices. In addition, MPLS and GMPLS will redefine packet and optical networks and provide network surviv-ability using protection and restoration techniques, which will enable the introduc-tion of new services together with enhanced methods to deliver existing services. By integrating these protocols into terabit switching and routing architecture, service providers will enable an intelligent, high-performance Internet that yields revenue opportunities with reduced operational costs. By using MPLS, scalable routing and intelligent optical switching, carriers can take large amounts of band-width and convert raw capacity into revenue-generating services.

Many of the optical crossconnects (OXCs) being deployed today use an extended version of MPLS to set up end-to-end lightpaths, called multiprotocol lambda switching (MPλS). MPλS is a kind of bandwidth-aware routing protocol that keeps track of the reserved and available wavelengths on each fiber pair. This routing protocol with RSVP or CR-LDP signaling allows for the rapid routing of lightpaths across the network. In addition, these optical networks also rely on the ability of MPLS to reroute these lightpaths in the event of a fiber failure to provide the survivability for the optical-service network. With the help of OXC, TSR implementation with extension to MPLS to set up end-to-end lightpaths in an all-optical switching network will happen in the future.

REFERENCES

1. T. S. Wang and S. Dixit, Optical interconnection networks for terabit packet switches, *Photonic Network Communications*, 2(1):61–71, Jan./Mar. 2000.
2. Z. Hass, The "staggering switch": An electrically controlled optical packet switch, *Journal of Lightwave Technology*, 11(5/6):925–936, May/June 1993.

3. D. Chiaroni et al., Sizeability analysis of a high-speed photonic packet switching architecture *Proc. 21st European Conference on Optical Communications, (ECOC'95)*, paper We.P.47, pp. 793–796, 1995.

4. C. Guillemot et al., KEOPS optical packet switch demonstrator: Architecture and testbed performance, *OFC'00 Technical Digest*, paper ThO1, pp. 204–206, 2000.

5. G. H. Duan et al., Analysis of ATM wavelength routing systems by exploring their similitude with space division switching, *Proc. ICC'96*, 3:1783–1787, 1996.

6. K. Sasayama et al., FRONTIERNET: Frequency-routing-type time-division interconnection network, *Journal of Lightwave Technology*, 15(3):417–429, Mar. 1997.

7. F. S. Choa and H. J. Chao, On the optically transparent WDM ATM multicast (3M) switches, *Fiber and Integrated Optics*, 15:109–123, 1996.

8. D. K. Hunter et al., WASPNET: A wavelength switched packet network, *IEEE Communications*, 37(3):120–129, Mar. 1999.

9. C. Raffaelli, Design of a core switch for an optical transparent packet network, *Photonic Network Communications*, 2(2):123–133, May 2000.

10. E. Arthurs et al., HYPASS: An optoelectronic hybrid packet switching system, *IEEE Journal on Selected Areas in Communications*, 6(9):1500–1510, Dec. 1988.

11. T. T. Lee et al., STAR-TRACK: A broadband optical multicast switch, *Bellcore Technical Memorandum Abstracts*, TM-ARH-013510, 1989.

12. A. Cisneros and C. A. Brackett, A large ATM switch based on memory switches and optical star couplers, *IEEE Journal on Selected Areas in Communications*, 9(8):1348–1360, Oct. 1991.

13. E. Munter et al., A high-capacity ATM switch based on advanced electronic and optical technologies, *IEEE Communications*, pp. 64–71, Nov. 1995.

14. Y. Nakahira et al., Evaluation of photonic ATM switch architecture: Proposal of a new switch architecture, *Proc. International Switching Symposium*, pp. 128–132, 1995.

15. N. Yamanaka et al., OPTIMA: Tb/s ATM switching system architecture, *Proc. IEEE ATM Workshop*, pp. 691–696, 1997.

16. H. J. Chao, J. S. Park, and T. S. Wang, Terabit/s ATM switch with optical interconnection network, *Proc. IEEE Lasers and Electro-Optics Society 1997 Annual Meeting*, Vol. 1, pp. 130–131, 1997.

17. Y. Maeno et al., A 2.56-Tb/s multiwavelength and scalable fabric for fast packet-switching networks, *IEEE Photonic Technology Letters*, 10(8):1180–1182, Aug. 1998.

18. S. Araki et al., A 2.56 Tb/s throughput packet/cell-based optical switch-fabric demonstrator, *Proc. ECOC'98*, 3:127–129, 1998.

19. H. J. Chao, X. Guo, C. Lam, and T. S. Wang, A terabit IP switch router using optoelectronic technology, *Journal of High Speed Networks*, pp. 35–38, Jan. 1999.

20. H. J. Chao, Saturn: A terabit packet switch using dual round-robin, *IEEE Communications*, 38(12):78–84, Dec. 2000.

21. J. Garcia-Haro and A. Jajszczyk, ATM shared-memory switching architecture, *IEEE Network*, 8(4):18–26, July/Aug. 1994.

22. C. Clos, A study of non-blocking switching networks, *Bell System Technical Journal*, 32:406–424, Mar. 1953.

23. F. Ebisawa et al., High-speed 32-channel optical wavelength selector using PLC hybrid integration, *OFC/IOOC '99 Technical Digest*, 3:18–20, 1999.

24. R. Kasahara et al., Fabrication of compact optical wavelength selector by integrating arrayed-waveguide gratings and optical gate array on a single PLC platform, in *ECOC'99*, 2:122–123, 1999.

25. C. Metz, Layer 2 over IP/MPLS, *IEEE Internet Computing*, 5(4):77–82, July/Aug. 2001.

26. A. Banerjee et al., Generalized multiprotocol label switching: An overview of signaling enhancements and recovery techniques, *IEEE Communications*, 39(7):144–151, July 2001.

27. B. Davie and Y. Rekhter, *MPLS Technology and Applications*, Morgan Kaufmann, San Francisco, 2000.

10 Optical Network Engineering

GEORGE N. ROUSKAS

North Carolina State University, Raleigh, North Carolina

10.1 INTRODUCTION

Over the last few years we have witnessed a wide deployment of point-to-point wavelength-division multiplexing (WDM) transmission technology in the Internet infrastructure. The corresponding massive increase in network bandwidth due to WDM has heightened the need for faster switching at the core of the network. At the same time, there has been a growing effort to enhance the Internet protocol (IP) to support traffic engineering [1,2] as well as different levels of quality of service (QoS) [3]. Label switching routers (LSRs) running multiprotocol label switching (MPLS) [4,5] are being deployed to address the issues of faster switching, QoS support, and traffic engineering. On the one hand, label switching simplifies the forwarding function, thereby making it possible to operate at higher data rates. On the other hand, MPLS enables the Internet architecture, built on the connectionless Internet protocol, to behave in a connection-oriented fashion that is more conducive to supporting QoS and traffic engineering.

The rapid advancement and evolution of optical technologies makes it possible to move beyond point-to-point WDM transmission systems to an all-optical backbone network that can take full advantage of the bandwidth available by eliminating the need for per-hop packet forwarding. Such a network consists of a number of optical cross-connects (OXCs) arranged in some arbitrary topology, and its main function is to provide interconnection to a number of IP/MPLS subnetworks. Each OXC can switch the optical signal coming in on a wavelength of an input fiber link to the same wavelength in an output fiber link. The OXC may also be equipped with converters that permit it to switch the optical signal on an incoming wavelength of an input fiber to some other wavelength on an output fiber link. The main mechanism of transport in such a network is the lightpath (also referred to as a λ-channel), an optical communication channel established over the network of OXCs, which may span a number of fiber links (physical hops). If no wavelength converters are used, a lightpath is associated with the same wavelength

IP over WDM: Building the Next-Generation Optical Internet, Edited by Sudhir Dixit.
ISBN 0-471-21248-2 © 2003 John Wiley & Sons, Inc.

on each hop. This is the well-known wavelength continuity constraint. Using converters, a different wavelength on each hop may be used to create a lightpath. Thus, a lightpath is an end-to-end optical connection established between two subnetworks attached to the optical backbone.

Currently, there is tremendous interest within both the industry and the research community in optical networks in which OXCs provide the switching functionality. The Internet Engineering Task Force (IETF) is investigating the use of generalized MPLS (GMPLS) [6] and related signaling protocols to set up and tear down lightpaths. GMPLS is an extension of MPLS that supports multiple types of switching, including switching based on wavelengths usually referred to as multi-protocol lambda switching (MPλ S). With GMPLS, the OXC backbone and the IP/MPLS subnetworks will share common functionality in the control plane, making it possible to integrate all-optical networks seamlessly within the overall Internet infrastructure. Also, the optical domain service interconnection (ODSI) initiative (which has completed its work) and the Optical Internetworking Forum (OIF) are concerned with the interface between an IP/MPLS subnetwork and the OXC to which it is attached as well as the interface between OXCs, and have several activities to address MPLS over WDM issues [7]. Optical networks have also been the subject of extensive research [8] investigating issues such as virtual topology design [9,10], call blocking performance [11,12], protection and restoration [13,14], routing algorithms and wavelength allocation policies [15–17], and the effect of wavelength conversion [18–20], among others.

Given the high cost of network resources and the critical nature of the new applications (especially those supporting business operations) that fuel the growth of the Internet, there has been increasing interest among network providers in traffic engineering techniques [2]. Network resource and traffic performance optimization will continue to be important issues, as providers introduce all-optical components in their networks. Since optical device technology has not yet reached the maturity level of electronic component technology, the transmission and switching devices that will need to be deployed to realize an all-optical network tend to be more expensive and bulky (i.e., require more storage space) than their electronic counterparts. Due to the extremely high data rates at which these networks are expected to operate (10 to 40 Gbps or beyond), a network malfunction or failure has the potential to severely affect critical applications. Also, a single fiber or wavelength may carry a large number of independent traffic streams, and a service outage may have wide implications in terms of the number of customers affected. Therefore, the application of network engineering methods to optimize the utilization of resources while meeting strict performance objectives for the traffic carried will be crucial for the emerging optical networks.

In this chapter we discuss the application of network and traffic engineering techniques to the design and operation of optical networks. The chapter is organized as follows. In Section 10.2 we introduce the basic elements of the optical network architecture and in Section 10.3 discuss the concepts of network and traffic engineering as they apply to optical networks. In Section 10.4 we describe off-line approaches for the design and capacity planning of optical networks and in Section 10.5 study on-line algorithms for dynamic provisioning of lightpaths. In Section 10.6 we discuss standardization activities under way for optical networks, with an emphasis on the control plane issues that are important for traffic engineering. Section 10.7 concludes the chapter.

10.2   OPTICAL NETWORK ARCHITECTURE

The architecture for wide-area WDM networks that is widely expected to form the basis for a future all-optical infrastructure is built on the concept of *wavelength routing*. A wavelength-routing network, shown in Fig. 10.1, consists of *optical cross-connects (OXCs)* connected by a set of fiber links to form an arbitrary mesh topology. The services that a wavelength-routed network offers to attached client subnetworks are in the form of *logical* connections implemented using *lightpaths*. Lightpaths are clear optical paths which may traverse a number of fiber links in the optical network. Information transmitted on a lightpath does not undergo any conversion to and from electrical form within the optical network, and thus the architecture of the OXCs can be very simple because they do not need to do any signal processing. Furthermore, since a lightpath behaves as a literally transparent "clear channel" between the source and destination subnetwork, there is nothing in the signal path to limit the throughput of the fibers.

The OXCs provide switching and routing functions for supporting the logical data connections between client subnetworks. An OXC takes in an optical signal at each of the wavelengths at an input port and can switch it to a particular output port, independent of the other wavelengths. An OXC with N input and N output ports capable of handling W wavelengths per port can be thought of as W independent $N \times N$ optical switches. These switches have to be preceded by a wavelength

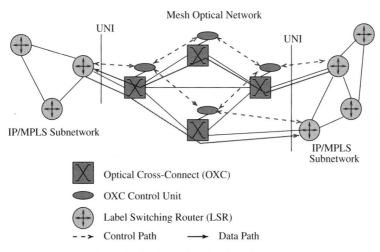

Figure 10.1   Optical network architecture.

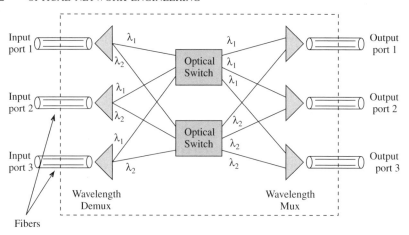

Figure 10.2   A 3 × 3 optical cross-connect (OXC) with two wavelengths per fiber.

demultiplexer and followed by a wavelength multiplexer to implement an OXC, as shown in Fig. 10.2. Thus, an OXC can cross-connect the different wavelengths from the input to the output, where the connection pattern of each wavelength is independent of the others. By configuring the OXCs appropriately along the physical path, logical connections (lightpaths) may be established between any pair of subnetworks.

As Fig. 10.1 illustrates, each OXC has an associated *electronic control unit* attached to one of its input/output ports. The control unit is responsible for control and management functions related to setting up and tearing down lightpaths; these functions are discussed in detail in Section 10.6. In particular, the control unit communicates directly with its OXC and is responsible for issuing configuration commands to the OXC to implement a desired set of lightpath connections; this communication takes place over a (possibly proprietary) interface that depends on the OXC technology. The control unit also communicates with the control units of adjacent OXCs or with attached client subnetworks over *single-hop lightpaths* as shown in Fig. 10.1. These lightpaths are typically implemented over administratively configured ports at each OXC and use a separate control wavelength at each fiber. Thus, we distinguish between the paths that data and control signals take in the optical network: Data lightpaths originate and terminate at client subnetworks and transparently traverse the OXCs, while control lightpaths are electronically terminated at the control unit of each OXC. Communication on the control lightpaths uses a standard signaling protocol (e.g., GMPLS), as well as other standard protocols necessary for carrying out important network functions, including label distribution, routing, and network state dissemination. Standardization efforts are crucial to the seamless integration of multivendor optical network technology and are discussed in Section 10.6.

Client subnetworks attach to the optical network via edge nodes which provide the interface between nonoptical devices and the optical core. This interface is denoted as UNI (user-to-network interface) in Fig. 10.1. The edge nodes act as the terminating points (sources and destinations) for the optical signal paths; the communication paths may continue outside the optical network in electrical form. In Fig. 10.1, only the label-switching routers (LSRs) of the two IP/MPLS subnetworks which are directly attached to an OXC implement the UNI and may originate or terminate lightpaths. For the remainder of this chapter we make the assumption that client subnetworks run the IP/MPLS protocols. This assumption reflects the IP-centric nature of the emerging control architecture for optical networks [21]. However, edge nodes supporting any network technology (including ATM switches and SONET/SDH devices) may connect to the optical network as long as an appropriate UNI is defined and implemented.

In addition to simply supporting logical connections between subnetworks, the optical backbone must protect clients from network impairments and the failure of any resources (including fiber links, optical transceivers, and OXCs) to ensure reliable operation and continuity of service. In this chapter we assume that the optical network can support the following levels of lightpath protection [22]:

1. *Dedicated protection.* A protected lightpath consists of a primary (working) route and a dedicated, diversely routed *backup* (protection) route. Under $1 + 1$ protection, the source transmits simultaneously on both the primary and backup path; upon a failure on the primary path, the destination simply switches to the backup path to receive the signal. Under $1 : 1$ protection, transmission occurs on the primary path only, while the backup path may carry low-priority traffic; upon a failure on the primary path, the source and destination of the protected traffic switch to the backup path, and the low-priority traffic is preempted. A $1 + 1$ or $1 : 1$ protected lightpath can recover from any failure on its working route.

2. *Shared protection.* A protected lightpath has a primary route and a diversely routed *shared* backup route. The backup route is shared among a set of protected lightpaths in such a way that any single failure on the primary route of any lightpath in the set can be restored by having the source and destination of the failed lightpath switch to the shared backup path.

3. *No protection.* The lightpath is not protected; that is, no spare capacity (backup route and wavelength) is set aside to protect from failures along the lightpath route. In the event of a failure, the logical connection is lost; however, the network may attempt to dynamically restore the lightpath by utilizing any available resources.

In [23] and [24], the concept of a lightpath was generalized into that of a *light tree*, which, like a lightpath, is a clear channel originating at a given source node and implemented with a single wavelength. But unlike a lightpath, a light tree has multiple destination nodes; hence it is a point-to-multipoint channel. The physical links implementing a light tree form a tree, rooted at the source node, rather than a path in the physical topology, hence the name. Light trees may be implemented by employing optical devices known as *power splitters* [25] at the OXCs. A power splitter has the ability to split an incoming signal, arriving at some wavelength λ, into up to m outgoing signals, $m \geq 2$; m is referred to as the *fanout* of the power splitter. Each of these m signals is then switched independently to a different output port of the OXC. Note that due to the splitting operation and associated losses, the optical signals resulting from the splitting of the original incoming signal must be amplified before leaving the OXC. Also, to ensure the quality of each outgoing signal, the fanout m of the power splitter may have to be limited to a small number. If the OXC is also capable of wavelength conversion, each of the m outgoing signals may be shifted, independently of the others, to a wavelength different than the incoming wavelength λ. Otherwise, all m outgoing signals must be on the same wavelength λ.

An attractive feature of light trees is the inherent capability for performing multicasting in the optical domain (as opposed to performing multicasting at a higher layer, e.g., the network layer, which requires electro-optic conversion). Such wavelength-routed light trees are useful for transporting high-bandwidth, real-time applications such as high-definition TV (HDTV). Therefore, OXCs equipped with power splitters will be referred to as *multicast-capable* OXCs (MC-OXCs). Note that just as with converter devices, incorporating power splitters within an OXC is expected to increase the network cost because of the need for power amplification and the difficulty of fabrication.

10.3 OPTICAL NETWORK AND TRAFFIC ENGINEERING

While the deployment of WDM technology networks in the Internet infrastructure is motivated primarily by the urgent need to accommodate the geometric growth rates of Internet traffic, simply expanding the information-carrying capacity of the network is not sufficient to provide high-quality service. As the Internet evolves into a mission-critical infrastructure and becomes increasingly commercial and competitive in nature, users will depend on (and indeed, expect or demand) reliable operation, expeditious movement of traffic, and survivability, while service providers will need to ensure efficient utilization of network resources and cost-effective operation. Consequently, emerging optical networks must be architected and engineered carefully so as to achieve both user- and provider-specific performance objectives.

In abstract terms, we can think of a network in general, and an optical network in particular, as having three components:

1. A *traffic demand*, either measured or estimated, which is usually expressed as a traffic matrix
2. A *set of constraints*, which represents the physical fiber layout, the link capacity, the OXCs, and other optical devices (including fiber amplifiers, wavelength converters, etc.) deployed
3. A *control policy*, which consists of the network protocols, policies, and mechanisms implemented at the OXC control modules

Building an optical network that efficiently and reliably satisfies a diverse range of service requests involves the application of *network and traffic engineering techniques* to determine optimal (or near-optimal) operating parameters for each of the three components above.

In essence, therefore, network and traffic engineering is a control problem. In general, depending on the nature of the network technology, control may be exercised at multiple time scales, and different mechanisms are applicable to different time scales [26]. Since optical WDM networks are circuit-switched in nature (i.e., lightpath holding times are assumed to be significantly longer than the round-trip time across the diameter of the optical network), we distinguish two time scales at which it is appropriate to apply control mechanisms:

1. The *connection-level* time scale is the time over which client subnetworks request, use, and tear-down logical connections (lightpaths). The optical network nodes and edge LSRs must implement a *signaling* mechanism for clients to make their requests and declare their traffic requirements. To ensure that the traffic requirements are met, the network must exercise *admission control* (i.e., it may be necessary to deny some lightpath requests). For requests that are allowed, *routing and wavelength assignment* must be performed at connection time. Also, *restoration* mechanisms to recover from network failures operate at this time scale.

2. If the network is persistently overloaded, the admission control algorithm will have to deny lightpath requests frequently and the clients will experience high blocking. The only solution in this situation is to increase network capacity. Since installing new OXCs and fiber trunks may take several months, network engineering techniques that deal with capacity shortage operate at this time scale. These techniques fall under the broad areas of *network design* and *capacity planning*. Also, *protection techniques* must be employed to set aside adequate spare capacity for lightpaths that must be protected from network failures.

Table 10.1 summarizes the control mechanisms appropriate for each of the two time scales.

TABLE 10.1   Control Time Scales in Optical Networks, and Associated Mechanisms

		Point Exercised	
Time Scale	Control Mechanism	OXC Control Unit	Edge LSR
Connection level	Signaling	Yes	Yes
(network operation	Admission control	Yes	No
phase)	RWA	Yes	No
	Restoration	Yes	No
Months or longer	Capacity planning	Yes	No
(network design phase)	Protection	Yes	No

We note that network and traffic engineering is by nature an adaptive process. Initially, the network topology is determined and a control policy is formulated; the policy depends not only on the constraints imposed by the network topology but also on factors such as the cost structure, the revenue model, and the performance objectives. During network operation, the traffic and network state are observed and analyzed continuously. The observation phase requires that a set of online performance and fault monitoring functions be implemented in the network. In the analysis phase, various techniques are applied to identify existing or potential bottlenecks that (may) affect network performance. The feedback from the analysis stage is used to update the network topology and/or reformulate the control policy in order to drive the network to a desirable operating range. The network and traffic engineering cycle then begins anew.

While capacity planning and dynamic provisioning of lightpaths operate at different time scales, the underlying control problem that must be addressed at both the network design and network operation phases is the routing and wavelength assignment (RWA) problem. Because of the prominent role of the RWA problem in the design and operation of optical WDM networks, it is discussed in detail next.

10.3.1 Routing and Wavelength Assignment

A unique feature of optical WDM networks is the tight coupling between routing and wavelength selection. As can be seen in Fig. 10.1, a lightpath is implemented by selecting a path of physical links between the source and destination edge nodes, and reserving a particular wavelength on each of these links for the lightpath. Thus, in establishing an optical connection we must deal with both routing (selecting a suitable path) and wavelength assignment (allocating an available wavelength for the connection). The resulting problem, referred to as the *routing and wavelength assignment* (RWA) *problem* [17], is significantly more difficult than the routing problem in electronic networks. The additional complexity arises from the fact that routing and wavelength assignment are subject to the following two constraints:

1. *Wavelength continuity constraint.* A lightpath must use the same wavelength on all the links along its path from source to destination edge node.
2. *Distinct wavelength constraint.* All lightpaths using the same link (fiber) must be allocated distinct wavelengths.

The RWA problem in optical networks is illustrated in Fig. 10.3, where it is assumed that each fiber supports two wavelengths. The effect of the wavelength continuity constraint is represented by replicating the network into as many copies as the number of wavelengths (in this case, two). If wavelength i is selected for a lightpath, the source and destination edge node communicate over the ith copy of the network. Thus, finding a path for a connection may potentially involve solving W routing problems for a network with W wavelengths, one for each copy of the network.

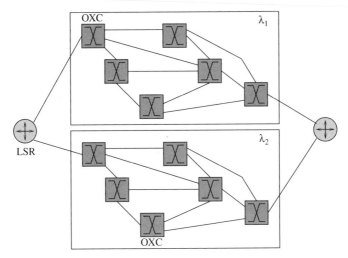

Figure 10.3   RWA problem with two wavelengths per fiber.

The wavelength continuity constraint may be relaxed if the OXCs are equipped with *wavelength converters* [18]. A wavelength converter is a single input/output device that converts the wavelength of an optical signal arriving at its input port to a different wavelength as the signal departs from its output port, but otherwise leaves the optical signal unchanged. In OXCs without a wavelength conversion capability, an incoming signal at port p_i on wavelength λ can be optically switched to any port p_j but must leave the OXC on the same wavelength λ. With wavelength converters, this signal could be optically switched to any port p_j on some other wavelength λ'. That is, wavelength conversion allows a lightpath to use different wavelengths along different physical links.

Different levels of wavelength conversion capability are possible. Figure 10.4 illustrates the differences for a single input and single output port; the case for multiple ports is more complicated but similar. *Full wavelength conversion*

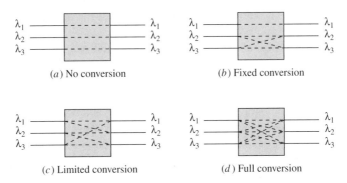

Figure 10.4   Wavelength conversion.

capability implies that any input wavelength may be converted to any other wavelength. *Limited wavelength conversion* [27] denotes that each input wavelength may be converted to any of a specific set of wavelengths, which is not the set of all wavelengths for at least one input wavelength. A special case of this is *fixed wavelength conversion*, where each input wavelength can be converted to exactly one other wavelength. If each wavelength is "converted" only to itself, we have no conversion.

The advantage of full wavelength conversion is that it removes the wavelength continuity constraint, making it possible to establish a lightpath as long as each link along the path from source to destination has a free wavelength (which could be different for different links). As a result, the RWA problem reduces to the classical routing problem, that is, finding a suitable path for each connection in the network. Referring to Fig. 10.3, full wavelength conversion collapses the W copies of the network into a single copy on which the routing problem is solved. On the other hand, with limited conversion, the RWA problem becomes more complex than with no conversion. To see why, note that employing limited conversion at the OXCs introduces links between *some* of the network copies of Fig. 10.3. For example, if wavelength λ_1 can be converted to wavelength λ_2 but not to wavelength λ_3, links must be introduced from each OXC in copy 1 of the network to the corresponding OXC in copy 2, but not to the corresponding OXC in copy 3. When selecting a path for the connection, at each OXC there is the option of remaining at the same network copy or moving to another one, depending on the conversion capability of the OXC. Since the number of alternatives increases exponentially with the number of OXCs that need to be traversed, the complexity of the RWA problem increases accordingly.

Wavelength conversion (full or limited) increases the routing choices for a given lightpath (i.e., makes more efficient use of wavelengths), resulting in better performance. Since converter devices increase network cost, a possible middle ground is to use *sparse conversion*, that is, to employ converters in some, but not all, OXCs in the network. In this case, a lightpath must use the same wavelength along each link in a segment of its path between OXCs equipped with converters, but it may use a different wavelength along the links of another such segment. It has been shown that implementing full conversion at a relatively small fraction of the OXCs in the network is sufficient to achieve almost all the benefits of conversion [11,19].

With the availability of MC-OXCs and the existence of multicast traffic demands, the problem of establishing light trees to satisfy these demands arises. We will call this problem the *multicast routing and wavelength assignment* (MC-RWA) problem. MC-RWA bears many similarities to the RWA problem discussed above. Specifically, the tight coupling between routing and wavelength assignment remains and even becomes stronger: In the absence of wavelength conversion the same wavelength must be used by the multicast connection not just along the links of a single path but along all the links of the light tree. Since the construction of optimal trees for routing multicast connections is by itself a difficult problem [28], the combined MC-RWA problem becomes even harder.

Routing and wavelength assignment is the fundamental control problem in optical WDM networks. Since the performance of a network depends not only on its physical resources (e.g., OXCs, converters, fibers links, number of wavelengths per fiber, etc.) but also on how it is controlled, the objective of an RWA algorithm is to achieve the best possible performance within the limits of physical constraints. The RWA (and MC-RWA) problem can be cast in numerous forms. The different variants of the problem, however, can be classified under one of two broad versions: a static RWA, whereby the traffic requirements are known in advance, and a dynamic RWA, in which a sequence of lightpath requests arrive in some random fashion. The static RWA problem arises naturally in the design and capacity planning phase of architecting an optical network and is discussed in Section 10.4. The dynamic RWA problem is encountered during the real-time network operation phase and involves the dynamic provisioning of lightpaths; this issue is addressed in Section 10.5.

10.4   OPTICAL NETWORK DESIGN AND CAPACITY PLANNING

If the traffic patterns in the network are reasonably well known in advance and any traffic variations take place over long time scales, the most effective technique for establishing optical connections (lightpaths) between client subnetworks is by formulating and solving a static RWA problem. Therefore, static RWA is appropriate for provisioning a set of semipermanent connections. Since these connections are assumed to remain in place for relatively long periods of time, it is worthwhile to attempt to optimize the way in which network resources (e.g., physical links and wavelengths) are assigned to each connection, even though optimization may require a considerable computational effort. Because off-line algorithms have knowledge of the entire set of demands (as opposed to online algorithms, which have no knowledge of future demands), they make more efficient use of network resources and project a lower overall capacity requirement.

In general, the optical network design and planning phase consists of three steps:

1. *Determination of user demand.* If a network already exists, user demand can be measured; otherwise, the demand must be estimated from the expected user population and the expected use patterns.
2. *Physical topology design.* Given the traffic demand, the physical topology of OXCs and fiber links to connect client subnetworks is determined.
3. *Virtual topology design.* At this step, a static RWA problem is formulated and solved to create lightpaths between client subnetworks to satisfy the traffic demand.

For the remainder of this section we assume that user demand is known, and we concentrate on the physical and virtual topology design problems and related issues.

10.4.1   Physical Topology Design

In this phase the network operator has a demand forecast and must decide on a topology to connect client subnetworks through OXCs. This step includes the sizing of links (e.g., determining the number of wavelength channels and the capacity of each channel) and OXCs (e.g., determining the number of ports) as well as the placement of resources such as amplifiers, wavelength converters, and power splitters. Moreover, to deal with link or OXC failures, it is desirable to ensure that there are at least two (or three) paths between any pair of OXCs in the network (i.e., that the graph corresponding to the physical topology of the optical network is two- or three-connected). Often, geographical or administrative considerations may impose further constraints on the physical topology.

If a network does not already exist, the physical topology must be designed from scratch. Obviously, the outcome of this step depends strongly on the accuracy of the demand forecast, and the potential for error is significant when designers have to guess the load in a new network. Therefore, many providers take a cautious approach by initially building a skeleton network and adding new resources as necessary by actual user demand. In this *incremental* network design, it is assumed that sets of user demands arrive over multiple time periods. Resources (e.g., OXCs, fiber links, wavelength channels) are added incrementally to satisfy each new set of demands, in a way that the additional capacity required is minimized.

A physical topology design problem was considered in [29]. Given a number of LSRs and a set of lightpaths to be set up among pairs of LSRs, the objective was to determine the two-connected physical topology with the minimum number of OXCs to establish all the lightpaths (this is a combined physical/virtual topology design problem in that the routing and wavelength assignment for the lightpaths is also determined). An iterative solution approach was considered, whereby a genetic algorithm was used to iterate over the space of physical topologies, and heuristics were employed for routing and wavelength assignment on a given physical topology (refer to Section 10.4.2 for details on RWA heuristics). The algorithm was applied to networks with up to 1000 LSRs and tens of thousands of lightpaths and provided insight into the capacity requirements for realistic optical networks. For example, it was shown that the number of OXCs increases much slower than the number of LSRs, and also that the number of OXCs increases only moderately as the number of lightpaths increases by a factor of 2 or 3. These results indicate that optical networks to interconnect a large number of LSRs can be built to provide rich connectivity with moderate cost.

Other studies related to capacity planning have looked into the problem of optimally placing network resources such as converters or power splitters (for multicast). The problem of converter placement was addressed in [11] and [30], and optimal [30] (for uniform traffic only) or near-optimal greedy [11] algorithms (for general traffic patterns) were developed. Although both studies established that a small number of converters (approximately 30% of the number of OXCs) is sufficient, the results in [11] demonstrate that the optimal placement of converters is extremely sensitive to the actual traffic pattern, and an incremental approach to deploying converters may not lead to optimal (or nearly optimal) results. The work in [31] considered the problem of optimally allocating the multicast-capable OXCs (MC-OXCs) to establish light trees, and a greedy heuristic was proposed. It was found that there is little performance improvement if more than 50% of the OXCs in the network are multicast-capable and that the optimal location of MC-OXCs depends on the traffic pattern.

Overall, the physical topology design problem is quite complex because the topology, the link and OXC capacities, and the number and location of optical devices such as converters and amplifiers strongly depend on the routing of lightpaths and the wavelength assignment strategy. If we make the problem less constrained, allowing the topology, routing, wavelength assignment, link capacity, and so on, to change, the problem becomes very hard because these parameters are coupled in complicated ways. In practice, the topology may be constrained by external factors, making the problem easier to deal with; for instance, the existence of a deployed fiber infrastructure may dictate the location of OXCs and the links between them. However, the area of physical topology design for optical networks remains a rich area for future research.

10.4.2 Virtual Topology Design

A solution to the static RWA problem consists of a set of long-lived lightpaths which create a *logical* (or *virtual*) topology among the edge nodes. This virtual topology is embedded onto the physical topology of optical fiber links and OXCs. Accordingly, the static RWA problem is often referred to as the *virtual topology design problem* [9]. In the virtual topology, there is a directed link from edge node s to edge node d if a lightpath originating at s and terminating at d is set up (refer also to Fig. 10.1), and edge node s is said to be "one hop away" from edge node d in the virtual topology, although the two nodes may be separated by a number of physical links. The type of virtual topology that can be created is usually constrained by the underlying physical topology. In particular, it is generally not possible to implement fully connected virtual topologies: For N edge nodes this would require each edge node to maintain $N - 1$ lightpaths and the optical network to support a total of $N(N - 1)$ lightpaths. Even for modest values of N, this degree of connectivity is beyond the reach of current optical technology, both in terms of the number of wavelengths that can be supported and in terms of the optical hardware (transmitters and receivers) required at each edge node.

In its most general form, the RWA problem is specified by providing the physical topology of the network and the traffic requirements. The physical topology corresponds to the deployment of cables in some existing fiber infrastructure and is given as a graph $G_p(V, E_p)$, where V is the set of OXCs and E_p is the set of fibers that interconnect them. The traffic requirements are specified in a traffic matrix $\mathbf{T} = [\rho p_{sd}]$, where ρp_{sd} is a measure of the long-term traffic flowing from source edge node s to destination edge node d [32]. Quantity ρ represents the (deterministic) total offered load to the network, while the p_{sd} parameters define the distribution of the offered traffic.

Routing and wavelength assignment are considered together as an optimization problem using integer programming formulations. Usually, the objective of the formulation is to minimize the maximum congestion level in the network subject to network resource constraints [9,10]. Although other objective functions are possible, such as minimizing the average weighted number of hops or minimizing the average packet delay, minimizing network congestion is preferable since it can lead to linear programming (ILP) formulations. Although we do not present the RWA problem formulation here, the interested reader may refer to [9], [10], and [32]. These formulations turn out to have extremely large numbers of variables and are intractable for large networks. This fact has motivated the development of heuristic approaches for finding good solutions efficiently.

Before we describe the various heuristic approaches, we note that the static RWA problem can be logically decomposed into four subproblems. The decomposition is approximate or inexact, in the sense that solving the subproblems in sequence and combining the solutions may not result in the optimal solution for the fully integrated problem, or some later subproblem may have no solution given the solution obtained for an earlier subproblem, so no solution to the original problem may be obtained. However, the decomposition provides insight into the structure of the RWA problem and is a first step toward the design of effective heuristics. Assuming no wavelength conversion, the subproblems are as follows:

1. *Topology subproblem.* Determine the logical topology to be imposed on the physical topology; that is, determine the lightpaths in terms of their source and destination edge nodes.

2. *Lightpath routing subproblem.* Determine the physical links of which each lightpath consists; that is, route the lightpaths over the physical topology.

3. *Wavelength assignment subproblem.* Determine the wavelength that each lightpath uses, that is, assign a wavelength to each lightpath in the logical topology so that wavelength restrictions are obeyed for each physical link.

4. *Traffic routing subproblem.* Route packet traffic between source and destination edge nodes over the logical topology obtained.

A large number of heuristic algorithms have been developed in the literature to solve the general static RWA problem discussed here or its many variants. Overall, however, the different heuristics can be classified into three broad categories: (1) algorithms that solve the overall ILP problem suboptimally, (2) algorithms that tackle only a subset of the four subproblems, and (3) algorithms that address the problem of embedding regular logical topologies onto the physical topology.

Suboptimal solutions can be obtained by applying classical tools developed for complex optimization problems directly to the ILP problem. One technique is to use LP relaxation followed by rounding [33]. In this case, the integer constraints are relaxed, creating a nonintegral problem that can be solved by some linear programming method, and then a rounding algorithm is applied to obtain a new solution which obeys the integer constraints. Alternatively, genetic algorithms or simulated annealing [34] can be applied to obtain locally optimal solutions. The main drawback of these approaches is that it is difficult to control the quality of the final solution for large networks: Simulated annealing is computationally expensive and thus it may not be possible to explore the state space adequately, while LP relaxation may lead to solutions from which it is difficult to apply rounding algorithms.

Another class of algorithms tackles the RWA problem by initially solving the first three subproblems listed above; traffic routing is then performed by employing well-known routing algorithms on the logical topology. One approach for solving the three subproblems is to maximize the amount of traffic that is carried on one-hop lightpaths (i.e., traffic that is routed from source to destination edge node directly on a lightpath). A greedy approach taken in [35] is to create lightpaths between edge nodes in order of decreasing traffic demands as long as the wavelength continuity and distinct wavelength constraints are satisfied. This algorithm starts with a logical topology with no links (lightpaths) and sequentially adds lightpaths as long as doing so does not violate any of the problem constraints. The reverse approach is also possible [36]: Starting with a fully connected logical topology, an algorithm removes the lightpath carrying the smallest traffic flows sequentially until no constraint is violated. At each step (i.e., after removing a lightpath), the traffic-routing subproblem is solved to find the lightpath with the smallest flow.

The third approach to RWA is to start with a given logical topology, thus avoiding direct solution of the first of the four subproblems listed above. Regular topologies are good candidates as logical topologies since they are well understood and results regarding bounds and averages (e.g., for hop lengths) are easier to derive. Algorithms for routing traffic on a regular topology are usually simple, so the traffic-routing subproblem can be solved trivially. Also, regular topologies possess inherent load-balancing characteristics which are important when the objective is to minimize the maximum congestion.

Once a regular topology is decided on as the one to implement the logical topology, it remains to decide which physical node will realize each given node in the regular topology (this is usually referred to as the *node mapping subproblem*) and which sequence of physical links will be used to realize each given edge (lightpath) in the regular topology (this *path mapping subproblem* is equivalent to the lightpath routing and wavelength assignment subproblems discussed earlier). This procedure is usually referred to as embedding a regular topology in the physical topology. Both the node and path mapping subproblems are intractable, and heuristics have been proposed in the literature [36,37]. For instance, a heuristic for mapping the nodes of shuffle topologies based on the gradient algorithm was developed in [37].

Given that all the algorithms for the RWA problem are based on heuristics, it is important to be able to characterize the quality of the solutions obtained. To this end, one must resort to comparing the solutions to known bounds on the optimal solution. A comprehensive discussion of bounds for the RWA problem and the theoretical considerations involved in deriving them can be found in [9]. A simulation-based comparison of the relative performance of the three classes of heuristic for the RWA problem is presented in [10]. The results indicate that the second class of algorithms discussed earlier achieve the best performance.

The study in [23] also focused on virtual topology design (i.e., static RWA) for point-to-point traffic but observed that since a light tree is a more general representation of a lightpath, the set of virtual topologies that can be implemented using light trees is a superset of the virtual topologies that can be implemented using only lightpaths. Thus, for any given virtual topology problem, an optimal solution using light trees is guaranteed to be at least as good and possibly an improvement over the optimal solution obtained using only lightpaths. Furthermore, it was demonstrated that by extending the lightpath concept to a light tree, the network performance (in terms of average packet hops) can be improved while the network cost (in terms of the number of optical transmitters/receivers required) decreases.

The static MC-RWA problem has been studied in [38] and [39]. The study in [38] focused on demonstrating the benefits of multicasting in wavelength-routed optical networks. Specifically, it was shown that using light trees (spanning the source and destination nodes) rather than individual parallel lightpaths (each connecting the source to an individual destination) requires fewer wavelengths and consumes a significantly lower amount of bandwidth. In [39] an ILP formulation that maximizes the total number of multicast connections was presented for the static MC-RWA problem. Rather than providing heuristic algorithms for solving the ILP, bounds on the objective function were presented by relaxing the integer constraints.

10.4.3   Design of Survivable Optical Networks

In Section 10.4.2 we considered the RWA problem under the assumption that lightpaths are not protected. To ensure that protected lightpaths survive network failures, it is important to develop solutions to the static RWA problem that take into account the requirements of lightpaths in terms of dedicated or shared protection (refer to Section 10.2). In other words, for each protected lightpath a diversely routed backup path (dedicated or shared) must also be computed and reserved during the network design phase. Clearly, the provisioning of backup lightpaths will increase the capacity requirements compared to a network that provides no protection. Therefore, a common objective in the design of survivable optical networks is to provide adequate protection while minimizing the total network capacity.

Optical network architectures are exposed to a wide range of risks due to either human activities (e.g., accidental fiber cuts or operational errors) or equipment malfunctions such as laser and OXC failures. To ensure survivability, the primary and backup route for a lightpath must be diverse so that a single failure will not affect both paths. Since several network links may pass though the same conduit or share the same right-of-way, two link-disjoint paths may not be immune to a single failure. The concept of a *shared risk link group* (SRLG) [40] can be used to express the risk relationship of optical channels with respect to a single failure. Specifically, each point of failure in the network (e.g., a fiber, a cable, or a conduit) is associated with an SRLG, which is the set of channels that would be affected by the failure. A particular link can belong to several SRLGs, and diverse routing of lightpaths implies that the primary and backup routes are SRLG-disjoint. Therefore, the availability of SRLG information is crucial to the design of survivable networks. In general, the different types of SRLGs are defined by the network operator, and the mapping of links into SRLGs must be configured manually since it may be impossible for network elements to self-discover this information. We also note that SRLG information does not change dynamically (i.e, it need not be updated unless changes are made to the network capacity or topology).

Protection from failures can be classified as either *local span (link) protection* or *path (end-to-end) protection*. In path protection, the ingress and egress OXCs of the failed lightpath coordinate to restore the signal on a predefined backup path that is SRLG-disjoint from the primary path. In local span protection, the OXCs closest to the failure coordinate to reroute the lightpath through alternate channels around the failure. These alternate channels are different for each failure and must be configured in advance and set aside for protection.

Given the SRLG information, the type of protection (local span or path protection), and the protection requirements of each lightpath (dedicated, shared, or no protection), the problem of designing a virtual topology that meets the lightpaths' protection requirements while minimizing total network capacity can be be modeled as an ILP and has been studied in [41] to [43]. Although we do not provide the formulation here, the reader is referred to [42] for three different formulations corresponding to dedicated path protection, shared path protection, and shared local span protection. The formulation is similar to the one discussed in Section 10.4.2 for unprotected lightpaths, with additional constraints to account for the fact that the primary and backup paths for a given lightpath are SRLG-disjoint, that two primary paths can share a link (channel) on their backup path only if they are SRLG-disjoint, and that a backup path consumes capacity if and only if there is a failure on the primary path. Consequently, similar solution approaches can be employed to obtain a nearly optimal set of primary and backup paths for the given set of lightpaths.

Finally, we note that the techniques described in this section can also be used in designing virtual private networks (VPNs) over the optical network [44]. A VPN is an overlay network implemented over a public network infrastructure using tunneling such that the client nodes at the endpoints of a tunnel appear to be connected by a point-to-point link of a specified bandwidth. VPN services are expected to be key drivers of new network technologies since they allow corporations to extend their services over geographically dispersed regions in a cost-effective manner. Providing VPN service over optical networks has the additional advantage of enhancing privacy and allaying security concerns due to the data transparency in the optical domain. From the point of view of the optical network provider, a VPN is simply a set of client edge nodes that need to be interconnected by lightpaths with certain protection requirements. The provider's objective is to maximize the total amount of VPN traffic by optimally utilizing the capacity of the optical network. This problem is similar to the virtual topology design problem we discussed in this section and can be addressed by the methodologies we have already described.

10.5   DYNAMIC LIGHTPATH PROVISIONING AND RESTORATION

During real-time network operation, edge nodes submit to the network requests for lightpaths to be set up as needed. Thus, connection requests are initiated in some random fashion. Depending on the state of the network at the time of a request, the available resources may or may not be sufficient to establish a lightpath between the corresponding source–destination edge node pairs. The network state consists of the physical path (route) and wavelength assignment for all active lightpaths. The state evolves randomly in time as new lightpaths are admitted and existing lightpaths are released. Thus, each time a request is made, an algorithm must be executed in real time to determine whether it is feasible to accommodate the request, and if so, to perform routing and wavelength assignment. If a request for a lightpath cannot be accepted because of lack of resources, it is blocked.

Because of the real-time nature of the problem, RWA algorithms in a dynamic traffic environment must be very simple. Since combined routing and wavelength assignment is a hard problem, a typical approach to designing efficient algorithms is to decouple the problem into two separate subproblems: the routing problem and the wavelength assignment problem. Consequently, most dynamic RWA algorithms for wavelength-routed networks consist of the following general steps:

1. Compute a number of candidate physical paths for each source–destination edge node pair and arrange them in a path list.
2. Order all wavelengths in a wavelength list.
3. Starting with the path and wavelength at the top of the corresponding list, search for a feasible path and wavelength for the lightpath requested.

The specific nature of a dynamic RWA algorithm is determined by the number of candidate paths and how they are computed, the order in which paths and wavelengths are listed, and the order in which the path and wavelength lists are accessed.

10.5.1   Route Computation

Let us first discuss the routing subproblem. If a *static algorithm* is used, the paths are computed and ordered independently of the network state. With an *adaptive algorithm*, on the other hand, the paths computed and their order may vary according to the current state of the network. A static algorithm is executed off-line and the computed paths are stored for later use, resulting in low latency during lightpath establishment. Adaptive algorithms are executed at the time a lightpath request arrives and require network nodes to exchange information regarding the network state. Lightpath setup delay may also increase, but in general, adaptive algorithms improve network performance.

The number of path choices for establishing an optical connection is another important parameter. A *fixed routing algorithm* is a static algorithm in which every source–destination edge node pair is assigned a single path. With this scheme, a connection is blocked if there is no wavelength available on the designated path at the time of the request. In *fixed-alternate routing*, a number $k, k > 1$, of paths are computed and ordered off-line for each source–destination edge node pair. When a request arrives, these paths are examined in the order specified and the first one with a free wavelength is used to establish the lightpath. The request is blocked if no wavelength is available in any of the k paths. Similarly, an adaptive routing algorithm may compute a single path or a number of alternative paths at the time of the request. A hybrid approach is to compute k paths off-line; however, the order in which the paths are considered is determined according to the network state at the time the connection request is made (e.g., least to most congested).

In most practical cases, the candidate paths for a request are considered in increasing order of path length (or path cost). *Path length* is typically defined as the sum of the weights (costs) assigned to each physical link along the path, and the weights are chosen according to some desirable routing criterion. Since weights can be assigned arbitrarily, they offer a wide range of possibilities for selecting path priorities. For example, in a static (fixed-alternate) routing algorithm, the weight of each link could be set to 1, or to the physical distance of the link. In the former case, the path list consists of the k minimum-hop paths, while in the latter the candidate paths are the k minimum-distance paths (where distance is defined as the geographic length). In an adaptive routing algorithm, link weights may reflect the *load* or *interference* on a link (i.e., the number of active lightpaths sharing the link). By assigning small weights to least loaded links, paths with larger number of free channels on their links rise to the head of the path list, resulting in a *least loaded routing algorithm*. Paths that are congested become "longer" and are moved farther down the list; this tends to avoid heavily loaded bottleneck links. Many other weighting functions are possible.

When path lengths are sums of of link weights, the k-shortest-path algorithm [45] can be used to compute candidate paths. Each path is checked in order of increasing length, and the first that is feasible is assigned the first free wavelength in the wavelength list. However, the k shortest paths constructed by this algorithm usually share links. Therefore, if one path in the list is not feasible, it is likely that other paths in the list with which it shares a link will also be infeasible. To reduce the risk of blocking, the k shortest paths can be computed so as to be pairwise link-disjoint. This can be accomplished as follows: When computing the ith shortest path, $i = 1, \ldots, k$, the links used by the first $i - 1$ paths are removed from the original network topology and Dijkstra's shortest-path algorithm [46] is applied to the resulting topology. This approach increases the chances of finding a feasible path for a connection request.

If the lightpath requires protection (dedicated or shared), a similar procedure can be used at connection setup time to determine two SRLG-disjoint paths, a primary (working) path and a backup (protection) one. The path cost for a lightpath with dedicated protection is the sum of the costs of the links along the primary and backup paths. Thus, the objective is to pick a pair of link-disjoint paths with the minimum combined cost. If the lightpath requires shared protection, the links of the backup path may be shared among other lightpaths. Note, however, that when determining the backup path for a given lightpath, selecting any link that is already shared by the backup paths of other lightpaths does not incur any additional cost (since the link has already been set aside for protection). Thus, the algorithm above can be used for determining the primary and backup paths by setting the weight of shared links in the backup path to zero.

An alternative to precomputed backup paths is to attempt to restore a lightpath after a failure, by dynamically determining a new route for the failed lightpath [14]. As with protection (see Section 10.4.3), restoration techniques can be classified as either *local span (link) restoration*, whereby the OXCs closest to the failure attempt to restore the lightpath on any available channels around the failure, or *end-to-end (path) restoration*, whereby the ingress OXC dynamically computes a new route to the egress OXC of the failed lightpath. This approach has the advantage that it incurs low overhead in the absence of failures. However, dynamic restoration has two disadvantages that make it unsuitable for lightpaths carrying critical traffic: There is no guarantee of successful restoration (e.g., due to lack of available resources at the time of recovery), and it takes significantly longer to restore traffic (e.g., due to the need to compute a new route on the fly) than when protection resources are provisioned in advance.

The problem of determining algorithms for routing multicast optical connections has also been studied in [39] and [47]. The problem of constructing trees for routing multicast connections was considered in [47] independent of wavelength assignment, under the assumption that not all OXCs are multicast capable (i.e., that there are a limited number of MC-OXCs in the network). Four new algorithms were developed for routing multicast connections under this *sparse light splitting* scenario. Although the algorithms differ slightly from each other, the main idea to accommodate sparse splitting is to start with the assumption that all OXCs in the network are multicast capable and use an existing algorithm to build an initial tree. Such a tree is infeasible if a non-multicast-capable OXC is a branching point. In this case, all but one of the branches out of this OXC are removed, and destination nodes in the removed branches have to join the tree at a MC-OXC. In [39], on the other hand, the MC-RWA problem was solved by decoupling the routing and wavelength assignment problems. A number of *alternate* trees are constructed for each multicast connection using existing routing algorithms. When a request for a connection arrives, the associated trees are considered in a fixed order. For each tree, wavelengths are also considered in a fixed order (i.e., the first-fit strategy discussed below). The connection is blocked if no free wavelength is found in any of the trees associated with the multicast connection.

We note that most of the literature (and the preceding discussion) has focused on the problem of obtaining paths that are optimal with respect to total path cost. In transparent optical networks, however, optical signals may suffer from physical layer impairments, including attenuation, chromatic dispersion, polarization mode dispersion (PMD), amplifier spontaneous emission (ASE), crosstalk, and various nonlinearities [40]. These impairments must be taken into account when choosing a physical path. In general, the effect of physical layer impairments may be translated into a set of constraints that the physical path must satisfy; for instance, the total signal attenuation along the physical path must be within a certain power budget to guarantee a minimum level of signal quality at the receiver. Therefore, a simple shortest-path-first (SPF) algorithm (e.g., Dijkstra's algorithm implemented by protocols such as OSPF [48]) may not be appropriate for computing physical paths within a transparent optical network. Rather, constraint-based routing techniques such as the one employed by the constraint-based shortest-path-first (CSPF) algorithm [5] are needed. These techniques compute paths by taking into account not only the link cost but also a set of constraints that the path must satisfy. A first step toward the design of constraint-based routing algorithms for optical networks has been taken in [40], where it was shown how to translate the PMD and ASE impairments into a set of linear constraints on the end-to-end physical path. However, additional work is required to advance our understanding of how routing is affected by physical layer considerations, and constraint-based routing remains an open research area [49].

10.5.2 Wavelength Assignment

Let us now discuss the wavelength assignment subproblem, which is concerned with the manner in which the wavelength list is ordered. For a given candidate path, wavelengths are considered in the order in which they appear in the list to find a free wavelength for the connection request. Again, we distinguish between the static and adaptive cases. In the static case, the wavelength ordering is fixed (e.g., the list is ordered by wavelength number). The idea behind this scheme, also referred to as *first fit*, is to pack all the in-use wavelengths toward the top of the list so that wavelengths toward the end of the list will have a higher probability of being available over long continuous paths. In the adaptive case, the ordering of wavelengths is typically based on *usage*, defined either as the number of links in the network in which a wavelength is currently used, or as the number of active connections using a wavelength. Under the *most used method*, the most used wavelengths are considered first (i.e., wavelength are considered in order of decreasing usage). The rationale behind this method is to reuse active wavelengths as much as possible before trying others, packing connections into fewer wavelengths and conserving the spare capacity of less-used wavelengths. This in turn makes it more likely to find wavelengths that satisfy the continuity requirement over long paths. Under the *least used method*, wavelengths are tried in the order of increasing usage. This scheme attempts to balance the load as equally as possible among all available wavelengths. However, least used assignment tends to "fragment" the availability of wavelengths, making it less likely that the same wavelength is available throughout the network for connections that traverse longer paths.

The most used and least used schemes introduce communication overhead because they require global network information in order to compute the usage of each wavelength. The first-fit scheme, on the other hand, requires no global information, and since it does not need to order wavelengths in real time, it has significantly lower computational requirements than either the most used or least used schemes. Another adaptive scheme that avoids the communication and computational overhead of most used and least used is *random wavelength assignment*. With this scheme, the set of wavelengths that are free on a particular path is first determined. Among the available wavelengths, one is chosen randomly (usually with uniform probability) and assigned to the requested lightpath.

We note that in networks in which all OXCs are capable of wavelength conversion, the wavelength assignment problem is trivial: Since a lightpath can be established as long as at least one wavelength is free at each link and different wavelengths can be used in different links, the order in which wavelengths are assigned is not important. On the other hand, when only a fraction of the OXCs employ converters (i.e., a sparse conversion scenario), a wavelength assignment scheme is again required to select a wavelength for each segment of a connection's path that originates and terminates at an OXC with converters. In this case, the same assignment policies discussed above for selecting a wavelength for the end-to-end path can also be used to select a wavelength for each path segment between OXCs with converters.

10.5.3 Performance of Dynamic RWA Algorithms

The performance of a dynamic RWA algorithm is generally measured in terms of the call blocking probability, that is, the probability that a lightpath cannot be established in the network due to lack of resources (e.g., link capacity or free wavelengths). Even in the case of simple network topologies (such as rings) or simple routing rules (such as fixed routing), the calculation of blocking probabilities in WDM networks is extremely difficult. In networks with arbitrary mesh topologies, and/or when using alternate or adaptive routing algorithms, the problem is even more complex. These complications arise from both the link load dependencies (due to interfering lightpaths) and the dependencies among the sets of active wavelengths in adjacent links (due to the wavelength continuity constraint). Nevertheless, the problem of computing blocking probabilities in wavelength-routed networks has been studied extensively in the literature, and approximate analytical techniques that capture the effects of link load and wavelength dependencies have been developed in [11], [16], and [19]. A detailed comparison of the performance of various wavelength assignment schemes in terms of call blocking probability can be found in [50].

Although important, average blocking probability (computed over all connection requests) does not always capture the full effect of a particular dynamic RWA algorithm on other aspects of network behavior: in particular, *fairness*. In this context, fairness refers to the variability in blocking probability experienced by lightpath requests between the various edge node pairs, such that lower variability is associated with a higher degree of fairness. In general, any network has the property that longer paths are likely to experience higher blocking than shorter ones. Consequently, the degree of fairness can be quantified by defining the *unfairness factor* as the ratio of the blocking probability on the longest path to that on the shortest path for a given RWA algorithm. Depending on the network topology and the RWA algorithm, this property may have a cascading effect that can result in an unfair treatment of the connections between more distant edge node pairs: Blocking of long lightpaths leaves more resources available for short lightpaths, so that the connections established in the network tend to be short ones. These shorter connections "fragment" the availability of wavelengths, and thus the problem of unfairness is more pronounced in networks without converters, since finding long paths that satisfy the wavelength continuity constraint is more difficult than without this constraint.

Several studies [11,16,19] have examined the influence of various parameters on blocking probability and fairness, and some of the general conclusions include the following:

- Wavelength conversion significantly affects fairness. Networks employing converters at all OXCs sometimes exhibit order-of-magnitude improvement in fairness (as reflected by the unfairness factor) compared to networks with no conversion capability, despite the fact that the improvement in overall blocking probability is significantly less pronounced. It has also been shown that equipping a relatively small fraction (typically, 20 to 30%) of all OXCs with converters is sufficient to achieve most of the fairness benefits, due to wavelength conversion.

- Alternative routing can significantly improve the network performance in terms of both overall blocking probability and fairness. In fact, having as few as three alternative paths for each connection may in some cases (depending on the network topology) achieve almost all the benefits (in terms of blocking and fairness) of having full wavelength conversion at each OXC with fixed routing.

- Wavelength assignment policies also play an important role, especially in terms of fairness. The random and least used schemes tend to "fragment" the wavelength availability, resulting in large unfairness factors (with least used having the worst performance). On the other hand, the most used assignment policy achieves the best performance in terms of fairness. The first-fit scheme exhibits a behavior very similar to most used in terms of both fairness and overall blocking probability, and has the additional advantage of being easier and less expensive to implement.

10.6   CONTROL PLANE ISSUES AND STANDARDIZATION ACTIVITIES

So far we have focused on the application of network design and traffic engineering principles to the control of traffic in optical networks with a view to achieving specific performance objectives, including efficient utilization of network resources and planning of network capacity. Equally important to an operational network are associated control plane issues involved in automating the process of lightpath establishment and in supporting the network design and traffic engineering functions. Currently, a number of standardization activities addressing the control plane aspects of optical networks are under way [51–53] within the Internet Engineering Task Force (IETF) [54], the Optical Domain Service Interconnection (ODSI) coalition [55], and the Optical Internetworking Forum (OIF) [56]. In this section we review the relevant standards activities and discuss how they fit within the traffic engineering framework; we note, however, that these are ongoing efforts and will probably evolve as the underlying technology matures and our collective understanding of optical networks advances.

Let us return to Fig. 10.1, which illustrates the manner in which client subnetworks (IP/MPLS networks in the figure) attach to the optical network of OXCs. The figure corresponds to the vision of a future optical network capable of providing a bandwidth-on-demand service by dynamically creating and tearing down lightpaths between client subnetworks. There are two broad issues that need to be addressed before such a vision is realized. First, a signaling mechanism is required at the user-to-network interface (UNI) between the client subnetworks and the optical network control plane. The signaling channel allows edge nodes to dynamically request bandwidth from the optical network, and supports important functions, including service discovery and provisioning capabilities, neighbor discovery and reachability information, address registration, and so on. Both the ODSI coalition [57] and the OIF [58] have developed specifications for the UNI; the OIF specifications are based on GMPLS.

Second, a set of signaling and control protocols must be defined within the optical network to support dynamic lightpath establishment and traffic engineering functionality; these protocols are implemented at the control module of each OXC. Currently, most work on defining control plane protocols in the optical network takes place under the auspices of IETF, reflecting a convergence of optical networking and IP communities to developing technology built around a single common framework (i.e., GMPLS) for controlling both IP and optical network elements [59]. There are three components of the control plane that are crucial to setting up lightpaths within the optical network and thus relevant to traffic engineering (refer to Fig. 10.5):

1. *Topology and resource discovery.* The main purpose of discovery mechanisms is to disseminate network state information, including resource use, network connectivity, link capacity availability, and special constraints.

2. *Route computation.* This component employs RWA algorithms and traffic engineering functions to select an appropriate route for a requested lightpath.

3. *Lightpath management.* Lightpath management is concerned with setup and teardown of lightpaths, as well as coordination of protection switching in case of failures.

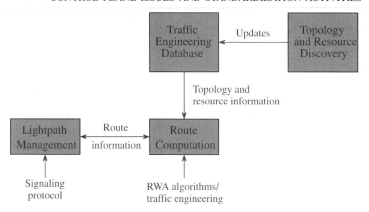

Figure 10.5   Control plane components.

The reader is referred to Chapter 14 for details on the control plane, and to Chapter 16 for details on standards activities. We now discuss briefly some of the standardization activities aimed at defining new protocols or extending existing ones to carry out these three functions.

Topology and resource discovery includes neighbor discovery, link monitoring, and state distribution. The link management protocol (LMP) [60] has been proposed to perform neighbor discovery and link monitoring. LMP is expected to run between neighboring OXC nodes and can be used to establish and maintain control channel connectivity, monitor and verify data link connectivity, and isolate link, fiber, or channel failures. Distribution of state information is typically carried out through link state routing protocols such as OSPF [48]. There are currently several efforts under way to extend OSPF to support GMPLS [61] and traffic engineering [62]. In particular, the link-state information that these protocols carry must be augmented to include optical resource information, including wavelength availability and bandwidth, SRLG information (discussed in Section 10.4.3), physical layer constraints (discussed in Section 10.5.1), and link protection information, among others. This information is then used to build and update the optical network traffic engineering database (see Fig. 10.5), which guides the route selection algorithm.

Once a lightpath is selected, a signaling protocol must be invoked to set up and manage the connection. Two protocols have currently been defined to signal a lightpath setup: RSVP-TE [63] and CR-LDP [64]. RSVP-TE is based on the resource reservation protocol (RSVP) [65], with appropriate extensions to support traffic engineering, whereas CR-LDP is an extension of the label distribution protocol (LDP) [66] augmented to handle constraint-based routing. The protocols are currently being extended to support GMPLS [67,68]. Besides signaling the path at connection time, both protocols can be used to handle the switchover to the protection path automatically once a failure in the working path has occurred.

As a final note, the control plane elements depicted in Fig. 10.5 are independent of each other and thus separable. This modularity allows each component to evolve independent of others or to be replaced with a new and improved protocol. As the optical networking and IP communities come together to define standards, the constraints and new realities (e.g., the explosion in the number of channels in the network) imposed by the optical layer and WDM technology will certainly affect our long-held assumptions regarding issues such as routing, control, discovery, and so on, which have been developed for mostly opaque electronic networks. As we carefully rethink these issues in the context of transparent (or almost-transparent) optical networks, protocol design will certainly evolve to better accommodate the new technology. Therefore, we expect that the control plane protocols will continue to be refined and/or replaced by new, more appropriate ones. The interested reader should check frequently with the activities within IETF and OIF for the most recent developments.

10.7   SUMMARY

We have discussed the issues arising in the design and operation of optical WDM networks. The emphasis has been on capacity planning and traffic engineering techniques as they apply to different time scales, from resource sizing and topology design to real-time provisioning of lightpaths. The underlying routing and wave-length assignment control problem has been studied extensively and is well understood within the research community. A number of control plane technologies in support of traffic engineering functions are also being developed at a rapid rate within the standards bodies. The next step is to fully incorporate RWA control algorithms and protocols into optical network elements, initially in field trials and later for deployment in commercial networks, in order to realize a seamless integration of optical networking within the overall Internet infrastructure.

REFERENCES

1. D. O. Awduche, MPLS and traffic engineering in IP networks, *IEEE Communications*, 37(12):42–47, Dec. 1999.
2. D. Awduche, J. Malcolm, J. Agogbua, M. O'Dell, and J. McManus, *Requirements for Traffic Engineering over MPLS*, RFC 2702, Sept. 1999.
3. S. Blake, D. Black, M. Carlson, E. Davies, Z. Wang, and W. Weiss, *An Architecture for Differentiated Services*, RFC 2475, Dec. 1998.
4. E. Rosen, A. Viswanathan, and R. Callon, *Multiprotocol Label Switching Architecture*, RFC 3031, Jan. 2001.
5. B. Davie and Y. Rekhter, *MPLS Technology and Applications*, Morgan Kaufmann, San Francisco, 2000.
6. P. Ashwood-Smith et al., Generalized MPLS: Signaling functional description, IETF draft, *draft-ietf-mpls-generalized-signaling-06.txt*, Apr. 2001.
7. D. H. Su and D. W. Griffith, Standards activities for MPLS over WDM networks, *Optical Networks*, 1(3), July 2000.

8. O. Gerstel, B. Li, A. McGuire, G. N. Rouskas, K. Sivalingam, and Z. Zhang Eds., Special issue on protocols and architectures for next generation optical WDM networks, *IEEE Journal on Selected Areas in Communications*, 18(10), Oct. 2000.

9. R. Dutta and G. N. Rouskas, A survey of virtual topology design algorithms for wavelength routed optical networks, *Optical Networks*, 1(1):73–89, Jan. 2000.

10. E. Leonardi, M. Mellia, and M. A. Marsan, Algorithms for the logical topology design in WDM all-optical networks, *Optical Networks*, 1(1):35–46, Jan. 2000.

11. Y. Zhu, G. N. Rouskas, and H. G. Perros, A path decomposition approach for computing blocking probabilities in wavelength routing networks, *IEEE/ACM Transactions on Networking*, 8(6):747–762, Dec. 2000.

12. L. Li and A. K. Somani, A new analytical model for multifiber WDM networks, *IEEE Journal on Selected Areas in Communications*, 18(10):2138–2145, Oct. 2000.

13. S. Ramamurthy and B. Mukherjee, Survivable WDM mesh networks, I: protection, *Proc. INFOCOM'99*, pp. 744–751, Mar. 1999.

14. S. Ramamurthy and B. Mukherjee, Survivable WDM mesh networks, II: restoration, *Proc. ICC'99*, pp. 2023–2030, June 1999.

15. A. Mokhtar and M. Azizoglu, Adaptive wavelength routing in all-optical networks, *IEEE/ACM Transactions on Networking*, 6(2):197–206, Apr. 1998.

16. E. Karasan and E. Ayanoglu, Effects of wavelength routing and selection algorithms on wavelength conversion gain in WDM optical networks, *IEEE/ACM Transactions on Networking*, 6(2):186–196, Apr. 1998.

17. H. Zang, J. P. Jue, and B. Mukherjee, A review of routing and wavelength assignment approaches for wavelength-routed optical WDM networks, *Optical Networks*, 1(1):47–60, Jan. 2000.

18. B. Ramamurty and B. Mukherjee, Wavelength conversion in WDM networking, *IEEE Journal on Selected Areas in Communications*, 16(7):1061–1073, Sept. 1998.

19. S. Subramaniam, M. Azizoglu, and A. Somani, All-optical networks with sparse wavelength conversion, *IEEE/ACM Transactions on Networking*, 4(4):544–557, Aug. 1996.

20. T. Tripathi and K. Sivarajan, Computing approximate blocking probabilities in wavelength routed all-optical networks with limited-range wavelength conversion, *Proc. INFOCOM'99*, pp. 329–336, Mar. 1999.

21. N. Ghani, Lambda-labeling: A framework for IP-over-WDM using MPLS, *Optical Networks*, 1(2):45–58, Apr. 2000.

22. O. Gerstel and R. Ramswami, Optical layer survivability: A service perspective, *IEEE Communications*, 38(3):104–113, Mar. 2000.

23. L. H. Sahasrabuddhe and B. Mukherjee, Light-trees: Optical multicasting for improved performance in wavelength-routed networks, *IEEE Communications*, 37(2):67–73, Feb. 1999.

24. D. Papadimitriou et al., Optical multicast in wavelength switched networks: architectural framework, IETF draft, *draft-poj-optical-multicast-01.txt*, July 2001.

25. B. Mukherjee, *Optical Communication Networking*, McGraw-Hill, New York, 1997.

26. S. Keshav, *An Engineering Approach to Computer Networking*, Addison-Wesley, Reading, MA, 1997.

27. V. Sharma and E. A. Varvarigos, Limited wavelength translation in all-optical WDM mesh networks, *Proc. INFOCOM'98*, pp. 893–901, Mar. 1999.

28. S. L. Hakimi, Steiner's problem in graphs and its implications, *Networks*, 1:113–133, 1971.

29. Y. Xin, G. N. Rouskas, and H. G. Perros, *On the Design of MPλS Networks*, Technical Report TR-01-07, North Carolina State University, Raleigh, NC, July 2001.

30. S. Subramaniam, M. Azizoglu, and A. K. Somani, On the optimal placement of wavelength converters in wavelength-routed networks, *Proc. INFOCOM'98*, pp. 902–909, Apr. 1998.

31. M. Ali and J. Deogun, Allocation of multicast nodes in wavelength-routed networks, *Proc. ICC 2001*, pp. 614–618, 2001.

32. R. Ramaswami and K. N. Sivarajan, Design of logical topologies for wavelength-routed optical networks, *IEEE Journal on Selected Areas in Communications*, 14(5):840–851, June 1996.

33. D. Banerjee and B. Mukherjee, A practical approach for routing and wavelength assignment in large wavelength-routed optical networks, *IEEE Journal on Selected Areas in Communications*, 14(5):903–908, June 1996.

34. B. Mukherjee et al., Some principles for designing a wide-area WDM optical network, *IEEE/ACM Transactions on Networking*, 4(5):684–696, Oct. 1996.

35. Z. Zhang and A. Acampora, A heuristic wavelength assignment algorithm for multihop WDM networks with wavelength routing and wavelength reuse, *IEEE/ACM Transactions on Networking*, 3(3):281–288, June 1995.

36. I. Chlamtac, A. Ganz, and G. Karmi, Lightnets: Topologies for high-speed optical networks, *Journal of Lightwave Technology*, 11:951–961, May/June 1993.

37. S. Banerjee and B. Mukherjee, Algorithms for optimized node placement in shufflenet-based multihop lightwave networks, *Proc. INFOCOM'93*, Mar. 1993.

38. R. Malli, X. Zhang, and C. Qiao, Benefit of multicasting in all-optical networks, *Proc. SPIE*, 3531:209–220, Nov. 1998.

39. G. Sahin and M. Azizoglu, Multicast routing and wavelength assignment in wide-area networks, *Proc. SPIE*, 3531:196–208, Nov. 1998.

40. J. Strand, A. L. Chiu, and R. Tkach, Issues for routing in the optical layer, *IEEE Communications*, pp. 81–96, Feb. 2001.

41. M. Alanyali and E. Ayanoglu, Provisioning algorithms for WDM optical networks, *IEEE/ACM Transactions on Networking*, 7(5):767–778, Oct. 1999.

42. S. Ramamurthy and B. Mukherjee, Survivable WDM mesh networks, I: protection, *Proc. INFOCOM'93*, pp. 744–751, Mar. 1999.

43. B. T. Doshi, S. Dravida, P. Harshavardhana, O. Hauser, and Y. Wang, Optical network design and restoration, *Bell Labs Technical Journal*, pp. 58–84, Jan./Mar. 1999.

44. Y. Qin, K. Sivalingam, and B. Li, Architecture and analysis for providing virtual private networks (VPNs) with QoS over optical WDM networks, *Optical Networks*, 2(2):57–65, Mar./Apr. 2001.

45. E. Lawler, *Combinatorial Optimization: Networks and Matroids*, Holt, Rinehart and Winston, New York, 1976.

46. D. Bertsekas and R. Gallager, *Data Networks*, Prentice Hall, Upper Saddle River, NJ, 1992.

47. X. Zhang, J. Y. Wei, and C. Qiao, Constrained multicast routing in WDM networks with sparse light splitting, *Journal of Lightwave Technology*, 18(12):1917–1927, Dec. 2000.

48. J. Moy, *OSPF Version 2*, RFC 2328, Apr. 1998.

49. A. Chiu, J. Luciani, A. Banerjee, and A. Fredette, Impairments and other constraints on optical layer routing, IETF draft, *draft-ietf-ipo-impairments-00.txt*, May 2001.

50. Y. Zhu, G. N. Rouskas, and H. G. Perros, A comparison of allocation policies in wavelength routing networks, *Photonic Network Communications*, 2(3):265–293, Aug. 2000.

51. Z. Zhang, J. Fu, D. Guo, and L. Zhang, Lightpath routing for intelligent optical networks, *IEEE Networks*, 15(4):28–35, July/Aug. 2001.

52. C. Assi, M. Ali, R. Kurtz, and D. Guo, Optical networking and real-time provisioning: An integrated vision for the next-generation internet, *IEEE Networks*, 15(4):36–45, July/Aug. 2001.

53. S. Sengupta and R. Ramamurthy, From network design to dynamic provisioning and restoration in optical cross-connect mesh networks: An architectural and algorithmic overview, *IEEE Networks*, 15(4):46–54, July/Aug. 2001.

54. Internet Engineering Task Force, *http://www.ietf.org*.

55. Optical Domain Service Interconnect, *http://www.odsi-coalition.com*.

56. Optical Internetworking Forum, *http://www.oiforum.com*.

57. G. Bernstein, R. Coltun, J. Moy, A. Sodder, and K. Arvind, *ODSI Functional Specification Version 1.4*, ODSI Coalition, Aug. 2000.

58. *User Network Interface (UNI) 1.0 Signaling Specification*, OIF2000.125.6, Sept. 2001.

59. B. Rajagopalan et al., IP over optical networks: A framework, IETF draft, *draft-many-ip-optical-framework-03.txt*, Mar. 2001.

60. J. P. Lang et al., Link management protocol (LMP), IETF draft, *draft-ietf-mpls-lmp-02.txt*, Sept. 2001.

61. K. Kompella et al., OSPF extensions in support of generalized MPLS, IETF draft, *draft-ietf-ccamp-ospf-gmpls-extensions-00.txt*, Sept. 2001.

62. D. Katz, D. Yeung, and K. Kompella, Traffic engineering extensions to OSPF, IETF draft, *draft-katz-yeung-ospf-traffic-06.txt*, Oct. 2001.

63. D. Awduche et al., *RSVP-TE: Extensions to RSVP for LSP tunnels*, IETF draft, *draft-ietf-mpls-rsvp-lsp-tunnel-08.txt*, Feb. 2001.

64. O. Aboul-Magd et al., Constraint-based LSP setup using LDP, IETF draft, *draft-ietf-mpls-cr-ldp-05.txt*, Feb. 2001.

65. R. Braden et al., *Resource Reservation Protocol: Version 1*, RFC 2205, Sept. 1997.

66. L. Andersson, P. Doolan, N. Feldman, A. Fredette, and B. Thomas, *LDP Specification*, RFC 3036, Jan. 2001.

67. P. Ashwood-Smith et al., Generalized MPLS signaling: RSVP-TE extensions, IETF draft, *draft-ietf-mpls-generalized-rsvp-te-05.txt*, Oct. 2001.

68. P. Ashwood-Smith et al., Generalized MPLS signaling: CR-LDP extensions, IETF draft, *draft-ietf-mpls-generalized-cr-ldp-04.txt*, July 2001.

11 Traffic Management for IP-over-WDM Networks

JAVIER ARACIL, DANIEL MORATO,* and MIKEL IZAL
Universidad Pública de Navarra, Pamplona, Spain

11.1 INTRODUCTION

The ever-increasing growth of the current Internet traffic volume is demanding a bandwidth that can only be satisfied by means of optical technology. However, although wavelength-division multiplexing (WDM) networks will bring an unprecedented amount of available bandwidth, traffic management techniques become necessary to make this bandwidth readily available to the end user. In this chapter we provide an overview of traffic management techniques for Internet protocol (IP)-over-WDM networks. More specifically, we address the following issues: (1) Why is traffic management necessary in IP-over-WDM networks? and (2) What are the traffic management techniques currently proposed for IP over WDM?

Concerning the first issue, we provide an intuitive rather than mathematically concise explanation about the specific features of Internet traffic that make IP traffic management particularly challenging. Internet traffic shows self-similarity features, and there is a lack of practical network dimensioning rules for IP networks. On the other hand, the traffic demands are highly non-stationary, and therefore the traffic demand matrix, which is the input for any traffic grooming algorithm, is time-dependent. Furthermore, traffic sources cannot be assumed to be homogeneous, since very few users produce a large fraction of traffic. In dynamic IP-over-WDM networks, we experience the problem of the multiservice nature of IP traffic. Since a myriad of new services and protocols are being introduced on a day-by-day basis, it turns out that the provision of quality of service on demand becomes complicated. We analyze the service and protocol breakdown of a real Internet link to show the complexity of the dynamic quality of service (QoS) allocation problem.

*Presently on leave at University of California–Berkeley.

IP over WDM: Building the Next-Generation Optical Internet, Edited by Sudhir Dixit.
ISBN 0-471-21248-2 © 2003 John Wiley & Sons, Inc.

Regarding existing and proposed traffic management techniques, we distinguish between static and dynamic WDM networks. Static WDM networks are dimensioned *beforehand* and the dynamic bandwidth allocation capabilities are very restricted. Such static bandwidth is not well suited to the bursty nature of Internet traffic and, as a result, traffic peaks require buffering, which is difficult to realize in the optical domain. Deflection routing techniques using the spare bandwidth provided by overflow lightpaths have been proposed as a technique to minimize packet loss and reduce optical buffering to a minimum. We investigate the advantages and disadvantages of such techniques, with emphasis on the particular features of IP overflow traffic. In the dynamic WDM networks scenario we study the optical burst switching paradigm [31] as a means to incorporate coarse packet-switching service to IP-over-WDM networks.

To end the chapter we focus on end-to-end issues to provide QoS to the end-user applications. Due to the availability of Tbps in a single fiber, the transfer control protocol (TCP) needs to be adapted to this high-speed scenario. Specifically, we analyze the TCP extensions for high speed, together with split TCP connections techniques, as a solution to provide QoS to the end user in a heterogeneous access-backbone network scenario.

11.1.1   Network Scenario

In what follows we assume three waves in the deployment of WDM technology: (1) first-generation or static lightpath networks, (2) second-generation or dynamic lightpath/optical burst switching networks, and (3) third-generation or photonic packet-switching networks.

Static WDM backbones will provide point-to-point wavelength speed channels (lightpaths). Such static lightpaths will be linking gigabit routers, as current ATM or frame relay permanent virtual circuits do. Such gigabit routers perform cell or packet forwarding in the electronic domain.

In a second-generation WDM network, the WDM layer provides dynamic allocation features, by offering on-demand lightpaths or coarse packet-switching solutions. The former provides a switched point-to-point connection service, while the latter provides burst switching service. Precisely, optical burst switching (OBS) [31] provides a transfer mode that is halfway between circuit and packet switching. In OBS, a reservation message is sent beforehand so that resources are reserved for the incoming burst, which carries several packets to the same destination. In doing so, a single signaling message serves to transfer several packets, thus maximizing transmission efficiency and avoiding circuit setup overhead.

The third generation of optical networks will provide photonic packet switching, thus eliminating the electronic bottleneck. As far as optical transmission is concerned, the challenge is to provide ultranarrow optical transmitters and receivers. Even more challenging is the development of all-optical routers that perform packet header processing in the optical domain. In fact, while packet header processing is very likely to remain in the electronic domain, all-optical packet routers based on optical codes are currently under development [38].

11.2   TRAFFIC MANAGEMENT IN IP NETWORKS: WHY?

There is a well-established theory for dimensioning telephone networks [9] based on the hypothesis of the call arrival process being Poisson and the call duration being well modeled as an exponential random variable. A blocking probability objective can be set, and by means of the Erlangian theory, telephone lines can be dimensioned to achieve the target degree of service.

IP over WDM networks and, in general, IP networks lack such dimensioning rules, for two reasons: (1) the traffic is no longer Poissonian but shows self-similar features, nonstationarity, and source heterogeneity; (2) there is very scarce network dimensioning theory for self-similar traffic. In this section we examine the specific features of self-similar traffic that are not present in other kinds of traffic, such as voice traffic. Such features provide the justification for traffic management techniques in IP-over-WDM networks.

11.2.1   Self-Similarity

In the recent past, voice and data traffic modeling has been based primarily on processes of independent increments, such as the Poisson process. Arrivals in disjoint time intervals are assumed to be independent since, intuitively, the fact that a user makes a phone call has no influence in other users making different calls. More formally, the call arrival process has a correlation that is equal to zero in disjoint time intervals. However, the independent increments property does not hold for Internet traffic. Not only is traffic in disjoint intervals correlated but the correlation decays *slowly*,* meaning that even if the time intervals under consideration are far apart from one another, the traffic is still correlated.

The traffic process is said to have a long memory or *long-range dependence*. In the specific case of Internet traffic, such long-range dependence is observed in the packet counting process, which represents the number of bytes in fixed duration (δ ms) intervals [20,29].

As a consequence of the slow decay of the autocorrelation function, the overflow probability in intermediate router queues increases heavily compared to a process with independent increments (Poisson). In [26] an experimental queueing analysis with long-range dependent traffic is presented, which compares an original Internet traffic trace with a shuffled version (i.e., with destroyed correlations). The results show a dramatic impact in server performance due to long-range dependence.

Figure 11.1 shows the results from trace-driven simulations of a single-server infinite buffer system with self-similar traffic. The self-similar trace is shuffled to show the effect of dependence in queueing performance. We note that the saturation breakpoint for a queueing system with self-similar input occurs at a lower utilization factor than that for the same system with Poissonian input.

---

*Correlation decays as a power law $\rho(k) \approx H(2H-1)k^{2H-2}$, k being the time lag.

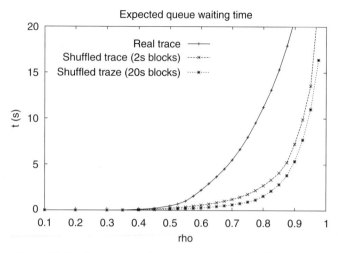

Figure 11.1   Queueing performance with a self-similar process.

Such a performance drop is due to the presence of bursts at any time scale. To assess such a phenomenon visually, Fig. 11.2 shows traffic in several time scales (10, 100, and 1000 ms) for a real traffic trace and a Poisson process. We observe that while the Poissonian traffic smooth out with the time scale toward the rate λ, the real traffic shows *burstiness at all time scales*. The self-similarity property can be explained in terms of aggregation of highly variable sources. First, most Internet traffic is due to TCP connections from the WWW service [4]. Second, WWW objects size and duration can both be well modeled with a Pareto random variable, with finite mean but infinite variance. The aggregation (multiplex) of such connections shows self-similarity properties [13,33]. As a result, self-similarity is an inherent property of Internet traffic, which is due to the superposition of a very large number of connections with heavy-tailed duration.

11.2.2 Demand Analysis

Optical networks will provide service to a large number of users requiring voice, video, and data services. To ensure a QoS to such users, there is a need for demand estimation, so that resources in the network can be dimensioned appropriately. In the telephone network it is well known that users are *homogeneous*, meaning that the traffic demand generated by a number of n users is simply equal to the sum of their individual demands: namely, $I = I'n$, with I' equal to the demand of an individual user.

The homogeneity assumption is most convenient since the operator will dimension the network, taking as a base a single user demand and extrapolating to the rest of the population. However, the traffic demand for the Internet radically differs from that of telephone networks, and the homogeneity assumption is no longer valid.

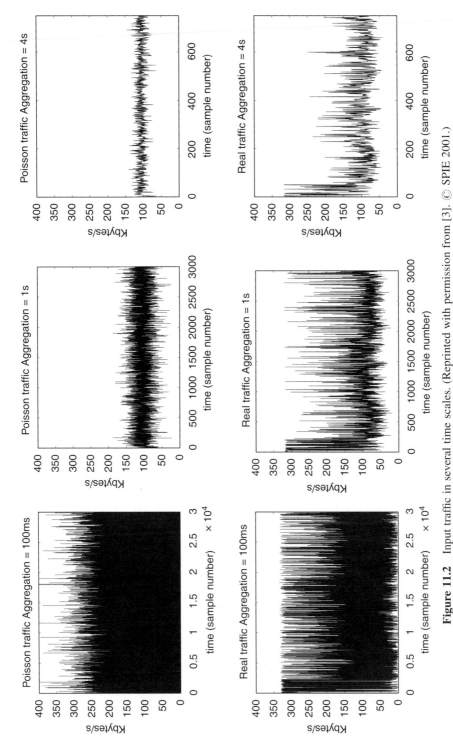

Figure 11.2 Input traffic in several time scales. (Reprinted with permission from [3]. © SPIE 2001.)

333

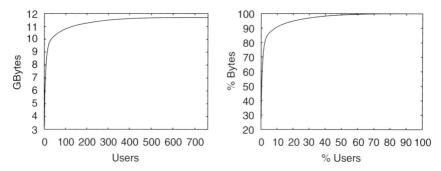

Figure 11.3   Traffic demand sorted by user. Left: Absolute value. Right: Percentage of bytes versus percentage of users.

We take a traffic trace from our university access link (see Section 11.5) and show, in Fig. 11.3 (left), the total volume of traffic generated by a group of n users, sorted according to volume of traffic. We observe that a very few users are producing a large volume of traffic. Figure 11.3 (right) shows the percentage of total volume of traffic in a working day versus the percentage of users generating such traffic. Less than 10% of users produce more than 90% of traffic, showing strong nonhomogeneity in the traffic demand. This result is in agreement with other measurements taken at the UC–Berkeley campus [14].

The results above show that a user population can be divided into three groups: bulk users, average users, and light users. We note that the appearance of a sole new bulk user can produce a sudden increase in network load. For instance, the top producer is responsible for 27.56% of the traffic. Should her machine be disconnected, the traffic load would decrease accordingly. As a conclusion, the existence of a group of top producers complicates matters for network dimensioning. For Internet traffic, the assumption that demand scales with the number of users does not hold. On the contrary, demand is highly dependent, with a very few users.

Furthermore, the dynamic behavior of top producers does not follow a general law. Figure 11.4 shows the accumulated number of TCP connections and bytes from the first and seventh top producers. While the former is a highly regular user, the latter produces a burst of traffic lasting several minutes and nearly no activity for the rest of the time.

11.2.3   Connection-Level Analysis

Optical networks will surely evolve to providing QoS on demand at the optical layer, in a forthcoming photonic packet-switching or lightpath-on-demand scenario. Nowadays, flow-switching mechanisms are being incorporated in IP routers to provide QoS at the connection level. To this end, an analysis of Internet traffic at the connection level is mandatory. First, we observe that Internet traffic is mostly due to TCP connections. Table 11.1 shows a breakdown per protocol of the total traffic in our university link. We note that TCP is dominant, followed at a significant distance

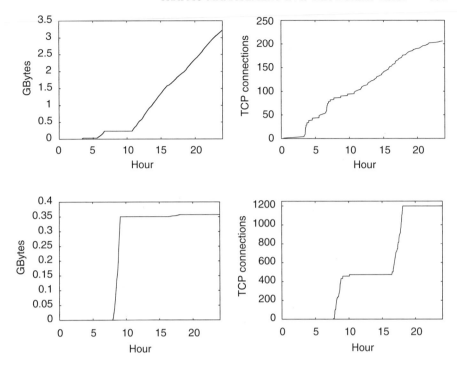

Figure 11.4   Demand versus time for the first (top) and seventh (bottom) producers. Left: TCP bytes. Right: TCP connections.

by UDP. On the other hand, TCP traffic is highly asymmetric outbound from server to client. Our traces show that 82.1% of the total TCP traffic corresponds to the server-to-client outbound. Next, we analyze what the service breakdown for TCP traffic.

11.2.3.1   Breakdown per Service   Figure 11.5 shows the number of bytes per TCP service. The WWW service, either through port 80 or variants due to proxies or secure transactions (port 443), is the most popular service in the Internet. We also note that there are a number of services which cannot be classified a-priori since they do not use well-known ports. This is an added difficulty with per-flow discrimination in flow-switching schemes.

TABLE 11.1   Traffic Breakdown per Protocol

Protocol	MBytes In	% Bytes In	MBytes Out	% Bytes Out
TCP	13,718	99.03	2835	89.38
UDP	109	0.8	317	10
ICMP	13.7	0.1	12.3	0.39
Other	10.2	0.07	7.4	0.23

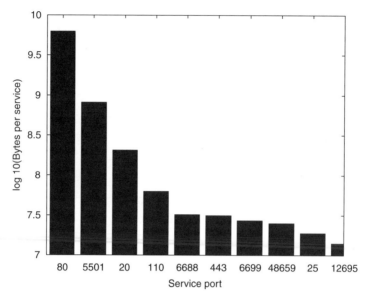

Figure 11.5   Breakdown per service.

11.2.3.2 Connection Size and Duration

Since most IP traffic is due to TCP connections to WWW servers, we focus our analysis in connection size and the duration of such TCP connections. Figures 11.6 and 11.7 show survival functions* of connection size (bytes) and duration (seconds) in log-log scales. We note that the

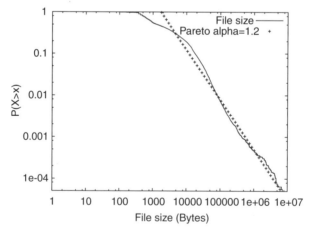

Figure 11.6   Survival function of file size. (Reprinted with permission from J. Aracil and D. Morato, Characterizing Internet load as a non-regular multiplex of TCP streams, *Proc. of 9th IEEE ICCCN Conference*, © 2000 IEEE.)

---

*A survival function provides the probability that a random variable X takes values larger than x [i.e., $P(X > x)$].

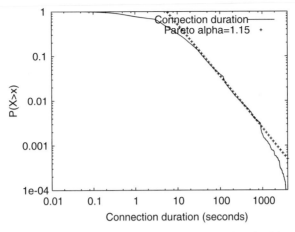

Figure 11.7 Survival function of connection duration. (Reprinted with permission from J. Aracil and D. Morato, Characterizing Internet load as a non-regular multiplex of TCP streams, *Proc. of 9th IEEE ICCCN Conference*, © 2000 IEEE.)

tail of the survival function fits well with that of a Pareto distribution with the parameter α. We plot the distribution tail least-squares regression line in both plots and estimate values of α of 1.15 and 1.2 for duration and size, respectively. Such values are in accordance with previous studies that report values of 1.1 and 1.2 [13].

File sizes in the Internet are heavy-tailed, due to the diverse nature of posted information, which ranges from small text files to very large video files [13]. Regarding duration, we note an even larger variability (lower α), which can be due to the dynamics of the TCP in presence of congestion, which will make connection duration grow larger if packet loss occurs.

Although connection size and duration are heavy-tailed, possibly causing the self-similarity features of the resulting traffic multiplex [13,33], we note that heavy-tailedness is a property of the *tail* of the connection and size distribution. Actually, most TCP connections are short in size and duration. Figure 11.8 shows a histogram of both size and duration of TCP connections. We note that 75% of the connections have less than 6 kilobytes and last less than 11 s. Such short-lasting flows

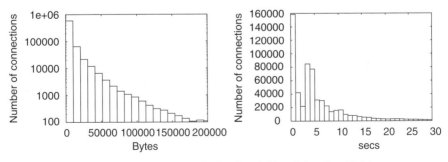

Figure 11.8 Histogram of TCP size (left) and duration (right).

complicate matters for per-flow bandwidth allocation since the resource reservation overhead is significant. As a conclusion, not only are high flexibility and granularity in bandwidth allocation required from the optical layer but also flow classification and discrimination capabilities, to distinguish which flows may be assigned separate resources.

11.3   IP TRAFFIC MANAGEMENT IN IP-OVER-WDM NETWORKS: HOW?

In this section we focus on the impact of IP traffic in WDM networks in first- (static lightpath) and second-generation (dynamic lightpath/OBS) WDM networks. Concerning the former, we explain the scenario and the traffic grooming problem. We also consider the use of overflow bandwidth to absorb traffic peaks. Concerning second-generation networks, we analyze the trade-off between burstiness and long-range dependence and the queueing performance implications. Then we consider signaling aspects, IP encapsulation over WDM, and label-switching solutions.

11.3.1   First-Generation WDM Networks

First-generation WDM networks provide static lightpaths between network end-points. The challenge is to provide a *virtual topology* that maximizes throughput and minimizes delay out of a physical topology consisting of a network topology with optical cross-connects linking fibers with a limited number of wavelengths per fiber.

It has been shown that the general optimization problem for the virtual topology is NP-complete [16]. Such an optimization problem takes as a parameter the traffic matrix, which is assumed to be constant, and as additional assumptions, packet interarrival times and packet service times at the nodes are independent and exponentially distributed, so that M/M/1 queueing results can be applied to each hop. We note that traffic stationarity and homogeneity become necessary conditions to assume that a traffic matrix is constant. A nonstationary traffic process provides an offered load that is time dependent. On the other hand, as shown in Section 11.2, a nonhomogeneous demand may induce large fluctuations in the traffic flows, since a sole bulk user produces a significant share of traffic. If, for instance, the top producer traffic changes destinations at a given time, the traffic matrix is severely affected. Furthermore, the independence assumption in packet arrival times is in contrast to the long-range dependence properties of Internet traffic.

A number of heuristic algorithms have been proposed to optimize the virtual topology of lightpaths (for further reference, see [24, part III]), assuming a constant traffic matrix. We note that even though such algorithms provide optimization of the physical topology, chances are that traffic bursts cannot be absorbed by the static lightpaths. As the buffering capabilities of the optical network are relatively small compared to their electronic counterpart, a number of proposals based on overflow or *deflection routing* have appeared.

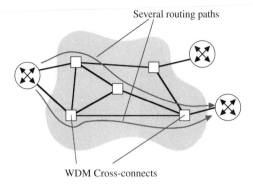

Several routing paths

WDM Cross-connects

Figure 11.9   WDM network.

Figure 11.9 presents a common scenario for first-generation optical networks. IP routers use the WDM layer as a link layer with multiple parallel channels, several of those being used for protection or overflow traffic, which leads to a network design with little buffering at the routers and a number of alternate paths to absorb traffic peaks. The same scenario is normally assumed in deflection routing networks, which are based on the principle of providing nearly no buffering at the network interconnection elements but several alternate paths between source and destination, so that the high-speed network becomes a distributed buffering system. The advantage is that buffer requirements at the routers are relaxed, thus simplifying the electronic design. In the WDM case, we note that the backup channels can be used to provide an alternate path for the overflow traffic as proposed in [5]. Rather than handling the traffic burstiness via buffering, leading to delay and packet loss in the electronic bottleneck, the backup channels can be used to absorb overflow traffic bursts. Interestingly, the availability of multiple parallel channels between source and destination routers resembles the telephone network scenario, in which a number of alternate paths (tandem switching) are available between end offices. The reason for providing multiple paths is not only for protection in case of failure of the direct link but also the availability of additional bandwidth for the peak hours.

The appearance of overflow traffic in the future optical Internet poses a scenario that has not been studied before, since standard electronic networks are based on buffering. Although the behavior of a single-server infinite queue with self-similar input has been well described [27,37], there is little literature on the study of overflow Internet traffic. On the contrary, due to the relevance of overflow traffic in circuit-switched networks, there is an extensive treatment of Poissonian overflow traffic. In the early stages of deployment of telephone networks, A. K. Erlang found that overflow traffic can no longer be regarded as Poissonian. In fact, overflow traffic burstiness is higher than in the Poissonian model. Equivalent Erlangian models for blocking probability calculations can be established [9] that incorporate the overflow load effect. The overflow traffic can be characterized by a Poissonian input with higher intensity, to account for the burstiness of the latter.

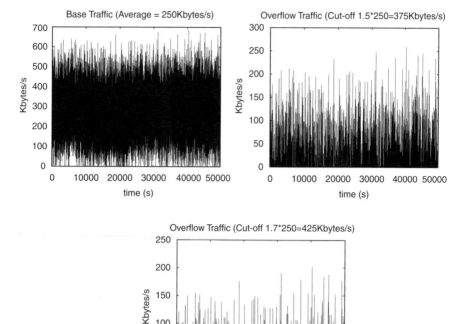

Figure 11.10   Original input and overflow traffic. (Reprinted with permission from [3].
© SPIE 2001.)

To illustrate the burstiness increase of overflow traffic, we plot several instance
of overflow versus total traffic for several cutoff values in Fig. 11.10. The figure
shows a significant burstiness increase in overflow traffic, which, consequently,
requires different treatment in terms of network and router dimensioning [3].

11.3.2   Second-Generation WDM Networks

Second-generation WDM networks will bring a higher degree of flexibility in
bandwidth allotment compared to first-generation static networks. Prior to photonic
packet-switching networks, which are difficult to realize with current technology
[39], dynamic lightpath networks make it possible to reconfigure the lightpaths'
virtual topology according to varying traffic demand conditions. However, the
provision of bandwidth on demand on a per-user or per-connection basis cannot be
achieved with an optical network providing lightpaths on demand, since they
provide a circuit-switching solution with channel capacity equal to the wavelength
capacity. As the next step, optical burst switching [31,32] provides a transfer mode

that is halfway between circuit switching and pure packet switching. At the edges of the optical network, packets are encapsulated in an optical burst, which contains a number of IP packets to the same destination. There is a minimum burst size due to physical limitations in the optical network, which is unable to cope with packets of arbitrary size (photonic packet switching).

Optical burst switching is based on the principle of "on the fly" resource reservation. A reservation message is sent along the path from origin to destination in order to set up the resources (bandwidth and buffers) for the incoming burst. A time interval after the reservation message has been sent, and without waiting for a confirmation (circuit switching), the burst is released from the origin node. As a result of the lack of confirmation, there is a dropping probability for the burst. Nevertheless, resources are statistically guaranteed and the circuit setup overhead is circumvented. Figure 11.11 shows the reservation and transmission procedure for optical burst switching.

On the other hand, optical burst switching allows for a differentiated quality of service by appropriately setting the value of the time interval between release of the resource reservation message and transmission of the optical burst [30,40]. By doing so, some bursts are prioritized over the rest, and thus they are granted a better QoS.

Even though the concept of OBS has attracted considerable research attention, there is scarce literature concerning practical implementations and impact in traffic engineering. A reference model for an OBS edge node is depicted in Fig. 11.12. Incoming packets to the optical cloud are demultiplexed according to their destination in separate queues. A timer is started with the first packet in a queue, and upon timeout expiration, the burst is assembled and relayed to the transmission queue, possibly requiring padding to reach the minimum burst size. Alternatively, a threshold-based trigger mechanism for burst transmission can be adopted, allowing for better throughput for elastic services.

The traffic engineering implications of the reference model in Fig. 11.12 can be summarized as follows [23]. First, we note that there is an increase in the traffic variability (marginal distribution variance coefficient) in short time scales, which is due to the grouping of packets in optical bursts. Furthermore, at short time scales,

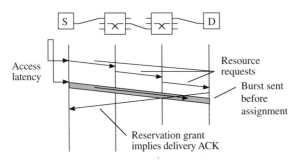

Figure 11.11   Optical burst switching.

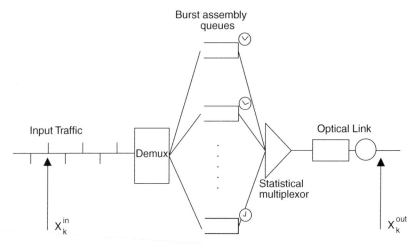

Figure 11.12   OBS reference model.

the process self-similarity is decreased, due to burst sequencing and shuffling at the output of the burst assembly queues. Nevertheless, at a long time scale, self-similarity remains the same. The beneficial effects of a self-similarity decrease at short time scales is compensated by the burstiness (traffic variability) increase at such time scales. Figure 11.13 shows an instance of the input and output traffic processes X^{in} and X^{out} in Fig. 11.12, plotted in several time scales, X^{in} being a fractional Gaussian noise, which proves accurate to model Internet traffic [27]. While significant traffic bursts can be observed at short time scales, we note that the process remains the same as the time scale increases, thus preserving the self-similarity features.

11.3.3   Signaling

Two different paradigms are being considered for integration of IP and WDM in the forthcoming next-generation Internet. The *overlay* model considers both IP and WDM networks as separate networks with different control planes (such as IP over ATM). The *peer model* considers that the IP and WDM network share the same control plane, so that IP routers have a complete view of the optical network logical topology. Figure 11.14 shows the overlay model compared to the peer model. The routers displayed in the figure act as *clients* from the optical network standpoint in the overlay model, since they use the optical network services to fulfill edge-to-edge connectivity requirements. Therefore, an optical user–network interface (UNI) becomes necessary at the network edges, as shown in the same figure. The Optical Internetworking Forum has produced a specification document containing a proposal for implementation of an optical UNI that interworks with existing protocols such as SONET/SDH [28].

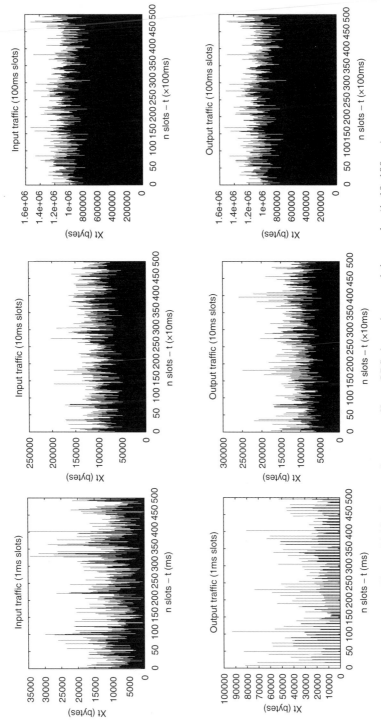

Figure 11.13 Input and output traffic to OBS shaper in several time scales (1, 10, 100 ms).

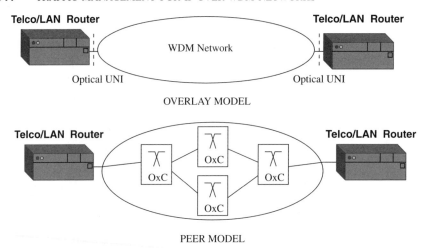

Figure 11.14 Overlay versus peer model.

To integrate IP and WDM layers in the peer model, a promising alternative is the use of multiprotocol label switching (MPLS), *with enhanced capabilities for optical networks*. The aim of MPLS is to provide a higher degree of flexibility to the network manager by allowing the use of label-switched paths (LSPs). Labels in MPLS are like VPI/VCI identifiers in ATM in the sense that they have local significance (links between routers) and are swapped at each hop along the LSP. They are implemented as fixed-length headers which are attached to the IP packet.

An LSP is thus functionally equivalent to a virtual circuit. On the other hand, a forward equivalence class (FEC) is defined as the set of packets that are forwarded in the same way with MPLS. The MPLS label is thus a short, fixed-length value carried in the packet header to identify an FEC. Two separate functional units can be distinguished in an MPLS-capable router: control and forwarding unit. The control unit uses standard routing protocols (OSPF/IS-IS) to build and maintain a forwarding table. When a new packet arrives, the forwarding unit makes a routing decision according to the forwarding table contents. However, the network operator may change the forwarding table in order to explicitly setup an LSP from origin to destination. This explicit routing decision capability provides extensive traffic engineering features and offers scope for quality of service differentiation and virtual private networking. Figure 11.15 shows an example of LSP (routers 1–2–5–6). Even though the link-state protocol mandates, for example, that the best route is 1–4–5–6, the network operator may decide, for load balancing purposes, to divert part of the traffic through routers 2–5–6. On the other hand, MPLS can work alongside with standard IP routing (longest-destination IP address prefix match). In our example, the packet may continue its way to the destination host, downstream from router 6, through a non-MPLS-capable subnetwork.

Explicit routes are established by means of two different signaling protocols: the resource reservation protocol (RSVP-TE) and constraint-based routing

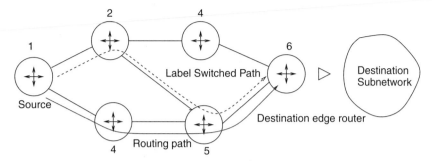

Figure 11.15   Label-switched path example.

label-distributed protocol. Both allow for strict or loose explicit routing with quality of service guarantees, since resource reservation along the route can be performed. However, the protocols are different in a number of ways; for instance, RSVP-TE uses TCP whereas CR-LDP uses IP/UDP. For an extensive discussion of MPLS signaling protocols, the reader is referred to [15] and to Chapters 13 and 15.

MPLS has evolved into a new standard that is tailored to the specific require-ments of optical networks: generalized MPLS (GMPLS) [6]. Since optical networks may switch entire fibers, wavelengths between fibers, or SONET/SDH containers, a FEC may be mapped in many different ways, not necessarily packet switched. For instance, a lightpath may be set up to carry a FEC. In this example there is no need to attach a label to the packet since the packet is routed end to end in an all-optical fashion. Therefore, the (nongeneralized) label concept is extended to the *general-ized* label concept. The label value is implicit since the transport medium identifies the LSP. Thus, the wavelength value becomes the label in a wavelength-routed LSP. Labels may be provided to specify wavelengths, wavebands (sets of wavelengths), time slots, or SONET/SDH channels. A SONET/SDH label, for example, is a sequence of five numbers known as S, U, K, L, and M. A packet coming from an MPLS (packet-switched) network may be transported in the next hop by a SONET/ SDH channel simply by removing the incoming label in the GMPLS router and relaying the packet to the SONET/SDH channel. A generalized label is functionally equivalent to the time-slot location in a plesiochronous network, where the slot location identifies a channel, with no need of additional explicit signaling (a packet header in MPLS).

GMPLS allows the data plane (in the optical domain) to perform packet forward-ing with the sole information of the packet label, thus ignoring the IP packet headers. By doing so, there is no need of packet conversion from the optical to the electronic domain and the electronic bottleneck is circumvented. The signaling for LSP setup is performed by means of extensions of both MPLS resource reservation protocols (RSVP-TE/CR-LDP) to allow a LSP to be routed explicitly through the optical core [7,8]. On the other hand, modifications have been made to GMPLS by adding a new link management protocol designed to address the specific features of the optical media and photonic switches, together with enhancements to the open shortest path first protocol (OSPF/IS-IS) to provide a generalized representation of the various link types in a network (fibers, protection fibers, etc.), which was not necessary in the packet-switched Internet. Furthermore, signaling messages are carried out of band in an overlay signaling network, to optimize resources and in contrast to MPLS. Since signaling messages are short in size and the traffic volume is not large compared to the data counterpart, the allocation of optical resources for signaling purposes seems wasteful. See Chapters 13 and 15 for more details on GMPLS.

As an example of the enhancements provided by GMPLS, consider the number of parallel channels in a DWDM network, which is expected to be much larger than in the electronic counterpart. Consequently, assigning one IP address to each of the links seems wasteful, due to the scarcity of IP address space. The solution to this problem is the use of unnumbered links, as detailed in [19]. The former example illustrates the need of adapting existing signaling protocols in the Internet (MPLS) to the new peer IP over WDM paradigm.

As for third-generation optical networks based on photonic packet switching, there are recent proposals for optical code multiprotocol label switching (OC-MPLS) based on optical code correlations [25]. The issue is how to read a packet header in the all-optical domain. Each bit of the packet header is mapped onto a different wavelength in a different time position, forming a sequence that can be decoded by correlation with all the entries in a code table. While the use of photonic label recognition has been demonstrated in practice the number of codes available is very scarce (8 bits/128 codes) [38], thus limiting applicability of the technique to long-haul networks.

11.3.4 Framing Aspects for IP over WDM

Concerning framing techniques for IP-over-WDM, the IP-over-SONET/SDH standard is foreseen as the most popular solution in the near future [10]. The rationale of IP over SONET/SDH is to simplify the protocol stack of IP over ATM, which in turn relies on SONET/SDH as the physical layer. By encapsulating the IP datagrams on top of SONET/SDH, the bandwidth efficiency is increased while the processing burden imposed by segmentation and reassembly procedures disappears. However, a link-layer protocol is still needed for packet delineation. Figure 11.16 shows the IP datagram in a layer 2 PDU (HDLC-framed PPP), which, in turn, is carried in a SONET/SDH STS-1 payload.

The current standards for layer 2 framing propose the use of HDLC-framed PPP as described in RFC 1662/2615 [34,35]. However, due to scrambling and reliability problems, the scalability is compromised beyond OC-48 [21]. In response to the need for higher speeds, other proposals for IP over SONET/SDH have appeared, such as the PPP over SDL (simplified data link) standard (RFC 2823) [11]. In PPP over SDL the link synchronization is achieved with an algorithm that is similar to I.432 ATM HEC delineation. Instead of searching for a flag (Ox7E), as is done in POS, the receiver calculates the CRC over a variable number of bytes until it "locks" to a frame. Then the receiver enters the SYNC state and packet delineation

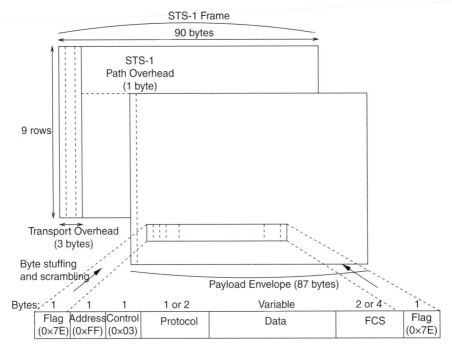

Figure 11.16   IP over SONET/SDH.

is thus achieved. In addition, data scrambling is performed with a self-synchronous $x^{43} + 1$ scrambler or an optional set–reset scrambler independent of user data, which makes it impossible for the malicious user to break SONET/SDH security.

We note that the performance of IP over SONET/SDH, in both POS and PPP over SDL, is highly dependent on the IP packet size. Assuming no byte stuffing (escaping Ox7E flags) and a 16-bit frame check sequence, the POS overhead is 7 bytes. For a SONET/SDH-layer offered rate of 2404 Mbps (OC-48), the user-perceived rate on top of TCP/UDP is 2035 Mbps for an IP packet size of 300 bytes.

Finally, there are ongoing efforts to provide lightweight framing of IP packets over DWDM using 10-gigabit Ethernet (IEEE 802.3ae task force) [36] and digital wrappers (ITU G.709) [1].

11.4   END-TO-END ISSUES

The success of the next-generation optical Internet will depend not only on optical technology, but also on the set of protocols that translate the availability of gigabit bandwidth into user-perceived quality of service. Figure 11.7 shows a reference model for WDM network architecture. The WDM wide/metropolitan area network serves as an Internet backbone for the various access networks. The geographical

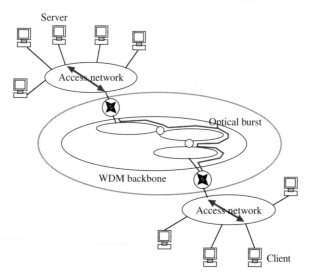

Figure 11.17 WDM network reference model. (Reprinted with permission from [17]. © 2001 SPIE.)

span of the WDM ring can be metropolitan or regional, covering areas up to 1000 miles. The WDM network input traffic comes from the multiplex of a large number of users (in the thousands) at each access network. Examples of access networks in our architecture are campus networks or Internet service provider networks. Access and WDM backbone will be linked by a domain border *gateway* that will perform the necessary internetworking functions. Such a domain border gateway will typically consist of a high-speed IP router.

The scenario shown in Fig. 11.17 is the most likely network configuration for future all-optical backbones, which will surely have to coexist with a number of significantly different technologies in the access network, such as Ethernets, wireless, HFC, and xDSL networks. The deployment of fiber optics to the end-user site can be incompatible with other user requirements, such as mobility, even though there is a trend toward providing high-speed access in the residential accesses. In any case, the access network will be subject to packet loss and delay due to congestion or physical layer conditions which are not likely to happen in the optical domain. On the contrary, the high-speed optical backbones will provide channels with high transmission rates (tens of Gbps) and extremely low bit error rates (in the range 10^{-15}).

Bridging the gap between access and backbone network becomes an open issue. A flat architecture based on a user end-to-end TCP/IP connection, although simple and straightforward, may not be a practical solution. The TCP slow-start algorithm severely constrains the use of the very large bandwidth available in the lightpath until the steady state is reached. Even in such a steady-state regime, the user's socket buffer may not be large enough to provide the storage capacity needed for the huge bandwidth-delay product of the lightpath. Furthermore, an empirical study conducted by the authors in a university network [4] showed that nearly 30% of the transaction latency was due to TCP connection setup time, which poses the burden of the round-trip time in a three-way handshake.

However, TCP provides congestion and flow control features needed in the access network, which lacks the ideal transmission conditions that are provided by the optical segment. We must notice that heterogeneous networks also exist in other scenarios, such as mobile and satellite communications. To adapt to the specific characteristics of each of the network segments' *split TCP connection models*, an evolutionary approach to the TCP end-to-end model has been proposed [2,12]. We note that TCP splitting is not an efficient solution for optical networks, since due to the wavelength speed, on the order of Gbps, the use of TCP in the optical segment can be questioned. For example, considering a 10-Gbps wavelength bandwidth and a 10-ms propagation delay in the optical backbone (2000 km), the bandwidth delay product equals 25 megabytes (MB). File sizes in the Internet are clearly smaller than such a bandwidth-delay product [4,22]. As a result, the connection is always slow starting unless the initial window size is very large [18]. On the other hand, since the paths from gateway to gateway in the optical backbone have different round-trip delays, we note that the bandwidth delay product is not constant. For example, a 1-ms deviation in round-trip time makes the bandwidth-delay product increase to 1.25 MB. Therefore, in this scenario it becomes difficult to optimize TCP windows to truly achieve transmission efficiency. Furthermore, the extremely low loss rate in the optical network makes retransmissions very unlikely to happen, and since the network can operate in a burst-switched mode in the optical layer, we note that there are no intermediate queues in which overflow occurs, thus making most TCP features unnecessary.

11.4.1 TCP for High-Speed and Split TCP Connections

As a first approach to solving the adaptation problem between access and backbone, the TCP connection can be *split* in the optical backbone edges. As a result, each (separate) TCP connection can be provided with TCP extensions tailored to the specific requirements of both access and backbone networks. More specifically, the backbone TCP connection uses TCP extensions for speed, as described in [18]. Such TCP extensions consist of a larger transmission window and no slow start.

A simulation model for the reference architecture of Fig. 11.17 is shown in Fig. 11.18. The ns^3 [41] simulator is selected as a simulation tool since an accurate TCP implementation is available. We choose a simple network topology consisting of an optical channel (1 Gbps) which connects several access routers located at the boundaries of the optical network, as shown in Fig. 11.18. Regarding the access network, two access links provide connectivity between the access routers and client and server, respectively. We simulate a number of network conditions with varying network parameters: namely, link capacities, propagation delay, and loss probability. The objective is to evaluate candidate transfer modes based on TCP, with the performance metric being connection throughput.

Figure 11.19 shows throughput versus file size for split-TCP (STCP-PPS) and end-to-end TCP (EE-PPS) connection with the access network parameters shown in

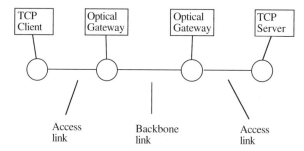

Figure 11.18   An ns model. (Reprinted with permission from [17]. © SPIE 2001.)

Table 11.2. On the other hand, results from a proposal of a transfer protocol using OBS [17], files over lightpaths (FOL), are also presented. In FOL, files are encapsulated in an optical burst in order to be transmitted across the optical network. We note that throughput grows with file size toward a value independent of access BW, and the less RTT, the more steady-state throughput. For small file sizes the connection duration is dominated by setup time and slow start, which does not allow the window size to reach a steady-state value. For large files the TCP reaches steady state and the throughput is equal to window size divided by round-trip time. Such behavior is expected in a large bandwidth-delay product network, in which connections are RTT-limited rather than bandwidth-limited.

We note that the use of split TCP provides a significant performance improvement. Thus, we expect that the forthcoming WDM networks will incorporate a connection adaptation mechanism in the edge routers so that the WDM bandwidth can be exploited fully.

11.4.2   Performance Evaluation of File Transfer (WWW) Services over WDM Networks

The results shown in Section 11.4.1 provide a comparison in error-free conditions: for instance, in a first-generation WDM network (static lightpath between routers). However, it turns out that second-generation WDM networks suffer blocking probability as a consequence of the limited number of wavelengths and burst dropping, due to limited queueing space in optical burst or photonic packet switches. In such conditions, split TCP becomes inefficient and alternate protocol design become necessary.

In FOL [17], files are encapsulated in optical bursts which are released through the optical backbone using a simple stop-and-wait protocol for error control. Assuming that the setup of an optical burst takes RTT/2, Figure 11.20 shows the achieved throughput with error probabilities lower than 0.1, for both FOL (different burst sizes) and Split TCP (STCP). We observe that TCP congestion avoidance severely limits transfer efficiency. *If loss probabilty is equal to 0.01 the throughput obtained with TCP is half the throughput obtained with a simple stop and wait*

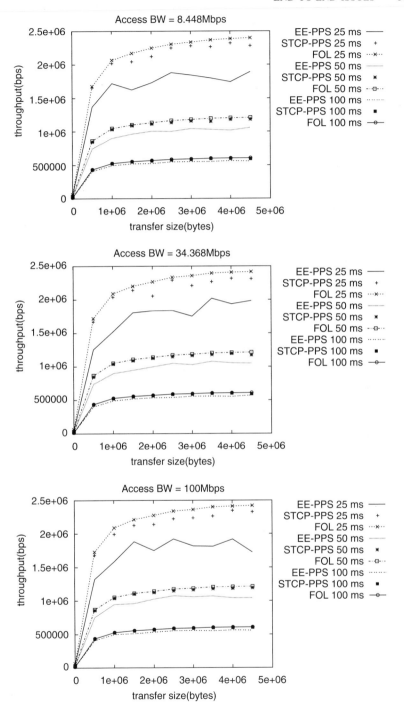

Figure 11.19   Average transfer throughput. (Reprinted with permission from [17]. © SPIE 2001.)

TABLE 11.2   Summary of Simulation Parameters

Parameter	Value
BW of backbone link	1 Gbps
Backbone link propagation delay	0–30 ms
BW of access link	8.4 M bps, 34.4 M bps, 100 M bps
Access link propagation delay	25 ms, 50 ms, 100 ms

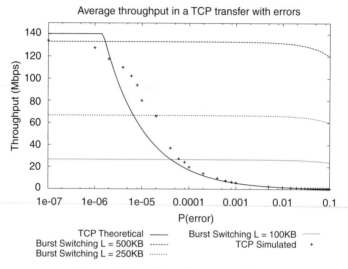

Figure 11.20   Throughput comparison.

protocol in FOL. This serves to illustrate that the throughput penalty imposed by the TCP congestion control mechanisms is rather significant.

The main difference between a simple FOL protocol and TCP is the way in which both protocols interpret congestion. Whereas TCP considers that loss is produced by queueing overflow, FOL is aware that loss is due to blocking. In a loss situation, TCP will lower the transmission window, which results in *no effect at all* since congestion is due to blocking, and the higher the blocking probability, the larger is the number of accesses to the optical network. Furthermore, since the BW × RTT product is extremely large, the TCP window size takes on a very high value. As a result, the slow-start or congestion avoidance phase that follows a packet loss take the longest time to complete.

11.5   SUMMARY

In this chapter we have presented the motivation, solutions, and open challenges for the emerging field of traffic management in all-optical networks. At the time of the writing of the chapter, there are a large number of open issues, including traffic management for IP traffic in dynamic WDM networks (dynamic lightpath and burst switching), design of an efficient integrated control plane for IP and WDM, and a proposal for a transport protocol that will translates the availability of optical bandwidth into user-perceived quality of service. We believe that the research effort in the next-generation optical Internet will be focused not only on the provision of a very large bandwidth but also in the flexibility, ease of use, and efficiency of optical networks.

Note: Measurement Scenario   Our traffic traces are obtained from the network configuration depicted in Fig. 11.21. The measurements presented were performed

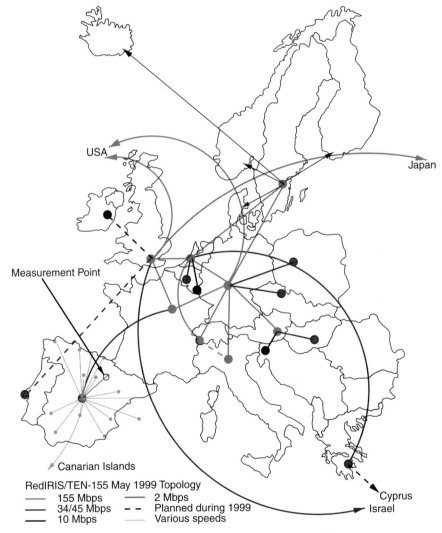

Figure 11.21   Network measurement scenario.

TABLE 11.3   Trace Characteristics

Start date	Mon 14/02/2000 00:00
End date	Mon 14/02/2000 24:00
TCP connections recorded	957053
IP packets analyzed	16375793

at the ATM permanent virtual circuit (PVC) that links the Public University of Navarra to the core router of the Spanish academic network (*RedIris* [42]) in Madrid. RedIris topology is a star of PVCs that connect the universities around the country to the central interconnection point in Madrid. From the central RedIris facilities in Madrid, a number of international links connect the Spanish ATM academic network to the outside Internet. The PVC measured uses bridged encapsulation and is terminated at both sides by IP routers. The peak cell rate (PCR) of the circuit is limited to 4 Mbps, and the transmission rate in the optical fiber is 155 Mbps.

We note that the scenario under analysis is a representative example of a number of very common network configurations. For example, most Spanish Internet service providers (ISPs) hire ATM PVC links to the operators to provide customers with access to the Internet. The same situation arises with corporate and academic networks, which are linked to the Internet through such IP-over-ATM links. On the other hand, measurements are not constrained by a predetermined set of destinations but represent a real example of a very large sample of users accessing random destinations in the Internet. Table 11.3 summarizes the main characteristics of the traffic trace presented in this chapter.

REFERENCES

1. ANSIT1X1.5/2001-064 Draft ITU-T Record G. 709, *Network Node Interface for the Optical Transport Network (OTN)*, Mar. 2001.

2. E. Amir, H. Balakrishnan, S. Seshan, and R. Katz, Improving TCP/IP performance over wireless networks, in *ACM MOBICOM'95*, Berkeley, CA, 1995.

3. J. Aracil, M. Izal, and D. Morato, Internet traffic shaping for IP over WDM links, *Optical Networks Magazine*, 2(1), Jan./Feb. 2001.

4. J. Aracil, D. Morato, and M. Izal, Analysis of internet services for IP over ATM links, *IEEE Communications Magazine*, Dec. 1999.

5. B. Arnaud, Architectural and engineering issues for building an optical Internet, *Proc. SPIE International Symposium on Voice, Video, and Data Communications—All-Optical Networking: Architecture, Control, and Management Issues*, Boston, Nov. 1998.

6. P. Ashwood-Smith et al., Generalized MPLS-signaling functional description, Internet draft, draft-ietf-mpls-generalized-cr-Ldp-07.txt, Nov. 2000.

7. A. Banerjee, J. Drake, J. Lang, B. Turner, D. Awduche, L. Berger, K. Kompella, and Y. Rekhter, Generalized multiprotocol label switching: An overview of signalling enhancements and recovery techniques, *IEEE Communications Magazine*, July 2001.

8. A. Banerjee, J. Drake, J. Lang, B. Turner, K. Kompella, and Y. Rekhter, Generalized multiprotocol label switching: An overview of routing and management enhancements, *IEEE Communications Magazine*, Jan. 2001.

9. J. Bellamy, *Digital Telephony*, Wiley, New York, 1991.

10. P. Bonenfant and A. Rodriguez-Moral, Framing techniques for IP over fiber, *IEEE Network*, July/Aug. 2001.

11. J. Carlson, P. Langner, E. Hernandez-Valencia, and J. Manchester, *PPP over Simple Data Link (SDL) Using SONET/SDH with ATM-Like Framing*, RFC 2823 (experimental), May 2000.

12. R. Cohen and I. Minei, High-speed internet access through unidirectional geostationary channels, *IEEE Journal on Selected Areas in Communications*, 17(2):345–359, Feb. 1999.

13. M. E. Crovella and A. Bestavros, Self-similarity in World Wide Web traffic: Evidence and possible causes, *IEEE/ACM Transactions on Networking*, 5(6):835–846, Dec. 1997.

14. R. Edell, N. McKeown, and P. Varaiya, Billing users and pricing for TCP, *IEEE Journal on Selected Areas in Communication*, 13(7), Sept. 1995.

15. E. W. Gray, *MPLS: Implementing the Technology*, Addison-Wesley, Reading, MA: 2001.

16. A. Ganz, I. Chlamtac, and G. Karmi, Lightnets: Topologies for high speed optical networks, *IEEE/OSA Journal of Lightwave Technology*, 11, May/June 1993.

17. M. Izal and J. Aracil, IP over WDM dynamic link layer: Challenges, open issues and comparison of files-over-lightpaths versus photonic packet switching, *Proc. SPIE OptiComm 2001*, Denver, CO, Aug. 2001.

18. V. Jacobson and R. Braden, *TCP Extensions for Long-Delay Paths*, RFC 1072, Oct. 1998.

19. K. Kompella and Y. Rekhter, Signalling unnumbered links in RSVP-TE, Internet draft, draft-ietf-mpls-cr-Ldp-ummum-03.txt, Sept. 2000.

20. W. E. Leland, M. S. Taqqu, W. Willinger, and D. V. Wilson, On the self-similar nature of Ethernet traffic, *IEEE/ACM Transactions on Networking*, 2(1):1–15, Jan. 1994.

21. J. Manchester, J. Anderson, B. Doshi, and S. Davida, IP over SONET, *IEEE Communications Magazine*, May 1998.

22. G. Miller, K. Thompson and R. Wilder, Wide-area Internet traffic patterns and characteristics, *IEEE Network*, pp. 10–23, Nov./Dec. 1997.

23. D. Morató, J. Aracil, M. Izal, E. Magana, and L. Diez-Marca, *Explaining the Impact of Optical Burst Switching in Traffic Self-Similarity*, Technical Report, Public University of Navarra, Navarra, Spain.

24. B. Mukherjee, *Optical Communication Netwoks*, McGrawHill, New York, 1997.

25. M. Murata and K. Kitayama, A perspective on photonic multiprotocol label switching, *IEEE Network*, July/Aug. 2001.

26. O. Narayan, A. Erramilli, and W. Willinger, Experimental queueing analysis with long-range dependent packet traffic, *IEEE/ACM Transactions on Networking*, 4(2):209–223, Apr. 1996.

27. I. Norros, On the use of fractional Brownian motion in the theory of connectionless networks, *IEEE Journal on Selected Areas in Communications*, 13(6):953–962, Aug. 1995.

28. Optical Internetworking Forum, OIF UNI 1.0- controlling optical networks, *http://www.oiforum.com*, 2001.

29. V. Paxson and S. Floyd, Wide area traffic: The failure of Poisson modeling, *IEEE/ACM Transactions on Networking*, 4(2):226–244, Apr. 1996.

30. C. Qiao, Labeled optical burst switching for IP-over-WDM integration, *IEEE Communications Magazine*, 38:104–114, Sept. 2000.

31. C. Qiao and M. Yoo, Optical burst switching (OBS): A new paradigm for an optical Internet, *Journal of High-Speed Networks*, 8(1), 1999.

32. C. Qiao and M. Yoo, Features and issues in optical burst switching, *Optical Networks Magazine*, 1:36–44, Apr. 2000.

33. R. Sherman, W. Willinger, M. S. Taqqu and Daniel V. Wilson, Self-similarity through high-variability: Statistical analysis of Ethernet LAN traffic at the source level, *IEEE/ACM Transactions on Networking*, 5(1), Feb. 1997.

34. W. Simpson and A. Malis, *PPP in HDLC-Like Framing*, RFC 1662, July 1994.

35. W. Simpson and A. Malis, *PPP over SONET/SDH*, RFC 2615, June 1999.

36. D. H. Su, Standards: The IEEE P802.3ae project for 10 Gb/s Ethernet, *Optical Networks Magazine*, 1(4), Oct. 2000.

37. B. Tsybakov and N. D. Georganas, On self-similar traffic in ATM queues: Definitions, overflow probability bound and cell delay distribution, *IEEE/ACM Transactions on Networking*, 5(3):397–409, June 1997.

38. N. Wada and K. Kitayama, Photonic IP routing using optical codes: 10 gbit/s optical packet transfer experiment, *Proc. Optical Fiber Communications Conference 2000*, Baltimore, Mar. 2000.

39. S. Yao, B. Mukherjee, and S. Dixit, Advances in photonic packet switching: An overview, *IEEE Communications Magazine*, 38(2):84–94, Feb. 2000.

40. M. Yoo, C. Qiao, and S. Dixit, Optical burst switching for service differentiation in the next generation optical internet, *IEEE Communications Magazine*, 39(2), Feb. 2001.

41. *http://www-mash.cs.berkeley.edu/ns*.

42. *http://www.rediris.es*.

12 IP- and Wavelength-Routing Networks

RANJAN GANGOPADHYAY

Indian Institute of Technology, Kharagpur, India

SUDHIR DIXIT

Nokia Research Center, Burlington, Massachusetts

12.1 INTRODUCTION

Considering the fact of Internet dominance, there has been a noticeable change in the traffic demand profile from voice dominated to data centric. Further, Internet traffic is more biased to long-reach connections in contrast to voice traffic, which exhibits short-distance connections. Internet traffic is also characterized by uncertainty in traffic distributions and traffic asymmetry. Currently designed backbone networks for the transport of voice circuits are becoming less attractive for Internet protocol (IP)-dominated traffic environments. Synchronous Optical Network/Synchronous Digital Hierarchy (SONET/SDH) ring architecture is also proving to be inefficient for transport of long-distance IP demands. Further, the availability of terabit IP router/switches at the edge of the backbone network minimizes the need for SONET/SDH layer multiplexing. On the contrary, the IP backbone does not provide the needed quality of service (QoS) and other important requirements, such as rapid provisioning of dynamic connection and traffic demands, automatic protection/restoration strategies, survivability, and so on.

The future evolutionary path of optical networking will try to leverage the benefits of the cross-connected wavelength-division multiplexing (WDM) technology to provide high-capacity connection throughout the entire network infrastructure and to orient the network operation from leased-line circuit-switched concepts to an IP-based bandwidth-on-demand environment. Essentially, it is the IP-over-WDM networking solution that will support predominantly IP traffic over WDM optical paths [1]. In this chapter we focus mainly on IP routing as well as wavelength-routing networks. A brief introduction to the current routing protocols

IP over WDM: Building the Next-Generation Optical Internet, Edited by Sudhir Dixit.
ISBN 0-471-21248-2 © 2003 John Wiley & Sons, Inc.

for the datagram networks is presented. We then provide wavelength-routing network architecture and address the routing and wavelength assignment (RAW) policy in WDM networks in detail. This provides a straightforward vehicle for the design and analysis of WDM networks with and without wavelength conversion. Finally, the concepts of IP-over-WDM integration, integrated routing, and waveband routing have been introduced. It is understood that readers will get a more comprehensive view of IP-over-WDM networking utilizing the concepts of TCP/IP, MPLS/MPλs/GMPLS, and related topics in Chapters 2, 4, 10, 14, and 15.

12.2 INTERNET PROTOCOL AND ROUTING

The Internet protocol [2] is a part of the transfer control protocol (TCP)/IP suite and has been treated as the foundation of the Internet and virtually all multivendor internetworks. It is functionally similar to the ISO standard connectionless network protocol. IP is basically specified in two parts:

1. The interface with the next-higher layer (e.g., TCP), specifying the services that IP provides
2. The actual protocol format and mechanism

The IP service model has been central to today's Internet and is indispensable to the *next-generation Internet* (NGI) [3]. Mainly, there are two types of service primitives at the interface to the next-higher layer. The send primitive used to request data transmission of a data unit and the delivery primitive is used by IP to notify a user on arrival of a datagram. The important features of IP services include its support to a multiplicity of delivery choices (multicast and unicast), its simple decentralized operation, and its ability to recover from a wide range of outages.

At present, two versions of IP protocol exist: IP version 4.0 (IPv4) and IP version 6 (IPv6). The very important elements of IP are described as follows:

- IP defines the basic data unit that can be sent through an Internet (i.e., IP specifies the format of the datagram to be sent).
- IP software carries out the routing function based on the IP address.
- IP contains a set of rules for how routers and hosts should handle the datagrams and how error messages should be generated when the datagrams have to be discarded.

The main features of the existing Internet protocol are as follows:

- There is a 32-bits-word header length with a minimum of 20 octets. Padding is used to make the header a multiple of 32 bits.
- Time to live (TTL) is measured in terms of router hops.
- Use of only a header checksum reduces computational complexities.

- A 32-bit source and destination address is used.
- A datagram must be an integer multiple of 8 bits. The maximum length of a datagram (data plus header) is 65,535 octets.

Traditionally, the Internet has supported best-effort service only where all the packets have equal access to network resources. The network layer is concerned with transporting packets to their proper destination using a packet header based on the entry in the routing table. The separation of routing (creating, manipulating, and updating of the routing table) from the actual forwarding process is an important design concept of the Internet. The Internet Engineering Task Force (IETF) has come up with several solutions, such as IntServ/ RSVP, DiffServ, to enable the quality of service (QoS) [2].

12.3   ROUTING IN DATAGRAM NETWORKS

Routing [4] is the process of finding a path from a source to a destination in the network. There are two basic categories of routing algorithms: static and adaptive. In the *static routing algorithm*, the routing table is initialized during system startup. Static routing is incapable of adapting to changes in the network and is therefore not suitable for practical use in data networks. *Adaptive routing algorithms* are able to adapt their routing information dynamically to current network characteristics. Adaptive routing algorithms can be further subdivided into either centralized or distributed. In *centralized algorithms*, routing decisions are made by a central entity in the network and then distributed to the participating users. Since complete knowledge about the network is available, the routing decisions in this case are reliable. However, the centralized entity represents a single point of failure. One drawback to the centralized approach is that it is not very scalable and usable in a very large network.

In the *distributed routing* approach, each router makes an independent routing decision based on the information available to it. Distributed routing is also efficient for applications in large networks. The two categories of distributed routing are the link-state algorithm and the distance vector algorithm. In a *link-state approach*, every node in the network must maintain complete network topology information. Each node then finds a route for a connection request in a distributed manner. When the state of the network changes, all the nodes must be informed by broadcast of update messages to all nodes in the network. Dijkstra's [4] algorithm is used frequently in the calculation of routing information. With this algorithm the network is regarded as a directed graph with the respective node as a root. Calculation of the shortest path to all systems in the network is carried out with this model of directed graph. In carrying out Dijkstra's algorithm, all nodes (intermediate systems) are marked as temporary and identified with infinitely high costs. The nodes marked as temporary are not part of the directed graph. The first step involves marking the root node as permanent. The root node is the node in which the calculation is carried out. During each step, for each new permanent node the cost is calculated from the root node to the neighboring nodes of this permanent node. The algorithm then selects the node with the lowest determined cost and labels it as permanent. Marking nodes as permanent indicates that these nodes are part of the directed graph. We present below the formal description of Dijkstra's algorithm. We view each node i as being labeled with an estimate D_i of the shortest path length from the root node 1. When the estimate becomes certain, we regard the node as being permanently labeled and keep track of this with a set of P permanently labeled nodes.

12.3.1 Dijkstra's Algorithm

Initialization:

$$P = \{1\}, D_1 = 0, \text{ and}$$
$$D_j = d_{1j}, \text{ for } j \neq 1.$$

Step 1 (Find the next-closest node)

Find $i \notin P$ such that

$$D_i = \min_{j \notin P} D_j$$

Set $P := P \cup \{i\}$. If P contains all nodes, then stop; the algorithm is complete.

Step 2 (Updating of labels)

For all $j \notin P$, set

$$D_j := [D_j, D_i + d_{ij}]$$

Go to step 1.

Dijkstra's algorithm requires a number of operations proportional to the number of nodes, N, at each step and the steps are iterated $N - 1$ times so that in the worst case the computation time is $O(N^2)$.

12.3.2 Distance Vector Algorithm

The distance vector algorithm makes the assumption that each network node is aware of the distance between the node itself and all other nodes in the network. The routing tables of distance vector algorithms contain an entry for each possible destination system in a network or in a domain. The routing table will initially use the value infinity for all nonneighboring systems. This indicates that the distance to a particular system is not known and that it therefore cannot be reached. When a router receives a distance vector, it adds the remaining distance of the link to the values received with the distance vector. This newly calculated value is compared to the current entry in the routing table. If the new value is lower, it is used to replace the old value and a new entry is constructed in the routing table. Consequently, the router has determined a shorter path between sender and receiver than known previously. The advantage of a distance vector algorithm lies in its simplicity for the fact that routers do not require a global view of the entire network, instead, they operate on the basis of the knowledge of distance vectors exchanged periodically with their neighbors. However, the suitability of a distance vector algorithm is limited for large networks since information carrying any change in network topology has to be propagated n times in the network through an exchange of distance vectors, with n representing the number of links on the longest path through the network. We provide below one implementation of the distance vector algorithm based on Bellman–Ford algorithm [4].

In the distance vector routing, a node tells its neighbors its distance to every other node in the network. Its implementation is the same as in the Bellman–Ford shortest-path algorithm. In such an implementation, each node is responsible for computing the distance D_i from itself to the destination. These initial estimates may be arbitrary except for node 1 (i.e., $D_1 = 0$: distance from itself to itself). Each node i advertises to its neighbors the current estimates of D_i that it has. This advertisement is done periodically or whenever the estimate changes. Whenever a node receives an estimate from a neighbor, it updates its own estimate according to the following Bellman's equation:

$$D_i = \min\{w_{ij} + D_j : 1 \leq j \leq N\}, \qquad 1 < i \leq N$$

where N is the number of nodes in the network and w_{ij} denotes the weight assigned to the link (i, j); $w_{ij} = +\infty$ if node i is not a neighbor of node j and $w_{ij} > 0$ for all i and j.

Specifically, if $D_i(t)$ is the estimate of D_i at time t and if $D_j(t - \tau_j)$ is the estimate of D_j from node j that was most recently advertised to node i, then $D_i(t) = \min\{w_y + D_j(t - \tau_j) : j \in N(i)\}$. The minimum above can be taken over only the neighboring nodes of i. The algorithm above ensures convergence of the nodes' estimate to the actual distance to the destination even if the nodes asynchronously update their distance vectors. However, problems may arise if the link weights change rapidly in response to changing traffic characteristics and the algorithm does not have a chance to converge. The worst-case time complexity of the Bellman ford algorithm is easily seen to be $O(N^3)$.

12.3.3 Deflection Routing

With deflection routing, each node attempts to route packets along a shortest path. A packet is deflected to any other available output link if it fails to exit (because of contention) to the preferred output link, which is in the shortest path to the destination. The deflection routing has the advantage that packet buffering can essentially be eliminated and a reduction in control bandwidth demand, as no update messages are present in the network.

12.3.4 Hierarchical Routing

The lookup tables for routing can get very large with a large number of nodes in a network. A hierarchical routing strategy [5] is adopted to reduce the size of the lookup tables as well as the complexity of route computation. In this hierarchical routing model a subset of network nodes are identified as backbone nodes. The nonbackbone nodes are partitioned into clusters of nodes called *domains*. Each backbone node belongs to a distinct domain and acts as the parent of all nodes in the domain. Routing is done separately at the domain level and in the backbone level. Because of this decoupling, the problem of route computation and packet forwarding can be simplified considerably in large networks.

A hierarchical addressing system is also very convenient in a hierarchical routing approach. For example, a node address can be grouped into two parts: the domain address and the address of the node within the domain. Thus packet forwarding at the backbone level needs to look at the domain part of the node address, and packet forwarding at the domain level may ignore the domain part of the node address.

12.3.5 Common Routing Protocols

Two important routing protocols that are in use in the Internet are the interior and exterior routing protocols [2,5]. Within the interior routing protocols are the routing information protocol (RIP) and the open shortest-path protocol (OSPF). The protocols commonly used for exterior routing are the exterior gateway protocol (EGP) and the border gateway protocol (BGP).

1. *RIP.* RIP [2] is a distance vector protocol that was the original routing protocol in the Arpanet. It uses a hop-count metric to measure the distance between the source and a destination. Each hop in a path is assigned a hop-count value, which is typically 1, and the infinity metric is defined to be 16. RIP sends routing update messages at regular intervals (30 s) and when the network topology changes. When a router receives a routing update that contains a new or changed destination network entry, the router adds one to the metric value indicated in the update and enters the network in the routing table. The IP address of the sender is used as the next hop. RIP prevents routing loops from continuing indefinitely by implementing a limit on the number of hops allowed in a path. If the metric value reaches 16, the network destination is considered unreachable. RIP is useful for small subnets, where its simplicity of implementation more than compensates its inadequacy in dealing with link failures.

2. *OSPF.* OSPF [2,5] is the interior routing protocol developed for IP networks. It is based on a link-state routing protocol that calls for the sending of link-state advertisements (LSAs) to all other routers within the same hierarchical area. Information on attached interfaces, metrics used, and other variables is included in OSPF LSAs. With the accumulated link-state information, OSPF routers use the SPF algorithm (Dijkstra's algorithm) to calculate the shortest path to each node.

3. *EGP.* EGP used in the Internet is based on a distance-vector routing protocol which allows the network administrator to pick their neighbors in order to enforce interautonomous system (AS) routing policies. It also uses address aggregation in routing tables to allow scaling [5].

4. *BGP.* BGP [2] is a replacement of EGP which is based on the path vector protocol, where distance vectors are annotated not only with the entire path used to compute each distance, but also with certain policy attributes. An exterior gateway can compute much better paths with BGP than with EGP by examining these attributes.

12.4 WAVELENGTH-ROUTING NETWORKS

A scalable WDM network can be constructed by taking several WDM links and connecting them at a node by a switching subsystem. These nodes (also wavelength routers) are interconnected by fibers to generate diverse and complex topologies. Each wavelength router makes its routing decision based on the input port and wavelength of the arriving signal. The same signal being carried by wavelength λ_1 (say) will be routed similarly by the intermediate-wavelength routers until it reaches the final destination port (see Fig. 12.1). Such an end-to-end connection is called a *light path* (*wavelength path*) of a wavelength-routed (WR) network [6]. At the same time, the same wavelength can be reused in some other part of the network for another end-to-end lightpath establishment as long as two or more lightpaths corresponding to the same wavelength do not flow through the same fiber. Such spatial reuse of wavelengths imparts higher scalability to WR networks than for broadcast-and-select WDM networks. If the end-to-end connection is established by several different wavelengths in cascade by providing a wavelength conversion facility, the connection is said to have been established by a semi-lightpath [or virtual wavelength path (VWP)] and the node is designated as bering of the wavelength interchanging (WI) type.

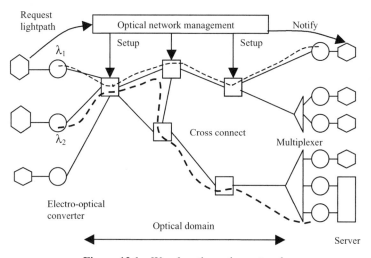

Figure 12.1 Wavelength-routing network.

Various architectural designs of the node implementation realized with optical cross-connects (OXCs) are possible [6]. An OXC provides the routing of individual channels coming from any of its input ports to any output port. Figure 12.2 depicts various possible architectures. The simple configuration (Fig. 12.2a) corresponds to fixed (or static) routing. A space-division switch is introduced in Fig. 12.2b to allow any wavelength from any input fiber to be routed to any output fiber not already

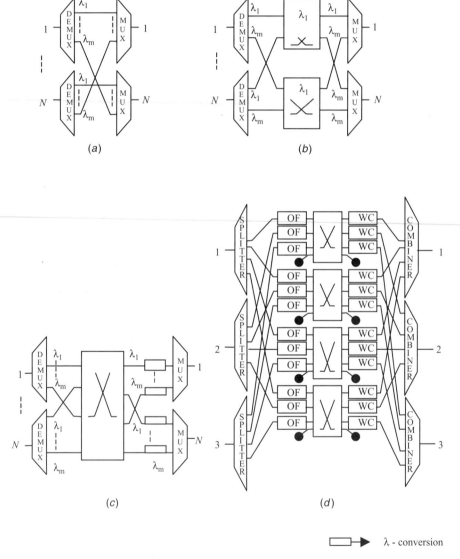

Figure 12.2 Optical cross-connect architectures: (a) static routing; (b) reconfigurable OXC; (c) nonblocking OXC; (d) nonblocking OXC with a reduced-size space switch.

using that wavelength. This makes the OXC rearrangeable. A strictly nonblocking realization of OXC is provided in Figure 12.2c, where the facility of wavelength translation allows two channels carried on two different fibers at the same wavelength to be routed onto a single outgoing link. This configuration, although introducing significant complexity to the routing node structure, permits better wavelength reuse. Finally, a strictly nonblocking OXC with lower dimension space switches can be realized as shown in Fig. 12.2d, where the channel selection is accomplished by a combination of passive power splitters and tunable filters (OFs).

12.4.1 Routing and Wavelength Assignment

One of the important requirements in the control plane of a wavelength-routed optical network (WRON) is to set up and take down optical connections, which are established by lightpaths between source and destination nodes. A lightpath is an all-optical path connecting a source and a destination such that every link of the path uses the same wavelength (wavelength continuity constraint). This is true when the OXCs in WRON do not have wavelength conversion capability. On the contrary, when the lightpath allocation is done link by link so that the end-to-end optical path is a chain of lightpaths of different wavelengths, it is called a *virtual wavelength path* (VWP) (or *semi-lightpath*) [7]. This is a consequence of the fact that the OXCs in WRON provide the additional facility of λ-conversion. We distinguish between static and dynamic lightpaths, depending on the nature of the traffic in the network. When the nature of the traffic pattern is static, a set of lightpaths is established all at once that remain in the network for a long period of time. Such static lightpath establishment is relevant in the initial design and planning stage and the objective is to maximize the network throughput (i.e., maximize the number of lightpaths established with given network resources).

For the dynamic traffic scenario where the traffic pattern changes rapidly, the network has to respond to traffic demands quickly and economically. In such a dynamic traffic case, a lightpath is set up for each connection request as it arrives, and the lightpath is released after some finite amount of time. This is called *dynamic lightpath establishment*. In this section we briefly address the dynamic path establishment in a WDM network. Given a constraint on the number of wavelengths, it is necessary to determine both the routes over which these light-paths are to be established and the wavelengths that are to be assigned to these routes. This problem is referred to as a *routing and wavelength assignment* (RWA) *problem*. The other important control and management protocols perform the tasks of exchanging the signaling information, resource reservation, and protection and restoration in case of occurrence of network failures.

12.4.2 RWA Algorithm

The RWA problem is viewed as a routing subproblem and wavelength assignment subproblem.

12.4.2.1 Routing Subproblem The routing subproblem is categorized as either static or adaptive and normally employs either global or local state information.

12.4.2.2 Fixed/Fixed–Alternate Path Routing In fixed routing, a single fixed route is predetermined for each source–destination path and stored in an ordered list at the source node routing table. As the connection request arrives, one route is selected from the precomputed routes [8].

12.4.2.3 Adaptive Routing By the adaptive routing approach, the likelihood of path connection can be improved by providing network global state information. This can be organized either centrally or in a distributed fashion. In a centralized algorithm, the network manager maintains complete network state information and is responsible for finding the route and setting up lightpaths for connection requests. In this case, a high degree of coordination among nodes is not needed, but the centralized entity is a possible single point of failure.

In *distributed adaptive routing* [9] based on the link-state approach, each node must maintain complete network state information. The information regarding any establishment or removal of a lightpath in a network must be broadcast to all the nodes for updating. This calls for significant control overhead. The second type of adaptive routing is based on distance vector or distributed routing, in which each node maintains a routing table for each destination and on each wavelength, giving indications of the next hop to the destination. The scheme requires updating the routing tables by sending routing updates by each node to its neighbors periodically or whenever the status of the node's outgoing link changes.

Another class of adaptive routing is *least-congested-path* (LCP) *routing*. A link with fewer available wavelengths is said to be *congested*. Congestion in a path is indicated by the most congested link in the path. Corresponding to each source–destination pair, a sequence of path is precomputed and the arriving request is routed on the least congested path. A variant of LCP exists that relies on only the neighborhood information and examines the first k links on each path rather than all links on all candidate paths.

12.5 LAYERED GRAPH APPROACH FOR RWA

Wavelength-routing networks offer the advantage of wavelength reuse and scalability and are suitable for wide area networks. As explained earlier, these networks consist of wavelength-routing nodes interconnected by optical fibers, to each of which are attached several access stations. A wavelength-routing node is capable of switching a signal based dynamically on the input port and the wavelength on which the signal arrives. Such a node can be implemented by a set of wavelength multiplexers and demultiplexers and a set of photonic switches (one per wavelength).

One of the key issues in the design of WDM networks is to solve the problem of routing and wavelength assignment (RWA). In this section we consider the network to operate in a circuit-switched mode in which the call requests arrive at a node according to a random point process and an optical circuit is established between a pair of stations for the random duration of the call. If wavelength conversion capability is absent in the routing node, the path establishment requires the wavelength continuity constraint to be fulfilled. This may lead to an increase in the call blocking probability. On the other hand, the added facility of wavelength conversion at a routing node may improve network performance, and the degree of improvement depends on many network-related parameters, such as network connectivity, number of wavelengths, number of transceivers/node used, and so on [10]. Nevertheless, this improvement is obtained at increased cost, due to wavelength converters.

12.5.1 Layered Graph Model

A network topology is defined as $N(R, A, L, W)$ for a given WDM network, where R is the set of wavelength router nodes, A the set of access nodes, L the set of undirected links, and W the set of available wavelengths per link. At a wavelength router $r \in R$, wavelengths arriving at its different input ports can be routed onward to any of its output ports, provided that all the wavelengths routed to the same output port are distinct. Each access node in A is attached to a wavelength router. The access nodes provide electro-optical conversions for supporting electronic packet- and circuit-switching operations. The links L in the network are assumed to be bidirectional. Each link consists of two unidirectional optical fibers, each carrying W wavelength-division-multiplexed channels.

The layered graph model [10,11] is a directed graph $G(V, E)$, consisting of V nodes and E edges obtained from a given network topology N as follows. Each node $i \in R$ in N is replicated $|W|$ times in G. These nodes are denoted as v_i^1, $v_i^2, \ldots, v_i^w \in V$. If a link $l \in L$ connects router i to router j, $i, j \in R$, in N, then node v_i^w is connected to node v_j^w by a directed edge, e_{ij}^w, $e_{ji}^w \in E$ for all $w \in W$. For an access node $a \in A$ attached to a router k, two nodes are created for each access node, one representing the traffic-generating part (i.e., source), the other, the traffic-absorbing part (i.e., destination). These two nodes are denoted as v_a^s and $v_a^d \in V$, respectively. The directed edges are formed from node v_a^s to nodes v_k^1, $v_k^2, \ldots, v_k^w \in V$, and from each of the nodes $v_k^1, v_k^2, \ldots, v_k^w \in V$ to v_a^d. Thus the number of edges in G is $E = 2L \times W$. Figure 12.4 shows an example of a layered graph that is obtained from the network topology shown in Fig. 12.3. It can be seen that if a set of paths in the layered graph is edge disjoint, all the paths in the set can be supported in the corresponding physical network topology. In the multigraph model, a set of edge disjoint paths do not guarantee that they can all be supported in the physical network topology, N.

12.5.2 Dynamic RWA Algorithm without Wavelength Conversion (Based on a Layered Graph)

The dynamic RWA problem can be solved simply by applying the shortest-path algorithm (e.g., Dijkstra's algorithm) for the lightpath requested in the layered

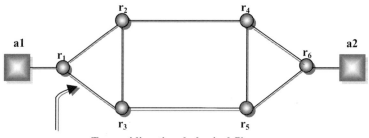

**Two unidirectional physical fibers
in opposite directions (two wavelengths per fibre)**

Figure 12.3   Optical network N.

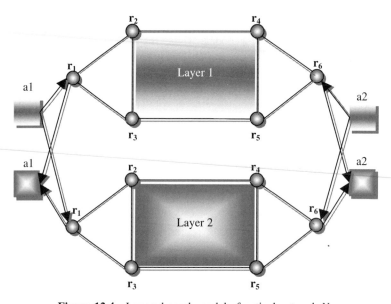

Figure 12.4   Layered graph model of optical network N.

graph. If the cost C_p of the shortest path is found to be finite, the request is accepted and the lightpath is set up along the shortest path; otherwise, the request is blocked. The weights on the links of G have to be updated whenever a lightpath is established or released. A formal description of the algorithm is given below.

1. Transform a given WDM optical network $N(R, A, L, W)$ into a layered graph $G(V, E)$. Let the cost of edge e in G be $c(e)$ for all $e \in E$.
2. Wait for a lightpath request. If it is a lightpath connection request, go to step 3. If it is a lightpath release request, go to step 4.
3. Find a shortest path p in G from the source to the destination node (e.g., by Dijkstra's algorithm). If the cost of the path $C_p = \infty$, reject the request;

otherwise, accept the request. Set up the connection along the shortest path. If the shortest path is via layer k, assign wavelength k to the lightpath. Update the cost of the edges on the path p to ∞. Go to step 2.

4. Update the cost of the edges e_i occupied by the lightpath to $c(e_i)$. Release the lightpath and then go to step 2.

12.5.3   Dynamic RWA with Wavelength Conversion (Based on a Layered Graph)

The RAW formulation with wavelength conversion can be effected by a shortest-path algorithm similar to the way a layered graph is created without a wavelength converter [10]. If the cost of the shortest path C_p is found to be finite, the request is accepted and the lightpath is set up along the shortest path; otherwise, the request is blocked. The weights on the links of G have to be updated whenever a lightpath is established or released. One needs in this case to find the shortest path between a source s (s connected to source router r_s and lightpath exiting at wavelength w) and the destination d (d connected to router r_d and lightpath exiting at wavelength w'), $w, w' \in W$, utilizing the augmented layered graph by putting additional edges between corresponding nodes in the layered graph reflecting wavelength conversion. Expectedly, the algorithm will require too much computation time for shortest-path computation when the network size is very large and/or the number of wavelengths is also large. A more computationally efficient algorithm is provided by the semilightpath (VWP) algorithm described in Section 12.6.

12.5.4   Simulation Results without Wavelength Conversion

To appreciate the application of RWA, we consider several network scenarios and provide results based on simulations carried out using the layered graph model, assuming Poisson traffic arrival [10].

1. *Irregular network Arpanet.* Figure 12.6 shows the blocking probability performance against mean call arrival rates for an Arpanet-like irregular network (Fig. 12.5) when the network uses six, eight, and 10 wavelengths. The number of access nodes is 24. For a mean arrival rate of 150, a minimum of eight wavelengths is required to achieve a blocking probability of 0.125.

2. *Ring network.* The call blocking probability performance of a 24-node ring network with six access nodes per router is shown in Fig. 12.7, where the ring performance is found to be poorer than that of Arpanet. The ring being a sparsely connected network, its path-length interference is considerably greater. To achieve the performance level of an irregular network, one either has to provide a larger number of wavelengths or increase network connectivity [10].

3. *Network with extra links.* The most critical parameter in a WRON is the physical topology onto which the logical demand has to be mapped, since it directly determines the lightpath allocation and hence the complexity of the WDM network and the wavelength requirement.

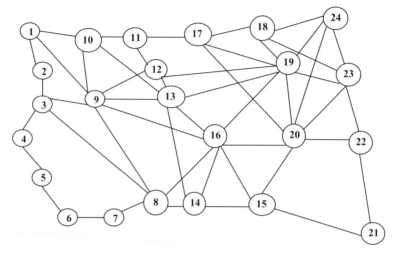

Figure 12.5   Arpanet topology.

In the present example we consider a network topology of N nodes connected by L bidirectional fibers. Further, from the network reliability point, any two subparts of the network must be connected by at least two links (i.e., the minimum degree of all nodes is two). The physical connectivity α is defined as the normalized number of bidirectional links with respect to a physically fully connected network of the same size [i.e., $\alpha = L/L\_FC = 2L/N(N-1)$]. In carrying out the simulation [10] based on a layered subgraph model, the arrival lightpath call requests are assumed to be Poisson with a mean rate r and exponential call holding time of mean 1. A call, once blocked, is removed from the list of pending lightpath requests. The traffic pattern is assumed to be distributed uniformly over all access node pairs.

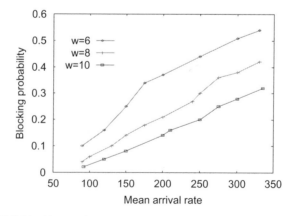

Figure 12.6   Call blocking performance of Arpanet using a layered graph without wavelength conversion.

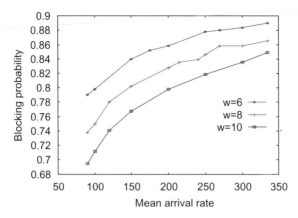

Figure 12.7   Call blocking performance of ring network with 24 router nodes.

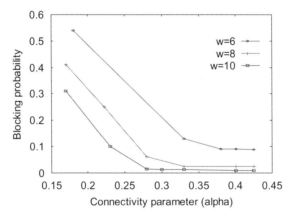

Figure 12.8   Call blocking performance of Arpanet-like network for various α values and wavelengths for a mean arrival rate $= 330$.

Figure 12.8 shows the blocking probability performance of Arpanet-type irregular networks with extra links for the desired value of α under different simulations as obtained by the layered graph approach. It is observed that the benefits of increase in wavelength is more at a higher value of α. The results would also permit allocation of the resources required to achieve a given blocking probability. If wavelength resource is scarce, it is necessary to increase the connectivity to achieve the required blocking probability.

12.5.5   Simulation Results with Wavelength Conversion

The layered graph model is extended to include the provision of wavelength conversion at all/selected routing nodes [10]. With the RWA algorithm with

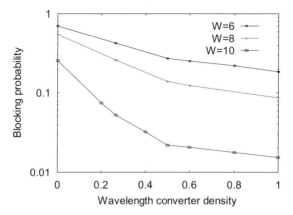

Figure 12.9   Blocking probability versus the wavelength interchanger density for a 20-node ring network at a load of 0.3 Erlang.

wavelength conversion, simulation studies are carried out on a 20-node router network with one access node per router. To accommodate wavelength conversion, we define the converter density q as the ratio of the number of routers having wavelength conversion capability to the total numbers of routers in the network.

Figure 12.9 shows how the blocking probability changes with wavelength converter density q for several values of number of wavelengths for a 20-node ring network for a network load of 0.3 Erlang. As can be seen, the wavelength conversion is more beneficial with more wavelengths. When the number of wavelengths is high, the blocking probability drops almost exponentially with increasing converter density up to a certain point and then begins to level off. This happens as blocking occurs due to lack of resources when the number of wavelengths is small, and therefore the capability of wavelength conversion helps marginally. When the number of wavelengths is increased, blocking occurs not due to a lack of resources but due to inefficiency in using these resources except with wavelength conversion. Finally, beyond a certain density of wavelength converters, a point of diminishing return is reached. This points out that having a wavelength converter at every router node is not necessary to improve performance.

12.6   VWP APPROACH FOR DESIGN OF WDM NETWORKS

The concept of VWP (i.e., a transmission path obtained by establishing and chaining together several lightpaths by introducing the capability of wavelength conversions at intermediate nodes) ensures improved network performance and network scalability. In the next section we describe the shortest-path algorithm for the wavelength graph (SPAWG) corresponding to the WDM network with λ-conversion [7]. The algorithm is applied to a WDM network of irregular physical topology such as Arpanet for deciding optimal placement of λ-converters as well as the optimal wavelength-changing facility at each wavelength-converting node satisfying the minimum network cost. The algorithm is quite general and can be used for efficient design of a wide-area WDM network of arbitrary topology. As a special case, the VWP algorithm is used to decide the best trade-off possible between space diversity and wavelength diversity in a multifiber, multiwavelength ring network.

12.6.1   Virtual Wavelength Path Algorithm

The VWP algorithm achieves fast routing of lightpaths and VWP between a given source and destination by minimizing the overall cost function that accounts for link cost (wavelength on a link) and wavelength conversion cost (doing wavelength conversion at a node when necessary). Essentially, the following steps are involved:

1. Construct the wavelength graph (WG) of the network. For example, Fig. 12.10 is the WG of a five-node unidirectional ring network.
2. Find a shortest path between a source and a destination using the shortest-path algorithm for WG (SPAWG).
3. Map the path found back to the network to obtain the minimum-cost VWP.

12.6.2   SPAWG Implementation

In obtaining a WG, arrange the total number of vertices $N = kn$, where k is the number of wavelengths and n is the number of nodes in the original network, in a matrix array with k rows and n columns. Draw a directed edge from node j to node h on the ith layer if there exists a link and assign weight $w(e, \lambda_i)$ to this edge. In column j, $j = 1, 2, 3, \ldots, n$, draw a directed edge from row i to row l, if at node j, wavelength conversion is available from λ_i to λ_l, and assign weight $c_j(\lambda_i, \lambda_l)$ to this edge. Such operations yield a direct 1 : 1 mapping of semi-lightpaths into the WG. To determine the shortest path (minimum-cost path) between two columns s and t in

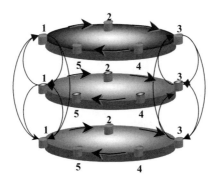

Figure 12.10 Wavelength graph for a five-node unidirectional ring network with full wavelength converters placed at nodes 1 and 3.

the WG, two nodes s' and t' are added in WG such that s' is connected to each vertex in column s via zero-weight edges and t' is connected to all vertices in column t with zero-weight edges. Clearly, the shortest path between s' and t' is the solution for the semi-lightpath in WG.

Next, the SPAWG algorithm [7] for finding the shortest path between two given vertices of a WG is presented. For easy notation the vertices of the WG are simply labeled by the numbers $1, \ldots, N$ and the weight of edge (i, j) is denoted by a_{ij}. If the edge does not exist, it has infinite weight. If two vertices, say i and j, are either in the same column or the same row, this relation is denoted by $i \sim j$. In the algorithm, three sets of numerical variables are used: a variable u_i, associated with each vertex; a variable R_i, associated with each row; and a variable C_j, associated with each column. The R_i and C_j variables will represent the row and column minimum values, respectively. For simplicity, the start point of the searched path is labeled 1, and the set-valued variables P and T represent permanently and tentatively labeled vertices, respectively. (Permanently labeled vertices are those for which computation has already terminated; that is, the variable u_i contains the weight of a shortest path from 1 to i. For tentatively labeled vertices, computation is still going on.)

12.6.3   SPAWG Algorithm

Step 1 (Initialization)
1. $u_1 = 0$.
2. If $i \sim 1$, then $u_i := a_{ij}$, else $u_i := \infty$ ($\forall\ i$).
3. $R_i := \min u_j$: j is in the ith row, $j \neq 1 (\forall\ i)$.
4. $C_j := \min u_i$: i is in the jth column, $i \neq 1 (\forall\ j)$.
5. $P := 1$; $T := 2, \ldots, N$.

Step 2 (Designation of a new permanent label)
1. Find the minimum of R_i, $C_j (\forall\ i, j)$.
2. Find an $h \in T$ with the minimum u_h in the row or column that gave the minimum above (ties are broken arbitrarily).
3. $T := T - h$; $P := P \cup h$.
4. If $T = \varnothing$, then stop.

Step 3 (Updating row and column minimum)
1. If h in step 2 is in row i and column j.
2. Then $R_i := \min u_j$: j is in the ith row, $j \in T$.
3. $C_j := \min u_i$: i is in the jth column, $i \in T$.
4. The minimum over an empty set is taken to be ∞.

Step 4 (Revision of tentative tables)
1. If h in step 2 is in row i and column j, then for all $l \in T$ in row i and in column j, set $u_1 := \min u_1, u_h + a_{h1}$. Go to step 2.

Like Dijkstra's algorithm, algorithm SPAWG computes the minimum weight distance from the root 1 to all other vertices. A dynamic traffic scenario is considered in which each node receives call requests that are Poisson distributed with mean arrival rate g and exponentially distributed call holding times with mean $1/\mu$. The destination nodes are chosen uniformly at random.

12.6.4   Simulation Results with Wavelength Conversion

The virtual wavelength path algorithm (SPAWG) has been found not only to accomplish simultaneous routing with wavelength assignment but can also be exploited in the design of WDM networks, as some of the following simulation results will indicate [10]. We consider some representative networks, such as single/multple-fiber ring, and the irregular topology Arpanet/NSFNET/MESH networks. We provide results for Poisson and non-Poisson traffic.

12.6.4.1   Arpanet   In deciding the optimal placement of λ-converters, the VWP algorithm first finds out the number of λ-conversions at each node while doing routing and wavelength assignment, assuming that all the nodes are provided with λ-converters. Only those nodes exhibiting the significant number of λ-conversions are decided for converter placement. For example, with a 24-node Arpanet with λ-converters placed at every node (i.e., converter density, $q = 1$), at a network load of 80 Erlangs, the nodes selected, numbered 8, 16, 9, 3, 19, 20, 7, 13, 4, 10, 12, and 17 (the ordering of nodes are in accordance with decreasing value of the number of λ-conversions taking place at the nodes) are the desired nodes where the λ-converters should be placed (corresponding to $q = 0.5$) (see Table 12.1). The blocking probability performance of a 24-node Arpanet with the heuristic placement of converters (number of wavelengths, $W = 8$) above is shown in Fig. 12.11. It has also been found that with a larger value of W, the corresponding heuristic

TABLE 12.1   Number of Wavelength Conversions at Selected Nodes for a 24-Node Arpanet with $W = 8$, $q = 0.5$, $c = 1$, and Load $= 80$ Erlang

Node Number	Number of Wavelength Conversions	Rank	Node Number	Number of Wavelength Conversions	Rank
1	0	13	13	358	8
2	0	14	14	0	18
3	857	4	15	0	19
4	311	9	16	1194	2
5	0	15	17	174	12
6	0	16	18	0	20
7	414	7	19	727	5
8	1202	1	20	700	6
9	1063	3	21	0	21
10	259	10	22	0	22
11	0	17	23	0	23
12	213	11	24	0	24

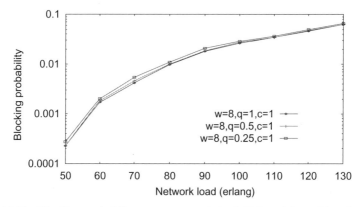

Figure 12.11 Blocking probability versus network load (Erlangs) for a 24-node Arpanet with heuristic placement of converters.

placement remains very efficient not only for a larger value of the network load but also over a greater variation of the load.

The VWP also provides the frequency of use of each wavelength at the wavelength-converting node (Table 12.2). Only those wavelengths of significant use may be selected for the partial-wavelength-changing facility. The results for the Arpanet show that with 50% converter density with λ-converters optimally placed by the VWP algorithm and within a 50% wavelength-changing facility ($c = 50\%$) can achieve performance almost equivalent to that with λ-converters placed at every node ($q = 1$) and with full wavelength-changing facility ($c = 100\%$) over a useful operating range of network loads.

12.6.4.2 Multifiber Ring For the case of a multifiber ring network (Fig. 12.12), the possible design trade-off between space diversity and wavelength diversity is best illustrated in the blocking probability performance of the multifiber multi-wavelength ring network (number of nodes = 20) in Fig. 12.13 under various

TABLE 12.2 Number of Wavelength Conversions of a Particular Wavelength at a Node

Wavelength Number	Node Number											
	3	4	7	8	9	10	12	13	16	17	19	20
1	140	38	55	182	169	57	41	52	163	26	123	94
2	137	49	63	185	186	73	57	96	192	38	149	124
3	152	49	59	186	189	51	58	90	213	53	147	120
4	127	40	51	173	177	48	26	63	195	31	143	126
5	112	42	58	152	134	16	20	33	161	18	94	94
6	7	33	49	117	95	10	5	17	130	4	53	57
7	57	35	39	117	64	3	4	6	86	2	15	48
8	53	25	40	90	49	1	2	1	54	2	3	28

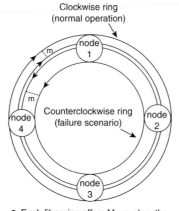

● Each fiber ring offers M wavelengths

Figure 12.12   Multifiber ring network.

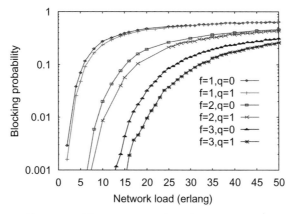

Figure 12.13   Blocking probability versus load for a 20-node multifiber ring network with $W = 6$ with/without wavelength converters; f is the number of fibers.

scenarios. A significant performance improvement results in the case of a multifiber ring network compared to a single-fiber ring network [10].

12.6.5   WDM Network Design with Non-Poisson Traffic

To study the performance of the VWP algorithm in WRON for non-Poisson traffic, we have adopted an interrupted Poisson process (IPP) model to characterize the non-Poisson traffic [12]. Figure 12.14 shows IPP generation with the required mean for the traffic load and random interruption of this Poisson source with another source that is also Poisson distributed. The mean of the interrupted Poisson source is selected so as to get the required peakedness factor (Z), defined as the ratio

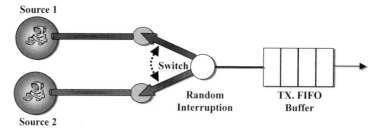

Source 1

Switch

Random
Interruption

TX. FIFO
Buffer

Source 2

Figure 12.14 IPP model for generation of non-Poisson traffic.

between the variance and the mean of the traffic. By carrying out very large number of simulations with the combined sources, the value of Z is found from the statistics. The traffic is identified as peaked, smooth, or Poisson, depending on a value of Z greater than, less than, or equal to 1, respectively. Here we have fixed the mean arrival rate (λ_1 and λ_2) and varied the probability of interrupt to get both smooth and peaky traffic.

12.6.5.1 MESH To bring out the superior performance of the VWP routing algorithm with other available routing schemes, we consider a 30-node mesh network (Fig. 12.15). In particular, we have considered the performance evaluation for both Poisson and non-Poisson traffic, and the results are depicted in Figs. 12.16 and 12.17. The results clearly establish the superior performance of the VWP (semi-lightpath) routing over other routing schemes discussed in [8].

12.6.5.2 NSFNET Finally, the blocking performance is also evaluated on another irregular topology, the NSFNET (see Fig. 12.18). The results are shown in Fig. 12.19, which shows that in the irregular topology network also, the performance obtained with the VWP algorithm is significantly better than the results obtained with other fixed/dynamic routing algorithms [8,12].

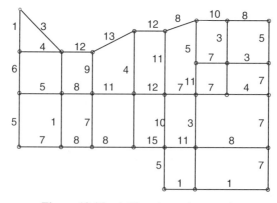

Figure 12.15 A 30-node mesh network.

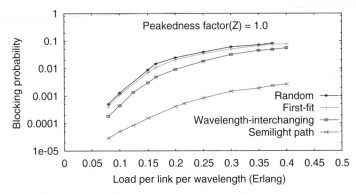

Figure 12.16 Blocking probability versus network load for a 30-node mesh network with different wavelength assignments ($W = 8$) under Poisson traffic.

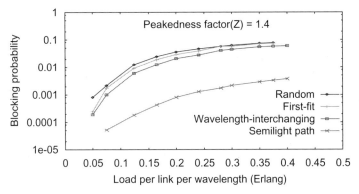

Figure 12.17 Blocking probability versus network load for a 30-node mesh network with different wavelength assignments ($W = 8$) under non-Poisson traffic.

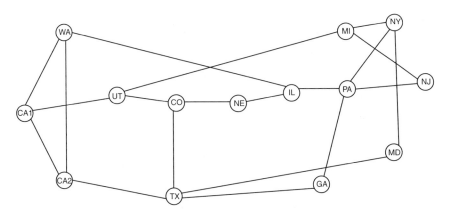

Figure 12.18 NFSNET topology.

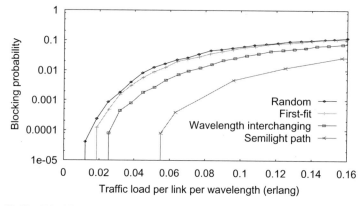

Figure 12.19 Blocking probability versus load for NSF network for several routing strategies.

In Fig. 12.20 we consider wavelength conversion that takes place at various nodes with random and optimal placement of converters [13,14] for a first-fit and VWP algorithms. The blocking probability with optimal placement of the converter is less compared with other placement of converters using first-fit and VWP algorithms with random placement of converters. The simulation results clearly indicate the benefit of wavelength conversion in reducing the blocking probability, the selection strategy of the required conversion density for a practical network, and the important feature (not shown here) that with λ-conversion, the blocking performance is virtually independent of the source–destination distance. Among the various RAW schemes considered for several WDM networks, the VWP or semi-lightpath algorithm for RAW is shown to provide superior performance in all the cases considered.

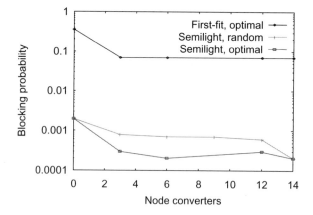

Figure 12.20 Blocking probability versus node conversion density for a 14-node NSFNET with optimal and random placement of converters.

12.7   MPLS/MPλS/GMPLS

The rapid proliferation of data-centric Internet applications and the dominant network layer protocol require a high-capacity optical transport network as well as optical internetworking systems that are versatile, reconfigurable, and capable of providing rapid provisioning and restoration. The meshed architecture of optical networks with optical cross-connects (OXCs) can achieve the required versatility and reconfigurability. The two-layer model mentioned earlier for the control plane for the purpose of network management achieves the extension of an IP control plane to connection-oriented technology, known as multiprotocol label switching (MPLS) [2], which allows a variable-length stack of four-octet labels in front of an IP packet.

An MPLS-enabled router called a *label-switched router* (LSR) forwards incoming packets using only the value of the fixed-length label on the top of the stack. This label, along with the port at which the packet was received, is used to determine the output port and the label for forwarding the packet. An LSR may also add or delete one or more labels to an IP packet as it progresses through it. Connections established using MPLS are referred to as label-switched paths (LSPs). The reader is referred to Chapters 2 and 14 for more details on MPLS.

The next level advance in the optical network control plane is to combine the MPLS with the network-level role of WDM technology and the switching functionalities of OXC for provisioning wavelength capabilities. This leads to multiprotocol lambda switching (MPλS). The key to realizing MPλS lies in establishing a logical topology (wavelength paths over the WDM physical network) in order to carry the IP packets utilizing the wavelength path [1]. The physical network consists of core routers (OXCs or optical add–drop multiplexers (OADMs)) interconnected by optical fiber links. The optical node may or may not have the wavelength conversion facility. The logical topology defined by the wavelength (no λ-conversion) forms the underlying network to the IP packet layer. In such a network the ingress/egress routers at the network boundary are the label switch routers (LSRs) (according to MPLS terminology).

A more general version of MPLS, GMPLS, is emerging which defines the extension of MPLS to cover TDM switching, λ-switching, and port switching. It also seeks to provide one single control protocol for integrating IP, SDH, and optical (DWDM) as well as to provide flexible provisioning at all layers (i.e., IP layer provisioning, SONET-level bandwidth provisioning, and DWDM λ-provisioning).

12.8   IP-OVER-WDM INTEGRATION

To reduce cost and simplify management, control, and signaling overhead, the present effort seeks a solution for integrating the IP layer directly onto the WDM layer. Such an integrated IP-over-WDM network model is shown in Fig. 12.21. Here, the clients (e.g., IP/MLPS routers) are connected to the optical core over a well-defined signaling and routing interface called the *user network interface*

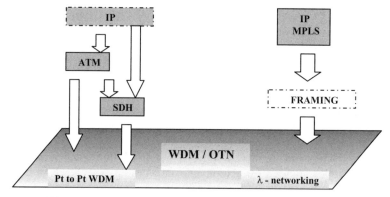

Figure 12.21   Layer view of multilayer networks. (From [1].)

(UNI). The optical core network consists of OXCs interconnected by optical links (fibers) in a general mesh topology as well as subnetworks (vendor specific) that can interact through the well-defined signaling and routing interface referred to as the network–network interface (NNI).

12.8.1   Interconnection Models

The two-layer model which is currently advocated for tighter integration between IP and optical layers offers a series of advantages over an existing multilayer architecture model (see Fig. 12.21).

As discussed in other chapters, MPLS and G-MPLS are the trends for the integrating structures between IP and optical layers. Following are the three models proposed [15]:

1. *Overlay model.* Under the overlay model, the IP domain acts as a client to the optical domain. IP/MPLS routing and signaling protocols are independent of the routing and signaling protocols of the optical layer. The topology distribution, path computation, and signaling protocols would have to be defined for the optical domain. The client networks are provided with no knowledge of the optical network topology or resources. The overlay model may be provisioned statically using a network management system

2. *Peer model.* Under the peer model, the IP/MPLS act as peers of the optical transport network (IP/MPLS routers and OXCs act as peers). A single routing protocol instance runs over both IP/MPLS and optical layers. A common IGP and OSPF or IS-IS [2] may be used to exchange topology information. A consequence of the peer model is that all the optical switches and routers have a common addressing scheme.

3. *Augmented model.* This model combines the best of the peer and overlay interconnection models. Under the augmented model, there are actually separate routing instances in the IP and optical domains, but the information from one

routing instance is passed through the other routing instance. For example, external IP addresses could be carried within the optical routing protocols to allow reachability information to be passed to IP clients.

12.9   INTEGRATED DYNAMIC IP AND WAVELENGTH ROUTING

One of the important needs for future IP-over-WDM networks is the dynamic setup of bandwidth-guaranteed paths between ingress–egress routers. In MPLS networks such paths are the label-switched paths (LSPs). Since a priori information about the potential LSP requests are not available, an online algorithm has to be used instead of an off-line algorithm. The conventional approach to routing of LSPs relies on independent routing at each layer (i.e., wavelength routing at the optical layer and the IP routing at IP/MPLS layer are done separately). To better utilize the network resources and to achieve efficiency, the use of integrated routing [16] is desirable. By integrated routing it is understood that the routing decision is made on the basis of knowledge of the combined topology and resource usage information at the IP and optical layers. The integrated routing should fulfil the following design objectives:

- Need for an online algorithm for the IP and IP over WDM
- Efficient network use
- Good rerouting performance upon link failure
- Computational efficiency
- Distributed implementation feasibility
- Use of knowledge of ingress–egress points of LSPs

The primary motivation for integrated routing is the use of one single control plane for MPLS and optical channel routing and the extension of the traffic engineering framework of MPLS. As in the IP layer, OSPF-like protocols can be run on both routers and OXCs for distribution of both link-state and resource usage information to all network elements. In optical layers, similar extension can be used to distribute wavelength use information for each link.

12.9.1   Network Model

Let us consider a network of n nodes interconnected by optical links. Each node is either a router, an OXC with λ-converters, or an OXC without a λ-converter. Let $R \underline{\underline{\Delta}}$ set of router nodes, $S \underline{\underline{\Delta}}$ set of OXCs with λ-conversion facility, and $T \underline{\underline{\Delta}}$ set of OXCs without λ-conversion facility. A router is capable of multiplexing or demultiplexing bandwidth with any granularity and doing wavelength conversion. At the nodes belonging to the set S, it is possible to do wavelength conversion and switching a wavelength from an input link to a wavelength at any output link. The nodes belonging to T can do only wavelength switching without wavelength conversion. A request for an LSP i is defined by a triple (o_i, t_i, b_i). The first field specifies the ingress router, the second field, t_i, signifies the egress router, and the third field, b_i, represents the amount of bandwidth required for LSP i. Having received a request for LSP setup, either at a router server or at an ingress router, the route server generates an explicit route, using the knowledge of the current topology and available capacities at both the IP and optical layers by running a link-state protocol with approximate extensions on all the network elements. The explicit route is then communicated back to the ingress router that uses a signaling protocol RSVP or LDP to set up a path to the egress router as well as reserving the bandwidth on each link on the path. In the following we describe an optimal route setup algorithm to ensure maximum accommodation of LSP requests without requiring any a priori knowledge regarding future arrivals using integrated IP and WR [16].

To explain integrated IP and optical networks we consider the example network shown in Fig. 12.22, where the shaded nodes 1 and 4 are the routers, node 2 is an OXC with wavelength conversion, and node 3 is an OXC without the capability of wavelength conversion. It is also assumed that each link carries two wavelengths, λ_1 and λ_2. A request from node 1 to node 2 can be routed along the path 1–3–4 using λ_1. Once this request is routed, node 3 cannot use λ_1 to route the traffic along the path 2–3–4, since node 3 is an OXC that can only switch wavelengths and cannot multiplex traffic from node 2 onto the unused λ_1. This direct connection from a node $r_1 \in R$ on a wavelength can be viewed as a logical link in the IP layer.

To compute routes taking into account the capabilities of different types of network elements, we provide the network representation of Fig. 12.23. Here, each node is expanded into a number of subnodes, one per wavelength. For OXC node without λ-conversion, each subnode is connected to a wavelength on each incoming and outgoing link. For OXC with λ-conversion and for routers, all the subnodes at each node are connected to a supernode as shown in Fig. 12.23.

For the MPLS network, let us assume that a demand $r \leq 1$ units is to be routed from an ingress node $a \in R$ to an egress node $b \in R$ following a path that generally consists of a sequence of routers and OXCs (shown as an example in Fig. 12.24). If r units of bandwidth is consumed by this demand, a capacity of $1 - r$ units of bandwidth is available for future demands. Since OXCs do not perform any

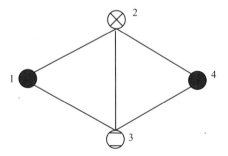

Figure 12.22   Example network.

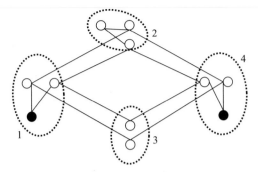

Figure 12.23   Network representation for integrated routing computation.

Figure 12. 24   LSP path through routers and OXCs.

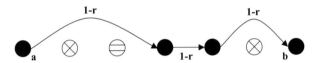

Figure 12.25   Cut-through arcs modeling path with routers and OXCs.

subgranularity switching or multiplexing, the availability of $(1 - r)$ units of bandwidth can be modeled by adding a cut-through arc between the routers and eliminating the links in the original graph (see Fig. 12.25). This cut-through is the logical link in the IP layer. For two adjacent routers, the residual capacity of the link connecting them is also reduced by r units. This reduced graph is then used for routing of future demands. Note that if the logical link reaches one unit of capacity, the logical link is eliminated, and the original physical (optical) link is reestablished with unit capacity.

It has to be emphasized that the integrated routing approach must seek a path corresponding to a given ingress–egress pair that will interfere least with potential future setup requests among other ingress–egress pairs. This is equivalent to maximizing the residual capacity or open capacity between the different ingress–egress pairs. This can be done by computing the maximum flow on the network with both logical IP links and optical links. The path that maximizes the open capacity is the one that maximizes the sum of the maxflows between all other ingress–egress routers. Since for a network the maximum flow is the minimum cut, those arcs of a particular min-cut are the critical links for an ingress–egress pair. If the capacity of a critical link decreases, the max-flow between the ingress–egress pair also decreases. One of the obvious strategies will be not to route the LSPs on

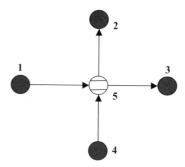

Figure 12. 26 Illustrative network.

critical links to the extent possible. To accommodate this policy, a new weighted graph is generated where the critical links have weights proportional to their criticality [16]. It is possible to assign weights to the critical links for different source–destination pairs (Fig. 12.27). The formation of the graph corresponding to Fig. 12.26 is illustrated in Fig. 12.27, where the weight (number of ingress–egress pairs for which the arc is critical) of the arc is indicated.

The algorithm for this integrated routing is termed the *maximum open capacity routing algorithm* (MOCA). Following are the formal descriptive steps of the MOCA [16]:

Input: A graph $G(N, L)$ with N nodes and L links, the cut-through arcs, and the residual capacities of all links; an ingress node a and an egress node b between which a flow of D units have to be routed.

Output: A route between a and b having a capacity of D units of bandwidth.

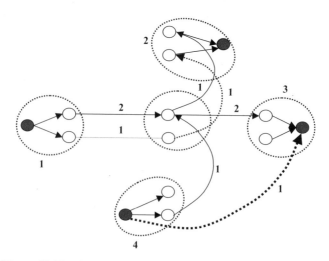

Figure 12.27   Graph (with residual capacities) after routing $4 \to 3$.

Algorithm

1. Compute the critical links for $(s, d) \in P \backslash (a, b)$.
2. Compute the weight of all the links in the network (including the cut-through arcs).
3. Eliminate all links that have residual capacities less than D and form a reduced graph.
4. Using Dijkstra's algorithm, compute the minimum weight path in the reduced graph.
5. Form the cut-through arcs with the appropriate residual capacities as well as updating the residual capacity of the router-to-router arcs along the routed path.

Detailed simulation results [16] indicate that the number of rejects for a given demand is far less with the MOCA than with integrated minimum hop (IMH) routing (where weights of all links are unity). It can further be concluded [16] that the MOCA is the form of integrated dynamic routing scheme that is robust to changing traffic patterns in the IP layer than a scheme that uses dynamic routing at the IP layer only and uses a static wavelength topology dictated by a priori assumed traffic distribution.

12.10   WAVEBAND ROUTING IN OPTICAL NETWORKS

To address the performance and scalability issues in IP-over-WDM networks, the concept of hierarchical path layers plays an important role in reducing the node cost in optical networks. For example, the number of optical–electrical–optical (O/E/O) ports required for processing individual wavelengths can be reduced significantly if wavelength aggregation into wavebands [17] is done and each waveband is switched optically as a single unit, requiring only two (input and output) ports of an optical switch in a node. However, cost-efficient implementation of optical path hierarchy demands efficient routing and scheduling algorithms. In this section we provide a classification of hierarchical routing models and present heuristic waveband routing and aggregation algorithms. Such waveband routing would be extremely important in future optical transport networking, where each fiber will be able to carry hundreds of wavelengths, demanding OXC structures to have fewer ports, but the individual port throughput should be maintained at a high value, thereby minimizing the cost of an individual OXC.

12.10.1   Hierarchical/Integrated Optical Node/Network Models

In a hierarchical optical network [18], some wavelength paths can be aggregated into a waveband path at one node and disaggregated into wavelength paths at another node. Such aggregation and disaggregation can be effected at a node that is equipped with both a waveband (transparent optical, O/O/O) and a wavelength

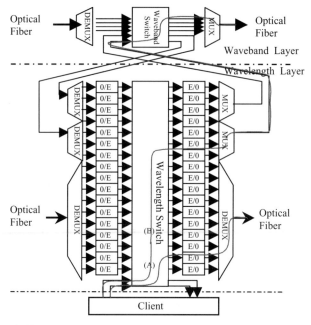

Figure 12.28 Integrated waveband/wavelength switching node. (From [18].)

(opaque O/E/O) switch. The general structure of such a hierarchical node is shown in Fig. 12.28.

In this node, a wavelength path input to the wavelength switch can be either routed directly to a neighboring node (shown in flow A) or aggregated into a wavelength path, which is then routed to a neighbor node via a wavelength switch (shown in flow B). The nodes in the network can be controlled centrally (central control plane) or in a distributed manner (distributed control plane). Both should take advantage of neighbor discovery, topology discovery, and route computation.

Figure 12.29 shows an example of homogeneous and heterogeneous network models. In a homogeneous network model, all the nodes are hierarchical, so a waveband path can be set up between any two nodes. In a heterogeneous network model, only a few nodes are hierarchical; others are nonhierarchical. Since the latter nodes do not have the waveband switch, a waveband path can only be set up between two hierarchical nodes.

12.10.2 Waveband Routing Models

For the purpose of routing hierarchical paths such as waveband and wavelength paths in optical network, routing models can be classified as follows [19]:

1. *Fixed-node aggregation or arbitrary node aggregation*. In a fixed-node aggregation model, a network has a limited number of integrated nodes; the

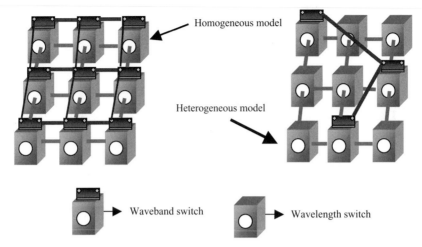

Figure 12.29   Homogeneous/heterogeneous network models.

remaining nodes do not have the waveband switches. The network is viewed to be composed of wavelength-layer subnetworks and a waveband-layer core network. In the arbitrary-node aggregation model, all the nodes in the network are integrated nodes. Hence the topology of a waveband-layer network is the same as that of a wavelength-layer network.

2. *Separated routing or integrated routing.* Under a separate routing network model, the routing of wavelength paths and waveband paths are performed independently. Each layer has its own routing protocol. In integrated routing model, a source node of a demand computes routes of waveband path. A common routing protocol runs on both layers. The status of the wavelength and waveband is advertised by a common routing protocol, so that the routing protocol is based on the total cost of wavelength and waveband resources. A further classification can be made on the following lines: (a) fixed-node aggregation/separate routing; (b) fixed-node aggregation/integrated routing; (c) arbitrary node aggregation, separate routing; and (d) arbitrary node aggregation/integrated routing

As an example, Fig. 12.30 depicts routing models (c) and (d). All nodes here are integrated nodes. In model (c), each node along the route from source to destination decides locally whether or not waveband paths should be created. In the figure, N0 creates a waveband path between N0 and N3, and node N6 creates a waveband path between N6 and N9. In routing model d, it is the source node that decides whether or not waveband paths should be created. This model achieves the highest effective use of resources but suffers from high complexity in the routing algorithm. The key feature of the hierarchical routing above is the potential cost benefit due to wavelength aggregation into wavebands reducing the number of O/E/O ports required for processing of individual wavelengths.

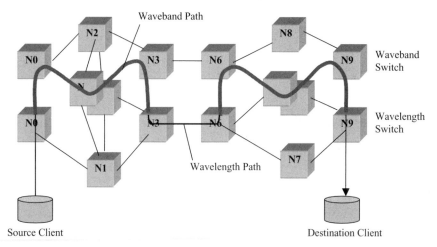

Figure 12.30 Arbitrary-node aggregation, separate routing, and arbitrary-node aggregation, integrated routing.

12.10.3 Hierarchical Online Routing Algorithm

For a homogeneous network it is assumed that some of the wavelengths can be aggregated into wavebands consisting of G wavelengths each. For the purpose of routing, let $L_i = [l_1^i, l_2^i, \ldots, l_n^i]$ represent a wavelength path, where l_k^i is the link ID used in the kth link of the wavelength path L_i. For each string L_i, a binary string $B_i = [b_1^i, b_2^i, \ldots, b_n^i]$ is defined. If $b_k^i = 0$, the link l_k^i of L_i can be aggregated into a waveband path. If $b_k^i = 1$, l_k^i of L_i is already a part of a waveband path.

We reproduce here a formal description of the algorithm from [18]. We set $S = \{L_k\}$, where L_k is the current wavelength path. We also set $b_m^k = 0$ for all m. Let N denote the length of the string L_k and let A_n^k denote a substring of L_k of length n, where $n \leq N$. A substring A_n^k of L_k is formed by selecting n adjacent elements $\{l_j^k, \ldots, l_{j+n}^k\}$ from L_k for which the corresponding $b_m^k = 0$ for all m from j to $j + n$.

The following steps are performed next.

Step 0:
  Initialize $n = N$.
Step 1:
  Select a (not previously selected) substring A_n^k of L_k.
  Assign $g = 1$.
Step 2:
  For all p from 1 to $k + 1$:
  Begin
    Find all substrings A_n^p belonging to L_p where $p \leq k$.
    If $(A_n^p = A_n^k)$, assign $g = g + 1$.

If ($g = G$): {G identical substrings found}

Begin {form the waveband}

For any l_j^i in matched substring or A_n^k

Begin

Assign $b_j^i = 1$.

End

End

End

Step 3:

If (all substrings A_n^k are selected)

Set $n = n - 1$.

If ($n < 2$)

Stop

Else

Go to step 1.

Figure 12.31 illustrates the algorithm for $G = 4$. Wavelength path LP4 is set up after paths LP1, LP2, and LP3. On LP4, the longest segment, 4–13 (LP4 itself), is chosen first. However, only one wavelength path (LP4) runs through the entire segment. By reducing the length of segments, a shorter segment 4–16 is selected:

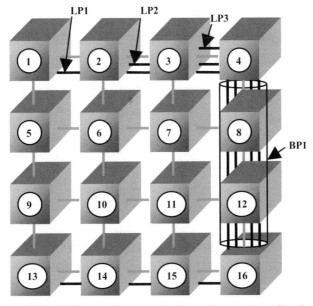

Figure 12.31   Online routing example for setting up a wavelength path.

Four wavelengths run through it. As a result, a new waveband path, BP1, is set up on segment 4–6, and LP1, LP2, LP3, and LP4 are aggregated into BP1.

Simulation results [18] based on the algorithm above indicate that the aggregation benefit depends largely on network topology, network size, and waveband granularity G. It is also found from simulation that the ratio of the number of wavelength ports (O/E/O) to the number of waveband (O/O/O) ports increases with waveband size G and it decreases with network size due to creation of longer waveband bands. In general, the online algorithm under separate routing discussed above has been found to deliver about 30% aggregation benefit for large networks with appropriate waveband granularity.

The concept of waveband routing can be extended to implement integrated routing as well as for heterogeneous networks. A further interesting work that can be pursued will relate to the development of an RWA algorithm that will optimize network resources in terms of wavelength requirements, wavelength conversions, and OXC switch port counts following the concept of integrated wavelength and waveband routing.

12.11 ADDITIONAL ISSUES IN OPTICAL ROUTING

12.11.1 Physical Layer Constraints

In most current work on optical routing, it is usually assumed that all the routes have adequate signal quality, and the impact of physical layer impairments is not considered at all. In reality, the optical networks suffer from many transmission-related degradation effects [10], such as group velocity dispersion, accumulated amplifier spontaneous emission (ASE), noise of erbium-doped fiber amplifiers (EDFAs), and various nonlinear effects, such as self-phase modulation (SPM), four-wave mixing (FWM) and cross-phase modulation (XPM), polarization mode dispersion (PMD), and imperfections due to nonideality of passive/active optical devices and components. The choice of both the physical route and wavelength assignment will depend on the nature of the physical layer impairment and on the traffic pattern to ensure lightpath provisioning with the desired quality of reception. The physical-layer information also increases the amount of information to be distributed in the LSA updates and may therefore increase distribution/settling times.

12.11.2 Diversity Routing Constraints

Routing in an optical network also has an important bearing in ensuring network survivability. *Survivability* means that the network has the ability to maintain an acceptable level of performance during network and/or equipment failures. For example, a failure in a fiber segment can lead to failure of all the lightpaths which traverse that failed fiber. Usually, protection and restoration strategies are used to guarantee the network survivability [6,17].

Protection depends on redundant network resources to rehabilitate the failed connections. Upon fault discovery, automated switching needs to be coordinated to use the protection entities (i.e., lightpath or span level), which are predetermined at setup time. Several practical protection schemes, such as $1 + 1$, $1 : 1$, and $1 : N$, are normally employed in practice (see Fig. 12.32). The interested reader is referred to Chapter 15 for more details on protection and survivability.

In $1 + 1$, the same information is sent through two paths and the better one is selected at the receiver. The receiver makes a blind switch when the working path's signal is poor. In $1 : 1$ protection, signal is sent only on the working path; a protection path is also set but can be used for priority signals that are preempted if the working path fails. A signaling channel is required to inform the transmitter to switch path if the receiver detects a failure in the working path. The $1 : N$ protection

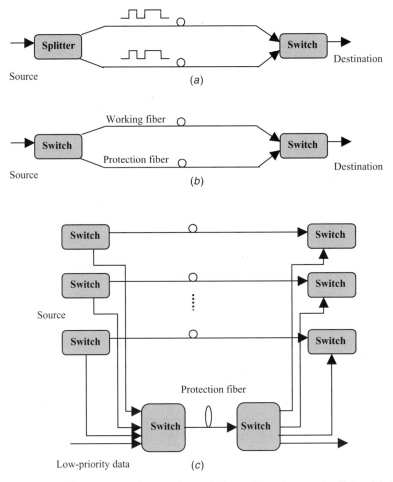

Figure 12.32 Different types of protection techniques for point-to-point links: (*a*) $1 + 1$ protection; (*b*) $1 : 1$ protection; (*c*) $1 : N$ protection.

is a generalization of a 1 : 1 protection scheme in which one protection fiber is shared among *N* working fibers. Like protection, restoration works to rehabilitate failed connections and supports end-to-end connections by computing routes dynamically after fault occurrence. To provide fast protection and restoration, diversity routing is used in practice. Two lightpaths are said to be *diverse* if they have no single point of failure. To support diversity routing, the concept of a shared risk link group (SRLG) has been introduced. SRLG information is used to denote all links, subject to a similar type of failure at a lower layer. In diverse lightpath computation, this information should be taken into account to ensure that two lightpaths are completely disjoint with respect to their SRLG values. This reduces the probability of simultaneous lightpath failures (e.g., between working and protection paths). SRLG is further aided by introducing the concept of a shared risk group (SRG) which covers information of nodes or domains. Both SRLG and SRG provide new opportunities and further possibilities for efficient routing in optical networks.

12.11.3   Wavelength Conversion Constraints

In all the optical networks where no wavelength conversion capability is available, wavelength continuity has to be maintained on an end-to-end basis (i.e., the same wavelength must be available on all links on the selected path). Such a λ-continuity constraint complicates the RWA computation as wavelength resource information must be disseminated in link-state advertisements (LSAs) for consideration in the routing process. This may introduce a scalability problem in the routing algorithm.

12.11.4   Interdomain Routing

For large optical networks, interdomain routing [5] becomes important, which must resolve issues such as constraints to flood topological information across the domains, exact type of information to be exchanged between domains, interdomain reachability between users, and so on. Currently, these issues are being studied actively, including the role of the border gateway protocol (BGP) for interdomain routing.

12.12   SUMMARY

The chapter provides basic introduction to an important network functionality, such as routing in IP datagram networks as well as in a wavelength-routed optical network. Two fundamental ways to route the packets in the IP network are to use distance vector and link-state routing. The corresponding algorithms (e.g., Bellman–Ford, Dijkstra) have been provided since these are the ones mostly used in practice. Some other routing schemes that have applications in IP networks have been also described. We then address the routing issue in the case of WDM networks. Routing in a WDM network has distinct differences from that of the Internet network. In a WDM network it is not the end-to-end path establishment alone but coupled with it the link-by-link wavelength assignment which has to be achieved. The process of simultaneous routing and wavelength assignment described as RWA is a key network functionality in a WDM network. Implementation of efficient RAW algorithm is therefore very crucial and depends on many network parameters, such as number of nodes, network connectivity, number of wavelengths, number of λ-converters, and so on.

To establish RAW algorithms, we first follow the layered-graph approach for the WDM network. The algorithmic description has been provided which accommodates both wavelength conversion and no wavelength conversion. To make RWA algorithms for WDM network more computationally efficient in the event of λ-conversion, the semi-lightpath/VWP algorithm has been introduced and its software implementation has been provided. Both the layered-graph and semi-lightpath algorithms have been shown to provide not only the least cost path, but they serve as an added tool for the design guidelines for the WDM network. The efficiency of the semi-lightpath (VWP) routing schemes has been exemplified in the results for several network topologies under both Poisson and non-Poisson traffic. It is interesting to note that the semi-lightpath routing always enjoys a superior performance compared to other routing schemes.

With the current focus moving toward a two-layer architecture that transports IP traffic directly over the optical network, the problem of RWA will assume new challenges to be addressed for effective bandwidth provisioning and protection/restoration schemes in this new architecture. The concept of MPLS/MPλS/GMPLS will have a key role in deciding control/signaling and automated path provisioning in the process of the IP-over-WDM integration effort. In this new scenario we have described the new dynamic integrated routing and waveband routing in optical networks. In integrated routing, the routing decision is made on the basis of knowledge of the combined topology and resource use information at the IP and optical layers. In particular, the maximum open capacity routing algorithm (MOCA) using IP and wavelength routing has been described which is robust to changing traffic patterns in the IP layer without requiring a priori knowledge about future traffic arrivals.

In waveband routing, several wavelengths are aggregated into a waveband and each waveband is switched optically as a single unit in the core OXC, reducing the number of OEO ports required for processing of individual wavelengths. Such traffic grooming strategies will be quite useful in future IP-over-WDM networks. How to allocate λ-group path and λ-path between interoptical layers of cross-connect nodes is an interesting open issue. More concerted activities are required to be undertaken on optimized path computation and wavelength allocation algorithms in integrated IP and aggregated λ-switching for efficient lightpath establishment, protection, and restoration purposes to enable practical realization of fully automated intelligent optical transport network.

REFERENCES

1. N. Ghani, S. Dixit, and T. Wang, On IP-over-WDM integration, *IEEE Communications*, pp. 72–84, Mar. 2000.

2. W. Stalling, *High Speed Networks and Internets*, 2nd ed., Prentice Hall, Upper Saddle River, NJ, 2001.

3. C. Assi, A. Shami, M. A. Ali, R. Kurtz, and D. Guo, Optical networking and real-time provisioning: An integrated vision for the next-generation Internet, *IEEE Network*, 54, July/Aug. 2001.

4. D. Bertsekas and R. Gallager, *Data Networks*, Prentice Hall, Upper Saddle River, NJ, 1987.

5. S. Keshav, *An Engineering Approach to Computer Networking*, Addison-Wesley, Reading, MA, 1997.

6. R. Ramaswami and K. N. Sivarajan, *Optical Networks: A Practical Perspective*, Morgan Kaufmann, San Francisco, 1998.

7. I. Chlamtac, A. Farago, and T. Zhang, Lightpath (wavelength) routing in large WDM networks, *IEEE Journal on Selected Areas in Communications*, 5: 909–913, 1996.

8. E. Karasan and E. Ayanoglu, Effects of wavelength routing and selection algorithms on wavelength conversion gain in WDM optical networks, *IEEE/ACM Transactions on Networking*, 6(2), 1994.

9. A. Mokhtar and M. Azizoglu, Adaptive wavelength routing in all-optical networks, *IEEE/ACM Transactions on Networking*, 6, 1998.

10. R. Gangopadhyay, *Design of Advanced Wavelength-Routed Optical Network, Final report (IIT-Kharagpur), European Commission, Dec. 1998.*

11. C. Chen and S. Banerjee, A new model for optimal routing and wavelength assignment in wavelength division multiplexed optical networks, *Proc. Infocom*, Vol. 2a.4.1-8, pp. 164–171, 1996.

12. R. Gangopadhyay, M. Rameshkumar, and A. K. Varma, Performance evaluation of WDM optical networks with/without wavelength converters for Poisson and non-Poisson traffic, *Proc. CPT-2002 Conference*, Tokyo, 2002.

13. S. Subramaniam, M. Azizoglu, and A. K. Somani, All optical networks with sparse wavelength conversion, *IEEE ACM Transactions on Networking*, Aug. 1996.

14. S. Subramaniam, M. Azizoglu and A. K. Somani, On optimal converter placement in wavelength routed networks, *IEEE ACM Transactions on Networking*, Oct. 1999.

15. S. Dixit and Y. Ye, Streamlining the Internet–fiber connection, *IEEE Spectrum*, 38(4): 52–57, Apr. 2001.

16. M. Kodialam and T. V. Lakshman, Integrated dynamic IP and wavelength routing in IP over WDM networks, Proc. Infocom, 2001.

17. T. E. Stern and K. Bala, *Multiwavelength Optical Networks*, Addison Wesley, Reading, MA, 1999.

18. R. Ismailov, S. Ganguly, Y. Suemura, I. Nishioka, Y. Maeno, and S. Araki, Waveband routing in optical networks, *Proc. ICC Conference*, New York, May 2002.

19. Y. Suemura, I. Nishioka, Y. Maeno, and S. Araki, Control of hierarchical paths in an optical network, *Proc. APCC Conference*, 2001.

13 Internetworking Optical Internet and Optical Burst Switching

MYUNGSIK YOO and YOUNGHAN KIM

Soongsil University, Seoul, Korea

13.1 INTRODUCTION

The unprecedented growth of Internet traffic for the past few years has accelerated the research and development in the optical Internet. One of the key contributions to optical Internet is the rapid advance in the optical technology. Due to the DWDM [dense wavelength-division multiplexing (WDM)] technology, it is possible to put more than 100 wavelengths in a single fiber, each of which is operating at 10 Gbps, even at 40 Gbps. In addition, continuing developments in the optical components, such as optical amplifier, optical switches, and wavelength converters, makes it possible to transmit Internet protocol (IP) traffic in optical form through WDM optical transmission systems [1,2].

Although optical Internet may have a huge network capacity using DWDM technology, it is still being debated how to provision the network resources (e.g., lightpaths). In particular, how frequently to provision the lightpaths in the optical (or WDM) domain is quite important in network management and is highly dependent on the optical switching technique in use [3,4]. Thus, it is worthwhile to take a look at the characteristics of existing switching paradigms in the optical domain, which are summarized in Table 13.1. The interested reader is referred to Chapter 6 for more details on optical packet switching. Optical packet switching (OPS) results in good network utilization since it can easily achieve a high degree of statistical multiplexing and effectively be amenable for traffic engineering. On the other hand, optical circuit switching (OCS) or wavelength routing results in relatively poor utilization, due to its infrequent reconfiguration (e.g., every few months or days) and dedicated use of wavelengths. However, OCS is much easier to implement than OPS since while the former requires only slow (operating at millisecond speeds) optical components, the latter requires very fast (in nano- or picoseconds) optical components. OPS even requires that more functions (e.g.,

IP over WDM: Building the Next-Generation Optical Internet, Edited by Sudhir Dixit.
ISBN 0-471-21248-2 © 2003 John Wiley & Sons, Inc.

TABLE 13.1   Characteristics of Optical Switching Techniques

Switching Technique	Utilization	Granularity	Implementation
Optical circuit switching (OCS)	Low	Coarse	Easy
Optical burst switching (OBS)	Moderate	Moderate	Moderate
Optical packet switching (OPS)	High	Fine	Difficult

synchronization, optical random access memory) be implemented, which is quite difficult with the optical technology available today.

Optical burst switching (OBS) achieves a good balance between OCS and OPC. The advantage of OBS is that it provides a workable solution to optical Internet. Specifically, OBS can achieve high utilization, as in the OPS, and at the same time, can be implemented with currently available optical technology, as in the OCS. Readers may refer to [5] to [16] for various technical aspects of the OBS. It is expected that wavelength-routing networks based on OCS will prevail for the near future, due to technical limitations on optics. However, the optical Internet will evolve to optical networks based on OBS and even on OPS in the long run. Therefore, it is worthy to discuss OBS technology from the optical Internet viewpoint. In this chapter we introduce the concept of optical burst switching and its applicability to internetworking with the optical Internet. In particular, in Section 13.2 we take a look at the characteristics of OBS from both the protocol and architectural aspects. In Section 13.3 we explain how quality of service (QoS) can be provisioned in the OBS networks. The survivability issues in the optical Internet are addressed in Section 13.4, where general restoration procedures based on the OBS protocols are introduced. We summarize this chapter in Section 13.5.

13.2 OVERVIEW OF OPTICAL BURST SWITCHING

13.2.1 Protocol Aspect of OBS

The distinction between OBS and OCS is that the former uses one-way reservation whereas the latter uses two-way reservation [2,15]. It is called *two-way reservation* when there must be a connection setup procedure before data transmission takes place. It is called *one-way reservation* when the data (which are called the *data burst*) follow the connection setup request immediately after some delay. This delay, called *offset time*, is explained later. Note that the connection setup request in the OBS is called the *control packet* or *burst control packet* (BCP) in the remainder of the chapter. Although OBS and OPS have similar characteristics (e.g., statistical multiplexing on the links), the distinction between the two is that the former has unique features, such as offset time and delayed reservation [2,15]. In addition, the payload in the OBS is much larger than that in the OPS. The payload unit in the OBS networks is referred as the data burst hereafter.

Operation of the OBS protocol is illustrated in Fig. 13.1a. When a source node S has a data burst to send, the BCP is generated and transmitted first. At each node, BCP takes time δ to be processed and makes the reservation for the following data

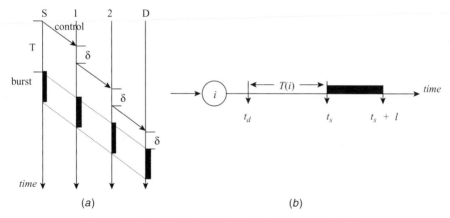

Figure 13.1   Offset time and delayed reservation in OBS.

burst. Meanwhile, the data burst, after waiting for an offset time, which is denoted as T, at the edge router, immediately follows the BCP. Since BCP has already set up the necessary configurations, including the resource reservation, data burst does not need to be delayed (buffered) for processing at the intermediate nodes (nodes 1 and 2) and just cuts through to the destination node D. Thus, OBS networks can provide the data plane with end-to-end transparency without going through any optical–electrical–optical (O/E/O) conversions.

The advantage of offset time is to decouple the BCP from the data burst in time, which makes it possible to eliminate buffering at the intermediate nodes. To ensure that the data burst does not overtake the BCP, the offset time should be long enough considering the processing time expected at each intermediate node and the number of hops to traverse by BCP [5,7,14]. How to determine offset time effectively is another research issue to be addressed.

Another feature of OBS is *delayed reservation* (DR), which makes it possible to multiplex data bursts statistically on the links. The DR is illustrated in Fig. 13.1*b*, where $T(i)$ is the offset time at node i, i is the transmission time of the data burst at node i, and t_a and t_s indicate the arrival time of BCP and its corresponding data burst, respectively. In DR, after being processed, BCP reserves the resources for a future time (which is the arrival time of the data burst), which can be obtained from the offset time $T(i)$. Also, the reservation is made for just enough time to complete the transmission (data burst duration l). In OBS, two parameters, the offset time and the mean data burst size, need to be selected carefully in the design step since they have a major impact on performance.

13.2.2   Architectural Aspect of OBS

Now, we discuss OBS technology from an architectural standpoint [6]. For simplicity, we focus on a single OBS domain,* where all nodes (or OBS routers) are well

---

OBS domain and *OBS network* are used interchangeably.

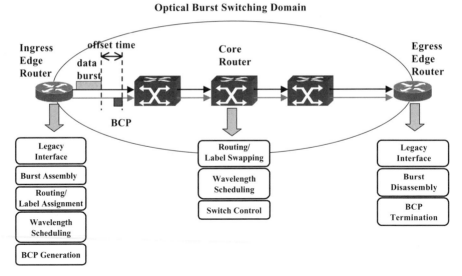

Figure 13.2   Functions of OBS routers.

aware of OBS protocols. Depending on the location in the OBS domain, OBS routers are classified as edge or core routers, as shown in Fig. 13.2. Note that edge routers should be equipped with the capabilities of both the ingress edge router (IER) and the egress edge router (EER). Thus, the edge router functions like an ingress router when there are inbound data to the OBS domain, whereas the edge router functions like the egress router when there are outbound data from the OBS domain. The OBS routers have different functions depending on where they are located (i.e., ingress, core, egress). In the following discussion we describe the functions and general architectures for each type of OBS router.

The general architecture of IER is shown in Fig. 13.3. The IER should provide the interface between the OBS network and other legacy networks. It also needs to provide the signaling interface between two networks, discussed in Section 13.2.3. When IER receives incoming data (e.g., IP packets, ATM cells, or voice streams) from line interface cards, the burst assembly process takes place in which multiple packets are packed into a data burst. We discuss the burst assembly process in detail in Section 13.2.4. The arriving packets are switched to the appropriate assembly queues based on their destination and QoS. The first packet arrival in an assembly queue initiates BCP generation, where some BCP fields,* such as burst size, offset time, and label, are determined and filled later when the burst assembly is done.

When the burst is assembled sufficiently to meet the length requirement, the BCP obtains the field information of burst size and offset time.† Based on the

---

*The structure of BCP and its fields need to be studied further.
†Deciding the average burst length and offset time is an important design issue in the OBS protocol, which is also related to QoS provisioning.

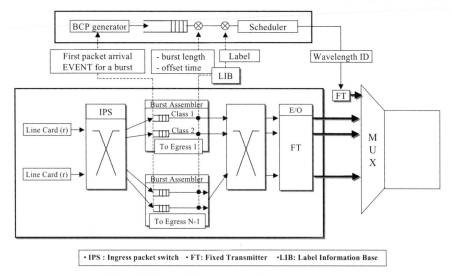

Figure 13.3   Architecture of an ingress edge router.

routing decision, a label is assigned to establish the label-switched path (LSP). If the LSP already exists (because it has been set up by the previous BCP), the label used previously is assigned. Otherwise (i.e., if a new LSP needs to be set up or the existing LSP needs to be changed), a new label is assigned. We need the reachability information in the routing process, discussed in Section 13.2.3. To set up the LSP, the downstream nodes perform the label swapping, where the inbound label is mapped into the outbound label at the local node. The label information in the BCP should be updated accordingly. How to distribute and assign the labels depends on label distribution protocols such as RSVP-TE and CR-LDP [17,18].

According to OBS protocols, the BCP is transmitted to the core OBS router an offset time earlier than its corresponding data burst. The wavelength assignment and scheduling are required in this step. There are two types of wavelengths (or channels) in the OBS domain: the control channels that carry the BCPs and called the *control channel group* (CCG), and the data burst channels that carry the data bursts and is called the *data burst channel group* (DCG). Thus, two wavelengths are assigned and scheduled for transmission of BCP and its data burst. The wavelength scheduling on DCG (for data bursts), which can enhance the utilization, is another interesting research area. A few scheduling algorithms have already been proposed, such as first fit, horizon scheduling, and void-filling scheduling [6,16]. It is noted that while the labels carry the path (LSP) information, the wavelengths are meaningful only at the local node. In this way, the downstream nodes can assign the wavelength dynamically, which binds with the label only for the duration of data burst.

An architecture for an EER is shown in Fig. 13.4. The main functions of EER are BCP termination and data burst disassembly. When a BCP arrives, EER processes it

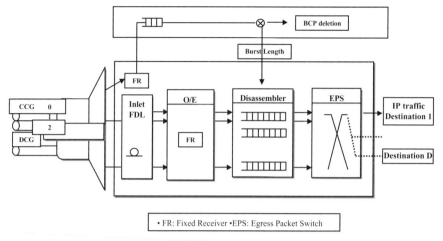

Figure 13.4   Architecture of an egress edge router.

and reserves the buffer space as required by the burst length field for the burst disassembly process. Upon arrival, the data burst, after being converted to the electrical signal, goes through the burst disassembly process. Then the disassembled packets are distributed to their destined port.

The core routers located inside the OBS network have the general architecture shown in Fig. 13.5. It consists of two parts: the switch control unit and the data burst unit. While the switch control unit is responsible for processing BCPs on the CCG, the data burst unit is responsible for switching data bursts to the destined output port, which is controlled by the switch control unit. Most functions in the

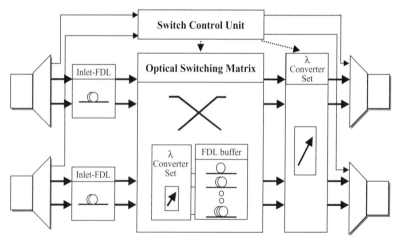

Figure 13.5   Architecture of a core router.

core router take place in the switch control unit, which includes label swapping for establishing the LSP, local wavelength scheduling, burst contention resolution, and generation of the control signal to the data burst part. The data burst unit consists roughly of demultiplexers, inlet fiber delay lines (FDLs) (for adjusting the timing jitter of offset time between BCP and data burst), optical switches, wavelength converters, FDL buffers (for contention resolution), and multiplexers. The components of the data burst unit are passively configured by the control signals from the switch control unit.

Before we close the discussion on architectural aspects, it is worthwhile mentioning some important design parameters, such as the capacity of the core router, the switching speed of the optical switch, and the average burst size. The capacity of a router is determined by the number of incoming fibers and the number of wavelengths available in each fiber. For example, a core router of 10 Tbps capacity needs 32 incoming fibers, each of which has 32 wavelengths operating at 10 Gbps per wavelength. Of course, since some wavelengths are dedicated to CCG, the router capacity is determined only by the number of wavelengths in the DCG.

The core router requires two switches; one in the switch control unit and the other in the data burst unit. Whereas the former switches BCPs on CCG, which can be implemented with a small electronic switch (depending on the number of wavelengths on CCG), the latter switches data bursts on DCG, which can be implemented with an optical switch. The dimension of the optical switch depends on the number of wavelengths in DCG. The architecture of the optical switch may be either a simple single stage if a large optical switch is available (e.g., MEMS crossbar switch [19]) or a multistage interconnection network (MIN) as in an ATM switch [20] if multiple small optical switches are used.

Next, consider the average burst size and the switching speed of an optical switch. The switching time of an optical switch is the delay it takes to change from one state to another state. The switching speed imposes the overhead on the switch throughput. In other words, the slower the switching speed, the worse is the throughput. However, the overhead caused by switching speed can be reduced if the data burst is long enough. Thus, for an optimum switch throughput, the average burst size should be selected appropriately depending on the switching speed of the optical switch in use.

13.2.3 Signaling in the Optical Internet

Until now, we have looked at the OBS from both the protocol and architectural standpoints. Now, we look into the internetworking aspect of OBS with the optical Internet. We define two different types of domains: IP domains and optical domains. An optical domain may be either the OBS domain, which is well aware of the OBS protocols, or the non-OBS domain, which is not. We exclude the non-OBS domain from our discussion, and the OBS and non-OBS domains are treated equally as far as the internetworking issue is concerned. We reiterate this when we discuss interdomain networking later in the chapter. We also assume that the control plane is running on top of each node and is responsible for exchanging routing and

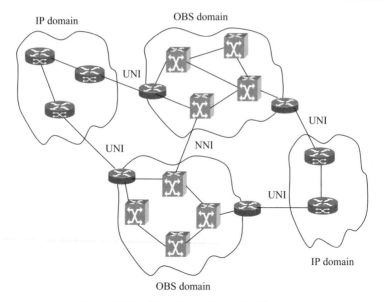

Figure 13.6   Signaling in the optical Internet.

signaling information. Standardization of a unified control plane for both IP and optical networks [21–23] is actively in progress. It is likely that the optical Internet will adopt the IP-centric control plane (e.g., G-MPLS). Next, we focus on signaling in the optical Internet and the signaling interface between different domains. We use the example of optical Internet shown in Fig. 13.6.

First, we define the autonomous system (AS) as a single domain that is administered by a single entity. In Fig. 13.6, each IP or OBS domain is an AS, for example. It is easy to understand internetworking in a single AS. Since all control planes are under a single administration, they are well aware of each other. Each node exchanges the routing and signaling information through its control plane, building up knowledge on the reachability within its own AS. Inside an AS, an extension of routing protocols such as OSPF or IS-IS can be used for information exchange.

When multiple ASs are involved, we need another approach to signaling and routing to get the reachability information across. The user-to-network interface (UNI) and network-to-network interface (NNI) signaling can be used for this purpose, where the former exchanges the information between the IP domain and the OBS domain, while the latter does the same between two OBS domains. As mentioned earlier, NNI signaling can be used between different optical domains (e.g., OBS and non-OBS domains) as far as both domains run unified control planes. When two domains run the control planes independently, the NNI signaling needs to be updated to interface the two, or the UNI signaling may be used.

There are generally two approaches mentioned in the literature when considering the architectural aspects of interdomain signaling [24]: the overlay model and peer model. In the overlay model, as in the IP over ATM approach, the IP domains are independent of the OBS domains. Thus, they perform routing and signaling in a totally different way. The overlay model is practical in a sense that the two domains may need to maintain the domain-specific characteristics and requirements. In this approach, since the control planes of two domains are independent of each other, IP domains obtain the reachability information across the OBS domains via UNI signaling, and vice versa. Two OBS domains exchange reachability information via NNI signaling. Although it is the practical approach, network management becomes complex since two independent domains need to be managed and maintained.

In the peer model, all domains (i.e., IP or OBS domains) run unified control planes. Thus, all domains are integrated into a single network from the control plane viewpoint. In the peer model, there is no difference between UNI and NNI. One domain obtains the reachability information from across other domains via NNI signaling. However, since IP domains need to be unified with the information specific to OBS domains, and vice versa, it is a more difficult approach than the overlay model. In either model, the label-switched path (LSP) can be established as long as reachability information is available via UNI/NNI signaling.

13.2.4 Burst Assembly

As we mentioned in Section 13.2.2, the burst assembly process takes place at the ingress edge router. The incoming data (e.g., IP packets) are assembled into a super packet called the *data burst*. In the following discussion, we look into the issues of burst assembly.

It is obvious that the IP packets would be the dominant traffic in the optical Internet. Unlike the traffic from traditional telephone networks, Internet traffic is quite difficult to predict. This is because Internet traffic shows self-similarity [25,26]. Self-similarity means that even with a high degree of multiplexing, the burstiness of traffic still exists in all time scales, which makes network resource dimensioning and traffic engineering more difficult. To show the difference between self-similar traffic and traditional telephone network traffic, the packets are generated using two distribution functions: exponential distribution and Pareto distribution. Figure 13.7 shows two different patterns of packet arrivals when the time scale of packet aggregation is changed from 1 unit to 10,000 units. As the time scale increases, while the traffic pattern when using the exponential distribution (shown on the left side in Fig. 13.7) smoothes out to the average value, the traffic pattern when using Pareto distribution (shown on the right side in Fig. 13.7), still shows a high variance above the average value. According to the variance-time plot [25,26], exponential distribution results in a Hurst parameter (H) of 0.5, whereas the Pareto distribution has H higher than 0.5 ($0.5 < H < 1.0$). One of the advantages of the burst assembly is that it may reduce the burstiness of Internet traffic [12].

As shown in Fig. 13.3, the incoming IP packets are classified and queued into an assembly buffer according to their destination and QoS requirements. In the following discussion, for simplicity, we consider the burst assembly with a single

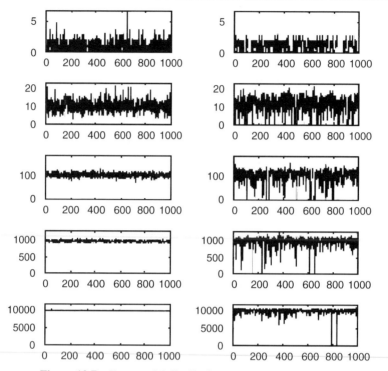

Figure 13.7 Exponential distribution and Pareto distribution.

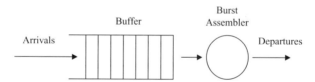

Figure 13.8 Burst assembly process.

assembly buffer, as shown in Fig. 13.8. The burst assembler at the end of assembly buffer runs a timer, which expires at the given time (i.e., the assemble time). Whenever the timer expires, the burst assembler takes the burst (multiple IP packets queued during a given period of time) out of the assemble buffer for transmission. The key parameter of the burst assembler is the assemble time, which controls the size of the data burst.* The distribution of assemble time could be deterministic or uniform random, called *constant assemble time* (CAT) and *variable assemble time* (VAT), respectively. The burst assembler waits for a constant time when using the CAT, whereas it waits for a random amount of time when using the VAT. For

---

*Since the assemble time determines the burst size, it should be large enough that the assembled data burst can meet the average burst size required in OBS protocol.

TABLE 13.2   **Performance of CAT Burst Assembly**

Assemble Time (units)	Variance of Interarrival Time	Mean Delay
0	38.3827	0.0
1×125	16.0982	0.318
3×125	3.9938	1.286
5×125	1.7955	2.506
10×125	0.5820	5.583
20×125	0.2492	11.623

simplicity, we take the CAT burst assembly as an example and address the optimization issue.

Now we are interested in optimizing the performance of a CAT burst assembly as we change the assemble time. For the performance evaluation, we consider two variables: the variance of interarrival time of data burst and the mean assembly delay. The interarrival time means the time interval between two consecutive data bursts. The variance of interarrival time indicates the regularity of data bursts appearing on the outgoing link, while the mean assembly delay indicates the delay that an IP packet suffered in the assembly buffer on average. The mean assembly delay is a relative value in the transmission time of a single IP packet. For a simple simulation, the IP packets are generated with Pareto distributed inter-arrival time and exponentially distributed packet size with mean of 100 bytes. It is assumed that a single IP packet arrives at the assembly buffer in on the average 125 units.

Table 13.2 shows the performance of a CAT burst assembly. According to the results, when the assembly time is zero, the burst assembly process does not take place, resulting in a large variance in the interarrival time. As the assembly time increases (in 125 units), the variance of the interarrival time of data bursts decreases. However, the table also shows that the mean assembly delay increases linearly as the assembly time increases. Thus, as far as the mean assembly delay is concerned, the assembly time should be minimized. Now we have two parameters, which behave opposite to each other. To optimize the performance of the CAT burst assembly, we need a compromise between the variance of interarrival time and the mean assembly delay. For optimum performance, the assembly time should be chosen to minimize both variables at the same time. For a complete understanding of the burst assembly process, more extensive study of a general case of burst assembly is required in which many other parameters are optimized.

13.3   QoS PROVISIONING WITH OBS

Given that some real-time applications (such as Internet telephony and video conferencing) require a higher QoS value than that for non-real-time applications such as electronic mail and general Web browsing, the QoS issue should be addressed. Although QoS concerns related to optical (or analog) transmission (such as dispersion, power, and signal-to-noise ratio) also need to be addressed, here we focus on how to transport the critical data in the OBS domain more reliably than the noncritical data. Unlike existing QoS schemes that differentiate services using the buffer, the QoS scheme introduced in the following discussion takes advantage of the offset time, which was explained in Section 13.2.1. That is why we call it the offset-time-based QoS scheme. For the offset-time-based QoS scheme, we introduce a new offset time, called the *extra offset time*. Note that the offset time introduced in Section 13.2.1, called the *base offset time*, is different from the extra offset time.

We now explain how the offset-time-based QoS scheme works [5,27]. In particular, we explain how class isolation (or service differentiation) can be achieved by using an extra offset time in both cases with and without using fiber delay lines (FDLs) at an OBS node. Note that one may distinguish the following two different contentions in reserving resources (wavelengths and FDLs): the intraclass contentions caused by requests belonging to the same class, and the interclass contentions caused by requests belonging to different classes. In what follows we focus on how to resolve interclass contentions using the offset-time-based QoS scheme.

For simplicity, we assume that there are only two classes of (OBS) services: class 0 and class 1, where class 1 has priority over class 0. In the offset-time-based QoS scheme, to give class 1 a higher priority for resource reservation, an extra offset time, denoted by t_o^1, is given to class 1 traffic (but not to class 0, i.e., $t_o^0 = 0$). In addition, we assume that the base offset time is negligible compared to the extra offset time, and refer to the latter hereafter as simply the offset time. Finally, without loss of generality, we also assume that a link has only one wavelength for data (and an additional wavelength for control).

13.3.1 Class Isolation without FDLs

In the following discussion, let t_a^i and t_s^i be the arrival and service-start times for a class i request denoted by req(i), respectively, and let l_i be the burst length requested by req(i), where $i = 0, 1$. Figure 13.9 illustrates why a class 1 request that is

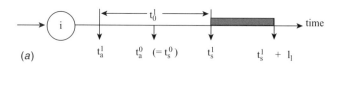

(a)

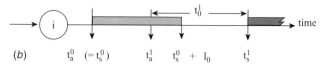

(b)

Figure 13.9 Class isolation without FDLs.

assigned an (extra) offset time obtains a higher priority for wavelength reservation than a class 0 request in the case of no FDLs. We assume that there is no burst (arrived earlier) in service when the first request arrives. Consider the following two situations, where contention among two classes of traffic is possible. In the first case, as illustrated in Fig. 13.9a, req(1) comes first and reserves a wavelength using DR, and req(0) comes afterward. Clearly, req(1) will succeed, but req(0) will be blocked if $t_a^0 < t_s^1$ but $t_a^0 + l_0 > t_s^1$, or if $t_s^1 < t_a^0 < t_s^1 + l_1$. In the second case, as in Fig. 13.9b, req(0) arrives first, followed by req(1). When $t_a^1 < t_a^0 + l_0$, req(1) would be blocked had no offset time been assigned to req(1) (i.e., $t_o^1 = 0$). However, such a blocking can be avoided by using a large enough offset time so that $t_s^1 = t_a^1 + t_o^1 > t_a^0 + l_0$. Given that t_a^1 may be only slightly behind t_a^0, t_o^1 needs to be larger than the maximum burst length over all bursts in class 0 in order for req(1) to avoid being blocked completely by req(0). With that much offset time, the blocking probability of (the bursts in) class 1 traffic becomes only a function of the offered load belonging to class 1 (i.e., independent of the offered load belonging to class 0). However, the blocking probability of class 0 is determined by the offered load belonging to both classes.

13.3.2 Class Isolation with FDLs

Although the offset-time-based QoS scheme does not mandate the use of FDLs, its QoS performance can be improved significantly even with limited FDLs, so as to resolve contentions for bandwidth among multiple bursts. For the case with FDLs, the variable B will be used to denote the maximum delay that an FDL (or the longest FDL) can provide. Thus, in the case of blocking, a burst can be delayed up to the maximum delay B.

Figure 13.10a and b illustrate class isolation at an OBS node equipped with FDLs where contention for both wavelength and FDL reservation may occur. In Fig. 13.10a, let us assume that when req(0) arrives at t_a^0 (t_s^0), the wavelength is in use by a burst that arrived earlier. Thus, the burst corresponding to req(0) has to be delayed (blocked) for t_b^0 units. Note that the value of t_b^0 ranges from 0 to B, and an FDL with an appropriate length that can provide a delay of t_b^0 is chosen. Accordingly, if $t_b^0 < B$, the FDL is reserved for a class 0 burst as shown in Fig. 13.10b (the burst will be dropped if t_b^0 exceeds B), and the wavelength will be reserved from $t_s^0 + t_b^0$ to $t_s^0 + t_b^0 + l_0$, as shown in Fig. 13.10a. Now assume that req(1) arrives later at t_a^1 (where $t_a^1 > t_a^0$) and tries to reserve the wavelength, req(1) will succeed in reserving the wavelength as long as the offset time is so long that $t_s^1 = t_a^1 + t_o^1 > t_a^0 + t_b^0 + l_0$. Note that had req(1) arrived earlier than req(0) in Fig. 13.10a, it is obvious that req(1) would not have interclass contention caused by req(0). This illustrates that class 1 can be isolated from class 0 when reserving wavelength because of the offset time. Of course, without the offset time, req(1) would be blocked for $t_a^0 + t_b^0 + l_0 - t_a^1$, and it would be entirely up to the use of FDLs to resolve this interclass contention.

Similarly, Fig. 13.10b illustrates class isolation in FDL reservation. More specifically, let us assume that req(0) has reserved the FDLs as described earlier,

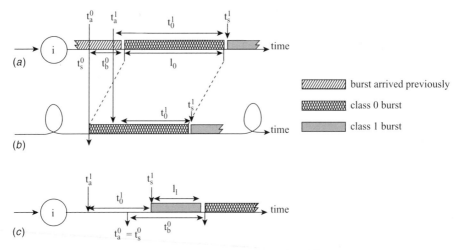

Figure 13.10   Class isolation with FDLs.

and because t_o^1 is not long enough, req(1) would be blocked in wavelength reservation and thus needs to reserve the FDLs. In such a case, req(1) will reserve the FDLs successfully if the offset time is still long enough to have $t_s^1 = t_a^1 + t_o^1 > t_a^0 + l_0$. Otherwise (i.e., if $t_s^1 < t_a^0 + l_0$), req(1) would contend with req(0) in reserving the FDL and would be dropped.

As shown in Fig. 13.10c, if req(1) comes first and reserves the wavelength based on t_s^1 and delayed reservation (DR), and req(0) comes afterward, req(1) is not affected by req(0). However, req(0) will be blocked either when $t_a^1 < t_a^0 < t_s^1$, but $t_a^0 + l_0 > t_s^1$, or when $t_s^1 < t_a^0 < t_s^1 + l_1$. Similarly, if req(1) arrives first, it can reserve the FDL first regardless of whether req(0) succeeds in reserving the FDL. As mentioned earlier, this implies that class 1 can be isolated from class 0 in reserving both wavelength and FDL by using an appropriate offset time, which explicitly gives class 1 a higher priority than class 0. As a result of having a low priority on resource reservation class 0, bursts will have a relatively high blocking and loss probability.

Although the offset-time-based QoS scheme does provide a good service differentiation even when buffering is not possible, it has some disadvantages that need to be explained. For example, the offset-time-based QoS scheme introduces the delay overhead caused by the extra offset time. Since the highest class suffers from the longest delay [5,27], QoS provisioning will be strictly restricted within a given delay budget. It also lacks controllability on the QoS, which can be enhanced by introducing the measurement-based QoS provisioning and assigning the appropriate weight to each QoS class. In addition, it is worth considering how the offset-time-based QoS scheme can be integrated with existing QoS domain such as DiffServ.

13.4 SURVIVABILITY ISSUE IN OBS NETWORKS

The survivability issue in the optical Internet becomes more and more important since a huge amount of data traffic is carried by a single fiber, which is susceptible to a simple network failure such as a fiber cut. Even, a single network element failure brings about multiple session errors in the higher layers, which has an impact on network service. There are many studies dealing with protection and restoration in the optical and MPLS networks. Readers are referred to [28] to [33] and Chapter 15 for more details. Here, we describe a restoration approach, which is based on OBS. We introduce network fault detection mechanisms that can be used in OBS restoration, then explain the OBS restoration procedures.

13.4.1 Network Failure Detection

In the optical Internet, a link between two nodes is composed of several optical fibers, each of which carries many wavelengths (or channels). The channels are divided into CCG (control channel group) for the transmission of BCPs and DCG (data channel group) for the transmission of data bursts. Thus, several types of optical network faults can exist depending on the location of the fault: data channel failure, control channel failure, fiber failure, link failure, and node failure. Figure 13.11 shows the different types of failure in the OBS domain.

When the network failure occurs, it is desirable that the restoration be done as quickly as possible. The restoration time is highly dependent on the fault detection time. The detection mechanisms in the higher layer, such as the one using the "hello" message at the IP layer, take relatively long time to detect a fault. Hence, it

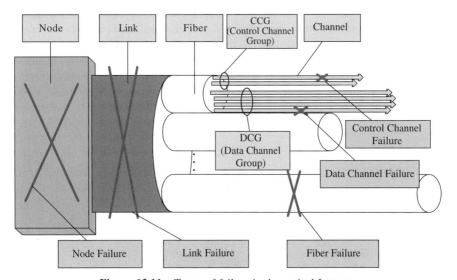

Figure 13.11 Types of failure in the optical Internet.

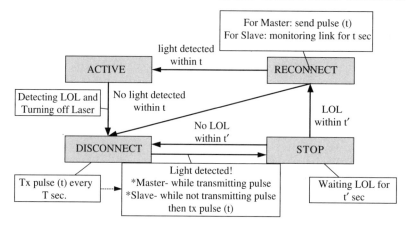

Figure 13.12 State diagram of open fiber control. (From [15, Fig. 10.18].)

is necessary to define the fault detection mechanism in the optical layer, which can detect the fault much faster.

First, the channel failure can be detected by monitoring the optical signal quality per wavelength with the help of an optical spectrum analyzer. Second, fiber failure can be detected by using an open fiber control protocol [34], which maintains the state information per fiber by monitoring the loss of light (LOL) as shown in Fig. 13.12. When the state of a specific fiber changes from "active" to "disconnect," it is a fiber failure. Third, the link failure can be detected implicitly by using the state information of all fibers belonging to a specific link. In other words, when all fibers in a link have the disconnect state, it is assumed to be a link failure. Finally, a node failure is detected by checking the following two conditions. Every node exchanges the control messages with its neighbor node and counts them periodically. If no control message is received in a certain period of time, it is assumed to be a seminode failure, in which case the states of the control channels belonging to this node are checked to determine whether or not the seminode failure has resulted from the control channel failure. If the channels are found to be normal, a node failure is confirmed.

13.4.2 OBS Restoration

The OBS-based restoration is classified into link- and path-level restorations. Link-level restoration is done at the local node, while path-level restoration is done globally between OPSL (OBS path switch LSR) and OPML (OBS path merge LSR). OPSL and OPML are two LSRs (label-switching routers) where the working and backup paths meet. For a path-level restoration, OPSL and OPML must be equipped with intelligent functions, and the alternative control channels for restoration must be preestablished for each working control channel. In addition, the path for the failure indication message, called O-RNT (OBS reverse notification

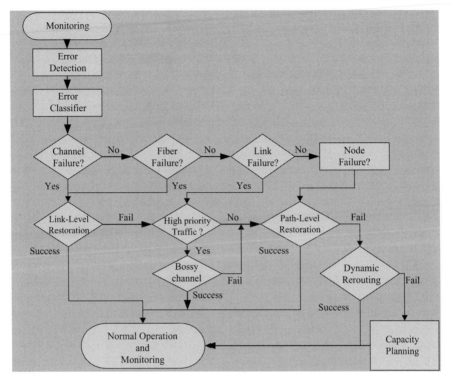

Figure 13.13 OBS restoration procedure.

tree), should be established in advance. The RNT means that the path is the same as the working path but is in the opposite direction.

The OBS restoration procedure can be explained in detail as follows. Every node in the OBS domain monitors the network status using the fault detection technique explained previously. When a fault is detected, an appropriate restoration technique is applied according to the type of the fault detected, as shown in Fig. 13.13. For channel failure and fiber failure, link-level restoration is tried, whereas for link failure and node failure, path-level restoration is tried. Note that for link failure, if the traffic has a high priority, restoration using the bossy channel may be used, which is described later.

For channel failure, a data channel and a control channel must be considered together. If a fault in either a control or data channel occurs and a corresponding backup channel is available, $1:N$ restoration can be applied successfully. Figure 13.14a shows a link-level restoration procedure when the control channel has a fault. Node b, which has detected the channel failure, replaces the control channel affected by this fault with the backup channel. This node updates the content of BCP-2* received from node a and activates the backup channel

*Note that in Fig. 13.14, BCP and data burst are numbered in pairs in the transmission sequence.

414

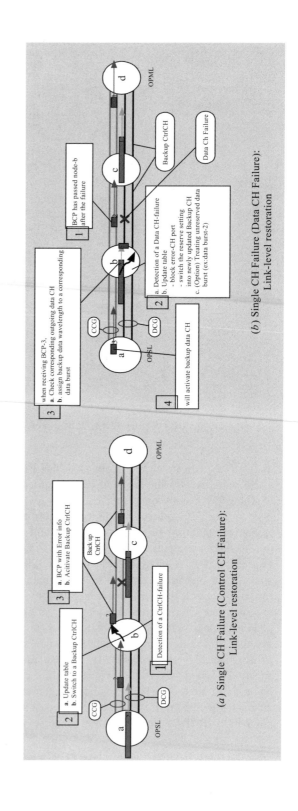

(a) Single CH Failure (Control CH Failure): Link-level restoration

(b) Single CH Failure (Data CH Failure): Link-level restoration

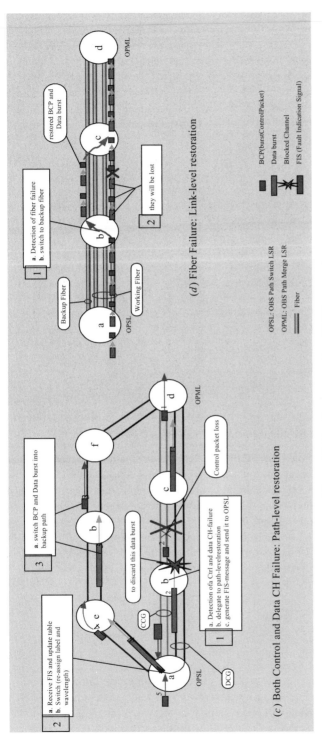

a. switch BCP and Data burst into
backup path

3

to discard this data burst

a. Detection of a Ctrl and data CH-failure
b. delegate to path-level restoration
c. generate FIS-message and send it to OPSL

Control packet loss

1

a. Receive FIS and update table
b. Switch (re-assign label and
wavelength)

2

CCG

DCG

OPSL

OPML

f

b

e

d

c

a

5

(c) Both Control and Data CH Failure: Path-level restoration

---

a. Detection of fiber failure
b. switch to backup fiber

1

they will be lost

2

Backup Fiber

Working Fiber

OPSL

OPML

restored BCP and
Data burst

a

b

c

d

(d) Fiber Failure: Link-level restoration

BCP(burstControlPacket)

Data burst

Blocked Channel

FIS (Fault Indication Signal)

OPSL: OBS Path Switch LSR
OPML: OBS Path Merge LSR
═══ Fiber

Figure 13.14 Some examples of OBS restoration.

by transmitting it. In this way, the data channel will hardly be affected by the failure.

In Fig. 13.14*b*, link-level restoration is executed to restore the data channel failure. In this case the data burst affected by the data channel failure (e.g., burst 2) has to be addressed. When both control and data channels fail at the same time, link-level restoration is applied first. If, however, a pair of these channels cannot be restored using link-level restoration, path-level restoration is attempted. Figure 13.14*c* illustrates path-level restoration where node *b* in Fig. 13.14*c* can restore the data channel but does not have a backup control channel available. As soon as a fault is detected, node *b* sends failure indication messages (FISs) upstream along the control channel to the OPSL. After receiving the FIS, the OPSL starts to switch all traffic on the working path to the backup path. The backup path will be activated by sending the first-arriving BCP (e.g., BCP-3 in Fig. 13.14*c*) to this backup control channel.

For a fiber fault, link-level restoration is executed. Figure 13.14*d* shows how all traffic on the working fiber is switched onto the backup fiber. However, if the backup fiber has already been occupied or had a fault at the same time as did the working fiber, it triggers path-level restoration immediately. For a link or node failure, path-level restoration applies directly. The link or node failure can cause a huge number of control channels (a large volume of traffic) to be restored. Thus, OPSL may split the restored traffic into several backup paths, which must be preplanned. If all the methods mentioned above fail, the backup path will be calculated dynamically based on the routing information. If the dynamic restoration is unsuccessful, the network capacity must be replanned to ensure survivability.

13.4.3   Bossy Channel Restoration and Unnecessary Reservation

When link-level restoration fails or link failure occurs, path-level restoration is used. Normally, path-level restoration requires that the local node detect the fault to send FIS to OPSL, and OPSL to initiate restoration, which causes low utilization and a long restoration time. Thus, if the traffic to be restored requires fast restoration, path-level restoration is inappropriate. To provide this kind of service, bossy channel restoration can be used. In bossy channel restoration, instead of sending FIS to OPSL, the node detecting the failure initiates restoration by establishing a new backup path using a preplanned path. Bossy channel restoration allows a preemption attribute in the BCP fields so that it can preempt the lower-priority channels when reserving labels and wavelengths. However, if all the resources are occupied by other bossy channel restorations, path-level restoration will be tried. In this way, traffic with a high priority can be restored much faster. Note that bossy channel restoration requires that each node have preplanned paths and reserves a small portion of labels.

Finally, we point out one of the problems in OBS restoration, called *unnecessary reservation*. As shown in Fig. 13.15*a*, when any type of fault occurs just after a BCP has passed the node, the BCP will make an unnecessary reservation of the resources at downstream nodes, which may cause other traffic to be blocked. To solve this

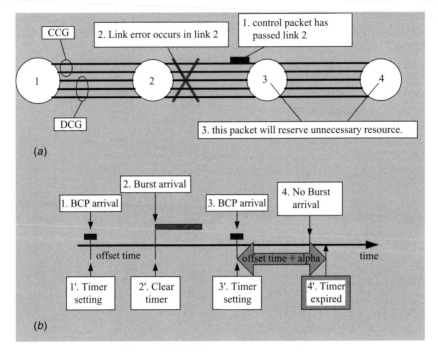

Figure 13.15   Unnecessary reservation.

problem, the unnecessary reservation should be released, as shown in Fig. 13.15*b*. When each node receives the BCP, it sets a timer and waits for an "offset time $+\alpha$" until receiving the corresponding data burst. If this data burst arrives within that period, the timer is reset. Otherwise, the reservation is assumed to be unnecessary and is released immediately.

13.5 SUMMARY

With the rapid advances in optical technologies, the optical Internet has firmly positioned itself as an infrastructure of choice for the next-generation Internet. The optical networks responsible for transporting Internet traffic play an important role in the optical Internet and are expected to evolve with optical switching technology. Although wavelength-routing networks based on OCS are the major thrust today due to limitations in optical technology, it is very likely that they will evolve into burst-switched or even packet-switched optical networks in the future. Due to its salient features, OBS achieves a good balance between OCS and OPS and provides an interim solution for the optical Internet. Thus, while the optical technology is being improved to implement OPS, OBS technology might be a practical choice for the optical Internet as an alternative to OCS. We have addressed many issues related to OBS in this chapter, including the protocol and architectural aspects, the signaling interface, the QoS provisioning, and the survivability. However, many issues in OBS technology remain open to research and implementation.

ACKNOWLEDGMENTS

This work was supported in part by the Korean Science and Engineering Foundation (KOSEF) through the OIRC project.

REFERENCES

1. E. Karasan and E. Ayanoglu, Performance of WDM transport networks, *IEEE Journal on Selected Areas in Communications*, 16:1081–1096, 1996.

2. K. Sivalingam and S. Subramaniam, *Optical WDM Networks: Principles and Practice*, Kluwer Academic, Dordrecht, The Netherlands, 2000.

3. N. Ghani, S. Dixit, and T. Wang, On IP-over-WDM integration, *IEEE Communications*, 38(3):72–84, Mar. 2000.

4. M. Listanti, V. Eramo, and R. Sabella, Architectural and technological issues for future optical Internet networks, *IEEE Communications*, 38(9):82–92, Sept. 2000.

5. M. Yoo, C. Qiao, and S. Dixit, QoS performance of optical burst switching in IP-over-WDM networks, *IEEE Journal on Selected Areas in Communications*, 18(10):2062–2071, Oct. 2000.

6. Y. Xiong, M. Vandenhoute, and H. C. Cankaya, Control architecture in optical burst-switched WDM networks, *IEEE Journal on Selected Areas in Communications*, 18(10):1838–1851, Oct. 2000.

7. M. Yoo, C. Qiao, and S. Dixit, Optical burst switching for service differentiation in the next generation optical Internet, *IEEE Communications*, 39(2):98–104, Feb. 2001.

8. J. Y. Wei and R. I. McFarland, Just-in-time signaling for WDM optical burst switching networks, *Journal of Lightwave Technology*, 18(12):2019–2037, Mar. 2001.

9. X. Lisong, H. G. Perros, and G. Rouskas, Techniques for optical packet switching and optical burst switching, *IEEE Communications*, 39(1):136–142, Jan. 2001.

10. S. Verma, H. Chaskar, and R. Ravikanth, Optical burst switching: A viable solution for terabit IP backbone, *IEEE Network*, 14(6):48–53, Nov./Dec. 2000.

11. C. Qiao, Labeled optical burst switching for IP-over-WDM integration, *IEEE Communications*, 38(9):104–114, Sept. 2000.

12. A. Ge, F. Callegati, and L. Tamil, On optical burst switching and self-similar traffic, *IEEE Communication Letters*, 4(3):98–100, Mar. 2000.

13. J. Turner, Terbit burst switching, *Journal of High Speed Networks*, 8(1):1–18, 1999.

14. C. Qiao and M. Yoo, Optical burst switching (OBS): A new paradigm for an optical Internet, *Journal of high speed networks*, 8(1):69–84, 1999.

15. C. Qiao and M. Yoo, Choice, features and issues in optical burst switching, *Optical Network*, 1(2):36–44, May 2000.

16. J. S. Turner, WDM burst switching for petabit data networks, *OFC'2000*, 2:47–49, 2000.

17. P. Ashwood-Smith et al., Generalized MPLS signaling: RSVP-TE extensions, IETF draft, draft-ietf-mpls-generalized-rsvp-te-06.txt, 29 Nov., 2001.

18. P. Ashwood-Smith et al., Generalized MPLS signaling: CR-LDP extensions, IETF draft, draft-ietf-mpls-generalized-cr-ldp-07.txt, Aug. 2002.

19. L. Y. Lin and E. L. Goldstein, MEMS for free-space optical switching, *LEOS'99*, 2:483–484, 1999.

20. R. Awdeh and H. T. Mouftah, Survey of ATM switch architecture, 27:1567–1613, Nov. 1995.

21. D. Awduche et al., Multi-protocol lambda switching: Combining MPLS traffic engineering control with optical crossconnects, IETF draft.

22. A. Bellato et al., G.709 optical transport networks GMPLS control framework, IETF draft.

23. B. Rajagopalan et al., A framework for generalized multi-protocol label switching (GMPLS), IETF draft.

24. B. Rajagopalan et al., IP over optical networks: Architectural aspects, *IEEE Communications*, pp. 94–102, Sept. 2000.

25. V. Paxon and S. Floyd, Wide area traffic: The failure of Poisson modeling, *IEEE Transactions on Networking*, 3(3):226–244, 1995.

26. W. Leland et al., On the self-similar nature of Ethernet traffic (extended version), *IEEE Transactions on Networking*, 2(1):1–15, 1994.

27. M. Yoo and C. Qiao, Supporting multiple classes of services in IP over WDM networks, *Proc. IEEE Globecom*, pp. 1023–1027, 1999.

28. A. Fumagalli and L. Valcarenghi, IP restoration vs. WDM protection: Is there an optimal choice? *IEEE Network*, 14(6):34–41, Nov./Dec. 2000.

29. G. Mohan et al., Lightpath restoration in WDM optical networks, *IEEE Network*, pp. 24–32, Nov. 2000.

30. Y. Ye et al., On joint protection/restoration in IP-centric DWDM-based optical transport networks, *IEEE Communications*, pp. 174–183, June 2000.

31. Z. Dongyun and S. Subramaniam, Survivability in optical networks, *IEEE Network*, 14(6):16–23, Nov./Dec. 2000.

32. T. M. Chen and T. H. Oh, Reliable services in MPLS, *IEEE Communications*, 37(12): pp. 58–62, Dec. 1999.

33. A. Banerjee et al., Generalized multi-protocol label switching: An overview of signaling enhancements and recovery techniques, *IEEE Communications*, pp. 144–151, July 2001.

34. R. Ramaswami and K. N. Sivarajan, *Optical Networks: A Practice Perspective*, Morgan Kaufmann, San Francisco, 1998.

14 IP-over-WDM Control and Signaling

CHUNSHENG XIN and SUDHIR DIXIT

Nokia Research Center, Burlington, Massachusetts

CHUNMING QIAO

State University of New York, Buffalo, New York

14.1 INTRODUCTION

During the last decade, optical network technology has evolved with leaps and bounds [1]. For one thing, optical network elements have advanced from the electronic terminals of early days, to passive stars, optical add–drop multiplexers, wavelength routers, and intelligent, reconfigurable optical cross-connects (OXCs). The first-generation OXCs have the electronic core and are called *opaque OXCs*. More recently, photonic OXCs, which are truly optical and called *transparent OXCs*, have been developed. As a result of this evolution, optical network elements are able to support more functions and features, enabling optical network topology to change from point-to-point to rings to meshes. Consequently, optical networks are becoming automatically switchable, intelligent, and controllable.

Although optical technology has made tremendous advances, at present it is difficult for OXCs to perform packet-by-packet processing at a speed to match optical transmission speed. Therefore, optical networks with mesh topology are being developed that set up end-to-end high-speed optical connections (called *lightpaths*), eliminating the need for intermediate data processing. Lightpaths also offer other benefits as well, such as protocol transparency. Since a wavelength-division multiplexing (WDM) optical network can offer a substantial number of lightpaths, current optical networks are based primarily on circuit-switching technology. Transmission of client traffic [e.g., Internet protocol (IP) traffic] on the optical network is achieved through a logical topology in the form of switched lightpaths. Each lightpath is used to provide the link connectivity for a pair of client nodes. Therefore, in WDM optical networks, a control mechanism is needed to

IP over WDM: Building the Next-Generation Optical Internet, Edited by Sudhir Dixit.
ISBN 0-471-21248-2 © 2003 John Wiley & Sons, Inc.

establish a lightpath and to tear it down. In this chapter we overview various control and signaling schemes in optical networks. In Section 14.2 we introduce the basic concepts of network control in the telecommunications and Internet worlds. In Section 14.3 we focus on IP-centric optical network control, which is under study by the Internet Engineering Task Force (IETF) and industry coalition organizations. In Section 14.4 the signaling protocols in the IP-centric control plane are discussed. In Section 14.5 we discuss optical internetworking, and in Section 14.6, describe an IP-centric control plane. Section 14.7 concludes the chapter.

14.2   NETWORK CONTROL

A critical challenge in an optical network is to provide the capability to instantiate and route lightpaths in real time or almost real time and to support a variety of protection and restoration capabilities required by the rapidly evolving Internet. This requires efficient control mechanisms to dynamically establish and tear down lightpaths. The control protocols have to deliver real-time connection provisioning and minimize blocking probability. Network control (NC) can be classified as centralized or distributed [2]. In centralized NC, the route computation/route control commands are implemented and issued from one place. Each node communicates with a central controller and the controller performs routing and signaling on behalf of all other nodes. In a distributed NC, each node maintains partial or full information about the network state and existing connections. Routing and signaling are performed independently at each node. Therefore, coordination between nodes is needed to alleviate the problem of contention.

Basically, there exist two different NC approaches for the telecommunication network and the Internet (IP network) [3,4]. In telecommunication networks, NC is tightly coupled with network management and is implemented as a part of a management system called network control and management (NC&M) within the telecommunication management network (TMN). On the other hand, the Internet separates management and control and focuses more on automatic control.

With WDM being extending into a network-layer technology, both the telecommunication and Internet NC approaches are being studied for their applications to optical NC. By reusing the existing know-how, a substantial cost and time saving can be made rather than developing and standardizing a completely new NC system. Additionally, telecommunication and Internet NC have undergone a long period of development and testing to become robust and reliable, making them well suited for adoption for optical network control.

14.2.1   Telecommunication NC&M

The traditional telecommunication NC&M is a well-structured, layered management system. It generally adopts a centralized approach to facilitate optimization of resource use, such as traffic engineering. The telecommunication NC&M can also adopt to the distributed approach, such as recent work on developing a distributed telecommunication NC&M-based control and management network for MONET

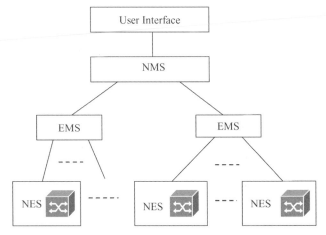

Figure 14.1   Telecommunication NC&M structure.

[5–7]. In the telecommunication NC&M, the network elements (NEs) are generally not aware of the entire network topology and focus on their own operation. The NC&M functions are separated into three layers: network element system (NES), element management system (EMS), and network management system (NMS), as shown in Fig. 14.1. NES controls and manages the local network element. EMS manages multiple NES, usually all NESs in a subnetwork. NMS manages the entire administration domain, which may contain multiple subnetworks. The interfaces between EMS and NES and between EMS and NMS are well defined. The operator interacts with NMS through a user interface (typically, a graphic user interface). This hierarchy is well known in the telecommunication community as the telecommunication management network (TMN). The lightpath is provisioned through the management system and the provisioning function resides in the connection management component of TMN. Due to the hierarchical structure of TMN, lightpath provisioning generally utilizes a hierarchical algorithm. The topology and resource discovery can be automatic but generally are coordinated by TMN. That is, typically, topology and resource usage information is exchanged between EMS and NES. Through the coordination of EMS, topology and resource use in a subnet are collected and sent to NMS. Thus, NMS has global topology and resource use information (which may be summarized). The route computation and signaling are also typically hierarchical, starting from NMS and recurring downward to the NES. For more details, the reader is referred to [5] to [8].

14.2.2   Internet Network Control and MPLS Traffic Engineering Control Plane

Since its inception, the Internet has employed a distributed NC paradigm and has focused on automation in provisioning, flexibility, and controllability. The Internet

NC consists of many loosely coupled protocols. The functionalities of topology discovery, resource discovery and management, and path computation and selection are achieved through routing protocols. Since the Internet is connectionless, it does not need the signaling function. Failure detection and recovery are also achieved through routing or rerouting, which is a very robust mechanism, although routing convergence could be a problem for critical traffic. The Internet NC is completely automatic except for some initial configurations of the routers.

Multiprotocol label switching (MPLS) [9] has been proposed by the IETF, the research organization of the Internet community, to enhance classic IP with virtual circuit-switching technology in the form of a label-switched path (LSP). The traffic engineering capability of MPLS attracts special attention from service providers. MPLS proposes a well-structured and flexible control framework for the Internet, which is also called an MPLS traffic engineering control plane (or MPLS control plane, for short) in the literature. The MPLS control plane consists of the following components: resource discovery, state information dissemination, path selection, and path management, as shown in Fig. 14.2. Resource discovery is a mechanism to exchange resource use information (including topology information) among network elements. State information dissemination is utilized to distribute relevant information concerning the state of the network, including topology and resource availability information. In the MPLS context, this is accomplished by extending conventional IP link-state interior gateway protocols (IGPs) to carry additional information in their link-state advertisements. The path selection component selects an appropriate route inside the MPLS network for explicit routing. Path management, which includes label distribution, path placement, path maintenance, and path revocation, is used to establish, maintain, and tear down LSPs in the MPLS domain. Path management is implemented through a signaling protocol, such as the resource

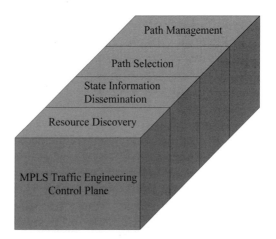

Figure 14.2 MPLS traffic engineering control plane.

reservation protocol (RSVP) extensions [10] or through constraint-based routing LSP setup (CR-LDP) [11]. These components of the MPLS traffic engineering control plane are separable, and independent of each other. They define the functionality primarily and are not restricted to specific protocols or algorithms. This is a very attractive feature because it allows an MPLS control plane to be implemented using a composition or synthesis of best-of-breed modules and facilitates parallel evolution of various components.

Due to many isomorphic relationships between the MPLS and optical networks, IETF has proposed extending the MPLS traffic engineering control plane into the optical layer, called the multiprotocol lambda switching (MPλS) control plane [12]. In the MPLS network, the label-switching router (LSR) uses the label-swapping paradigm to transfer a labeled packet from an input port to an output port. In the optical network, the OXC uses a switch matrix to switch the data stream (associated with a lightpath) from an input port to an output port. Both the LSR and OXC need a control plane to discover, distribute, and maintain network state information and to instantiate and maintain the connections (LSPs/lightpaths) under various traffic engineering rules and policies. The explicit LSP and lightpath exhibit certain common characteristics. Both of them are the abstractions of unidirectional point-to-point virtual path connections. The payload carried by both the LSP and the lightpath is transparent along their respective paths. They can be parameterized to stipulate their performance, behavioral, and survivability requirements from the network. There are also certain similarities in the allocation of labels to the LSP and in the allocation of wavelength to the lightpath.

Certainly, the OXC and LSR have certain differing characteristics. The LSR can perform certain packet-level operations in the data plane, whereas the OXC cannot. The switching operation of LSR depends on the explicit label information carried in the data packets, while the switching operation of OXC depends on the wavelength or optical channel transporting the data packets (i.e., the OXC switches the individual wavelength or the optical channel and does not need any information from the data packets). The label-stacking concept in MPLS is also not available (and not needed) in the optical domain.

This extension work of MPLS is going through an additional step to support various types of switching technologies of the underlying networking gears and is called generalized multiprotocol label switching (GMPLS) [13,14]. Currently, the following switching technologies are being considered.

1. *Packet switching.* The forwarding capability is packet based. The networking gear is typically an IP router.
2. *Layer 2 switching.* The networking gears (typically, Ethernet, ATM, or frame relay switches) can forward data on a cell or frame base.
3. *Time-division multiplexing* (or time-slot switching). The networking gears forward data based on the time slot and the data are encapsulated into the time slots. The Synchronons Optical Network/Synchronons Digital Heirarchy (SONET/SDH) digital cross-connect (DCS) and add–drop multiplexer (ADM) are examples of such gears.

4. *Lambda switching.* λ-switching is performed by OXCs.
5. *Fiber switching.* The optical switching granularity is in the unit of a fiber, not a wavelength. The networking gears are fiber-switch-capable OXCs.

The MPλS control plane focuses on λ-switching, and GMPLS encompasses almost the full range of networking switching technologies and attempts to apply a unified control plane for all such networks. This extension brings some complexity and restrictions to the original MPLS. However, the tremendous benefit brought by GMPLS is obvious, such as network management simplification, cost reduction, and effective integration, and internetworking of different networks. In GMPLS, an LSP can be set up only between the networking gears (more specifically, the interfaces) of the same switching type. Except for the label-stacking capability available in MPLS for packet-switching LSPs, a natural hierarchy of LSP stacking exists in GMPLS for LSPs of different switching type (i.e., packet-switching LSPs over layer 2 switching over time-slot switching over λ-switching over fiber-switching LSPs, as shown in Fig. 14.3). Of course, this hierarchy is not strict. For example, the packet-switching LSP can be stacked directly over a time-slot switching circuit or even the λ-switching lightpath. In the remainder of the chapter we focus on the MPλS and the optical network control part of GMPLS. However, whenever the specific GMPLS concerns rises, we give a brief introduction to GMPLS in terms of those concerns.

In addition to IETF, there are other industry forums working in this area, such as the Optical Domain Service Interconnect (ODSI) and the Optical Internetworking Forum (OIF). ODSI was initiated by a group of service providers and equipment vendors to rapidly develop a user-to-network interface (UNI) for service provisioning and delivery between the optical core network and the client network (e.g., the IP network). Similar to ODSI, the current focus of OIF is also on the UNI interface. Network control inside the optical core has not been addressed yet by either of the organizations. On the other hand, the IETF proposes to extend the MPLS control framework across both the UNI and the optical core network to form a unified control plane over the optical network and client networks (indeed, the client networks have a peer relationship with the optical network and, thus, essentially, are

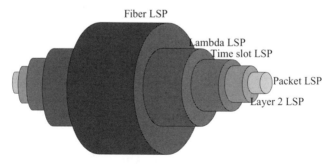

Figure 14.3 GMPLS label-stacking hierarchy.

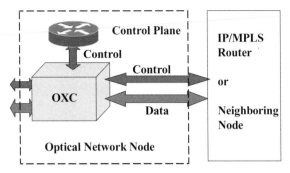

Figure 14.4   Optical node model.

no longer the clients of the optical core network). The goal of this chapter is to describe control and signaling inside the optical network. Therefore, the remainder of this chapter focuses mainly on work being done in the IETF.

14.2.2.1   Optical Node Model   In this chapter we assume that an optical node consists of an OXC and an optical network control plane, as shown in Fig. 14.4. Between two neighboring nodes, there is a preconfigured control channel, which may be in-band or out-of-band, to provide IP connectivity for transmission of control messages such as routing or signaling messages. The switching function in an OXC is controlled by appropriately configuring the cross-connect fabric. Conceptually, this is like setting up a cross-connect table whose entries have the form ⟨input port, input wavelength, output port, output wavelength⟩, indicating that the data stream incoming from the input wavelength at the input port will be switched to output wavelength in the output port.

14.3   MPλS/GMPLS CONTROL PLANE FOR OPTICAL NETWORKS

It is believed that IP traffic will dominate other traffic (e.g., voice) in the future data networks. In fact, many service providers have reported that in their networks, IP traffic has already surpassed legacy voice traffic. Therefore, the optical network is becoming a transport network for IP traffic (i.e., IP over optical), as IP over ATM did when ATM emerged. Partly due to the lessons learned from mapping IP networks over ATM, there is a consensus in the research community and industry that an IP-centric optical control plane is the best choice to make the optical network serve better as an IP transport network. Qiao and Yoo [15] were the early proponents of IP-centric control over optical networks. Recently, there have been many ongoing research efforts in the context of MPλS/GMPLS in this area [12,16–18]. The MPλS/GMPLS control plane for optical networks contains routing, signaling, constraint-based routing shortest path first (CSPF), and restoration management, as shown in Fig. 14.5. Next we discuss briefly the components noted above and some issues of the MPλS/GMPLS control plane.

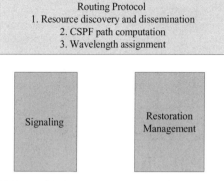

Figure 14.5 MPλS/GMPLS control plane for optical networks.

14.3.1 Resource Discovery and Link-State Information Dissemination

Lightpath routing in optical networks requires topology and resource use information. Hence, it is important for each optical node to broadcast the local resource use and neighbor connectivity information to other nodes so that each node has global topology and resource information (which may be summarized). In the MPλS/ GMPLS control plane, the routing protocol is used to address this issue. Two standard IGP protocols, open shortest path first (OSPF) [19] or intermediate system to intermediate system (IS-IS) [20], are adopted with extensions [21,22] as the routing protocol candidates in the MPλS/GMPLS control plane. Based on the manual initial configuration, the routing protocol can employ a neighbor discovery mechanism to find the neighbor connectivity and resource information, such as the number of ports, the peering nodes/ports, the number of wavelengths per fiber, and the channel capacity. Such information will be attached to a link-state advertisement by the routing protocol to disseminate to the network.

Neighbor discovery is used by a node to discover the state of local links connecting to all neighbors and is a crucial part of the routing protocol. Neighbor discovery generally requires some manual initial configuration and an automated protocol running between adjacent nodes when the nodes are in operation. The neighbor discovery typically requires in-band communication on the bearer channels. For opaque OXCs with SONET termination, one instance of neighbor discovery protocol is needed on every pair of connecting ports between adjacent nodes. The SONET overhead bytes are used as the communication channel. For the transparent OXC, neighbor discovery generally utilizes a separate protocol such as the link management protocol (LMP) [23].

The optical network raises new problem for link addressing and link-state advertisement. In MPLS networks, there are typically a few links between two LSRs. In general, the link is uniquely addressed through the IP addresses of the two end interfaces on the routers. In optical networks, there may be many parallel fiber links, with each fiber containing tens or even hundreds of wavelength channels, between two optical nodes to support the huge bandwidth in the core network. Each wavelength channel may need to be considered as a "link" from the routing point of view. Therefore, there can be thousands of links between two OXCs, which is much larger than the number of links between two routers in MPLS. This raises a scalability problem for link addressing and link-state advertisement. Allocating a unique IPv4 address for each interface (tranceiver) is impractical, due to the address space exhaustion problem. Even for IPv6, which has a huge address space, the link-state advertisement problem still exists. The overhead would be a prohibitively high to flood link-state information for thousands of links between each neighboring node pair. Furthermore, the link-state database would be formidably large, so that the processing and storage would be extremely difficult, if not impossible. GMPLS utilizes two technologies to address this problem: unnumbered links and link bundling. An unnumbered link does not have a unique IP address for itself. It can be addressed through a local "selector" which is unique for the end node (OXC). Thus an unnumbered link can be addressed globally through the globally unique end-node ID (LSR ID) plus the local selector. Link bundling technology is used to dramatically reduce link-state advertisement overhead and the size of the link-state database. The link attributes of multiple wavelength channels or even fiber links of similar characteristics can be aggregated or summarized to represent a logical bundled link. Thus link-state updates propagate the resource information of those "bundled" links, such as end nodes, available/allocated wavelengths, channel capacity, and attributes. Such information is propagated periodically throughout the same IGP administration domain. Thus, the link-state database at each node has the up-to-date network topology and resource use information required for path computation in lightpath routing. The link bundling does have some restrictions. For the links to be bundled into one link bundle, they must originate and terminate at the same end-node pair and have the same link type, traffic engineering property, and administration policy characteristics.

14.3.2 CSPF Path Computation

In an MPLS traffic engineering control plane, a path is computed using the shortest path first (SPF) algorithm, subject to some specified resource and policy constraints to achieve the traffic engineering objective. The algorithm is called the *constraint-based SPF* (CSPF) [24]. Such a path computation is NP-complete and the heuristics have to be used. The constraints are typically the bandwidth required for the lightpath, perhaps along with administrative and policy constraints. The objective of path computation could be to minimize the total capacity required for routing lightpaths.

In optical networks, two new concerns arise for the CSPF algorithm: link bundling and restoration path computation. Due to the link bundling, some information will be lost, making it difficult to compute a restoration path that requires physical diversity from the primary path. Therefore, the link bundling has to keep enough information for restoration path computation, which typically employs shared risk link group (SRLG) information. An SRLG is an administration group associated with some optical resources that probably share common vulnerability to a single failure [25]. For example, all fibers in the same conduit can be assigned with one SRLG identifier because they all will probably fail under a conduit cut. A link can be in one or more SRLGs. To keep the link SRLG information, a link bundle generally contains only the links belonging to the same set of SRLGs. Consequently, when a primary path includes a specific link bundle as one hop, no matter which specific link will be used in the bundle during signaling, the SRLG information regarding the link can be recognized since all links share the same set of SRLGs.

14.3.3 Wavelength Assignment

In optical networks, the transparent OXC generally can only cross-connect a wavelength channel in the input port/fiber to the same wavelength in the output port/fiber. This is called the *wavelength continuity constraint* in the optical network. However, for opaque OXCs, it is easy to cross-connect a wavelength in the input port to a different wavelength in the output port due to its 3R (regeneration, retiming, and reshaping) operations. This is called *wavelength conversion*. The recent optical technology advance even enables the transparent OXC to have the wavelength conversion capability, although with a high cost. In the optical network with wavelength conversion capability, the wavelength assignment is a trivial problem. When receiving a lightpath setup message, every node along the path can independently select an available wavelength in the downstream link and record it in the setup message.

In an optical network without wavelength conversion, the wavelength assignment has to select the same wavelength along the entire path. The wavelength assignment can be taken as a constraint to the CSPF algorithm to compute a lightpath (a path plus the assigned wavelength), or can be performed independently at the source node through some heuristics, such as first fit and random, after the path has been computed. Depending on the information availability in the link-state database, there are two scenarios. If the source node has complete wavelength use information throughout the network, wavelength assignment can be done at the source. If the source node has only summarized wavelength use information, wavelength assignment needs the coordination of all nodes along the path. In this *summarized usage scenario*, a random wavelength assignment may be used (i.e., randomly picking a wavelength and then sending out the lightpath setup message). When the source receives a negative acknowledgment (NAK), the source randomly picks another wavelength and resends the lightpath setup message. This cycle repeats until successful lightpath establishment or a preset threshold number of retries has been reached. The method may be very inefficient when the wavelength use is high in a network.

A more feasible approach is dynamic wavelength reservation which is tightly coupled with the signaling [26]. The basic idea is as follows. The source sends out the lightpath setup message, which includes a vector to record wavelength availability. The vector initially records the available wavelengths in the first hop in the path. Upon receiving the setup message, every node along the path checks the available wavelengths for the outgoing link. If a recorded wavelength in the vector is not available in the outgoing link, this wavelength is removed from the vector. This procedure continues until the destination node receives the setup message. Now the destination node has complete knowledge of available wavelengths along the entire path and uses a heuristic to pick any one of them to assign to the lightpath. The lightpath setup message then traverses the reverse direction of the path toward the source node. The path establishment, such as OXC table config-uration, can be performed at this time. Alternatively, the intermediate node only relays the setup message to the source, and the source issues the second setup message, which contains the assigned wavelength now. There are many variants of this method. The reader is referred to [26] and to Chapter 12 for more compre-hensive information.

An optical network may be partially wavelength conversion capable, which means that only some nodes possess the wavelength conversion capability or that the wavelength conversion has some limitations (e.g., the wavelength conversion can be performed from one range to another range, not from a wavelength to any other wavelength). This situation further complicates the wavelength assignment problem. However, the approaches in the optical network without wavelength conversion can be adapted to solve the problem.

14.3.4   Restoration Management

Survivability is a critical issue in the optical network considering the huge band-width of a lightpath. Generally, the IP layer also provides robust failure restoration. But the restoration time scale, generally on the order of tens of seconds or even minutes, is not acceptable for mission-critical traffic. Although the MPLS restora-tion scheme substantially reduces the time scale, it is still not as fast as the optical layer restoration, which takes only tens of milliseconds. In addition, optical layer restoration has many other merits compared with MPLS/IP layer restoration. For example, optical layer restoration operates on the scale of a lightpath or even a link. In terms of messaging overhead, this is very efficient and saves lots of restoration messages because a link or a lightpath in the optical network may contain many LSPs. On the other hand, some client layers, such as SONET, provide high-quality restoration. Such client layers typically require the optical layer not provide restoration, to avoid interference. Hence the management and coordination of restoration among multiple layers is an important issue for the control plane.

The optical protection/restoration mechanisms can be classified into path protection or link protection. Link protection is local protection for a link provided by the two end nodes of a link. Path protection considers protection in terms of lightpaths and can be further classified into two approaches. The first approach, called *disjoint path protection*, employs an edge-disjoint protection path to reroute the client traffic when the working (primary) path carrying the traffic fails. The second approach, called *link-dependent path protection*, uses different (and typi-cally non-edge-disjoint) protection paths to protect the working path, depending on where the failure happens in the working path. The protection path can be precomputed or dynamically selected when a failure happens. There are two major types of *disjoint path protection*: $1 + 1$ and $1 : 1$. The $1 + 1$ protection is called *dedicated protection* because for each working lightpath, a dedicated protection lightpath needs to be set up (generally, the client traffic goes to both the working and protection paths and it is the destination's decision to receive from either the working or the protection path). There are two types of $1 : 1$ protection. In the first case, a dedicated protection path is required, but the protection path can carry low-priority traffic. In the second case, the protection path is not dedicated and multiple protection lightpaths can share the same λ as long as there is no shared link among their associated working lightpaths. The $1 : 1$ protection can be extended into $M : N$ protection, which uses N protection lightpaths to protect the traffic on M working lightpaths.

To recover from a failure, restoration management contains three functional components: failure detection, failure notification, and failure restoration. A link or node failure can be detected by the hardware when lower-layer impairments such as loss of signal occur, or by the higher (IP) layer via the link-probing mechanism. The link-probing mechanism typically employs an approach as in the OSPF *hello* message mechanism or the TCP keep-alive mechanism. When there is no message coming from the peering node within a prenegotiated period, or there is no acknowledge (ACK) message coming from the peering node after a node sends a certain number of keep-alive messages, the link is assumed to be in the fault status. The failure notification utilizes the signaling protocol to broadcast the link failure to the traffic source node, or even throughout the network. To reduce the large number of messages flying in the network during failure notification, the propagation mechanism has to be designed carefully, such as using a broadcast tree. The failure restoration phase contains the restoration path computation (if not precomputed), establishment (if not a $1 + 1$ scheme), and traffic rerouting from the primary path to the restoration path.

14.3.5   Signaling

In path signaling, a critical issue is network element addressing. To specify a signaling path and send the signaling messages to the correct intermediate and destination nodes, the network element has to have a unique address. In addition, to perform the correct path signaling, many other entities need to be addressed, such as optical links, bundled links, optical channels (i.e., wavelength) and subchannels (in TDM). In IP networks, a router generally has several ports and every port is assigned an IP address. In addition, the router is assigned a unique router ID in the routing domain. In optical networks, the number of entities needing to be addressed is several orders of magnitude bigger than that in the IP network. It is neither practical nor necessary to assign a unique IP address to every entity. For example, the routing of a lightpath does not need to depend on the termination point information of the intermediate node. Therefore, in an MPλS control plane, it is reasonable to assign unique IP addresses only to the OXCs. Other entities are addressed through "selectors" associated with the OXC address. The port selector identifies a port in a given OXC and the channel selector identifies a channel in a given port of a given OXC. The subchannel selector further identifies the subchannel in a given channel. Such a selector hierarchy uniquely addresses any network entities in a network.

Signaling is a distributed path establishment operation across the optical network. Once the path and wavelength are selected, all OXCs along the path must establish appropriate cross-connects in a coordinated fashion. After the appropriate configuration of the OXCs' cross-connect tables, the end-to-end lightpath becomes available to transmit client traffic. The end-to-end lightpath signaling includes the following three major operations: lightpath setup, teardown, and abort. The lightpath setup and teardown operations are generally initiated from the source node and the lightpath abort is typically initiated by a node detecting the failure.

The lightpath setup operation conceptually requires three messages: SETUP, SETUP_ACK, and SETUP_NAK, illustrated in Fig. 14.6. The SRLG information is used for shared protection path reservation. Figure 14.7 shows the generic lightpath setup operation by the coordinated nodes along the path. Between the source and destination nodes, there are two intermediate nodes, INT_A and INT_B. There are two phases for lightpath setup signaling: lightpath reservation and lightpath commitment. When the source node (which may be a router) receives a call request, it first computes an explicit path and assigns a wavelength. Then the source node reserves the wavelength resource in the outgoing link and invokes the path signaling procedure by sending a SETUP message to the next hop along the path. The lightpath signaling goes into the reservation phase. The SETUP message proceeds to INT_A and INT_A checks the resource availability. If the assigned wavelength in the SETUP message is available, INT_A reserves it and sends the

type	ID	path	λ	Prot_type	Other parameters

Message type
1. SETUP
2. SETUP_ACK
3. SETUP_NAK

Prot type
1. "1+1"
2. "1:1"
3. others

ID: domain unique identifier for a lightpath

Other parameters:
1. Traffic engineering constraints
2. SRLG information

Figure 14.6 Sample lightpath setup messages.

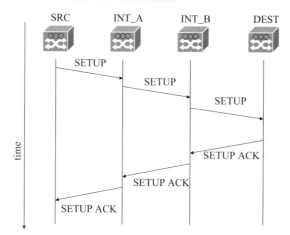

Figure 14.7   Lightpath setup operation.

SETUP message to INT_B. INT_B performs the same operations as INT_A, and as a result, the SETUP message proceeds to the destination node. The destination determines whether or not to accept this lightpath setup. The destination does not need to reserve the resource. However there may be other constraints to be considered when deciding to accept a lightpath setup (e.g., the administrative policy). When the destination decides to accept the lightpath setup, it sends a SETUP_ACK message back to INT_B in the reverse direction of the path. At this stage, the lightpath setup goes into the second phase: the lightpath commitment phase. INT_B looks up the resource reservation table for this lightpath and issues a command to configure the OXC to connect the incoming wavelength channel from the incoming link to the outgoing wavelength channel in the outgoing link. Note that if this is a network without wavelength conversion, the incoming and outgoing channels should be the same. At the same time, the SETUP_ACK message is passed to INT_A and INT_A performs the same operations. The SETUP_ACK finally proceeds to the source node. Now the lightpath from the originating port of the source to the terminating port of the destination has been connected completely from end to end and is ready to be used to transmit traffic. At each node, some data structures are required to maintain the lightpath information. For example, a lightpath table may be needed to record the lightpath ID, incoming/outgoing port/channel, SRLG information, and so on.

However, the lightpath setup operation may fail at some place, for many reasons. The path computation and wavelength assignment may fail at the source itself, and consequently, the signaling procedure does not initiate at all. During the lightpath reservation phase, the wavelength resource may not be available (reserved by another lightpath setup a little earlier). In this case, the intermediate node sends back a SETUP_NAK message to the source in the reverse direction of the signaling path and all nodes along the reverse path release the reserved wavelength resource.

During the lightpath commitment phase, there may be a failure such that the reservation table does not include this lightpath. In this situation, a lightpath abort message is sent to the upstream and downstream nodes toward the source and destination.

The lightpath teardown operation is similar to the lightpath setup. The first phase is the resource release phase, where each node marks the used wavelength by the lightpath as available. There are some other processing operations, such as the SRLG information clear and cross-connect table clear. The OXC itself does not need to be configured during the teardown operation. When the destination receives the TEARDOWN message and sends a TEARDOWN_ACK message in the reverse path, each node only relays the message toward the source.

14.3.6   GMPLS Signaling Functional Requirements

In GMPLS, the two end nodes (LSRs) of an LSP should have the same switching capability (e.g., both are OXCs). GMPLS extends MPLS signaling in many aspects (e.g., supporting nonpacket/noncell switching links, label suggestion capability by the upstream node, restricting label selection range of the downstream node, bidirectional LSP setup, label control over a specific interface, supporting administration policy, separation of control and data channels, and inclusion of technology-specific parameters).

In GMPLS, the *generalized label request* from upstream to downstream should carry the LSP encoding type, required link switching type, and payload identifier. The LSP encoding type indicates the encoding of LSP, such as lambda (lightpath) or packet (traditional LSP). The payload identifier defines the client layer that will use this LSP. GMPLS also defines a bandwidth encoding to match the discrete bandwidth allocation capability, such as in SONET. The *generalized label* is defined with enough flexibility to have the power to represent a label for different switching types, such as a fiber, a waveband, a wavelength, a frame relay DLCI, and so on. For waveband switching, the waveband ID, start, and end labels are needed to define the *generalized label*. The *suggested label* mechanism is used to expedite the LSP setup. When the upstream node provides a suggested label to the downstream, it can begin to configure its switching fabric as soon as possible, assuming that the downstream will select the label suggested. In optical networks with long-configuration-time OXCs, this can substantially reduce the LSP setup delay compared with waiting for the label from downstream and then starting to configure the switching fabric. Of course, if the downstream node has selected a different label and later distributed to the upstream, the upstream node has to reconfigure the switching fabric according to the new label supplied by the downstream. The *label set* is used by the upstream to restrict label selection of the downstream to a set of acceptable labels. A *label set* consists of one or multiple subobjects or TLVs. Each subobject/TLV can contain a set of labels or range of labels to be excluded or included for label selection. The *label set* mechanism is useful in many cases, such as lightpath (GMPLS LSP) setup in an optical network without wavelength conversion.

GMPLS supports bidirectional LSP setup and provides a contention resolution mechanism when two bidirectional LSP setup requests coming from the opposite direction contend for the same label. GMPLS also enhances the notification of label error by supplying a set of acceptable labels in the notification message. The *explicit label* control offers the capability of explicit label selection on a specific hop on an explicit route. GMPLS signaling should be able to carry the link protection information and administrative status information by defining new objects or TLVs.

In GMPLS, data channels and control channels may be separate (which is usually the case in optical networks). Therefore, signaling should have the capability to identify specific data channels in signaling messages transmitted on the control channel. An object or TLV needs to be defined to indicate the specific data link(s) through using the IP address (packet-switching link) or LSR ID plus a locally unique selector (e.g., for wavelength channels). The separation of data and control channels results in a more complex situation for fault handling. There are three failure scenarios: the data channel failure, control channel failure, or both. The GMPLS signaling for fault handling should minimize the loss. For example, if only the control channel fails, the LSPs on the data channel should not be disrupted (such as rerouting), and when the control channel is restored, the state information should be synchronized between the upstream and downstream nodes.

14.3.7   Traffic Engineering Extension

The MPLS discussed in this chapter is assumed traffic engineering capable [24]. A traffic engineering link in MPLS is a link on which an IGP (e.g., OSPF) routing adjacency is established. Such a link has two metrics, the regular link metric used in traditional IP routing, and the traffic engineering link metric used for constraint-based routing. GMPLS complicates the concept of the traffic engineering link. First, some links may have traffic engineering metrics, but the routing adjacency cannot be established on such links (e.g., those links are SONET links). Second, an LSP (e.g., lightpath) may be taken as a traffic engineering link, but the routing adjacency does not need to be established directly between the two end nodes of the LSP. Third, a link bundle can be advertised as a traffic engineering link. As a result, the one-to-one mapping of a routing adjacency and a traffic engineering link in the MPLS is broken in GMPLS. Therefore, in GMPLS, a traffic engineering link is a logical link with traffic engineering properties. The management of a traffic engineering link can be conducted through LMP.

14.3.8   Adjacencies

In GMPLS, there are three types of adjacencies: routing, signaling, and forwarding. The *routing adjancency* represents an adjacency relationship between two nodes with regard to the routing protocol (e.g., OSPF), and two such nodes are said to be *neighbors* of the routing protocol. The *signaling adjacency* represents a peering relationship of two nodes established by the signaling protocol (e.g., CR-LDP or RSVP). Both routing and signaling adjacencies are not necessarily established between physically connected nodes, although this is generally the case. The *forwarding adjacency* is a traffic engineering link that transits three or more GMPLS nodes in the same instance of a GMPLS control plane. A forwarding adjacency can be represented as unnumbered or numbered links, or even a bundle of LSPs with the same end nodes. There is not one-to-one mapping between any two of the three adjacencies. The routing adjacency is generally not established over the traffic engineering links of a forwarding adjacency. However, if a forwarding adjacency traffic engineering link is used as a tunnel to establish LSPs over it, a signaling adjacency needs to be set up over the traffic engineering link between its two end nodes.

14.3.9 Link Management Protocol

In GMPLS networks, the meaning of a link can be, for example, a packet-switched link (e.g., a serial link between two routers), a SONET circuit, a wavelength, a fiber, an LSP in the same layer, a bundled link, or a bundled LSPs with the same type. Therefore, due to the extreme complexity of link management in optical networks, *link management protocol* (LMP) is defined as a companion of GMPLS. In the context of optical networks, LMP can run between OXCs and, between an OXC and an optical line system (e.g., WDM multiplexer). LMP is a protocol supporting control channel management, link connectivity verification, link property correlation, and fault management between adjacent nodes.

LMP control channel management is used to establish and maintain control channels between two nodes to exchange routing, signaling, and other management messages. An *LMP adjacency* can be formed between two nodes supporting the same LMP capabilities. LMP assumes that the control channel is manually configured. However, the LMP peers of an LMP adjacency can negotiate the configuration of channel capabilities. The control channel may be in-band (coupled with a data-bearing channel) or out-of-band through a separate control network. Multiple control channels can be active simultaneously for one LMP adjacency, and each channel negotiates its capability parameters separately and maintains the connectivity through exchanging hello messages. The optional LMP link connectivity verification is used to verify the physical connectivity of data-bearing links and to exchange the link identifiers used in signaling. The links to be verified must be opaque. Verification is conducted through sending test messages in-band over the data-bearing links. LMP link property correlation is used to aggregate multiple data-bearing links into one bundled link and to exchange, correlate, or change traffic engineering parameters of those links. This allows adding component links into a link bundle or changing the properties of a specific link. LMP fault management supports a fault localization process. A downstream LMP peer that detects data link failures will send an LMP message to its upstream peer, which can then correlate the failure with the corresponding input ports to determine if the failure is between the two nodes.

14.4 SIGNALING PROTOCOL

In Section 14.3.5 we described a generic signaling procedure to set up a lightpath in the optical network. However, a practical and robust signaling protocol in the MPλS/GMPLS control plane needs many more functions and capabilities. To have a complete understanding of the issues and procedures in signaling, it is necessary to introduce a standardized signaling protocol. In the MPLS control plane, there are two major signaling protocols: the label distribution protocol (LDP) [27] and RSVP [10]. Both protocols have been extended to support traffic engineering [24], which has been generalized as constraint-based routing and named *constraint-based routing LDP* (CR-LDP) [11] and *RSVP with traffic engineering extension* (RSVP-TE) [28], respectively. Now, CR-LDP and RSVP-TE are being enhanced to support optical network signaling and GMPLS control plane signaling. We introduce those protocols in this section.

14.4.1 LDP Signaling Protocol in the MPLS Control Plane

The formal description of LDP has been defined in [27], which is a large document containing about 130 pages. Hence, it is impossible to cover it here in depth. Instead, we highlight the main mechanisms, procedures, and messages to perform the signaling functionality. Although critical, issues such as security, loop detection, and fault notification/processing are omitted. LDP is the set of procedures and messages by which LSRs establish LSPs through a network by mapping network-layer routing information directly to data-link-layer switched paths. These LSPs may have an endpoint at a directly attached neighbor (comparable to IP hop-by-hop forwarding) or may have an endpoint at a network egress node, enabling switching via all intermediate nodes. LDP associates a forwarding equivalence class (FEC) with each LSP it creates. The FEC associated with an LSP specifies which packets are mapped to that LSP. LSPs are extended through a network as each LSR "splices" incoming labels for a FEC to the outgoing label assigned to the next hop for the given FEC.

14.4.1.1 LDP Peer Discovery and Session Establishment   Before two LSRs can communicate using LDP, a LDP session has to be established between them and, after the session establishment, the two LSRs are called *LDP peers*. An LSR employs discovery mechanisms to discover the potential LDP peers. There are two discovery mechanisms: a basic discovery mechanism and an extended discovery mechanism. The basic discovery mechanism is used to discover potential LDP peers that are connected directly at the link level, and the extended discovery mechanism is utilized to locate potential LDP peers that are not directly connected at the link level. In both mechanisms, the LSR sends the hello messages (link hellos/targeted hellos) to the potential peers, and the hello messages carry the label space information between the LSR and its peer.

After the hello message exchange between an LSR and its potential LDP peer, the LSR can establish an LDP session with the potential peer through a two-step procedure: transport connection establishment and session initialization. The LDP session is established over the transport layer (i.e., TCP layer) and a TCP connection has to be set up at first between the two LSRs. The address information required for the TCP connection establishment is carried in the hello messages through a transport address optional object or simply in the source address of the hello message. The LSR compares its own address with the peer's address. If its own address is greater, the LSR plays an active role in session establishment and initiates a TCP connection to the peer. Otherwise, it plays a passive role and waits on the well-known LDP port for the peer's TCP connection request. The peer conducts the same check and determines its role in the session initialization.

After LSR1 and LSR2 establish a transport connection, they negotiate session parameters by exchanging LDP initialization messages. The parameters negotiated include an LDP protocol version, label distribution method, timer values, label range, and so on. It is convenient to describe LDP session negotiation behavior in terms of a state machine. The LDP state machine is defined with five possible states and the behavior is presented in Fig. 14.8 as a state transition diagram.

14.4.1.2 Label Advertisement Mode and LSP Control Mode LDP has two types of label advertisement modes: *downstream on demand*, by which an LSR distributes a label-FEC binding in response to an explicit request from another LSR (typically, upstream LSR), and *downstream unsolicited*, where an LSR can distribute label-FEC bindings to LSRs that have not explicitly requested them. For any given LDP session, each LSR must be aware of the label advertisement

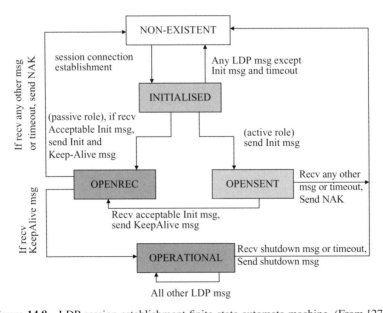

Figure 14.8 LDP session establishment finite-state automata machine. (From [27].)

0 1 2 3 4 5 6 7 8 9 0 1 2 3 4 5	6 7 8 9 0 1 2 3 4 5 6 7 8 9 0 1
U F Type	Length
Value	

Figure 14.9   TLV encoding.

mode used by its peer, and the LSRs exchange the label advertisement mode during the initialization. Each interface on an LSR can be configured to operate in either downstream unsolicited or downstream on demand mode.

The LSP setup behavior of an LSR is determined by the LSP control mode. LDP defines two LSP control modes: independent and ordered. With *independent control*, each LSR may distribute label mappings to its neighbors any time it desires. For example, when operating in independent control mode, an LSR may answer requests for label mapping immediately, without waiting for a label mapping from the next hop. With *ordered control*, an LSR may initiate the transmission of a label mapping only for a FEC for which it has a label mapping for the FEC from the next hop, or for which the LSR is the egress. When an LSR is not the egress and there exists no mapping, it must wait for the label mapping from the downstream before mapping the FEC and passing corresponding labels to upstream LSRs. That is why it is called *ordered control*.

14.4.1.3 LDP TLV Encoding   LDP defines messages and TLVs in four areas: peer discovery, session management, label distribution, and notification of errors and advisory information. A type-length-value (TLV) scheme is used to encode most information carried in LDP messages. An LDP TLV is encoded as in Fig. 14.9. The U bit represents an "unknown" TLV bit. Upon receipt of an unknown TLV, if U is clear, a notification must be returned to the message originator and the entire message must be ignored; if U is set, the unknown TLV is silently ignored and the rest of the message is processed as if the unknown TLV does not exist. The F bit represents a "forward" unknown TLV bit. This bit applies only when the U bit is set and the LDP message containing the unknown TLV is to be forwarded. If F is clear, the unknown TLV is not forwarded with the containing message; if F is set, the unknown TLV is forwarded with the containing message. The value field is an octet string that encodes information to be interpreted as specified by the type field. The value field itself may contain TLV encoding. The TLV encoding scheme is very general. In principle, everything appearing in an LDP PDU could be encoded as a TLV.

1. *FEC TLV.* Labels are bound to FECs. An FEC is a list of one or more FEC elements. An FEC element is a wildcard FEC, an address prefix FEC, or a host address FEC. The wildcard FEC is typically used in the label withdraw or release messages to withdraw or release all FECs associated with a label. The address prefix FEC encodes an address prefix such as an IPv4 address prefix, and the host address FEC encodes a single host address. The FEC TLV is illustrated in Fig. 14.10.

```
0 1 2 3 4 5 6 7 8 9 0 1 2 3 4 5 6 7 8 9 0 1 2 3 4 5 6 7 8 9 0 1
```
0	0	FEC(0x0100)	Length
FEC Element 1			
...			
FEC Element n			

Figure 14.10   FEC TLV.

```
0 1  2 3 4 5 6 7 8 9 0 1 2 3 4 5  6 7 8 9 0 1 2 3 4 5 6 7 8 9 0 1
```
0	0	Generic Label (0x0200)	Length
Label			

Figure 14.11   Label TLV.

2. *Label TLV.* A label is encoded through a label TLV. Figure 14.11 shows the label TLV encoding of a generic label. A generic label is a 20-bit number in a four-octet field. Other labels, such as the ATM label (VPI/VCI) or frame relay label (DLCI), can also be encoded in a label TLV.

14.4.1.4 LDP Messages and Processing Procedure   LDP message exchange is accomplished by sending LDP protocol data units (PDUs) over the LDP session TCP connection. Each LDP PDU includes an LDP header followed by one or more LDP messages. The LDP PDU header is illustrated in Fig. 14.12. The LDP Identifier contains six octets that uniquely identify the label space of the sending LSR for which this PDU applies. The first four octets are the 32-bit router ID assigned to the LSR, and the last two octets identify a label space within the LSR.

All LDP messages have the format shown in Fig. 14.13. The U bit is the "unknown" message bit. Upon receipt of an unknown message, if U is clear, a notification is returned to the message originator. If U is set, the unknown message is silently ignored. The message ID is a 32-bit value used to identify this message.

```
0 1 2 3 4 5 6 7 8 9 0 1 2 3 4 5 6 7 8 9 0 1 2 3 4 5 6 7 8 9 0 1
```
Version	PDU Length
LDP Identifier	

Figure 14.12   LDP PDU header.

```
0 1 2 3 4 5 6 7 8 9 0 1 2 3 4 5  6 7 8 9 0 1 2 3 4 5 6 7 8 9 0 1
```
U	Message Type	Message Length
Message ID		
Mandatory Parameters (TLVs)		
Optional Parameters (TLVs)		

Figure 14.13   LDP message format.

0 1 2 3 4 5 6 7 8 9 0 1 2 3 4 5	6 7 8 9 0 1 2 3 4 5 6 7 8 9 0 1
0 | Label Request (0x0401) |	Message Length
Message ID	
FEC TLV	
Optional Parameters	

Figure 14.14 Label request message.

The mandatory parameters are a variable-length set of required message parameters. The optional parameters are a variable-length set of optional message
parameters. LDP defines 11 messages. The hello, initialization, and keep alive
messages are used for the LDP peer discovery and LDP session establishment.
The address and address withdraw messages are used by an LSR to advertise or
withdraw the interface address to the LDP peer. The label withdraw message is used
for an LSR to signal the LDP peer not to continue to use the specific FEC-label
mappings the LSR has advertised previously. The label release message is used by
an LSR to signal the LDP peer that the LSR no longer needs the specific FEC-label
mappings requested previously from and advertised by the peer. The label request/
mapping messages are used to set up the LSPs by the LSRs along the path in a
coordinated manner. The label abort message is used by an LSR to abort a label
request message in some situations. We will now introduce the label request/
mapping messages and how to set up an LSP by using them. For other messages,
the reader is referred to [27].

1. *Label request message.* The label request message, shown in Fig. 14.14, is
used by an upstream LSR to explicitly request the downstream LSR to assign and
advertise a label for an FEC. The possible optional parameters are the hop count
TLV and path vector TLV used for loop detection. Typically, when an LSR does not
have a label-FEC mapping from the next hop (specified in the forwarding table) for
a given FEC, the LSR will send a label request message to the next hop (which
should be an LDP peer of this LSR). The next-hop LSR receiving the label request
message should respond via a label mapping message with the requested label-FEC
binding, or with a notification message indicating the reason for rejecting the label
request message. When the FEC TLV includes a prefix FEC element or a host
address FEC element, the receiving LSR uses its routing table to determine its
response. Unless its routing table includes an entry that exactly matches the requested prefix or host address, the LSR must respond with a no route notification
message.

2. *Label mapping message.* The label mapping message, illustrated in Fig. 14.15,
is used by an LSR to advertise FEC-label bindings to the LDP peer. FEC TLV
specifies the FEC component, and label TLV specifies the label component in the
FEC-label mapping. The possible optional TLVs are label request message ID TLV,
hop count TLV, and path vector TLV. If this label mapping message is a response to
a label request message, it must include the label request message ID TLV. Hop
count and path vector TLVs are used for loop detection.

0 1 2 3 4 5 6 7 8 9 0 1 2 3 4 5	6 7 8 9 0 1 2 3 4 5 6 7 8 9 0 1
0 | Label Mapping (0x0400)	Message Length
Message ID	
FEC TLV	
Label TLV	
Optional Parameters	

Figure 14.15   Label mapping message.

14.4.1.5 LSP Setup with Ordered Control and Downstream on Demand Mode.

In this section we illustrate how an LSP is set up via label request/mapping messages. In our example, all LSRs are assumed to use ordered LSP control mode and downstream on demand label advertisement mode. As shown in Fig. 14.16, we assume that there is a routing path for the 23-bit address prefix FEC, 128.205.35.0, between an ingress LSR and an egress LSR in an MPLS domain. The FEC is a supernet attached to the egress LSR. We assume that this is a newly found FEC, and no LSRs have label mapping assigned for this FEC. LDP is a routing protocol–driven signaling protocol. Hence, when the ingress LSR finds the new FEC, it sends a label request message to its routing next hop for this FEC, which is LSR1 in our

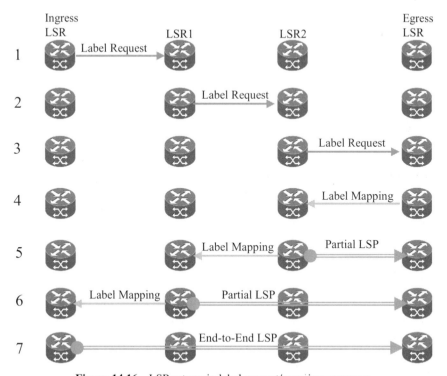

Figure 14.16   LSP setup via label request/mapping messages.

example. LSR1 checks its forwarding table and label information base (LIB). LSR1 does not find any label binding to the FEC 128.205.35.0. The ordered control requires that before LSR1 receives a label-FEC mapping from LSR2, LSR1 cannot send a label mapping message to the ingress LSR. LDP requires that an LSR operating in downstream on demand advertisement mode cannot assume that the downstream will send a label-FEC mapping even if the downstream is operating in downstream unsolicited mode. Therefore, due to this requirement, LSR1 should not just wait and expect the downstream to issue a label mapping message automatically. Instead, LSR1 has to send a label request message to its downstream (LSR2) to request a label-FEC mapping, as illustrated in step 2 of Fig. 14.9. Now LSR2 receives a label request message from LSR1 and behaves similarly as LSR1 since LSR2 also does not have the label mapping and is operating in ordered LSP control and downstream on demand mode. Finally, the label request message proceeds to the egress LSR. The label request message propagation stops at this point. The egress LSR picks an available label in its label space to bind to the FEC and sends the label-FEC binding in a label mapping message to the upstream LSR2. LSR2 records this mapping in LIB and uses this label for the future forwarding operation. Then LSR2 picks an available label from its label space to bind to the FEC on its uplink and sends this label-FEC binding in a label mapping message to LSR1. LSR1 performs similar operations as LSR2, and eventually, the ingress LSR receives a label mapping message corresponding to the label request message issued by itself at the beginning. The ingress LSR records the label-FEC binding in a LIB and uses the label for future forwarding. Now an end-to-end LSP to carry the traffic of an FEC is completely established. The ingress LSR can classify the incoming packets of the FEC 128.205.35.0 and use the LSP to transport them.

14.4.1.6 *LSP Setup with Other LSP Control and Advertisement Mode Combinations* Replacing the ordered control with the independent control mode in the example above does not complicate the signaling procedure much. When an LSR is configured as an independent LSP control node, it can respond to the label request message immediately (e.g., when LSR1 is operating in independent control mode). Then, when LSR1 receives a label request message from the ingress LSR, LSR1 still send the label request message to LSR2 to request the label-FEC binding due to the downstream on demand mode requirement. However, LSR1 does not need to wait for the label mapping message from LSR2. It can just pick an available label and send the label-FEC binding in a label mapping message to the ingress LSR immediately. There is a risk that the ingress LSR will begin to send packets on the LSP segment established between the ingress LSR and LSR1 before LSR1 receives the label-FEC mapping from LSR2. If this happens, LSR1 just extracts the packets and uses the regular layer 3 routing and forwarding to process those packets. After LSR1 receives the label-FEC binding, an end-to-end LSP is completely established and the packets can go through the entire LSP toward the egress LSR.

If an LSR is operating in downstream unsolicited advertisement mode, it can distribute a label-FEC mapping through a label mapping message as it desires. Suppose that LSR2 is operating in downstream unsolicited advertisement mode; it is possible that LSR2 sends a label mapping message binding FEC 128.205.35.0 with an available label to LSR1 when it finds that this is a new FEC. After some time, when the ingress LSR sends a label mapping message to LSR1, LSR1 finds that its LIB already has the label-FEC binding. LSR1 will immediately send a label mapping message to the ingress LSR whether it is operating in independent or ordered LSP control. As for how the LSR2 gets the label-FEC binding from the egress LSR, there are two possibilities. The first possibility is that LSR2 sends a label request message to the egress LSR when it finds that the FEC 128.205.35.0 is a new FEC and the LSR2 receives a label mapping message from the downstream (in either independent or ordered LSP control mode). Another possibility is that the downstream of LSR2 (in this example, it is the egress LSR) is also operating in downstream unsolicited advertisement mode and already sent a label mapping message to LSR2.

Note that in our example, we assume that the ingress LSR initiates the LSP setup. However, in LDP, any LSR can initiate a label request message as long as it finds a new FEC, an FEC next-hop change, or other situation. The label request message can stop anywhere considering the different configurations for the label advertisement mode and the LSP control mode in an LSR. For example, the label request message can stop just at downstream (next hop) of the message originator, and as a result an LSP segment is set up between the two adjacent LSRs. Moreover, any LSR can send a label mapping message to the upstream as it desires to set up an LSP segment between them if the LSR is configured in downstream unsolicited mode. Thus along the routing path of an FEC, there may be many such LSP segments established at the same time or different times. An LSP segment will grow by one hop when a node joins to become the new originating LSR by sending a label request message to the originating LSR of the segment, or to become the new termination LSR by sending a label mapping message to the termination LSR. Two LSP segments will be spliced into one LSP segment when an LSR is the termination LSR of one segment and the originating LSR of another. Eventually, the LSP segments will be spliced into an end-to-end LSP from the ingress LSR to the egress LSR.

14.4.2   CR-LDP

The LSP setup using the LDP signaling protocol is driven completely by the routing protocol. The LSP setup is initiated automatically when the LSRs discover a new FEC. Additionally, the LSP establishment through LDP needs only the routing information (e.g., the FEC and the next hop for the FEC). An important service offered by MPLS technology is to support constraint-based routing, a mechanism to offer the traffic engineering capability to a network. Figure 14.17 illustrates the constraint-based routing procedure. The routing protocol with traffic engineering extension collects the link-state information with the traffic engineering parameters into a traffic engineering database. The user supplies constraints to the CSPF algorithm and CSPF utilizes the traffic engineering database to compute an explicit path to meet the constraints. The explicit path is then passed to the signaling

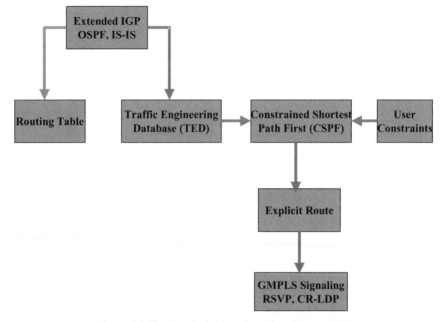

Figure 14.17   Constraint-based routing. (From [24].)

protocol for path establishment. The constraint-based routing requires that the LSP can be set up subject to some constraints, such as explicit route constraints, and/or QoS constraints. Such requirements can be met by extending LDP to support the constraint-based routing LSP (CR-LSP) setup [11]. The extension introduces some new TLV encoding for the constraints and procedures for LDP messages carrying those TLVs. The resulting protocol is called CR-LDP. The CR-LSP setup is typically initiated by the ingress LSR.

14.4.2.1   CR-LDP TLVs   CR-LDP defines the following TLVs: explicit route TLV (ER TLV), explicit route hop TLV (ER-Hop TLV), traffic parameters TLV, preemption TLV, LSPID TLV, route pinning TLV, resource class TLV, and CR-LSP FEC TLV. Those TLVs are also called CR-TLVs. This section introduces ER TLV, ER-hop TLV, and LSPID TLV. For others, the reader is referred to [11].

1. *ER-hop TLV.* Each ER-hop may identify a group of nodes in the constraint-based route. Hence, an ER-hop is also called an *abstract node*. The ER-hop TLV is illustrated in Fig. 14.18. CR-LDP defines four types of ER-hops: IPv4 prefix, IPv6

0 1	2 3 4 5 6 7 8 9 0 1 2 3 4 5	6 7 8 9 0 1 2 3 4 5 6 7 8 9 0 1
0 0	ER-Hop Type	Length
L	Contents	

Figure 14.18   ER-hop TLV.

0	1	2 3 4 5 6 7 8 9 0 1 2 3 4 5	6 7 8 9 0 1 2 3 4 5 6 7 8 9 0 1
0	0	ER-TLV Type (0x0800)	Length
		ER-Hop TLV 1	
		...	
		ER-Hop TLV n	

Figure 14.19   ER TLV.

prefix, autonomous system (AS), and CR-LSP ID (LSPID). An ER-hop of AS indicates that the abstract node is the entire AS domain. An ER-hop of LSPID indicates this ER-hop is using a tunnel LSP as one hop, and the tunnel LSP is specified by LSPID. The L bit indicates the attribute of the ER-hop. If the L bit is set, this is a loose ER-hop. Otherwise, this is a strict ER-hop. The loose or strict ER-hop is interpreted relative to their prior ER-hop. The path between a strict ER-hop and its prior ER-hop must include only network nodes from the strict ER-hop and its prior ER-hop. The path between a loose ER-hop and its prior ER-hop may include other network nodes, which are not part of the strict ER-hop or its prior ER-hop. The contents field contains a network node or abstract node.

2. *Explicit route TLV.* The ER-TLV is an object that specifies the path to be taken by the LSP being established. It is composed of one or more ER-hop TLVs (Fig. 14.19).

14.4.2.2 *CR-LDP Modification to LDP Messages*   CR-LDP extends label request/mapping messages to set up CR-LSP. Figures 14.20 and 14.21 show the extensions encoding to label request/mapping messages. When processing the label request message carrying CR-TLVs, the LSR has to operate in downstream on-demand label advertisement mode and ordered LSP control mode, although the

0	1 2 3 4 5 6 7 8 9 0 1 2 3 4 5	6 7 8 9 0 1 2 3 4 5 6 7 8 9 0 1
0	Label Request (0x0401)	Message Length
	Message ID	
	FEC TLV	
	LSPID TLV	
	ER TLV (optional)	
	Other CR-TLVs (optional)	

Figure 14.20   Extension to label request message.

0	1 2 3 4 5 6 7 8 9 0 1 2 3 4 5	6 7 8 9 0 1 2 3 4 5 6 7 8 9 0 1
0	Label Mapping (0x0400)	Message Length
	Message ID	
	FEC TLV	
	Label TLV	
	Label Request Message ID TLV	
	LSPID TLV (optional)	
	Other Optional TLVs	

Figure 14.21   Extension to label mapping message.

LSR may be configured to operate in other modes for regular LDP messages. The label mapping message must include a single label TLV. Hence a label mapping message is transmitted by a downstream LSR to an upstream LSR under one of the following conditions:

1. The LSR is the egress node of the CR-LSP and an upstream mapping has been requested.
2. The LSR receives a mapping from its downstream LSR for a CR-LSP for which an upstream request is still pending.

14.4.2.3   CR-LSP Setup through CR-LDP   CR-LDP depends on the following minimal LDP behaviors:

1. Basic and/or extended discovery mechanisms
2. Use of the label request/mapping messages in downstream on-demand label advertisement mode with ordered LSP control
3. Use of the notification message
4. Use of withdraw and release messages
5. Loop detection (in the case of loosely routed segments of a CR-LSP) mechanisms

In addition, CR-LDP augments the following functionality to LDP:

1. The label request message used to set up a CR-LSP includes one or more CR-TLVs. For instance, the label request message may include the ER-TLV.
2. An LSR implicitly infers ordered control from the existence of one or more CR-TLVs in the label request message. This means that the LSR can still be configured for independent control for LSPs established as a result of dynamic routing. However, when a label request message includes one or more of the CR-TLVs, ordered control is used to set up the CR-LSP. Note that this is also true for the loosely routed parts of a CR-LSP.
3. New status codes are defined to handle error notification for failure of established paths specified in the CR-TLVs.

Due to the second functionality augment, the CR-LSP setup is very similar to the procedures in Section 14.4.1.5, where the downstream on-demand advertisement and ordered LSP control mode are used in each LSR. The main difference is the CR-TLVs, especially ER TLV and ER-hop TLV, processing procedures augmented by CR-LDP (for detailed description of the procedures, see [11]). Furthermore, the CR-LSP setup is typically issued by the ingress LSR and the path computation is achieved in the constraint-based routing context through CSPF. Unlike the LSP setup in Section 14.4.1.5, which is initiated by dynamic routing, the initiation of a CR-LSP is due to the traffic engineering requirement.

14.4.3   Optical Extension to CR-LDP

The extension to CR-LDP for GMPLS signaling is currently under study [31,32]. Fan et al.'s work [30] is an early effort to examine CR-LDP and RSVP-TE to set up a lightpath in the optical network. Since this is work in progress, we will not go into the detailed encoding or specific procedures. Instead, we provide only a general overview.

Due to the fundamental differences in technologies, some procedures in CR-LDP signaling are not applicable in optical networks. For example, in IP networks, the FEC is bound to a label and the label is used for packet forwarding. All packets belonging to the same FEC are thus forwarded using the label bound to the FEC. This is not applicable in the optical domain. In IP networks, label request/mapping messages are used to distribute the label-FEC binding between the LDP/CR-LDP peers, but in the optical network, the two messages are used to distribute the port assignment information between neighboring nodes. The label request message is used by an upstream node to request a port assignment (or with the wavelength assignment) from the downstream node for establishing a lightpath. A label request message is generally initiated from the ingress node and the message may need to carry many parameters, such as the ingress and egress node address, output and input ports at the endpoints, assigned wavelength, bandwidth requirement, protection/restoration scheme, preemption parameter, and so on. The label mapping message is used for a downstream node to distribute a selected input port for establishing a lightpath to the upstream node. When a node receives a label mapping message, it should be able to derive the output port from the downstream input port from the port mapping information obtained from the neighbor discovery mechanism. Then it selects an input port and connects the input port to the output port via configuring the cross-connect table. Finally, it distributes the input port selected in a label mapping message to the upstream node. Using the downstream on-demand advertisement mode and ordered LSP control mode, the label request message propagates from the ingress node to the egress node. The egress node creates a label mapping message with the assigned input port. The label mapping message then propagates from the egress to the ingress and distributes the assigned port at each hop. A lightpath is committed when all cross-connect tables are configured appropriately.

There are several possible extensions to be added to the CR-LDP to accomplish the lightpath setup described above:

1. Introduction of signaling port ID TLV to specify the port assigned. Such a "label" must be assigned in a coordinated manner by the neighboring nodes since the port of a downstream node is connected to a specific port of an upstream node.
2. Introduction of optical switched path (lightpath) identifier to identify a lightpath being established.
3. Introduction of endpoint TLVs to specify the port level endpoints of the lightpath.

4. Introduction of restoration parameter TLV to specify the restoration scheme.
5. Recording the precise route of the establishing lightpath through the label mapping message.
6. Bidirectional lightpath setup.

14.4.4 GMPLS Extension to CR-LDP

The GMPLS extension to CR-LDP defines the CR-LDP encoding of the TLVs and messages (e.g., the generalized label request) discussed in the GMPLS signaling functional requirements, and the corresponding processing for those TLVs and messages. Readers are referred to [31] for details.

14.4.5 Resource Reservation Protocol

The resource reservation protocol (RSVP) [10] was designed originally for the integrated service architecture (IntServ) [33] to reserves resources in Internet. It is used by a host to request QoS from the network for particular application data flows. The router uses RSVP to deliver the QoS request to the nodes along the path of a flow, and to set up and maintain the corresponding states in all nodes to guarantee the QoS requested. RSVP is defined to request the QoS for unidirectional data flows. Different from many signaling protocols, the receiver, not the sender, is responsible for the resource reservation in RSVP. The resource reservation initiated from the receiver propagates all the way up to the sender. RSVP is also designed to support both unicast and multicast traffic effectively, which is different from LDP, where only unicast traffic is supported.

The QoS is implemented for a given data flow using the traffic control, including a packet classifier, admission control, and a packet scheduler (or a link-layer similar mechanism). The packet classifier classifies the packets into QoS classes and the packet scheduler is used to achieve the promised QoS for the packets of a data flow. The admission control is used to determine the resource availability for a new resource reservation request. This admission mechanism, together with policy control, which checks the administrative permission of a user to make the reservation, decides whether a new request should be accepted or rejected. Only when both the admission and policy control checks succeed will the request be accepted. An important point here is that both the QoS and policy control parameters are opaque to RSVP. RSVP is only responsible for transporting such information to the traffic control and policy control modules in all nodes along the path.

Considering the dynamic characteristic of the multicast group membership, RSVP uses a refresh mechanism to maintain the state information in all nodes, called a *soft state* (i.e., the reservation state information is maintained through sending the periodical refresh reservation messages).

14.4.5.1 RSVP Session RSVP defines a session (Fig. 14.22) as a data flow with particular destination and transport layer protocol (e.g., UDP/TCP). This can be defined formally as a triple ⟨dest address, protocol ID, dest ports⟩. The dest address

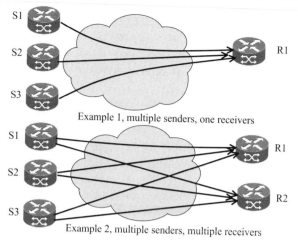

Example 1, multiple senders, one receivers

Example 2, multiple senders, multiple receivers

Figure 14.22   RSVP session examples.

can be a multicast address; the dest ports can include zero or more ports of the transport layer protocol specified by "protocol ID."

14.4.5.2 Reservation Styles   RSVP defines three reservation styles: fixed filter (FF), shared explicit (SE), and wildcard filter (WF). A WF style creates a single reservation for all senders. The reservation request propagates toward all senders. When there is a new sender joining, this single reservation extends to this new sender automatically. The FF style creates a distinct reservation for data packets from a particular sender. The SE style creates one reservation for a set of selected senders. The WF and SE styles can be used by multicast applications. RSVP prohibits the merging of reservations with different styles. For example, a reservation request with FF style may never be merged with a request with SE style to form a new request.

14.4.5.3 RSVP Messages   A RSVP message consists of a common header, and a variable number of objects with a variable length. The common header is shown in Fig. 14.23. "Flags" is currently reserved for future use. RSVP defines seven types of messages: Path, Resv, PathErr, ResvErr, PathTear, ResvTear, and ResvConf. "Send TTL" represents the time to live, which is copied from the IP packet TTL field. "Length" is the total length of the message, including the common header and message body. Every object consists of one 32-bit header, and the object content can be empty or multiples of 32-bit units, as shown in Fig. 14.24. "Class-Num"

Version	Flags	Message Type	RSVP Checksum
Send TTL		Reserved	Length

Figure 14.23   RSVP message common header.

Length in bytes	Class-Num	C-Type
Object Contents		

Figure 14.24   Object format.

represents the object class (the detailed value is defined in [10]). Several important classes are SESSION, RSVP_HOP, STYLE, FLOWS_SPEC, SENDER_ TEM-PLATE, and FILTERSPEC. "C-Type" is the object type inside a given class. For example, the encoding of the object within some class may need to be different for IPv4 and IPv6, and C-Type can be used to differentiate such difference within one class. The transmission of RSVP messages can be encapsulated in raw IP packets or UDP datagrams. The Path and Resv messages are illustrated in Figs. 14.25 and 14.26. For other messages, the readers are refered to [10].

1. *Path message.* The SESSION object specifies the session of this request (i.e., ⟨dest address, protocol ID, dest ports⟩). The PHOP object encodes the address of the previous hop sending this Path message. The TIME_VALUE object includes the expected refresh interval of this Path message. The optional objects may contain the policy data and/or sender descriptor. The sender descriptor includes a SENDER_ TEMPLATE object, a SENDER_TSPEC object, and an optional ADSPEC object.

The Path message is sent by the sender or an upstream node initially to establish or refresh the state information of a data flow. The SENDER_TEMPLATE object identifies the sender in the form of a FILTERSPEC object, and the SENDER_ TSPEC object defines the traffic characteristic of the data flow. The Path message should traverse the same path as the data packets in the network and may cross a non-RSVP cloud (the routers are RSVP incapable in the cloud). When an intermediate node receives a Path message, it establishes the Path state, including the information in the PHOP object, SENDER_TEMPLATE (FILTERSPEC)

Common Header
SESSION Object
PHOP Object
TIME_VALUES Object
Optional Objects

Figure 14.25   Path message.

Common Header
SESSION Object
NHOP Object
TIME_VALUES Object
STYLE Object
Flow Descriptors
Optional Objects

Figure 14.26   Resv message.

object, sender Tspec object, and so on. The Path message is then forwarded to the next hop specified by the routing protocol. If the SESSION object of this Path message is a multicast session, this message is forwarded to all the next hops in the multicast tree.

2. *RESV messages.* The SESSION and TIME_VALUES objects have the same meaning as in the Path message. The NHOP object is the downstream hop that sends this Resv message. The STYLE object encodes the reservation style. The flow descriptor contains the FLOWSPEC and/or FILTERSPEC objects, depending on the STYLE object. The optional objects many contain a SCOPE object, a policy data, and/or a RESV_CONFIRM object. The SCOPE object is used primarily to prevent looping of Resv message.

The Resv message is sent by the receiver toward senders in a session to reserve resource or to refresh the reservation for a data flow. A Resv message carries a reservation request and traverses the reverse path of the data packets using the PHOP information in the Path state which has been established by the Path message of this data flow. A reservation request consists of a flow descriptor. The FLOWSPEC object in the flow descriptor specifies QoS parameters to be set in the packet scheduler. It generally includes a service class and two values defined by IntServ, an Rspec representing the desired QoS, and a Tspec describing the data flow. The FILTERSPEC object in the flow descriptor specifies what kind of data packets to receive the QoS, and together with a specific RSVP session, defines the data flow to receive the QoS. The FILTERSPEC object is used to set parameters in the packet classifier. When an intermediate node receives a reservation request, it first attempts to make the reservation on a link by passing the parameters of the request to the admission control and policy control. If the reservation is made successfully, the reservation state is created or updated by supplying the FLOW-SPEC to the packet scheduler and FILERSPEC to the packet classifier. The Resv message attempts to merge with other Resv messages carrying reservation requests originated from other receivers in the multicast application. If the STYLE object is FF or SE, the merged Resv message is forwarded to all previous hops in those Path states whose SENDER_TEMPLATE objects match the FILTER_SPEC object in this Resv message. The Resv message forwarded upstream may be modified by the traffic control (e.g., the FLOWSPEC object may be modified). The basic reservation model is *one pass*. The receiver cannot ensure that the reservation has been established successfully along the entire path. Even requesting a confirmation message does not guarantee a successful end-to-end setup, due to the possible merging of requests in the intermediate nodes.

14.4.5.4 *Resource Reservation Using Path and RESV Messages*   Figure 14.27 is an example of the RSVP resource reservation procedure using Path and Resv messages for a unicast data flow. The RSVP session includes only one sender and one receiver. The Path message is processed hop by hop and a Path state is created in each intermediate node. After the Path message arrives at the receiver, a Resv message is sent upstream toward the sender. The Resv message is forwarded hop by

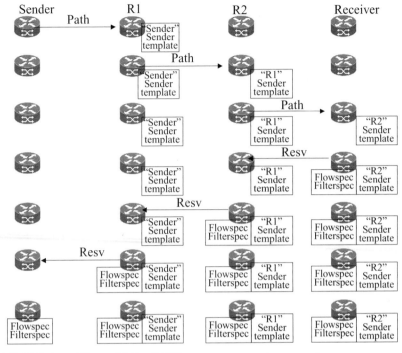

Figure 14.27 RSVP resource reservation example.

hop using the previous hop information stored in the Path state. When the Resv message has been processed by the sender, all nodes created the reservation states for this Resv message by supplying the FLOWSPEC and FILTERSPEC objects to the packet scheduler and packet classifier, respectively. Now the data packet can receive the QoS along the entire path. If there are multiple senders in a session, there may be multiple Path states for this session in each intermediate node along the multipoint-to-point path. If there are multiple receivers in a session (multicast session), the reservation parameters (e.g., FLOWSPEC objects) need to be merged to form a new Resv message at the branch nodes in the multicast tree.

14.4.6 RSVP-TE

RSVP-TE is a RSVP extension to set up LSP tunnels in MPLS with traffic engineering extension. In a network, if the nodes support both RSVP and MPLS, the labels can be associated with RSVP data flows. More specifically, the reservation state established for a data flow can be associated with a label in each node. The LSP defined by the labels associated with the data flow is called an *LSP tunnel*. Thus with some extension, RSVP can effectively support the setup of LSP tunnels. Furthermore, RSVP can offer another advantage for the LSP tunnel setup (i.e., the resource reservation for the LSP tunnel using FLOWSPEC and

TABLE 14.1 RSVP-TE Objects

Object	Property	Applicable Messages
LABEL_REQUEST	Mandatory	Path
LABEL	Mandatory	Resv
EXPLICIT ROUTE	Optional	Path
RECORD_ROUTE	Optional	Path, Resv
SESSION_ATTRIBUTE	Optional	Path

FILTERSPEC). Multiple LSP tunnels are called *traffic engineering tunnels* if the tunnels are set up to service the traffic engineering of a traffic trunk (e.g., the rerouting or spreading of the traffic trunk).

In RSVP-TE, reservation-style WF is not used because this style requires a single multipoint to the point path from all senders to one receiver, which is not practical for setting set up a LSP tunnel, where different senders probably require a different path to the receiver.

14.4.6.1 RSVP-TE Objects RSVP-TE defines five new objects and extends the Path and Resv messages to include those objects. Table 14.1 shows the objects and the messages that can include those objects. Additionally, new objects, called LST_TUNNEL_IPv4 and LSP_TUNNEL_IPv6, for the three classes SESSION, SENDER_TEMPLATE, and FILTERSPEC, have been defined to support LSP tunnel features. A SESSION object is used to represent a traffic engineering tunnel. A SENDER_TEMPLATE object (used to identify the sender) or a FILTERSPEC object contains an LSP ID. The LSP tunnel is defined by the SENDER_ TEMPLATE object together with the SESSION object.

There are three encodings for a LABEL_REQUEST object to request a generic label, an ATM label (VPI/VCI), or a frame relay label (DLCI), respectively. The LABEL_REQUEST object also indicates the network layer protocol that will use the LSP tunnel to carry data. When a Path message including a LABEL_REQUEST object is processed, this object is stored in the Path state. The LABEL object can encode a generic label, an ATM label, or a frame relay label. This object should be included in a Resv message to distribute the label from the receiver toward the sender. The label in the LABEL object is allocated by the downstream node.

The EXPLICIT_ROUTE object consists of a set of subobjects and is used to specify an explicit route for the LSP tunnel. A subobject is essentially one hop, or an abstract node in the explicit route. The subobject includes one bit, indicating that this hop is a loose or strictly explicit routing hop, the type and length of the subobject, and the abstract node address, which can be an IPv4 address, IPv6 address, or an autonomous system number. The semantics and processing of an explicit route in MPLS signaling has been discussed in CR-LDP and will be neglected here. The RECORD_ROUTE object is used to record the hop-by-hop path of the LSP tunnel, and its structure is similar to the EXPLICIT_ROUTE object. The SESSION_ATTRIBUTE object contains session attributes such as setup priority and holding priority for the LSP tunnel setup.

The RSVP can operate across a non-RSVP cloud as discussed in Section 14.4.5. However, during LSP tunnel setup, a non-RSVP-TE node (it is either a RSVP node that cannot recognize those newly defined objects or a non-RSVP node) receiving a message including those newly defined objects must return an error message (e.g., PathErr) to the message source. This is because MPLS LSP tunnel setup essentially requires all nodes along the path supporting those features.

14.4.6.2   *LSP Tunnel Setup Using RSVP-TE*   RSVP-TE sets up an LSP tunnel similar to that for CR-LDP, which utilizes downstream on demand label distribution and ordered control. The operation of Path and Resv messages in RSVP is in fact an ordered control since the Resv message needs the previous hop information in the Path state established by a previous Path message. Note that such an operation mode also restricts label distribution to be downstream on demand at least at the first setup (not the refresh) of the LSP tunnel.

The LSP tunnel setup is similar to that illustrated in Fig. 14.27, except that the Path state includes the LABEL_REQUEST object and the reservation state has been replaced with the label allocation and incoming/outgoing label mapping. Initially, the sender creates a Path message with a session type of LSP_TUNNEL_IPv4 or LSP_TUNNEL_IPv6, and a LABEL_REQUEST object specifying the label type/ range and the network layer protocol using this LSP tunnel to carry the data flow. Optionally, an EXPLICITE_ROUTE object can be attached with the Path message to specify the path for the LSP tunnel. Similarly, an optional RECORD_ROUTE object can be attached to the Path message to record the LSP tunnel path during the setup. A SESSION_ATTRIBUTE object may also be inserted into the Path message to provide some attribute for this LSP tunnel. Each intermediate node receiving this Path message creates a Path state including the LABEL_REQUEST object and forwards the Path message to the downstream node. When this message arrives at the receiver, a label is allocated according to the specification (e.g., label range) of the LABEL_REQUEST object and encoded in a LABEL object. A Resv message is created including this LABEL object and sent to the upstream nodes based on the previous hop information in the Path message. Note that a Resv message should not include a LABEL object if the corresponding Path message does not contain a LABEL_REQUEST object. The Resv message propagates toward the sender. Each intermediate node receiving this Resv message extracts the label allocated by the downstream node and uses it as the outgoing label for the traffic carried by this LSP tunnel. A new local label is also allocated by the intermediate node according to the LABEL_REQUEST stored in the Path state and replaces the extracted label in the LABEL object. The tuple ⟨local label, extracted downstream label⟩ will be used in the future label-swapping operation. The modified Resv message is forwarded to the upstream node. When the Resv message reaches the sender, the LSP tunnel is established successfully. If any intermediate node cannot recognize the RSVP-TE extended objects, or the label allocation fails, or the reservation fails in the case of a

QoS request, the LSP tunnel setup fails and a PathErr or ResvErr message is sent out indicating the error.

14.4.7 GMPLS Extension of RSVP-TE

RSVP-TE is being extended to support GMPLS signaling [29]. Five label-related objects have been defined: generalized LABEL_REQUEST, generalized LABEL, WAVEBAND_SWITCHING, SUGGESTED_LABEL, and LABEL_SETS. The WAVEBAND_SWITCHING object is used to support waveband switching. The SUGGESTED_LABEL object is used for the sender or upstream node to suggest a label for the downstream node. The LABEL_SETS object is used to restrict the label selection for the downstream node. A Path message can include multiple LABEL_SETS objects, which can specify which label is not acceptable, which label can be used, and so on.

The GMPLS extension of RSVP-TE also supports the bidirectional LSP tunnel setup. Two objects and one Notify message have been defined for the notification of label error and expedition of failure notification. The LABEL ERO and LABEL RRO subobjects are defined for the explicit label control, which extends the explicit route feature further to provide finer control of the LSP tunnel. The ADMIN_ STATUS object is defined to provide the administration information of a LSP tunnel. Some other objects are defined for fault handling, protection, administrative status information, and so on.

14.4.8 Other Signaling Approaches

Signaling has also been an active area of work in other coalition bodies, such as OIF, and ODSI. OIF has adopted the MPLS signaling protocols for UNI signaling to facilitate the optical network control evolution toward a unified control plane across the client and optical core networks. The UNI 1.0 specification [34] over SONET interface has been standardized. ODSI has defined and implemented a TCP-based signaling protocol for UNI signaling.

14.4.9 Signaling Protocols Comparison

Although each signaling protocol has its own pros and cons, it is sometimes helpful to carry out a comparative study on them. The objective here is not to prefer one approach over others but to better understand the similarities and differences. The focus here is on CR-LDP and RSVP approaches only since they are studied popularly in the IETF community, thus making them the de facto standards at this time. In addition, CR-LDP is a TCP-based signaling protocol and hence shares many common features with other TCP-based signaling approaches.

First, the two protocols are similar in various aspects. The label request/mapping messages support the same label distribution functions and follow signaling procedures similar to those of PATH/RESV messages. The function of both the TLV in CR-LDP and object in RSVP is to encode the signaling parameters. Both protocols support QoS signaling and are flexible in bandwidth reservation granularity. Although the two protocols exhibit more similarities than distinctions, next we compare them briefly in terms of scalability, complexity, reliability, and QoS model.

14.4.9.1 Scalability
RSVP was initially proposed to support integrated services (IntServ) architecture in the Internet. The main objective of RSVP is to reserve bandwidth in the network for a network traffic flow to support QoS. For RSVP, the traffic flow can be as small as a traffic stream between two applications on two hosts, and thus the number of flows supported can easily reach a huge number (e.g., millions). With RSVP, an intermediate node in the network needs to maintain the state information for each passing-through traffic flow. With the possible huge number of flows, it may be prohibitively difficult for an intermediate node to maintain state information in terms of both storage and processing. On the other hand, CR-LDP is used to set up an LSP that carries a traffic trunk in the MPLS domain (generally, core network) with a relatively large bandwidth requirement. In this sense, CR-LDP is said to have better scalability. However, the potential scalability problem of RSVP can be avoided if the use of RSVP is restricted for LSP setup in the MPLS domain. Therefore, the scalability problem of RSVP may not be a practical problem in the MPLS/MPλS/GMPLS domain.

14.4.9.2 Complexity
CR-LDP is based on the TCP for reliable transport of CR-LDP messages. The TCP connection initialization introduces some overhead to the CR-LDP initialization. CR-LDP peers have to go through an initialization procedure to establish a CR-LDP session in order to exchange signaling messages such as label request/mapping messages. CR-LDP also defines many control modes, such as label retention, LSP control and label broadcast modes, and many maintenance messages. Altogether, CR-LDP is a complex protocol. On the other hand, RSVP is based on UDP and is called a lightweight protocol for unicast traffic (although it is complex for multicast traffic). RSVP employs a soft-state mechanism which requires source–destination nodes to send PATH/RESV messages periodically to refresh flow state information in the intermediate nodes. Without refreshment, flow state information is discarded and the traffic flow is broken at the intermediate node. This introduces some maintenance overhead. However, the soft state also brings an advantage that in case of a failure (e.g., link failure) in some segment of a flow, the flow state information becomes obsolete after the refreshment interval and the flow is terminated automatically. In addition, the periodic transmission of refresh messages enables RSVP to recover from message loss.

14.4.9.3 Reliability and Setup Time
CR-LDP is a reliable protocol since it operates over the TCP connection. RSVP, on the other hand, is an unreliable protocol because it is based on the UDP connection. To recover from message loss depends on the soft-state refresh mechanism for the message retransmission. RSVP assumes that the message loss happens only occasionally. If this is not the case, RSVP may face a serious efficiency issue. When a message is lost, retransmission

can be initiated only after one refresh interval. There is a subtlety here. If the refresh interval is configured as large, this will result in substantial delay for the setup. If it is configured as small, the refresh overhead may be prohibitively high. Therefore, the refresh interval has to be chosen carefully.

14.4.9.4 QoS Model Support CR-LDP supports a QoS model based on ATM, which uses the peak rate, committed rate, peak burst size, committed burst size, and excess burst size to characterize the traffic. RSVP supports the IntServ QoS model, which is defined by a service class and the service parameters within the given service class. The values specifying these two levels are together referred to as a traffic specification (Tspec). The ATM QoS model has been used by the service providers in ATM networks. The IntServ QoS model has also been implemented and tested in real networks. It is hard to say which one is better at this time.

14.5 OPTICAL INTERNETWORKING AND SIGNALING ACROSS THE NETWORK BOUNDARY

The internetworking in IP networks offers powerful means for service creation and ubiquitous connectivity for end users. The IP networks in the same administrative domain are connected to form autonomous systems (ASs) and the ASs are connected to provide end-to-end transit service. Similarly, the optical internetworking is also a must to build interdomain optical networks. Figure 14.28 shows an optical internetworking model consisting of interconnected optical networks attached by IP clouds. The IP clouds are connected through dynamically established lightpaths which may cross one or more optical networks. Each optical network generally represents an administration domain and has a mesh topology to

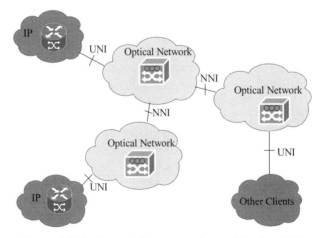

Figure 14.28 Network interconnection model using NNI.

improve the resource use. Interconnection between optical networks is assumed to be through compatible physical interfaces that is typically based on a peer-to-peer model. A lightpath from an ingress port in an OXC to an egress port in a remote OXC, which may be in another network, is established by setting up suitable cross-connects in the ingress, the egress, and intermediate OXCs such that a continuous physical path exists from the ingress to the egress port.

With such an optical Internet, it must be possible to provision and restore lightpaths dynamically across optical networks. Therefore, a standard addressing scheme is required to uniquely identify lightpath endpoints (node and/or port) in the optical Internet. A standard routing protocol is required to determine the reachability of endpoints in the optical Internet. A standard signaling protocol is required for provisioning lightpaths across the network-to-network interface (NNI) in the optical Internet. A standard restoration procedure is required to restore lightpaths across NNI. If the NNI represents a boundary of the administration domain, it is also required to support the policies that affect the flow of control information across networks. The IP-centric (MPλS/GMPLS) control plane for optical networks can be extended to satisfy the functional requirements of optical internetworking. Routing and signaling interaction between optical networks can be standardized across the NNI. Currently, BGP is being studied for optical interdomain routing and even signaling. This work is still in its preliminary phase.

If there is a standard signaling protocol across the entire optical Internet, the end-to-end signaling across NNI will be similar to the signaling in a single optical network. If some networks have different signaling protocols inside them, the signaling will have a two-layer structure. The higher layer, the *external signaling protocol* (ESP), is used for signaling across NNI and has to be standard. The lower layer, the *internal signaling protocol* (ISP), is used for signaling inside a single network and may be distinct in different optical networks. Figure 14.29 illustrates an end-to-end signaling procedure across an NNI in the optical Internet. If the ISP is different from ESP in an optical network, the border node (OXC) needs the ability to translate between the ISP message and ESP message. The core (non-border) node does not require such a capability. In Fig. 14.29, an optical node (named SRC), which is typically an edge node with UNI to the client, initiates a lightpath setup in the optical network A. The ISP of SRC then constructs an ISP signaling message carrying the endpoint (destination node and/or the port) information, and other optional parameters, such as the assigned wavelength. This ISP signaling message propagates to border node 1, which then translates this message into an ESP signaling message and sends it to the peering border node (border node 2) in optical network B. Border nodes 1 and 2 uses a standard signaling message and thus they can talk with each other without difficulty. Border node 2 then translates the ESP message to the ISP message in optical network B, where the ISP may differ from the ISP in optical network A. The ISP message passes one intermediate node in optical network B, where the message is processed using ISP procedures and then proceeds to border node 3. The NNI signaling interaction between border nodes 1 and 2 repeats over here between border nodes 3 and 4. The signaling message is translated by border node 4 into its own ISP

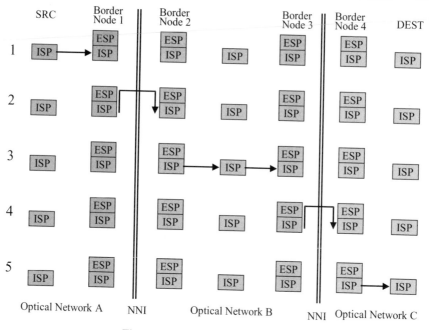

Figure 14.29   Signaling across NNI.

message and proceeds to the destination node (named DEST). DEST uses the ISP procedures in optical network C to process the message as if this message is from a node in the same network (i.e., optical network C). The propagation and processing of signaling message in the reverse direction is similar. Hence through a standard signaling protocol across NNI, it is possible to set up a lightpath across the network boundary. CR-LDP and RSVP-TE with appropriate extensions can be used for ESP in the optical Internet. The ISP can be a proprietary signaling protocol coming from the vendor or a standard protocol such as CR-LDP. Note that if an optical network uses the same protocol as the ISP and ESP, the border node does not need the translation capability.

14.6 SAMPLE IP-CENTRIC CONTROL PLANE FOR OPTICAL NETWORKS

In this section we describe a sample IP-centric control plane for the optical network [35]. The sample IP-centric control plane (for convenience, hereafter called the control plane) considers a mesh optical network which is used to connect high-speed IP/MPLS client routers in client networks, as shown in Fig. 14.30. Each optical node consists of a transparent OXC and a control plane. The control plane may be integrated into the same box as the OXC or a separate router that is used to control the OXC. Between two neighboring OXCs, a dedicated out-of-band

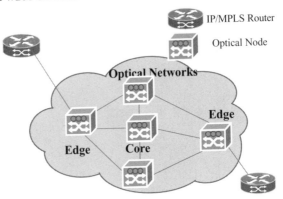

Figure 14.30 Mesh optical network.

wavelength is preconfigured as the IP connectivity and a reliable TCP socket connection is set up as the *control channel* to transmit the control messages. The control messages are processed and relayed in a hop-by-hop fashion.

As shown in Fig. 14.31, the sample control plane consists of four modules: the resource management module (RMM), connection module (CM), protection/restoration module (PRM), and main module (MM). RMM is used for routing and wavelength assignment (RWA), topology and resource discovery, and quality of service (QoS) support. CM is used for lightpath signaling and maintenance. In the sample control plane, a flexible and efficient signaling protocol based on TCP is implemented for CM. Considering the paramount importance of the survivability in optical networks, the sample control plane contains an independent PRM for fault monitoring and fast protection/restoration. The objective of the MM is to receive incoming messages and work closely with other modules to process the requests. With the functional modules above, the control plane enables the WDM layer to

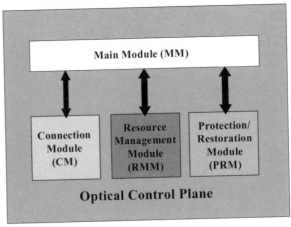

Figure 14.31 Sample IP-centric control plane structure.

extend into the network layer, providing network-layer intelligence to the optical network so that it becomes flexible, scalable, and resilient.

14.6.1 Main Module

The main function of the main module (MM) consists of initializing the control plane, waiting for client requests or incoming messages (from neighboring control planes) and invoking other modules to process the message or request. When a control plane is started initially or rebooted (after a failure), the control plane establishes the *control channels* to neighboring control planes (the neighboring information is manually configured) and its clients. It also creates all tables and data structures. Then the control plane queries neighboring control planes to update some of these tables. After initialization, the MM will accept requests from clients or the neighboring control planes and invoke other modules to process the requests.

14.6.2 Connection Module

The main functions of the connection module (CM) are lightpath signaling and maintenance. At each node, a *lightpath table* (LT) is maintained by CM to manage all lightpaths (originating, passing through, and terminating) over the OXC. Figure 14.32 shows an entry of LT. The *lightpath identifier* (lightpath ID) includes the source node ID, destination node ID, and a sequence number. The sequence number distinguishes the lightpaths originating at a given node. Thus a lightpath ID uniquely identifies a lightpath in the entire network. The "status" attribute indicates the state of a lightpath (creating, reserved, active, or deleted). The "QoS type" attribute indicates the service class of the traffic. The "input/output port ID" represents the ID of the incoming/outgoing port of the lightpath in the OXC (the port ID is only unique in each node). The "λ ID" indicates the assigned wavelength of this lightpath.

The signaling of a lightpath is performed in a hop-by-hop fashion. When MM receives a client request, it transfers the request to PRM, where the QoS information is extracted from the request. Then PRM invokes RMM for RWA with the QoS parameter. For each path, a *connection request message* is formed, including the lightpath ID, lightpath type (primary or protection), routing path, assigned wavelength, and QoS type. If this is a protection path of the protection-sensitive service, the SRLG list of its primary path is also included in the message. This message is processed at each hop and then forwarded to the next hop toward the destination

Lightpath ID			Status	QoS type	Input port ID	Output port ID	λ ID
SRC Node ID	DEST Node ID	SEQ NUM					

Figure 14.32 Lightpath table entry.

node. The destination node then sends an acknowledgment (ACK) back to the source node along the path. If the lightpath setup is blocked at any hop due to the resource conflict, a negative acknowledgment (NAK) is sent back to the source node.

At each hop, the processing of lightpath signaling can be divided into two parts: the resource reservation/release and the lightpath state transfer. The resource reservation is responsible for reserving the wavelength resource, and the resource release is basically an inverse procedure of the resource reservation. The following procedure illustrates the resource reservation (the wavelength status is to be discussed in RMM):

1. Determine the input/output ports by the path information.
2. If the QoS type is best-effort and the assigned wavelength is "available," set the wavelength status to "used and preemptible."
3. If the QoS type is mission critical:
   (a) If the assigned wavelength is "available," the wavelength status is set to "used and non-preemptible."
   (b) If the assigned wavelength is "used and preemptible," abort the existing lightpath on this wavelength. The wavelength status is set to "used and nonpreemptible."
4. If the QoS is protection sensitive:
   (a) If it is a primary lightpath and the assigned wavelength is "available," set the wavelength status to "used and nonpreemptible."
   (b) If it is a protection lightpath type:
      (1) If the assigned wavelength is "available," set its status to "reserved."
      (2) If the assigned wavelength is "reserved," compare the SRLG list in the message with the SRLG list of the assigned wavelength ("λ_i SRLG list"). If there is no SRLG conflict, add the SRLG list in the message to the SRLG list of the assigned wavelength.

If the resource reservation fails (in the wavelength status checking or SRLG list checking), a NAK, including the information in the *connection request message*, is sent back to the upstream node. Otherwise, CM begins to process the lightpath state transfer. The lightpath state transfer processing is also invoked when CM receives an ACK or NAK for a lightpath from the downstream node. Figure 14.33 shows the state transfer of a lightpath. After the resource reservation succeeds, CM allocates an entry in the LT and properly sets all attributes in the entry, with the "status" attribute set to "creating." When CM receives an ACK for a lightpath from the downstream node, the "status" attribute is set to "active" or "reserved," depending on whether it is a primary lightpath or a protection lightpath. A protection lightpath in the status of "reserved" becomes "active" when PRM has detected a failure on the primary lightpath and invokes the protection process. When CM receives a NAK for a lightpath from the downstream node during the setup procedure, or gets

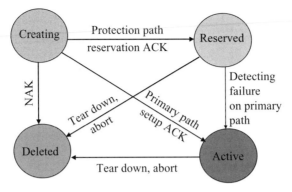

Figure 14.33 Lightpath state transfer graph.

a teardown or abort message after the lightpath has been established, the resource will be released and the associated entry in LT will be cleared.

14.6.3 Resource Management Module

The main functions of the resource management module (RMM) are resource discovery and maintenance, QoS support, and RWA. In RMM, the topology and resource availability information is stored in a database consisting of some tables. The active neighbor information (two neighboring nodes are defined as active neighbors if the physical link between them is in the working condition) between a given node and its neighbors is detected by the neighbor discovery mechanism and stored in a *local connectivity vector* (LCV), whose ith element represents the link cost of the active connectivity to the ith node. The link cost "0" indicates that the link is in fault or there is no link between the two nodes. A *topology connectivity matrix* (TCM) is used to store the network topology (in OSPF, this is stored in the link-state database). Figure 14.34 shows a TCM of an optical network with N

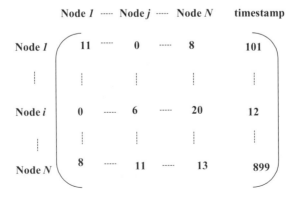

Figure 14.34 Topology connectivity matrix.

Port *1*	Node ID	λ_1 status	...	λ_j status	...	λ_w status
		λ_1 SRLG list		λ_j SRLG list		λ_w SRLG list
...	...	...	...	...	...	...
Port *i*	Node ID	λ_1 status	...	λ_j status	...	λ_w status
		λ_1 SRLG list		λ_j SRLG list		λ_w SRLG list
...	...	...	...	...	...	...
Port *M*	Node ID	λ_1 status	...	λ_j status	...	λ_w status
		λ_1 SRLG list		λ_j SRLG list		λ_w SRLG list

Figure 14.35   Local resource table.

nodes. The that row is the LCV of node i plus a time stamp (a sequence number) that is used to recognize the out-of-date LCV during the flooding procedure.

In RMM, the local resource availability is stored in a *local resource table* (LRT). Figure 14.35 shows the LRT of a node with M ports and W wavelengths per fiber. In the figure, the "node ID" (a unique ID to address the optical node) represents the ID of the peering node connected to that port. The "λ_i status" indicates the state of the ith wavelength in the fiber attached to the port. A wavelength can be in one of the following five states: "used and preemptable," "used and nonpreemptable," "reserved," "available," or "faulty." The "used and preemptable" state indicates that this wavelength is being used by a low-priority lightpath and can be preempted by a high-priority lightpath. The "reserved" state indicates that this wavelength is used by one or multiple protection path(s). The "λ_i SRLG list" stores the SRLG information of the primary paths whose protection paths have reserved this wavelength. It is valid only if the "λ_i status" is "reserved." RMM also maintains a *global resource table* (GRT) consisting of LRTs of all nodes (in OSPF, this is stored in the link-state database).

RMM adopts a neighbor discovery mechanism similar to that in OSPF [19]. To discover and maintain active neighbor information, each control plane sends *hello messages* periodically on all outgoing links. If a control plane does not receive a *hello message* from a neighboring control channel within some initially negotiated holding time, this link is considered as failed. The PRM will be invoked to process the fault. On the other hand, if a control plane receives a *hello message* from a failed link, the link is considered as restored. When the LCV is updated due to a link cost change, or a link failure, or a restoration, it is broadcast to the network. Resource discovery is also based on OSPF and employs a flooding mechanism similar to that in OSPF to broadcast to the entire network the LCV and LRT of a given node. Each node builds its own TCM and GRT using the received LCVs and

LRTs. Generally, when there is a change or the update timer expires, the LCV is encapsulated in a *topology update message* and is flooded in the network. For the LRT, the wavelength status needs to be broadcast. However, the LRT is generally large and impossible to encapsulate in a single message. In addition, it is not necessary to broadcast the entire LRT whenever there is a small change, such as the status change of one wavelength of a port. Hence, alternatively, only the wavelength status that has changed since the last broadcast needs to be flooded in the network. The status change of one wavelength is specified by the tuple ⟨node ID, port ID, λ status⟩. Generally, to reduce the message processing overhead, multiple tuples (i.e., multiple wavelength status changes) are encapsulated in a *resource update message*. The "λ_i SRLG list" will not be encapsulated into the *resource update message* unless the "λ_i status" is "reserved." As in the OSPF flooding mechanism, the control plane employs the sequence number (time stamp) from the *topology update message* and *resource update message* to resolve the looping problem.

RMM utilizes different RWA algorithms to support QoS. The current RWA algorithms have not considered the wavelength conversion. However, they can easily be extended to consider this situation. The explicit source routing is realized in the control plane, and the shortest path first (SPF) algorithm is used to compute the end-to-end routing path based on TCM. The computed routing path is cached for subsequent requests to save the per request routing computation overhead. The first-fit policy is used to assign an available wavelength for a lightpath based on GRT. In the control plane, three QoS classes have been defined: best effort mission critical, and protection sensitive. The *best-effort service* only requires restoration and may be preempted by the mission critical traffic. For a client request in this service class, RMM computes an end-to-end path using SPF and assigns a wavelength that is in the "available" status along the path. The *mission critical service* needs a dedicated protection path and the establishment of either a primary or protection lightpath may preempt an existing lightpath of the best-effort service class. For the mission critical client request, the primary path is computed first and an edge-disjointed path is computed as the protection path. Then for each path, RMM assigns a wavelength that is in the status of "available" or "used and preemptible" along that path.

The *protection sensitive service* requires only the shared protection, and once the primary and protection lightpath of this service type have been established, neither of them can be preempted. On the other hand, the establishment of both primary and protection lightpaths cannot preempt any existing lightpaths. The path computation for the protection sensitive service class is similar to that for the mission critical service class. The wavelength assignment for the primary path is also similar except that the assigned wavelength should be in the status of "available." However, the wavelength assignment for the protection path is more complex because wavelength sharing needs more consideration. With protection sensitive traffic, two protection paths can share a wavelength if their primary paths are link disjointed. To maintain such information, the control plane assigns a unique SRLG ID to each link. Thus, there is an SRLG list for a primary path. For those protection paths that share a wavelength (in the status of "reserved"), the SRLG lists of their primary paths are stored in the "λ_i SRLG list" of this wavelength. When there is a change in the λ_i SRLG list (e.g., adding an SRLG ID or deleting an SRLG ID), it is broadcast to the network in the *resource update message*. Such information is collected into the GRT by each node. Thus during the wavelength assignment for a protection path, RMM can check the conflict between the SRLG list of its primary path and the λ_i SRLG list of a "reserved" wavelength to determine if this wavelength can be assigned. Therefore, by maintaining a λ_i SRLG list for each reserved wavelength, RMM can assign a reserved wavelength to a protection path such that multiple protection paths can share the same wavelength.

14.6.4   Protection/Restoration Module

The protection/restoration module (PRM) provides the functions of setup coordination of the primary and protection lightpaths, fault detection, and notification. We consider end-to-end protection/restoration where a link disjointed path is selected for protection/restoration. Such end-to-end protection/restoration can provide fast recovery without locating the faulty link. The protection method includes fiber-level protection and channel-level protection. We will consider channel-level protection, for it will result in high network utilization. Because the mesh network is rich in connectivity, the spare resources can be shared among multiple protection paths as long as their corresponding working paths are link disjoint.

When a client request is received by the MM, it is transferred to the PRM. Then the PRM invokes the RMM for RWA with the QoS parameter extracted from the request. For each path, a *connection request message* is formed and the CM is invoked for signaling by the PRM. If the client request requires protection, signaling of the primary and protection paths will be invoked in parallel by the PRM. Hence the PRM (of the source control plane) needs to take care of the asynchronous feedbacks (i.e., ACK or NAK) of the two lightpaths. Only if both the primary and protection lightpaths have been set up, the CM sends an ACK to the client. Otherwise, when one of the two lightpaths has failed (getting NAK) on the setup, the CM sends a NAK to the client, and, if the other lightpath has been set up, that lightpath needs to be torn down.

The PRM is also responsible for detecting the failures and initiating a protection/ restoration process. Fault detection can be done by hardware to detect low-layer impairments such as loss of signal, or by the higher layer via link-probing mechanisms. In the control plane, the neighbor discovery mechanism is used to detect the failure of a link via periodically exchanging *hello messages* as discussed in the section on the RMM. The node failure can be divided into OXC failure or control plane failure. OXC failure is detected by the control plane through the hardware mechanism. The control plane failure is detected indirectly through the neighbor discovery mechanism. When a control plane fails, the neighboring control planes will not be able to receive *hello messages* from the failed control plane. As a result, every neighboring control plane broadcasts a *topology update message* to the network. After receiving all of these *topology update messages*, every control plane learns of the failed control plane.

After detecting a failure, the control plane will send out a failure notification for each lightpath affected by transmitting a *failure indication signal* (FIS) toward the source node. This notification is relayed hop by hop to the upstream control plane. Once the source control plane receives the FIS, it will check the QoS attribute of the lightpath. If it is restoration, the source node will calculate and signal a restoration path and then invoke rerouting to recover the working traffic. If the QoS attribute is protection sensitive, the source node immediately invokes setup signaling for the protection lightpath previously reserved. For mission critical traffic, the destination node can detect the failure of the primary lightpath and turn automatically to the protection path.

14.7 SUMMARY

In this chapter we have discussed control and signaling in optical networks. The focus was on IP-centric optical network control and signaling, which is under study in standards organizations (e.g., OIF and IETF). The optical internetworking and interdomain signaling was introduced. A sample control plane was also presented to explain further the concepts and mechanisms discussed in this chapter.

The extension and standardization of optical network control and signaling is still in progress. Currently, the biggest obstacle for automatic control and provisioning in the optical network is interoperability problem among the equipment from different vendors and/or the network boundary. To facilitate interoperability, it is critical that signaling and even some aspects of routing be standardized in the optical network. The evolution of optical networks toward automatic control and signaling will promise real-time, on-demand provisioning capability, which is needed greatly by both service providers and customers.

REFERENCES

1. B. Mukherjee, WDM optical communication networks: Progress and challenges, *IEEE Journal or Selected Areas in Communications*, 18(10):1810–1824, Oct. 2000.

2. C. Xin, Y. Ye, S. Dixit, and C. Qiao, An emulation-based comparative study of centralized and distributed control schemes for optical networks, *SPIE ITCom 2001*, Aug. 19–24, Denver; *Proc. SPIE* 4527:65–72, 2001.

3. O. Gerstel, Optical layer signaling: How much is really needed? *IEEE Communications*, pp. 154–160, Oct. 2000.

4. G. Berstein, J. Yates, and D. Saha, IP-centric control and management of optical transport networks, *IEEE Communications*, pp. 161–167, Oct. 2000.

5. J. Wei et al., Network control and management of reconfigurable WDM all-optical network, *IEEE NOMS'98*, 3:880–889, 1998.

6. B. Wilson et al., Multiwavelength optical network management and control, *Journal of Lightwave Technology*, 18(12):2038–2057, Oct. 2000.

7. J. Wei et al., Connection management for multiwavelength optical networking, *IEEE Journal on Selected Areas in Communications*, 16(7):1097–1108, Sept. 1998.

8. M. Maeda, Management and control of transparent optical networks, *IEEE Journal on Selected Areas in Communications*, 16(7):1008–1023, Sept. 1998.

9. E. Rosen et al., *Multiprotocol Label Switching Architecture*, IETF RFC 3031, Jan. 2001.

10. R. Braden et al., *Resource ReSerVation Protocol (RSVP), Version 1 Functional Specification*, IETF RFC 2205, Sept. 1997.

11. B. Jamoussi et al., Constraint-based LSP setup using LDP, Internet draft, *http://www. IETF.org/internet-drafts/draft-ietf-mpls-crldp-05.txt*, a work in progress.

12. D. Awduche et al., Multi-protocol lambda switching: Combining MPLS traffic engineering control with optical crossconnects, *IEEE Communications*, pp. 111–116, Mar. 2001.

13. E. Mannie et al., Generalized multi-protocol label switching (GMPLS) architecture, Internet draft, *http://www.IETF.org/internet-drafts/draft-ietf-ccamp-gmpls-architecture-02.txt*, a work in progress.

14. A. Banerjee et al., Generalized multiprotocol label switching: An overview of routing and management enhancements, *IEEE Communications*, pp. 144–150, Jan. 2001.

15. C. Qiao and M. Yoo, Optical burst switching: A new paradigm for an optical Internet, *Journal of High Speed Networks*, special issue on optical networking, 8(1):69–84, 1999.

16. B. Rajagopalan et al., IP over optical networks: A framework, Internet draft, *http://www.IETF.org/internet-drafts/draft-many-ip-optical-framework-01.txt*, a work in progress.

17. Z. Tang et al., Extensions to CR-LDP for path establishment in optical networks, Internet draft, *http://www.IETF.org/internet-drafts/draft-tang-crldp-optical-00.txt*, a work in progress.

18. Y. Ye et al., Design hybrid protection scheme for survivable wavelength-routed optical transport networks, *Proc. European Conference on Networks and Optical Communication*, pp. 101–108, June 2000.

19. J. Moy, *OSPF Version 2*, IETF RFC 2328, Apr. 1998.

20. R. Callon, *Use of OSI IS-IS for Routing in TCP/IP and Dual Environments*, IETF RFC 1195, Dec. 1990.

21. K. Kompella et al., IS-IS extensions in support of generalized MPLS, Internet draft, *http://www.IETF.org/internet-drafts/draft-ietf-isis-gmpls-extensions-04.txt*, a work in progress.

22. K. Kompella et al., OSPF extensions in support of generalized MPLS, Internet draft, *http://www.IETF.org/internet-drafts/draft-ietf-ccamp-ospf-gmpls-extensions-00.txt*, a work in progress.

23. J. P. Lang et al., Link management protocol, Internet draft, *http://www.IETF.org/internet-drafts/draft-ietf-mpls-lmp-03.txt*, a work in progress.

24. D. Awduche et al., *Requirement for Traffic Engineering over MPLS*, RFC 2702, Sep. 1999.

25. J. Strand, A. Chiu, and R. Tkach, Issues for routing in the optical layer, *IEEE Communications*, pp. 81–87, Feb. 2001.

26. Y. Mei and C. Qiao, Efficient distributed control protocols for WDM all-optical networks, *Proc. International Conference on Computer Communications and Networks (IC3N)*, pp. 101–108, Sept. 1997.

27. L. Andersson et al., *LDP Specification*, IETF RFC 3036, Jan. 2001.

28. D. Awduche et al., RSVP-TE: Extensions to RSVP for LSP tunnels, Internet draft, *http://www.IETF.org/internet-drafts/draft-ietf-mpls-rsvp-lsp-tunnel-08.txt*, a work in progress.

29. P. Ashwood-Smith et al., Generalized MPLS signaling: RSVP-TE extensions, Internet draft, *http://www.IETF.org/internet-drafts/draft-ietf-mpls-generalized-rsvp-te-05.txt*, a work in progress.

30. Y. Fan et al., Extensions to CR-LDP and RSVP-TE for optical path setup, Internet draft, *http://www.IETF.org/internet-drafts/draft-fan-mpls-lambda-signaling-00.txt*, a work in progress.

31. P. Ashwood-Smith et al., Generalized MPLS signaling: CR-LDP extensions, Internet draft, *http://www.IETF.org/internet-drafts/draft-ietf-mpls-generalized-cr-ldp-04.txt*, a work in progress.

32. P. Ashwood-Smith et al., Generalized MPLS: Signaling functional description, Internet draft, *http://www.IETF.org/internet-drafts/draft-ietf-mpls-generalized-signaling-02.txt*, a work in progress.

33. R. Braden et al., Integrated services in the Internet architecture: An overview, IETF RFC 1633, Jun. 1994.

34. User network interface (UNI) 1.0 signaling specification, *http://www.oiforum.com/public/documents/OIF-UNI-01.0.pdf*.

35. C. Xin et al., On an IP-centric optical control plane, *IEEE Communications*, pp. 88–93, Sept. 2001.

15 Survivability in IP-over-WDM Networks

YINGHUA YE and SUDHIR DIXIT

Nokia Research Center, Burlington, Massachusetts

15.1 INTRODUCTION

Internet protocol (IP) traffic has surpassed voice traffic and has already become the dominant traffic. Consequently, the multilayer network architecture [IP/ATM (Asynchronous Transfer Mode)/SONET (Synchronous Optical Network/WDM (Wavelength-Division Multiplexing)] which currently transports IP traffic is undergoing a delayering and moving into IP-optimized two-layer architecture (IP directly over WDM), which has to stand up to the same level of availability as the multilayer network architecture (i.e., 99.999%, or "five 9's" availability, which translates to approximately 5.25 minutes of downtime per year). In Section 15.1.1, an example is given to illustrate why such a high level of availability is called for. Network availability is assured through a combination of equipment reliability and the implementation of various network survivability approaches. Network equipment reliability is ensured via built-in redundancy. Survivability refers to the ability of the network to transfer the interrupted services onto spare network capacity to circumvent a point of failure in the network [1]. Such failures can involve fiber cuts and node-related failure. Fiber cuts are due primarily to human error or natural disasters leading to link failures, and node failures may be due to component failure within equipment or to entire node shutdown caused by accidents such as fire, flood, or earthquake. Node equipment or component failure can be dealt with by standby redundancy. However, whole node shutdown is difficult to recover from since it requires huge spare capacity. Nevertheless, the occurrence of such events is rare and can be handled by applying techniques of recovery from link failures by rerouting traffic around the failed node(s). The monitoring and maintenance of links is more difficult than nodes since links generally span a long distance with variable geographical conditions, such as mountains, under water, or in a heavily built-up

IP over WDM: Building the Next-Generation Optical Internet, Edited by Sudhir Dixit.
ISBN 0-471-21248-2 © 2003 John Wiley & Sons, Inc.

city. Thus, it is more difficult to locate the failure exactly and repair the fiber cut. In this chapter we restrict our discussion to survivability from link failure.

15.1.1   Importance of Survivability in IP-over-WDM Networks

Figure 15.1 shows a three-node network. Here we assume that the node is an optical node and to simplify the computation, wavelength conversion is allowed in each node. Furthermore, the shortest path is adopted, and each node has the same traffic load, where the traffic load is the product of the expected arrival rate and expected holding time. The blocking probability can be formulated using the Erlang B loss formula [2], which is

$$E(W, \rho) = \frac{\rho^W / W!}{\sum_{k=0}^{W} \rho^k / k!}$$

where W is the number of wavelengths per link and ρ is the offered load in each node. In our example we assume that W is 100 wavelengths, and the offered load is 70 Erlangs; then the network will experience 0.137% blocking probability. However, if failure happens between nodes 1 and 2, the traffic between nodes 1 and 2 will reroute to node 3 (shown in Fig. 15.1), finally getting to node 2. The offered load on link 1–3 and link 3–2 will increase to 140 Erlangs. Using the Erlang B loss equation, the blocking probability for links 1–3 and 3–2 will be 30.12%. Using link-dependent assumption, the end-to-end blocking probability is $1 - (1 - 30.12)^2 = 51.17$! Clearly, it would be unacceptable that even a single link failure would result in the end-to-end blocking probability jumping from 0.137% to 51.17%. Many incoming calls will be blocked and many existing calls passing link 1–2 will be dropped due to the limited capacity of link 1–3. Thus, the data loss is catastrophic, further degrading the quality of service to the end user to the worst extent possible. The final result is the loss of revenue to the operator. To maintain the end-to-end quality of service (QoS) guarantees to customers according to the agreed-upon SLA

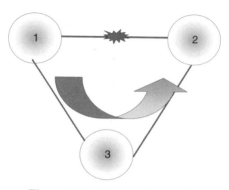

Figure 15.1   Three-node network.

(service level agreement) in the most cost-effective manner, various survivability schemes will need to be considered based on trade-offs between resource utilization, restoration time, complexity, and so on. From this example, we conclude that survivable network design is a requisite for future ultrahigh-bandwidth optical Internet.

In multilayer architecture, a high level of network availability is made possible by the SONET/SDH (Synchronous Optical Network/Synchronous Digital Hierarchy) layer, which is by far the most developed transport layer available today. To maintain the sub-50-ms resilience possible in SONET/SDH, the industry needs to devise a migration strategy from SONET/SDH to IP, and a target solution to enable IP transport directly over WDM. In a two-layer network architecture, eliminating the SONET layer and associated stand-alone network elements requires moving survivability and traffic grooming functions to either higher (IP, ATM) and/or lower (WDM) layers. Since IP and WDM have their own survivability mechanisms [3,4], it becomes very important to examine the survivability architecture, understand the pros and cons of performing recovery at each layer, and assess how the interactions between multilayer survivability affect overall network performance. Therefore, in Section 15.2, we present a network architecture of the next-generation transport infrastructure. In Section 15.3 we discuss the various survivability strategies that can be implemented to accommodate any particular technology. In IP-over-WDM networks, each layer provides survivability capabilities at various time scales, ranging from a few tens of milliseconds to minutes, at different bandwidth granularities from packet level to the whole wavelength level. In Section 15.4 we summarize three basic diverse routing algorithms and in Section 15.5 present details on survivability capabilities in different layers, give an example to illustrate how different survivability mechanisms react to a fiber cut, and introduce survivability schemes in generalized multiprotocol label switching (GMPLS)–capable networks. It is well known that the speed with which a fault is detected and notified directly affects the speed of service recovery. Therefore, in Section 15.6 we describe the various fault detection techniques in different layers, and in Section 15.7 we examine some restoration signaling issues in mesh networks. In Section 15.8 we look at the issues of survivability in future packet-based optical networks. Finally, Section 15.9 summarizes this chapter.

15.2 IP-OVER-WDM ARCHITECTURE

The explosive growth in data traffic and the emerging WDM technologies have had a profound impact on network architecture design. Here IP has become the revenue-generating convergence layer, and therefore the future network should be optimized to carry IP traffic. On the other hand, advances in WDM technologies, such as the optical cross-connects (OXCs), fiber amplifiers, single-mode fiber (SMF), tunable lasers, and tunable transceivers, are accelerating the migration from point-to-point transmission to mesh-based intelligent optical networking. Mesh type of networks well known as their rich network connectivity have proved to be very efficient against network faults. In the early 1990s, digital cross-connect system (DCS)–based mesh networks were studied extensively; however, due to the slow switching of DCS, applications of DCS-based mesh networks were confined to research. With the maturity of optical technologies, optical switches are now capable of providing switching speeds on the order of a few milliseconds to microseconds. These advances have made mesh networks a reality and the cost of end-to-end optical solutions acceptable. Today, much IP traffic is carried over WDM infrastructure, but with several other technologies separating them. These technologies [e.g., SONET/SDH, ATM, frame relay (FR)], provide features such as traffic grooming, performance monitoring, survivability, and standardized interfaces. What needs to be determined is whether those services can be provided adequately by a network architecture in which IP rides directly over WDM, and if so, what that network might look like. The widely accepted network model is that high-speed routers are interconnected by intelligent, reconfigurable optical networks [5]. Figure 15.2 shows one of the potential node architectures in IP over WDM networks. IP routers [i.e., label-switched routers (LSRs) as they are called in MPLS] are attached to multilayer OXCs and connected to their peers over dynamically switched lightpaths. The multilayer OXCs may consist of fiber-level cross-connects (FXCs) and wavelength-selective cross-connects (WSXCs) or wavelength interchange cross-connects (WIXCs) if wavelength conversion is allowed. OXC is controlled by the control plane, which is an IP-based controller that employs IP protocols to operate the connected OXC. The control plane may be integrated into the same box as the OXC, or it could be a separate router which is used to control the OXC. Between two neighboring OXCs, a dedicated out-of-band wavelength is preconfigured to provide IP connectivity.

Much effort has been directed to developing a multiprotocol label switching (MPLS) control plane that will support quality of service (QoS) and traffic

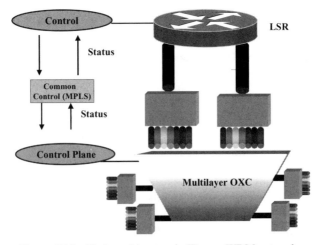

Figure 15.2   Node architecture in IP-over-WDM networks.

engineering in IP networks [1]. There are now proposals to extend MPLS to support multiple switching types [e.g., packet switching, time-division multiplexed (TDM) switching, λ-switching, and fiber switching]. This extension is called GMPLS. To better support IP over WDM integration in the GMPLS control plane, open shortest path first (OSPF) or intermediate system to intermediate system (IS-IS) is used as the interior gateway protocol (IGP) for neighbor discovery and link-state update. Constraint-based routing algorithms are developed to perform path computation for the connection requests, while the specific routing algorithms may be proprietary. Resource reservation protocol (RSVP) or constraint-based routing label distribution protocol (CR-LDP) is used as the signaling protocol to establish a path.

GMPLS enables IP routers to dynamically request bandwidth services from the transport layer using existing control plane techniques. Through the components of GMPLS such as neighbor discovery and link-state update, IP routers (i.e., LSRs) at the edge of the optical network have information about the network and will use this information to manage the available wavelengths or bandwidth in the optical layer. GMPLS supports two basic network architectures, overlay and peer to peer, for designing a dynamically provisionable and survivable optical network [6]. The reader is referred to Chapter 14 for more details on GMPLS.

In the *overlay model*, there are two separate control planes: one in the core optical network and the other in the LSRs. The interaction between the two control planes, although minimal, is through a user network interface (UNI). The LSRs see the lightpaths that are either dynamically signaled or statically configured across the core optical network, without seeing any of the network's internal topology (Fig. 15.3). When an LSR requests the wavelength service from its attached OXC, the attached OXC is responsible for processing the request, computing the lightpath, and establishing that lightpath to the other OXC, to which the destination router is attached. The disadvantage of this model is that connections among LSRs

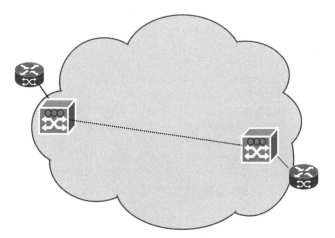

Figure 15.3   Overlay model.

are also used by the routing protocols, producing an excessive amount of control message traffic, which in turn limits the number of LSRs that can participate in the network. Nevertheless, the overlay model is easily deployed commercially while enabling carriers to use proprietary solutions as a competitive tool. It is anticipated that the overlay model will be the first step toward supporting IP-over-WDM integration. The survivability in the overlay model is provided by adopting separate survivability schemes in optical and IP domains, or multilayer escalated surviv- ability strategies (discussed in Section 15.5.5.1), where the necessary information between the two layers is realized by UNI. Since most services currently still require bandwidth of less than one full wavelength, traffic grooming is a key requirement for the success of an IP-over-WDM network. In the overlay model, traffic grooming can only be done in either the IP router for end-to-end connection (i.e., one lightpath) or OADM in the optical domain (discussed in Section 15.5.4).

In the *peer-to-peer model*, a single instance of a control plane spans both the core optical network and the surrounding edge devices (i.e., LSRs), allowing the edge devices to see the topology of the core network (Fig. 15.4). The lightpath is used exclusively for the purpose of data forwarding. As far as the routing protocols are concerned, each LSR is adjacent to the optical switch to which it is attached rather than to the other LSRs. That allows the routing protocols to scale to a much larger network. It is possible in a peer network to hide some or all of the internal topology of the optical network, if desired. That could be done by establishing permanent virtual circuit (PVC)–like connections or forwarding adjacencies through the cloud and sharing them in peer-to-peer topology exchanges with the routers. Survivability in the peer model has three options: MPLS protection/ restoration, MPLambdaS-based optical layer survivability, and integrated protection/

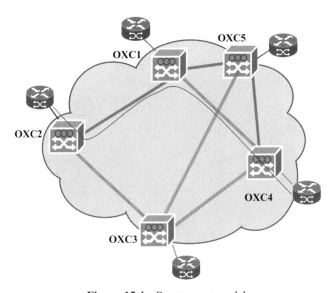

Figure 15.4   Peer-to-peer model.

restoration (discussed in Section 15.5.5.2). The peer-to-peer model is more flexible and effective than the overlay model. Traffic grooming (sometimes called *traffic aggregation*) can be coordinated by both the IP and WDM layers. Detailed schemes are examined in Section 15.5.5.2.

15.3 SURVIVABILITY STRATEGIES

Survivability can be classified as protection and restoration [7,8]. In most cases, protection mechanisms are designed against a single failure event, whereas restoration can be used against multifailure events. Further distinctions between protection and restoration can be made based on other factors, such as restoration speed, capacity overbuild, and degree of deterministic behavior.

15.3.1 Protection

Protection is preprovisioned failure recovery, where a secondary path is computed and reserved at the same time as the primary path; otherwise, the connection request is rejected. Simultaneous reservation of a secondary path ensures fast restoration by eliminating path computation when the failure actually occurs. Defects are used as triggers, resulting in fast detection time (e.g., physical media faults can be detected within 10 ms [9]). There are two forms of protection mechanism: dedicated and shared. *Dedicated protection* uses 100% of the protection capacity overbuilt for dedication to the primary connection (see Fig. 15.5). For example, for connection 1 from node 4 to node 6, the primary path uses path 4–5–6, and the secondary (i.e., dedicated protection) path uses path 4–1–2–6 while using the same wavelength, λ_2. Since there are two connections in the network, and their secondary paths share the same link between nodes 2 and 6, two wavelengths are needed to set up the connections.

The most commonly deployed protection architectures are $1 + 1$ and $1 : 1$. In $1 + 1$ *architecture*, data are transmitted in both primary and secondary paths,

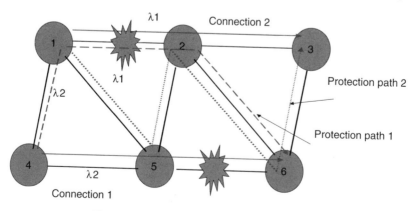

Figure 15.5 Example of dedicated protection.

where the destination picks up the better-quality signal. The $1+1$ architecture scheme does not support extra traffic since the primary and secondary paths carry the same traffic simultaneously. To prevent data loss, the source node should delay transmitting data on the secondary path for a certain amount of time, depending on difference in the propagation delays between the primary and secondary paths, plus the fault detection time [10]. There are two types of $1:1$ *protection*. In one type, a dedicated protection path is required, but the protection path is allocated to carry low-priority traffic under normal circumstances. In the second case, the protection path is not dedicated and multiple protection lightpaths can share the same resources in the protection path as long as there is no shared link among their associated working lightpaths. The $1:1$ protection scheme can be extended to $M:N$ protection, which uses N protection lightpaths to protect the traffic on M working lightpaths (i.e., if there is a failure on any one of the primary resources, data are switched to the backup resource). Apparently, $1+1$ architecture is faster and simpler than $1:1$ architecture but at the expense of lower network utilization.

The dedicated backup reservation method has the advantage of shorter restoration time since the resources are reserved for the secondary path when establishing the primary path. However, this method reserves excessive resources. For better resource utilization, multiplexing techniques [11] have been studied extensively (i.e., if multiple primary paths do not get affected simultaneously, their secondary paths can share the resources). In this case, the resources along the secondary paths are allocated in real time for one of the secondary paths whose primary path is affected by a fault; however, the backup capability of the other connections is no longer preserved. Therefore, proper *reversion* (e.g., moving traffic from the secondary path back to corresponding primary path after failure has been restored) should be considered. The details are discussed in Section 15.7. Since the *shared protection scheme* requires that the primary paths whose secondary paths share the resources cannot incur the same risk, the *shared risk link group* (SRLG) concept has

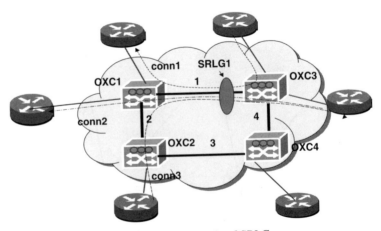

Figure 15.6 Example of SRLG.

been developed to help identify risk associations among various resources [12,13]. Figure 15.6 illustrates the SRLG concept. Each physical link between OXCs determines a SRLG; in particular, SRLG1 (associated with link 1) consists of *{conn 1, conn 2, conn 3}* and represents the shared risk of interrupting both connections if link 1 fails.

The SRLG concept is used to ensure that the primary and secondary paths of one call are not affected by the same failure, and to identify the primary paths that may not be capable of sharing the resources for their secondary paths. Shared protection offers the advantage of improving network utilization. However, to capture SRLG, more network status information is needed, and extended signaling protocol should be developed to disseminate SRLGs. Besides, a signaling protocol (such as RSVP or CR-LDP) is also involved to set up the protection path after the failure occurs. Therefore, the recovery time in a shared protection scheme is longer than in a dedicated scheme.

Figure 15.7 illustrates how the shared protection scheme works at the optical layer. We still use the same connections as in Fig. 15.5. The primary path (4–5–6) of connection 1 and the primary path (1–2–3) of connection 2 do not share any link or node, but the secondary paths of the corresponding connections do share the link between nodes 2 and 6. Since in our network design, we only try to avoid single-link failure, the secondary paths of the two connections can share the same wavelength on the link between nodes 2 and 6. Hence, only one wavelength is needed for setting up two connections. From this example we can see clearly that significant reductions in network resources can be achieved by sharing spare capacity across multiple independent failures.

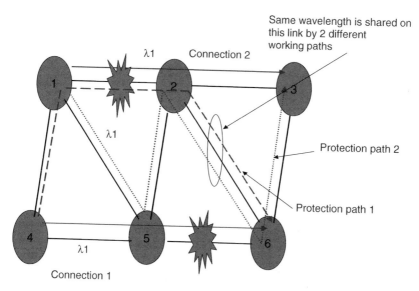

Figure 15.7   Shared protection scheme in optical layer.

15.3.2   Restoration

Restoration is after-the-fact dynamic signaled recovery, where the secondary path is computed and set up after fault alarms have triggered. The restoration mechanism was based on a centralized approach implemented within an external management system [9]. The centralized approach requires network management to be implemented at one place. This approach suffers from the problem of scalability, because when the network is large and the traffic load is high, the processing power at the central management site can become a bottleneck, and request processing may incur a large delay. In addition, the central controller is susceptible to a single point of failure. Distributed restoration mechanisms have been proposed resulting in faster restoration, where each individual node is fully or partially responsible for signaling, routing, and restoration, and the scalability problem is resolved. There are two basic distributed restoration techniques: end-to-end path restoration and link restoration. In *end-to-end path restoration*, the source and destination of each connection traversing the failed link or node participate in distributed algorithms to dynamically discover the second path. The fault has to be located and the source and destination must be notified of the failure so that the source can find an alternative path, avoiding the fault location. *Link restoration* is also called *local repair*, where the end nodes of the failed link or node compute and establish the alternative path. In both restoration approaches, if an alternative path cannot be discovered for a broken connection, that connection is blocked. Therefore, restoration cannot guarantee recovery and it takes much longer (in seconds) to restore service. The restoration mechanism is commonly deployed in mesh-based topologies. Generally, meshes are more efficient in utilizing capacity due to the possible sharing of spare resources. Failure detection and notification are very critical to achieving fast restoration. Since the basic principles of protection and restoration schemes are similar, for example, signaling and alternative path routing algorithms, they can be applied equally well in both schemes. In the remainder of the chapter, we focus on protection schemes.

15.4   SURVIVABLE ROUTING ALGORITHMS

In survivable networks, the fundamental and important issue is how to select reliable and efficient paths for each connection request. A survivable routing algorithm (i.e., survivable path selection) further divides into two subproblems: diverse routing and resource assignment. For example, in optical networks, survivable lightpath routing consists of diverse path routing and wavelength assignment (RWA). Compared with higher-layer (i.e., IP) path selection, the algorithms for diverse routing can be applied at both the optical and IP layers. However, the resource allocation for IP layer connection is slightly different from the wavelength assignment for the lightpath, where RWA adds more constraints in resource assignment, such as without wavelength conversion, the wavelength continuity needs to be maintained and traffic granularity is coarser. In the following, we take the RWA of a lightpath as an example to introduce the general concepts of how to realize routing and resource allocation.

15.4.1   Overview of Routing and Wavelength Assignment Approaches

The various survivable routing algorithms have been extensively studied in [9] and [14]. Depending on the different traffic patterns, many researchers have proposed different approaches to solve the RWA problem. For example, the static lightpath establishment (SLE) approach has been proposed for static traffic and the dynamic lightpath establishment (DLE) approach for dynamic traffic. For SLE, the RWA problem can be described as follows: Given the physical topology and a set of end-to-end lightpath demands, determine a path and wavelength for each demand with minimum resources. (Path-routing algorithms are introduced later in this section.) Solving the wavelength assignment problem can be equivalent to the graph-coloring problem, which has been studied extensively with existing algorithms [15]. In [14] an integer linear programming (ILP) formulation has been proposed to formulate the RWA problem that minimizes the wavelength utilization for a static traffic demand under different link and path protection schemes. With dynamic traffic demand, the connection request (s-d pair) comes dynamically in time, not all of a sudden. The established lightpath can also be torn down dynamically. Hence, in DLE, there is no optimization similar to that in SLE, and it may lead to higher blocking probability [14]. Anand and Qiao [16] proposed a dynamic strategy to rearrange protection paths to increase network capacity availability; hence the network can accommodate more connections. The lightpath routing in DLE for an s-d pair can just use some precomputed fixed paths or employ graph-theoretic approaches, such as shortest path first (SPF), to compute a path dynamically. The wavelength assignment uses heuristics such as first fit, random, and so on [17]. The DLE approach can be used whenever a connection request is received, and there are two choices: hop-by-hop routing or explicit routing. Hop-by-hop routing can make more accurate next-hop decisions because each node can consult its local resource information, which may be unavailable to other nodes. However, hop-by-hop routing requires path calculation at each node and needs loop-prevention mechanisms. Hop-by-hop routing is also more difficult when constraints other than additive link metrics are taken into account. The interested reader is referred to Chapter 12 for a more comprehensive overview.

15.4.2   Path Routing

Generally, routing in communication networks becomes a graph-theoretic problem; the nodes and links of a network are transformed into vertices and edges of a simple graph $G(V, E)$, where V is a set of nodes and E stands for a set of edges, each edge representing a pair of opposite directed links. The length of the edge reflects the cost of link. In Section 15.4.3 we discuss how to define the cost function of link and path costs [18].

15.4.3 Cost Function

15.4.3.1 Link Cost   There are many definitions of link cost in the literature, which can generally be classified into topology based, resource use based, or topology and resource use integrated. The topology-based link cost definitions include hop count, link mileage, reliability, and so on. The resource use–based link cost definitions include available bandwidth or wavelengths, available bandwidth/wavelengths weighted by the total bandwidth/wavelengths, and so on. The integrated link cost definitions generally integrate both topology information and resource use information into the cost. In most scenarios, the hop counts and available resources are combined and weighted to form the link cost. An integrated link cost function $l(i,j)$ for a link between nodes i and j is as follows:

$$l(i,j) = W_u \times f(R_{\text{used}}, R_{\text{total}}) + W_h \times 1$$

where R_{used} is the total amount of used resources, R_{total} the total amount of resources, and W_u and W_h are weights. The number "1" represents the hop count cost (i.e., a link represents "1" hop), and $f(x, y)$ is a function to control the effect of used resources and total resources in the cost function. One can simply use a function such as $f(x, y) = ax/y$, where a is a coefficient. However, more complex functions, such as log, polynomial, and exponential, can be used to introduce nonlinear effects to the cost. Through W_u and W_h, $l(i,j)$ can control the relative significance of the hop count or the resource use information. Via $f(x, y)$, $l(i,j)$ can control the cost increment/decrement rate with the changed resource use information.

15.4.3.2 Path Cost   Generally, for the working path or dedicated protection path, the path cost is the sum of the link cost at each hop. However, for the shared protection path, there are two possible scenarios. In the first scenario, similar to the case of the working path, the protection path cost is defined as the sum of the link costs. However, in the second scenario, if shared protection is considered, the protection path cost is more related to resource consumption than to link cost. Therefore, during the path selection process, if the protection path shares resources in an individual link with other protection path(s), the contribution of this specific link cost to the path cost should be reduced significantly. The working path cost CW and the protection path cost CP are defined as

$$\text{CW}_p = \sum_{i,j \in p} l(i,j)$$

$$\text{CP}_p = \sum_{i,j \in p} \phi(l(i,j), \alpha)$$

$\phi(l(i,j), \alpha)$ equals to $l(i,j)$ when p has no λ sharing with other protection path(s). Otherwise, $0 \leq \phi(l(i,j), \alpha) < l(i,j)$. In our study we use a linear function as the

$\phi(l(i,j), \alpha)$; that is,

$$\phi(l(i, j), \alpha) = \alpha \times l(i, j)$$

$$\alpha = \begin{cases} \alpha(n) & \text{if the link is already used by other protection paths} \\ 1 & \text{otherwise} \end{cases}$$

where $0 \le \alpha \le 1$ and α is called the *sharing control weight* (SCW) because it is used to control the link cost of the shared link.

15.4.4 Survivable Routing Algorithms

Thus, survivable routing involves finding the shortest pair of edge disjoint or vertex disjoint paths between a single pair of vertices in graph *G*. The best known and most efficient shortest-path algorithm is Dijkstra's algorithm [19]. Three survivable routing approaches are discussed below.

15.4.4.1 Conventional Survivable Routing The conventional survivable routing method adopts a two-step algorithm; it selects a route with minimum cost for the working path (in the rest of this chapter, route and path terms are used inter-changeably) at first, removes the links along the working path, and then selects a route with minimum cost for the protection path. Although both the working and protection paths are selected separately with minimum costs, the sum of the working and protection path costs may not be minimal. Figure 15.8 gives an example of a two-step survivable routing algorithm for setting up a connection between node A and node H. Route 1 (A–C–F–G–H) is selected as the working path with minimal cost 4, and route 2 (A–B–E–H) is obtained with minimal cost

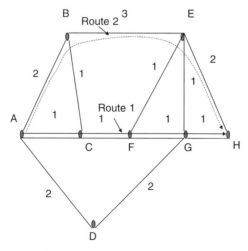

Figure 15.8 Conventional two-step survivable routing.

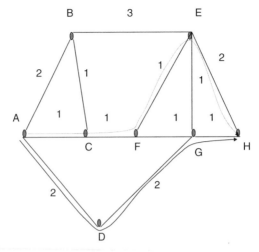

Figure 15.9   Optimal routes for connection A to H.

$2 + 3 + 2 = 7$ as the protection path after removing all the links of the working path. Thus, the total cost of connection from A to B is $4 + 7 = 11$. However, routes 1 and 2 are not the optimal choice, since the sum of costs of the disjointed path pair A–C–F–E–H, and A–D–G–H shown in Fig. 15.9 (i.e., $5 + 5 = 10$) is less than the sum of cost of routes 1 and 2. Furthermore, in the worst case, it may not find a secondary path by using the conventional two-step routing approach.

15.4.4.2 Suurballe's Algorithms   To overcome the shortcomings of the conventional approach, Suurballe's algorithms were developed to search for the shortest pairs of disjointed paths [20]. In Suurballe's algorithms, Dijkstra's algorithm is used first to construct the shortest-path tree, rooted by the source node. Let $d(i)$ denote the shortest distance from source node to node i. Second, the graph transformation is conducted (i.e., $G \rightarrow G'$), where the cost of link $l(i,j)$ in G is transferred into $l'(i,j)$ by $l'(i,j) = d(i) + l(i,j) - d(j)$. Third, the graph G' is modified (i.e., all links along the shortest path that are directed toward the source node are removed). Then Dijkstra's algorithm is run again to find the shortest path from the source node to the destination node in the modified graph G'. Finally, the interlacing part of the shortest path from the source node and the destination node in original G and the shortest path from the modified G' are removed. This results in the desired pair of paths. Suurballe's algorithms require determination of the shortest paths from source node to all other nodes and transformation of modified G'. Unfortunately, these tasks take time to finish, and as the number of links in the network increases, this time can be substantial. Furthermore, for the shared protection case, in the second scenario, the cost function for the working path is different from that for the protection path. However, Suurballe's algorithms cannot be applied here since they simultaneously obtain a pair of shortest paths, where the cost functions of the pair of paths have to be the same.

15.4.4.3 *Joint Working and Protection Routing Approach*   To simplify computation and make the disjoint path algorithm more practical, a joint lightpath
selection approach known as *joint path selection* (JPS) has been proposed [18].
The motivation here is that if the sum of the costs is minimized, the network
resource utilization is optimized at least for each new call, and better performance
can be expected. JPS selects the working (W) and protection (P) paths between a
source node and a destination node using the following heuristic:

1. Compute K candidate routes with W_i representing the ith $(1 \le i \le K)$ route
   and CW_i as the cost of W_i. Note that in a given topology, it is possible that,
   between two specific nodes, the number of all possible routes is less than K.
   In such cases, just compute all possible routes.
2. For $W_i(1 \le i \le K)$, compute an edge-disjoint route, represented as P_i and the
   cost of P_i denoted as CP_i.
3. Find h such that $CW_h + CP_h = \min(CW_i + CP_i),\ 1 \le i \le K$.
4. Select W_h as the working path and P_h as the protection path.

15.5 SURVIVABILITY LAYER CONSIDERATIONS

15.5.1 IP Layer

15.5.1.1 *Conventional IP Protection and Restoration Approaches*   The conventional IP networks are connectionless; each router dynamically manages the
routing table, which is established by executing routing protocols such as OSPF or
IS-IS. Usually, for router to router, the routing protocol computes a set of *least cost
paths* based on the network topology and resource information, and the path cost
can be defined as the function of the link cost, which can be the hop count, the link
delay, and so on. For a given destination, each router may have several cost-
associated entries in the local routing table. Each entry points to the local egress
port for the next router along the route to the destination. Whenever there is any
change, the topology update message must be propagated to every router in the
network; then every router computes the new path and fills out the related entries in
its routing table. Therefore, a packet is routed on a hop-by-hop basis from source to
destination (i.e., when a packet arrives at an intermediate router, the router inspects
its destination IP address, accesses the routing table to check the egress port, and
forwards the packet to the next router along the path). Protection and restoration in
an IP layer are essentially best effort in nature. IP dynamic rerouting and load
balancing in multiple diverse paths can be used against failure. Figure 15.10
illustrates how an IP layer reacts against a link failure by adopting dynamic
rerouting. Let us assume that router A sends traffic to router D along route 1, and
router B detects the link failure between router B and router D, at which time
it rapidly determines an alternative entry from the local routing table, rerouting
the traffic to router C along route 2 until a new route 3 is set up completely. At the

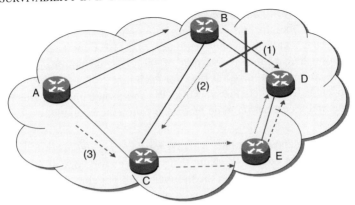

Figure 15.10 IP layer rerouting approach.

same time, it sends out the topology updates to notify the network that the link between B and D has failed. When router A receives the update messages, it updates its topology and resource database, calculates the new optimal route 3 to router D (i.e., A–C–E–D) installs a new forwarding table in the router, and finally, the traffic is totally switched to route 3. On load balancing, the IP layer distributes the traffic over multiple paths, and some routing protocols (such as OSPF and IS-IS) explicitly support this capability. Thus, spreading the traffic over multiple paths guarantees that the source router can still send some packets over at least one path to the destination router in the event of a single point of failure (i.e., single link or node failure). When a better route becomes available afterward, the source router has the option to select the new route.

Protection and restoration at the IP layer may take up to several minutes. The inherent reasons for slow recovery are as follows. First, the network topology takes several seconds to converge, depending on the network size and topology; this may sometimes result in incomplete or inaccurate topology and resource information, leading to a bad route selection and/or packets looping around the network or dropping. Second, the detection time in the IP layer is quite long because "hello" messages are used periodically to check the health status of the link. Only after several consecutive "hello" messages get lost does the router conclude that the link is down, and therefore fault detection time is directly related to how frequently a "hello" message is sent. The slow reaction time to fault determination might cause transient service disruptions on the order of seconds and reductions in QoS when restored [4]. Another major problem in IP layer protection is that of *failure propagation*: that is, one physical cut can expand to tens of thousands of simultaneous logical link failures at the IP layer. It is to be noted that the survivable logical topologies are often designed to withstand only a single logical link failure, and it is likely that the primary path and the secondary path share the same physical link since the IP layer is transparent to how the lightpaths are routed at the physical layer. In the worst-case scenario, a single link/node failure may lead to partitioning the virtual layer into many disjointed islands. Hence, with conventional IP rerouting, it is almost impossible to achieve full-service recovery against lower-layer failures.

15.5.1.2 MPLS-Based Recovery

MPLS enhances network performance by supporting traffic engineering and QoS by separating forwarding (based on MPLS label) from constraint-based network routing. (The reader is referred to Chapter 14 for details on MPLS, MPλS, and GMPLS.) It is expected to provide greater fault protection and faster restoration than in conventional IP approaches. In MPLS-capable networks, since the control and data planes are separated, switching traffic to secondary path is not dependent on IP routing convergence. Therefore, restoration speed can be improved significantly to the extent that the IETF (Internet Engineering Task Force) considers MPLS recovery mechanisms to be capable of providing SONET/SDH-like fast recovery. But thus far, the MPLS-based recovery approach has not been standardized. An additional advantage of MPLS is that the constraint-based explicit routing can provide flexible and differentiated reliability services. The IETF has defined two recovery models: protection switching and fast rerouting [8,21].

In protection switching, a primary label-switched path (LSP) and the corresponding secondary LSP have to be established at the same time. Here, $1 + 1, 1 : 1$, $1 : N$, or $M : N$ protection types can be applied. There are also two basic subtypes of protection switching: link protection (i.e., local repair) and path protection (global or end-to-end repair). In link protection, several secondary LSPs may need to be configured for one primary LSP, depending on the specific faulty link that each secondary LSP protects. For example, in Fig. 15.11, to protect primary LSP (A–B–D), two secondary LSPs are needed against link AB failure and link BD failure, respectively. In the event of link BD failure, label-switched router (LSR) B is an immediate upstream node of the fault and initiates the recovery process. The LSR B that is responsible for diverting the traffic to backup LSP is called the PSL (path-switched LSR). When the LSR B detects that the link BD is down, the concept of label stack is applied [22]; an extra label is pushed into the packet at LSR B, and popped off at LSR D, also called PML (path merge LSR), where the protection traffic is merged back onto the primary path.

In path protection, the protection is end to end, since in most cases the recovery path can be link or node disjointed with its primary path. Thus, only one secondary LSP can protect against all link and node failures on the primary path. The end-to-end path protection methods also enhance the network reliability and availability since they utilize the global view of network resources. However, PSL is usually far from the failure and needs to be notified via appropriate signaling protocols. Thus, there may be some delay in the switchover to secondary LSP. In the example shown in Fig. 15.11, LSR B detects the fault and sends fault notification to the ingress router A (i.e., PSL); then LSR A diverts the traffic to the existing (path protection based) backup channel (A–C–E–D).

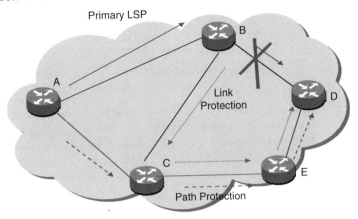

Figure 15.11   Protection in MPLS networks.

Compared to protection switching, the fast rerouting model can make use of network resources more efficiently since it allocates resources for recovery LSP right after a fault occurs. This model can be further classified into established-on-demand type and prequalified type. In an established-on-demand protection type, after a fault occurs, the PSL has to follow two steps: (1) compute a new path based on real-time network status, and (2) set up a backup LSP. The restoration time for an established-on-demand model includes fault notification time, path selection time, and path setup time. The prequalified rerouting is a compromise between established-on-demand rerouting and protection switching. It has higher network utilization than protection switching but slower restoration. Furthermore, it can achieve faster restoration but at lower efficiency than established-on-demand approach. This is because in prequalified rerouting, the secondary path is calculated at the primary LSP setup time, but the backup LSP is set up after the occurrence of the fault. However, at the time the fault occurs, the backup LSP may not be the optimal path compared to the one found in the rerouting method.

Nevertheless, MPLS recovery is still hampered by failure detection. It should be noted that even a single fiber cut or node failure will affect multiple LSPs, which means that hundreds of ingress routers of LSPs should be notified, requiring intensive processing if MPLS signaling is used. Multiple LSP failures also lead to even more LSRs to update the topological and forwarding information, causing the network to become potentially unstable while raising the issue of signaling scalability.

Overall, IP layer survivability mechanisms require minimal capacity overbuild and provide very flexible protection options in meshed topology, but they need careful IP dimensioning to ensure that spare capacity in layer 3 is available via adding larger/additional pipes and interfaces on routers; otherwise, these approaches may result in poor QoS, due to network congestion.

15.5.2   WDM Layer

WDM lacks traffic granularity and operates at the aggregate lightpath level. There are two main protection schemes in the optical domain: a fiber-based protection scheme a and lightpath-based protection scheme. Fiber-based protection is employed at the optical multiplex section level by performing fiber protection switching. The protection granularity is one fiber capacity. Its applications include linear (point-to-point) and ring architectures. Each working fiber is backed by another disjointed protection fiber path in the ring [e.g., automatic protection switching (APS)]. This fiber-based protection scheme requires only simple control and management mechanisms, and provides sub-50-ms protection speeds. However, beyond providing simple fast traffic restoration, network utilization is also very important. A major drawback of fiber-based protection is its very low network utilization.

The lightpath-based protection scheme is used primarily in OXC-based mesh architectures deployed at the wavelength level, and it is similar to the IP layer protection. It has three variants:

1. *Link protection*, which reroutes the disrupted traffic between the end nodes of the failed link. This approach offers the fastest failure detection and tries to recover traffic locally at the expense of efficiency (more hops, more bandwidth, more end-to-end delay).

2. *General path protection*, which reroutes the broken traffic between the two end nodes of the affected path and takes advantage of the spare capacity all over the network. It has a lower spare capacity requirement (as expected) than does link protection, but takes longer to detect failure and to propagate the error notification to a responsible node, which can be the source node or the intermediate node, and to set up the alternative path.

3. *Disjoint-link path protection*, which is a special case of path protection, adds the link-disjoint constraint to path selection and can restore traffic immediately once it discovers the path failure. The source doesnot need to know the exact location of the failure, as it is capable of protecting against multiple simultaneous failures on the working lightpath.

General path protection includes link-dependent protection and disjoint-link protection. Figures 15.12 and 15.13 show the difference between a link-dependent path protection scheme and a disjoint-link path protection scheme. In Fig. 15.12, the primary path is OXC2–OXC1–OXC4, and the secondary path selects OXC2–OXC1–OXC3–OXC4. The primary and secondary paths share part of the resources from the link between OXC1 and OXC2, so it may be more effective in some cases. The link-dependent approach also offers more chances to accept calls compared with disjointed path protection (shown in Fig. 15.13), since it alleviates the constraint of finding the secondary path, and in disjointed path protection there may be times when no resources are available, since in optical networks, wavelength continuity must be obeyed without wavelength converters.

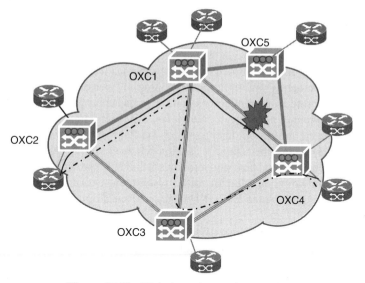

Figure 15.12 Link-dependent path protection.

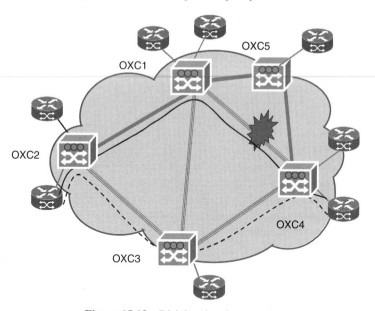

Figure 15.13 Disjointed path protection.

The commonly used survivability mechanisms in WDM network are listed in Table 15.1. The currently deployed survivable architectures are dedicated linear protection and shared ring protection. Optical multiplex section protection (OMSP) and optical channel subnetwork connection protection (OCH-SNCP) belong to dedicated linear protection category, and shared protection architecture is applicable

TABLE 15.1 Various Survivability Mechanisms

Survivability Mechanism	Recovery Speed (ms)	Capacity Overbuild	Flexibility	Standardization
OMSP	50	1 + 1 or 1 : 1	Linear topology	Standardized
OCH-SNCP	50	1 + 1 or 1 : 1	Linear or ring topology	Standardized
OMS-SPRing	50	1 : 1 shared less than 100%	Linear or ring topology	In progress
OCH-SPRing	50	1 : 1 shared less than 100%	Linear, ring, or mesh topology (but not optimized)	In progress
Distributed restoration	< 100	Minimal capacity overbuild	Linear, ring, and optimal solutions for mesh	Time frame questionable

Source: Data from [9].

only in a ring or multiring environment. The latter includes an optical multiplex-section shared protection ring (OMS-SPRing) and optical channel shared protection ring (OCH-SPRing). Other optical channel shared protection schemes are still in the study phase. Optical restoration can flexibly use both fiber- and channel-level restoration mechanisms.

15.5.3 Illustrative Example

An IP/WDM network can employ the following three options according to protection granularity: fiber protection (at the OMS level), channel protection (at the OCh level), and higher-layer protection (at the IP layer). To better understand the difference among various granularity survivability schemes, we take a simple network (Fig. 15.14) as an example to show how these three protection options interwork after a single link failure has been detected [1]. The sample network consists of four OXCs and seven LSRs. The physical connections are shown in Fig. 15.14a. The virtual topology presents the connections among the routers (Fig. 15.14b), where one router-to-router link is the lightpath in the WDM layer and may span several physical links. The routing and wavelength assignment algorithm doesn't consider wavelength conversion since it is quite expensive and the technology is not yet mature. Therefore, the constraint that the wavelength is continuous from the end-to-end path has to be considered. Given that there is a fiber cut between OXC1 and OXC4 and the two lightpaths, which connect LSR4 with LSR5 and LSR1 with LSR5, respectively, are affected, if we use the fiber protection option, once OXC1 detects a failure, it triggers the protection mechanism and consults its own database, thus setting the fiber protection path (OXC1–OXC3–OXC4) (Fig. 15.14c). OXC1 therefore sends notification messages to OXC3 and OXC4, and modifies the fiber interface element in the entry of the

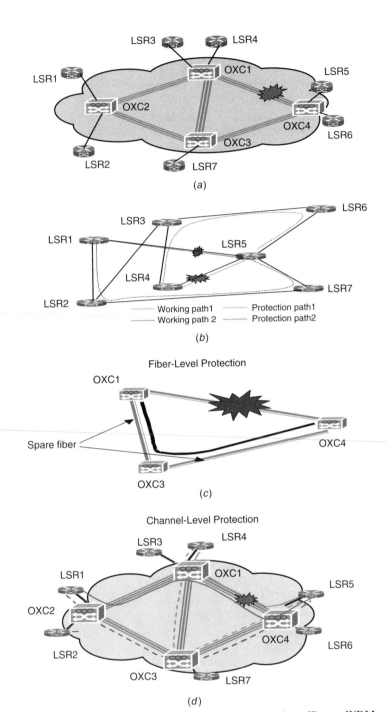

Figure 15.14 Example demonstrating joint protection scheme in an IP-over-WDM network. (*a*) The cable cut shows in the physical topology; (*b*) protection is implemented in virtual topology (router–router connectivities); (*c*) fiber-level protection needs spare fibers to reroute all the traffic carried by the failed cable; (*d*) channel-level protection reroutes the traffic carried by the failed lightpaths among the working fibers.

494

wavelength-forwarding table while keeping the wavelength element untouched. Through fiber switching, the fiber protection route is set up. In OXC1, the traffic is switched to the protection fiber, and in OXC4, the traffic is switched back to the normal status.

If lightpath-level protection scheme is used, then to recover the traffic carried by the two lightpaths, the protection paths have to be found and reserved at the same time as the working lightpaths. The protection path selection process is the same as the working path selection process, except that it adds additional constraints (e.g., the working lightpath cannot share the same failed link as the protection lightpath) to the routing protocol that runs over the link-state database. The protection lightpath may use a different wavelength than the working lightpath wavelength provided that wavelength agility is allowed. After OXC1 detects failure, it sends failure notification messages, encapsulated in TLVs (type–length–value), to the protection-switched OXCs, which are responsible for switching traffic from working lightpaths to protection lightpaths. In our example, OXC2 is a protection-switched OXC. A lightpath from LSR1 to LSR5 expressly bypasses (i.e., cuts through) OXC1 and OXC4 without getting into the attached routers. Once OXC2 gets alarms, it will activate the protection path OXC2–OXC3–OXC4 through proper signaling. A lightpath from LSR4 to LSR5 will be protected by an alternative path, OXC1–OXC3–OXC4 (Fig. 15.14d).

The two optical protection schemes above do not affect the virtual topology since the lightpaths provide the link for IP connection and router-to-router links are not changed. If IP layer protection is chosen, the traffic is rerouted on the virtual topology and no new direct connection needs to be established. Here, since there is lightpath disruption, which results in disconnection between LSR1 and LSR5 and LSR4 and LSR5, the virtual topology is changed. The IP layer protection employs the existing lightpaths to reroute traffic. The LSRs process the incoming traffic on a packet-by-packet basis (where the traffic can also be buffered). The edge LSR knows the virtual topology along with the resources utilized and selects a protection path by running explicit constraint-based routing. For LSR-to-LSR links, physical link diversity is a major consideration. Due to the multiplexing of wavelengths, it can be difficult to ensure that a physical link failure does not affect two or more logical links, and it takes time to complete routing convergence in the IP layer. Hence, to implement rapid restoration in the IP layer without waiting for routing convergence, the edge LSR has to remove the unqualified links (e.g., lightpaths) when it computes the secondary path, and then applies the shortest path algorithm to find an alternative path. Since the protection uses the existing lightpaths, it may not provide 100% survivability. To provide maximum survivability, the intermediate LSRs may drop some low-priority traffic (e.g., small bandwidth LSPs), which can be distinguished from the label of a packet, and for the traffic disrupted, the protection first considers the high-priority traffic. For this, the traffic has to undergo optical–electrical–optical (O/E/O) conversion at the intermediate LSRs. In Fig. 15.14b, traffic between LSR1 and LSR5 can be protected by the alternative LSP, which proceeds as LSR1–LSR2–LSR7–LSR5. The traffic between LSR4 and LSR5 can be restored by the path LSR4–LSR3–LSR6–LSR5. In IP/MPLS, traffic engineering allows for multipath routing, label stacking, and label splitting. The protection granularity is flexible. For example, if a large-granularity LSP cannot be protected, it can be demultiplexed into many small-granularity LSPs. Then each small LSP can have its own protection path, and each working LSP may have multiple protection paths where one of them carries a part of the protected traffic of the working LSP.

15.5.4   Traffic Grooming for Low-Rate Streams in Survivable WDM Mesh Networks

With the deployment of WDM, there exists a mismatch between the transmission capacity of one wavelength and the users' requested bandwidth per connection. To make the network workable and cost-effective, the necessary traffic grooming function has to be implemented. *Traffic grooming* refers to the act of multiplexing, demultiplexing, and switching low-rate traffic streams into high-capacity lightpaths. In WDM networks, traffic grooming is performed by OADMs (optical add–drop multiplexers). A traffic stream occupying a set of time slots on a wavelength on a fiber can be switched to a different set of time slots on a different wavelength on another fiber. Thiagarajan and Somani [11] addressed the problem of dynamically establishing dependable low-rate traffic stream connections in WDM mesh optical networks. In [11], the route on which the traffic stream passes is called the *traffic stream path* (TSP). A TSP can travel more than one lightpath, and each lightpath can carry many multiplexed low-rate streams, each of which can be a primary or backup TSP. In addition, backup multiplexing technique is used to allow backup resources to be shared among backup TSPs going through a link or a wavelength. Thiagarajan and Somani [11] also proposed two schemes for grooming traffic streams onto wavelengths: mixed primary–backup grooming policy (MGP) and segregated primary–backup grooming policy (SGP). In MGP, both primary and backup TSPs can be groomed onto the same wavelength, and in SGP, the wavelength can consist of either primary or backup traffic streams but not both. Figure 15.15 shows an example of MGP and SGP on a link with three wavelengths. The capacity on a wavelength can essentially be grouped into three types: capacity used by the primary TSPs or *primary capacity* (PC), capacity used by the backup TSPs or *backup capacity* (BC), and unused or *free capacity* (FC). Study results show that SGP is useful for topologies with good connectivity and a good amount of traffic switching and mixing at the nodes. On the other hand, MGP is useful for topologies such as ring, with high load correlation and low connectivity. Up to now, we have presented the survivable options in different network layers; in the next section, these options are used flexibly against various network failures.

15.5.5   Multilayer Survivability Mechanisms

Since the IP and WDM layers both are capable of providing survivability, the network can employ survivability in one or both layers. To provide multilayer survivability, there are two main approaches based on different network models. In

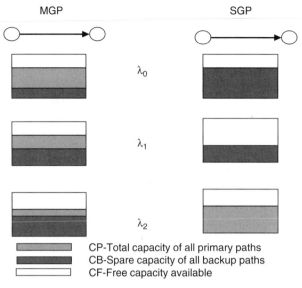

MGP   SGP

CP-Total capacity of all primary paths
CB-Spare capacity of all backup paths
CF-Free capacity available

Figure 15.15   Traffic stream grooming policies in a link.

the client–server or augmented model, the IP and WDM layers are unaware of each other; therefore, some coordination between the two layers has to be defined to provide the required survivability. The escalation strategies are used to allow the IP and WDM layers to cooperate efficiently. In the peer-to-peer network model, the introduction of a uniform control plane for IP/WDM internetworking brings about a great opportunity to ensure that certain network survivability goals are met in a most efficient and flexible manner for the various services supported by networks. In the following subsections, two principal approaches are described.

15.5.5.1 Escalation Strategies   Since the IP and WDM layers provide similar survivability functionality, reduced resource utilization and routing instabilities may result [23]. To overcome this problem, in escalation strategies, coordination between the IP and WDM layers is implemented by hold-off timers. Hold-off timers inform the other layer to hold until the recovery result at the active layer is known. Two overall escalation strategies are commonly deployed, bottom-up and top-down [24]. The bottom-up strategy starts at the layer closest to the failure and escalates toward the upper layer upon expiration of the hold-off timer (shown in Fig. 15.16). If the physical layer detects a fault, it will trigger fiber layer protection; subsequently, if the services cannot totally recover within a given maximum restoration time, optical channel layer survivability mechanisms will be triggered, and so on. In the worst case, the IP layer survivability mechanism will be activated. The bottom-up strategy provides simple recovery because of the coarser switching granularity of the lower layers and offers very fast recovery time since the recovery is close to the failure. Figure 15.17 shows the process of the top-down strategy when the lower

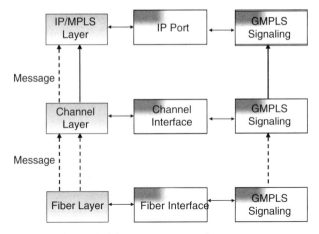

Figure 15.16 Bottom-up escalate strategy.

layer detects a fault. It sends a message to the upper layer, then the upper (most) layer activates its own survivability mechanisms. If the upper layer (e.g., the IP layer or channel layer) cannot find the spare resources, it will escalate a recovery request downward to the lower layer(s) via GMPLS signaling. The top-down escalation strategy allows different recovery actions to take place for different reliability requirements on services and strategy offers higher-capacity efficiency than that of the bottom-up strategy, but it is also very complex and slow, since one failure in the lower layer may lead to many failures in the upper layers.

15.5.5.2 Integrated Survivability Strategies An intelligent open optical network can be built only if the above-mentioned vertically layered model migrates to a horizontal model where all network elements work as peers to dynamically

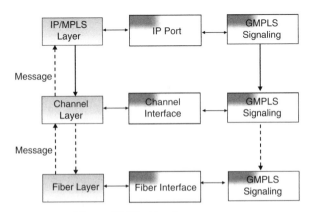

Figure 15.17 Top-down strategy.

establish optical paths through the network and participate in network performance monitoring and management [25]. The IETF and OIF have already addressed the interworking of routers and OXCs through GMPLS [26]. Advances in GMPLS have provided enhanced survivability capabilities (e.g., performance monitoring and protection/restoration), have supported traffic engineering functions at both the IP and WDM layers, and have made it possible effectively to share spare resources. Issues of unstable routing and fast end-to-end survivability have been addressed. After the control plane detects or is notified that a fault has occurred, it knows which layer has the fault. If it is an IP layer fault (such as an IP interface failure), it invokes the survivability mechanisms only at the IP layer. To combat the large granularity failure at the WDM layer, the more effective way is to use an integrated approach [27], which takes advantage of network state information from both layers to provide survivability. In IP-over-WDM networks, there are at least three QoS classes that need to be supported from the availability viewpoint: mission critical, protection sensitive, and best effort [28]. Mission critical service may require dedicated protection, protection-sensitive service can be backed up by shared protection, and best-effort service only requires restoration. Next, we give a brief overview of integrated protection approaches in IP-over-WDM networks and present ways the shared and dedicated concepts can be applied into integrated approaches [27].

Overview of Integrated Protection Schemes   The objective of integrated protection schemes is to integrate IP layer routing with WDM layer routing and wavelength assignment and to allow the IP and WDM layers simultaneously to perform their survivability tasks against a large granularity fault. These schemes consider providing joint protection based on both optical physical topology and virtual topology. The integrated protection schemes incorporate network state information (such as the cost information in physical links and the bandwidth available information on each lightpath) into protection path allocation. The purpose of these schemes is to improve network utilization by reducing the number of lightpaths for protection resource and to obtain benefits by dynamically provisioning low-rate traffic streams at the IP/MPLS layer in future IP-centric WDM-based optical networks. Several low-rate data flows are statistically multiplexed (groomed) onto one lightpath at the IP/LSR router. (Note that LSR is an MPLS-capable router.) Then conventional dynamic lightpath provisioning schemes at the physical WDM layer, where the bandwidth of a connection request is assumed to be a full wavelength capacity, are extended to allow provisioning of sub-λ connection flow requests at the IP/MPLS layer. One of the basic properties of these schemes is that the bandwidth allocated for each primary and secondary path for different connections can be groomed onto the same lightpath. Note that the primary and secondary paths for each connection must be link disjointed. Another important feature of integrated protection is that the differentiated reliability requirements associated with different applications can be provided by a protection factor α of the connection [i.e., $\alpha f(K)$ represents the fraction of traffic that needs to be protected, $0 \le \alpha \le 1$].

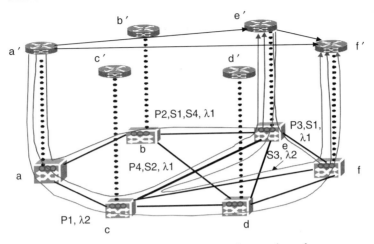

Figure 15.18 Application of integrated protection scheme.

Figure 15.18 shows how to apply the proposed protection scheme to network design. The upper layer represents the logical connectivity between IP routers at an instance of time, and the lower layer shows the physical layer topology and how the lightpaths are routed in the WDM layer. Connections are labeled to show primary and secondary routes for each flow. Here, we assume that the traffic is well aggregated. In Fig. 15.18, P1, P2, P3, and P4 represent primary paths for the flows, and S1, S2, S3, and S4 represent secondary paths for the same flows. We assume that flow 1 arrives at router a' and is destined to router f'. First, the primary path P1 needs to be allocated. Based on the routing table at a', f' is not connected to a', and there is not enough bandwidth available from the logical topology, so router a' requests a lightpath from the physical topology to f'. A lightpath (a'–a–c–d–f–f') on wavelength $\lambda 2$ is set up, and router a' adds one more entry to its routing table between a' and f' and reserves the required bandwidth for flow 1. Second, to provide resilience against any single link failure at the optical layer, router a' needs to find a route S1 that is link disjoint with the primary route for flow 1 since we assume that the protection path bandwidth $\alpha f(k)$ is not as large as the working path bandwidth $f(k)$, and there are two existing connections between a'–e' and e'–f' in logical topology. Here, the connection between a'–e' used by P2 and S4 corresponds to lightpath a'–a–b–e–e' with $\lambda 1$, and the connection e'–f' used by P3 is for lightpath e'–e–f–f' with $\lambda 1$. A multihop route [a'–e', e'–f'] on the logical topology has available bandwidth for S1. Since P1 doesn't physically share a link with the connections a'–e' and e'–f', therefore multihop a'–e', e'–f' will be chosen as the route for S1. In the case of any link failure along P1, the failure is detected at the physical layer and notification messages are propagated to router a', where the latter will switch the traffic from one port to another port. Traffic will be processed at router e', and e' will forward S1 to f' via the existing connection e'–f'. In the next section we present and discuss the numerical results of the proposed approach under different constraints.

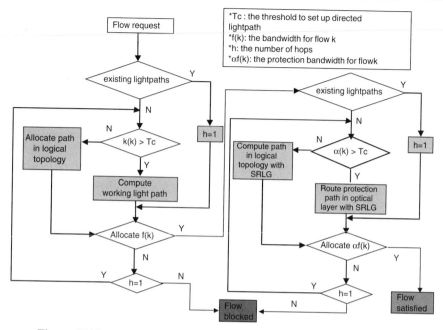

Figure 15.19   Flow chart of proposed integrated shared protection approach.

Shared Integrated Protection Scheme   Figure 15.19 is a flowchart of the dynamic integrated shared protection approach. Upon receiving a flow request, the LSR begins a path selection process for the working path. First, LSR allocates the flow to existing lightpaths between the source and destination nodes. If lightpaths with enough available capacity exist, a working path is set up as the primary active path, the existing bandwidth entry is updated by the additional amount allocated to the new request, and updated messages are sent out to the network via IGP. Concurrently, the protection path allocation is started. If no existing lightpath is available, the bandwidth requirement is checked to see whether a direct lightpath is needed. If necessary, the RWA process is invoked, where the routing in the optical domain is based on MPLS constraint-based routing, which computes paths that satisfy certain specifications subject to certain constraints. For example, if the OXCs are not equipped with wavelength conversion capability, a common wavelength must be located on each physical link along the entire route (single hop). It is implemented by running a path selection algorithm on a traffic engineering (TE) database in the WDM layer, in which the information is obtained by the interior gateway protocol. If the lightpath requested cannot be found, the flow request is blocked. If it isn't necessary to set up a direct lightpath, a path selection algorithm is run on the TE database for the IP layer topology. Once the primary path is allocated, the protection path selection almost repeats the same process as above. Here the proposed approach supports QoS via providing differentiated reliability services, where $\alpha f(k)$ represents the fraction of traffic that needs to be protected, $0 \leq \alpha \leq 1$. The rest of traffic can be restored by a restoration process. This is different from conventional protection schemes, where either full protection or none is chosen.

Note that the backup path must be link disjoint from the primary working path selected, where the backup statistic multiplexing technology can be applied [11]. During the protection path allocation process, we look first at whether there is a disjointed existing lightpath, and whether there is a backup bandwidth reserved along this lightpath for other flows. If this is the case, we check if the primary paths of these flows are in the SRLG as the primary path of the current flow. If not, the backup bandwidth could be shared to protect the current flow. In this case, if the bandwidth required is larger than the one already reserved, the already reserved bandwidth is replaced by the new required bandwidth. If they are in the same SRLG, the extra bandwidth for the current flow will need to be reserved.

Second, if the source LSR cannot locate a link-disjoint backup path from existing lightpaths (i.e., in Fig. 15.19, $h = 1$), or there are existing lightpaths but these lightpaths don't have available bandwidth that matches the requested backup bandwidth, the router performs the following steps: (1) Run the path selection algorithm with constraints; for example, we limit the number of logical hops to two to achieve fast restoration. (Note that the two logical hops must not be shared by any part of the primary path of the flow from physical layer.) (2) If step 1 fails, request a lightpath from the WDM layer with SRLG requirement and make sure that the route computed by the ingress OXC is physically completely link disjoint from the primary path, and begin the reservation process while updating the network status; otherwise, drop the flow request and release all the reserved resources.

Dedicated Integrated Protection Schemes   Provisioning a connection request by the dedicated protection scheme implies that a flow of data is routed successfully if both an active path and an alternate link and node-disjoint backup path are set up at the same time. In this case, provisioning of all requests (both active and backup) is fully accomplished at the IP/MPLS layer. Here we consider two variants of integrated protection schemes: The first one, called IP-over-WDM protection scheme I, only allows the primary path to traverse a single-hop route (single lightpath with wavelength continuity constraint), but the secondary path can use two-hop routing (i.e., two lightpaths can be used to transport the protection traffic). However, the two logical hops must not share any part of the primary path of the flow of the physical layer. The second scheme (i.e., integrated IP-over-WDM protection scheme II) is a modified version of the first scheme. Here it is assumed that both primary and secondary paths are allowed to be multihop on the logical topology. If a working path for a traffic stream connection spans two lightpaths ($\{w\} = \{w1\} \cup \{w2\}$), and if the backup path is routed on two different lightpaths ($\{b\} = \{b1\} \cup \{b2\}$), they should satisfy the following property to ensure fault tolerance: $\{b1\} \cap \{w\} = \emptyset$ and $\{b2\} \cap \{w\} = \emptyset$. The working path selection process is the same as in the shared protection scheme, the difference being that in the protection path allocation, no SRLG information is needed, and the protection paths do not share any spare resources. Therefore, the dedicated protection schemes require less complex signaling and management and can achieve rapid restoration.

15.6   FAULT DETECTION AND NOTIFICATION

Fault detection and notification play important roles in realizing accurate and rapid restoration. The effective fault management has to be sensitive to hardware and software failures and provide the appropriate mechanisms for their notification.

15.6.1   Fault Detection

The fault detection and isolation are crucial to achieve rapid service recovery. The first task for all survivability mechanisms is to detect the fault. Fault detection can be done by performance monitoring to detect the WDM layer impairments or by the IP layer via link-probing mechanisms. Most current optical networks adopt O/E or frame monitoring schemes, such as reusing SONET/SDH framing with B1 byte or J0 monitoring, loss of signal/loss of frame alarm information, or a digital wrapper approach with forward/reverse defect indicator bytes [3]. These opaque monitoring approaches, which require monitoring of every data channel (wavelength) and electronic termination of data channels at each intermediate node, lead to high system cost. This further limits network scalability and inhibits evolution to all-optical networks. This is why digital wrapper technologies are not very workable solutions. To overcome these limitations, optical monitoring schemes [7] have been proposed to detect failure by analyzing optical parameters such as power-level monitoring, optical signal-to-noise ratio (O-SNR), and Q factors. The major difficulty is how to correlate these performance measurement metrics with the quality of the signal in the electrical domain. Another alternative is to perform the end-to-end channel fault detection for which performance monitoring is implemented at the receiver side of the optical channel. For protection, an edge-disjointed protection path (from working path) is needed.

The IP layer relies on the link management protocol (LMP) between neighboring routers to detect the health of the path. Probing messages such as keep-alive or hello messages are exchanged periodically between the peering routers [28]; for example, if k consecutive probing messages are not received from a peer router, the path will be considered as faulty. Using this approach to detect a fault will take subseconds, which is unacceptable for voice transport services. The IP layer can also provide the path-degraded indicator LOP (loss of packets), to indicate whether the connection between the peering routers is up or down, and whether the quality of the connection is acceptable. LOF (loss of frame) occurs when there is excessive packet discarding at a router interface due either to label mismatch or to time-to-live (TTL) errors [8].

15.6.2 Fault Notification

Upon detecting the failure, alarms will be generated to initiate the recovery process. To avoid multiple alarms stemming from the same failure, LMP [28] provides a set of diagnostic mechanisms to localize the failure location through a *Failure indication signal* (FIS) notification message. This message may be used to indicate that a single data channel has failed, multiple data channels have failed, or an entire TE link has failed. Once a fault is detected, the downstream node that first detects the fault should send out a notification of the fault by transmitting FIS (bundling together the notification of all failed data links) to those of its upstream nodes that were sending traffic on the working path that is affected by the fault. The upstream nodes should correlate the failure to see if the failure is also detected locally for the corresponding LSP(s). If the failure has not been detected on the input of the upstream node or internally, the upstream node will decide whether to localize the failure or continue to notify its upstream neighbors. Usually, for local repair approaches, the fault has to be localized; for end-to-end path protection approaches, fault notification is required by source node or destination node to initiate the restoration process; the notification is relayed hop by hop by each subsequent related node to its upstream (or downstream) neighbor until it eventually reaches the source (or destination) node. In the optical domain, there are two possible options to notify failure. One is to develop enhancements to the existing RSVP-TE/ CR-LDP LSP protection (survivability) signaling proposals and to tailor them for optical lightpath LSP protection. The other would be to develop completely new, dedicated protection-switching protocol: namely, an optical APS (O-APS) protocol. This approach considers that FIS should be transmitted with high priority to ensure that it propagates rapidly toward the affected nodes(s). In an MPLS-capable IP domain, depending on how fault notification is configured in the LSRs of an MPLS domain, the FIS could be sent as a layer 2 or layer 3 packet [13]. The use of a layer 2–based notification requires a layer 2 path direct to the node. An example of a FIS could be the liveness message sent by a downstream LSR to its upstream neighbor, with an optional fault notification field set, or it can be denoted implicitly by a teardown message. Alternatively, it could be a separate fault notification packet, and the intermediate LSR should identify which of its incoming links (upstream LSRs) to propagate the FIS on [7]. In the case of $1 + 1$ protection, the FIS should also be sent downstream to the destination where the recovery action is taken.

15.7 SIGNALING PROTOCOL MECHANISM

To support fast restoration, the signaling network, which consists of the control channels, has to be robust and efficient to the extent that a failure should not affect its fault recovery capabilities. In this section we present different signaling protocols which support fast recovery according to GMPLS specification and examine how these capabilities are used to implement different survivability mechanisms using GMPLS signaling. Since IP-based control protocols have been accepted extensively in GMPLS for provisioning of primary and secondary paths, it is natural to reuse IP-based protocols as much as possible to achieve fast restoration [29]. The restoration protocol has to support the following operations:

1. *Secondary path reservation.* In shared protection, when the connection is established, the secondary path of the connection is not set up but is reserved along the shared protection capacity pool without cross-connecting the channels. It is only after a failure event has occurred that it is activated. The nodes along the secondary path need to know the information about the primary path, so the reservations can be shared among SRLG disjointed failures along the primary path. Therefore, a new mechanism is needed to distinguish the path setup, reservation of shared resources, and the allocation of shared resources to a particular connection [30].

2. *Secondary path setup.* Dedicated protection schemes such as $1 + 1$ and $1 : 1$, don't need the secondary path reservation process, and they set up the primary path and secondary path simultaneously via the normal path provisioning process. Shared protection schemes have special requirements for signaling to implement path establishment. Depending on the initiating node (e.g., either source node or destination node), several message types need to be used and messages sent along the reserved secondary path. If the source node initiates the setup process, the message with ⟨connection ID, input port, wavelength, output port, wavelength, etc.⟩ will be sent over the reserved secondary path to activate the relevant OXCs to perform the required cross-connections. In the case of slow-switching technologies, the control message will move forward to the next OXC without waiting for the completion of switching in current OXC. Once the setup message reaches the destination node, the acknowledgment message is sent back to the source node. On the other hand, if the destination node decides to activate the reserved secondary path, a procedure is used, but the new set of message types need to be defined for *setup message and acknowledgment.* Meanwhile, the setup process can be triggered simultaneously. When the control messages reach an intermediate node, the node will send acknowledgment messages to the end nodes.

3. *Error handling.* The restoration signaling protocol has to have the capability to report errors during the path setup process. For example, when there is contention for spare resources that leads to path establishment failure, the resources already allocated must be released immediately.

4. *Reversion function.* The reverting capability means that after failure has been fixed, the traffic is moved back to the primary path. This is not necessary for dedicated protection schemes. However, generally, routing of the protection path is always longer than that of the primary path. Furthermore, reversion will be very critical for shared protection schemes, since the resources in the secondary path of the affected connection are shared by many other primary paths, and other connections can be better protected after the connection affected has been moved back to its primary path. It is important to have mechanisms that allow traffic reversion to be conducted without disrupting services. A *bridge and roll approach* has been proposed. In this approach [30,31], the source node initially starts to send

traffic on both the primary and secondary paths and then the source node sends a bridge and roll request message to the destination, requesting that the destination select the primary path signal. Upon completing the bridge and roll at the destination, the destination sends a *notification* message to the source, confirming completion of the bridge and roll operation, letting the source node stop sending traffic on the secondary path.

   5. *Path deletion operation.* The signaling requirements are similar to those for the teardown function in normal status provisioning, except that in shared protection schemes, information about the primary path needs to be sent along with the teardown message so that each related node can update its own SRLG.

So far, we have discussed some basic functions that should be included in any signaling protocol that supports restoration. Regarding the implementation of these functions, there are two different approaches realizing fault recovery capabilities by adopting IP-based protocols. One approach is to enhance existing RSVP/CR-LDP protocols to support restoration [30,32,33], and the other is to define a new lightweight protocol for restoration [31]. Each approach has pros and cons. The first approach can simplify the standards-based approaches, but it cannot give different priorities to different RSVP/CR-LDP messages [31]. However, for a network to be able to react against failure very rapidly and to further guarantee service-level agreement, restoration-signaling messages should have priority over all other control messages. On the other hand, using the second approach will complicate the standardization process. Therefore, a network operator should decide which application to use based on their specific requirements.

15.8   SURVIVABILITY IN FUTURE IP-OVER-OPTICAL NETWORKS

The current much-touted IP-over-WDM networks actually consist of IP routers and circuit-based optical transport networks, which provide virtual connectivity for IP routers. This network architecture inevitably suffers from various shortcomings [34]: for example, sophisticated traffic aggregation has to be done at the network edge to implement subwavelength traffic demand, and the lightpath provisioning time scale also seriously weakens the effectiveness of IP-over-WDM integration. Furthermore, as more IP routers are added into the network, the number of wavelengths required to provide full-mesh connectivity as well as the size of OXCs may increase dramatically, resulting in very expensive solutions. To avoid this full-mesh connectivity problem, a very complex dynamic control and provisioning logical network topology (consisting of lightpaths) will be required. Since all optical packet switches are capable of switching at various granularities, they effectively bridge the gap between IP packet-level switching and optical circuit–based wavelength-level switching. That is, the optical packet switches do not read the actual bits in the payload data and can stay in a given switching state for an arbitrary period of time. Such switches can function as LSR and OXC, making them theoretically perfect for at least IP over packet switching.

Survivability is inherently important in optical packet-switching networks. Several IP layer schemes, which may include conventional rerouting schemes and MPLS-based schemes, can be extended to route the working path and the protection path. Whereas in IP layer protection schemes, the ongoing packets to the affected node are buffered by electronic random access memory (ERAM), in optical packet networks no such equivalent optical memory is available. Similar to resolving the contention problem in optical packet switching [35], there are two basic approaches to addressing the buffer issue [i.e., fiber delay line (FDL)–based time buffering and deflection routing]. In the first approach, FDLs used together with the switches provide certain fixed delays, but this approach is less efficient than its equivalent ERAM. Once the fault is detected, the affected node will put the ongoing packets in FDLs and find an alternative route to avoid the fault. Since obtaining the alternative route may take some time to process, many packets may need to be buffered, and a large number of FDLs may be required, making the node more complex. In the second approach to deflection routing, the ongoing packets are deflected to the other port rather than the one desired. A part of the network is used as the buffer to store these packets. The other node, which receives these packets, will check their destination and perform hop-by-hop routing to route the packets to their destination. As described in [35], the deflection routing approach actually introduces some extra link propagation delays, which lead to out-of-order packets.

Optical burst switching (OBS) is an alternative approach between circuit switching and packet switching. It has intermediate granularity compared with circuit switching and packet switching. In OBS, multiple IP packets are collected to form a burst, and control information (e.g., source and destination addresses, routing information) is sent as either a header (if in-band signaling is used) or as a separate control packet (carried over a separate control wavelength) along the route that the burst would take. The control message is sent ahead of the corresponding data, and it is used to configure the switch and to make sure that the port is ready when the burst actually arrives at the switch. Thus, a burst will cut through the intermediate nodes without being buffered. Through adjusting an offset time between the control packet and the data burst, various classes of services can be offered. The schemes for routing the primary and secondary paths in GMPLS (possibly with resource sharing) can be applied to OBS. Figure 15.20 demonstrates an example of how OBS can handle faults by using deflection routing. When an intermediate node detects a fault, it deflects the control packet to another port, and the burst will follow the path set up by the control packet.

Since OBS allows for statistical multiplexing between bursts and there is no need to get an acknowledgment from the switches along the route, it is expected that the efficiency would be better in OBS than in circuit-based lightpath routing. For example, new protection schemes such as $1 + N$ or $1 : N$ may become possible [i.e., one primary path can be protected by n secondary paths, each of which carries only a fraction of the working traffic (bursts)] [34]. However, OBS might not be favorable for small IP packets, since it takes numerous small packets to form a burst. The interested reader is referred to Chapter 13 for more details on protection and restoration in OBS.

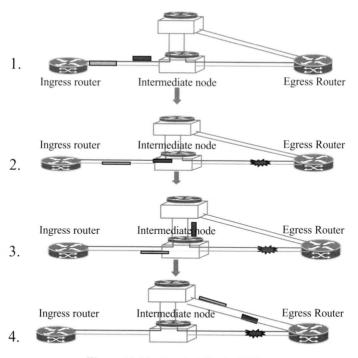

Figure 15.20   Survivability in OBS.

Overall, the major problem with all optical packet switching is that enabling technologies are still in their infancy. For example, to switch packets at a multi-gigabits per second speed, the switch must have a nanosecond switching time. The current popular microelectromechanical systems (MEMSs) can achieve only milli-second switching time. Some other potential candidates to performing switching in the nanosecond range include LiNbO3-based switch elements and semiconductor optical amplifiers (SOAs), but they are cost-prohibitive and are still under development in the laboratories.

15.9 SUMMARY

With the arrival of optical Internet, the importance of survivability has never been greater. In this chapter we studied the network architecture of the next-generation transport infrastructure (i.e., IP-over-WDM network architecture), reviewed the features and characteristics of the general survivability architectures, and described three diverse routing algorithms. The details on survivability capabilities in different layers were described, and how to realize traffic grooming of low-rate streams in meshed optical network was introduced. Several implementation proposals were examined based on the various IP-over-WDM network models while combining the IP and WDM layers. Since survivability mechanisms are activated by fault detection and notification, which are extremely important to improve the recovery time in any survivability approach, we provided an overview of fault detection and notification approaches in IP-over-WDM networks. In addition, we discussed the operational requirements of the corresponding signaling protocol for implementing the restoration protocol. Finally, we presented a brief overview of survivability in future packet-based optical networks (e.g., OBS). Overall, successful deployment of survivability should consider the requirements of recovery time scale, flexibility, resource efficiency, and reliability for multiservices supported by IP-over-WDM networks.

REFERENCES

1. Y. Ye, S. Dixit, and M. Ali, On joint protection/restoration in IP-centric DWDM-based optical transport networks, *IEEE Communications*, 38:174–183, June 2000.

2. D. Medhi, network reliability and fault tolerance, *Encyclopedia of Electrical and electronics Engineering*, 1999.

3. S. Dixit and Y. Ye, Streamlining the Internet–fiber connection, *IEEE Spectrum*, 38:52–57, Apr. 2001.

4. R. Iraschko and W. Grover, A highly efficient path-restoration protocol for management of optical network transport integrity, *IEEE Journal on Selected Areas in Communications*, 18(5):779–793, May 2000.

5. B. Rajagopalan et al., IP over optical networks: Architecture aspects, *IEEE Communications*, pp. 94–102, Sept. 2000.

6. L. Ceuppens, Multiprotocol lambda switching comes together, *Lightwave, http://lw.pennnet.com/home.cfm*, Aug. 2000.

7. N. Ghani and S. Dixit, Channel provisioning for higher-layer protocols in WDM networks, *Proc. SPIE All Optical Networking Conference: Architecture, Control, and Management Issues*, Sept. 1999.

8. V. Sharma et al., Framework for MPLS-based recovery, IETF draft, *draft-ietf-mpls-recovery-frmwrk-03.txt*, July 2001.

9. J. Meijen, E. Varma, R. Wu, and Y. Wang, *Multi-layer Survivability*, white paper, Lucent Technologies, 1999.

10. H. Zang and B. Mukherjee, Connection management for survivable wavelength-routed WDM mesh networks, *SPIE Optical Networks*, 2(4):17–28, July/Aug. 2000.

11. S. Thiagarajan and A. K. Somani, Traffic grooming for survivable WDM mesh networks, *Proc. OPTICOMM 2001*, Aug. 2001.

12. D. Papadimitriou et al., Inference of shared risk link groups, IETF draft, *draft-many-inference-srlg-00.txt*, Nov. 2001.

13. S. Kini et al., Shared backup label switched path restoration, IETF Draft, *draft-kini-restoration-shared-backup-01.txt*, May 2001.

14. S. Ramamurthy and B. Mukherjee, Survivable WDM mesh networks, I: Protection, *Proc. IEEE INFOCOM 1999*, pp. 744–751, 1999.

15. R. Ramaswami and K. N. Sivarajan, *Optical Networks: A Practical Perspective*, Morgan Kaufmann, San Francisco, 1999.

16. V. Anand and C. Qiao, Dynamic establishment of protection paths in WDM networks, *Proc. ICCCN 2000*, pp. 198–204, Oct. 2000.

17. A. Mokhtar and M. Azizoglu, Adaptive wavelength routing in all-optical networks, *IEEE/ACM Transactions on Networking*, 6:197–206, Apr. 1998.

18. C. Xin, Y. Ye, S. Dixit, and C. Qiao, A joint lightpath routing approach in survivable optical networks, *Proc. APOC'01*, Nov. 2001.

19. R. Bhandari, *Survivable Networks: Algorithms for Diverse Routing*, Kluwer Academic, Dordrecht, The Netherlands, 1999.

20. J. W. Suurballe and R. E. Tarjan, A quick method for finding shortest pairs of disjoint paths, *Networks*, 14:325–336, 1984.

21. D. Awaduche et al., Extensions to RSVP for LSP tunnels, IETF draft, *draft-ietf-mpls-rsvp-lsp-tunnel-00*, Nov. 1998.

22. C. Mez, IP protection and restoration, *IEEE Internet Computing*, pp. 97–102, Mar./Apr. 2000.

23. P. Demeester et al., Resilience in multilayer networks, *IEEE Communications*, 37:70–76, Aug. 1999.

24. D. Papadimitriou, R. Hartani, and D. Basak, *Optical Restoration and Escalation Strategies*, OIF 2001.252, 2001.

25. L. Ceuppens et al., Performance monitoring in photonic networks in support of MPL (ambda)S, IETF draft, *draft-ceuppens-mpls-optical-00.txt*, Mar. 2000.

26. E. Mannie et al., Generalized multi-protocol label switching (GMPLS) architecture, IETF draft, *draft-ietf-ccamp-gmpls-architecture-00.txt*, June 2001.

27. Y. Ye, C. Assi, S. Dixit and M. Ali, A simple dynamic integrated provisioning/protection scheme in IP over WDM networks, *IEEE Communications*, 39:174–182, Nov. 2001.

28. J. P. Lang et al., Link management protocol (LMP), IETF draft, *draft-ietf-ccamp-lmp-00.txt*, Mar. 2001.

29. C. Xin et al., On an IP-centric optical control plane, *IEEE Communications*, 39:88–93, Sept. 2001.

30. G. Li et al., RSVP-TE extensions for shared mesh restoration in transport networks, IETF draft, *draft-li-shared-mesh-restoration-00.txt*, July 2001.

31. B. Rajagopalan et al., Signaling for protection and restoration in optical mesh networks, IETF draft, *draft-bala-protection-restoration-signaling-00.txt*, Aug. 2001.

32. K. Owens et al., Extensions to RSVP-TE for MPLS path protection, IETF draft, *draft-chang-mpls-rsvpte-path-protection-ext-01.txt*, Nov. 2000.

33. K. Owens et al., Extensions to CR-LDP for MPLS path protection, IETF draft, *draft-owens-mpls-crldp-path-protection-ext-01.txt*, Nov. 2000.

34. C. Qiao, Label optical burst switching for IP-over-WDM integration, *IEEE Communications*, pp. 104–114, Sept. 2000.

35. S. Yao, S. J. Yoo, B. Mukherjee, and S. Dixit, All-optical packet switching for metropolitan area networks: Opportunities and challenges, *IEEE Communications*, pp. 142–148, Mar. 2001.

16 Optical Internetworking Models and Standards Directions

YANG CAO and RICHARD BARRY

Sycamore Networks, Chelmsford, Massachusetts

16.1 INTRODUCTION

To keep pace with technology advances and rapidly changing customer requirements, networks are continuing to evolve. The emergence of new broadband applications and the increased demand for core network capacity is forcing the evolution of one of the most critical architectural aspects of high-speed networks—how to internetwork with the intelligent optical layer.

Traditionally, the optical layer of the network has been viewed as a simple transport medium. Its primary function has been to provide a connection between electrical layer devices such as Internet protocol (IP) routers and asynchronous transfer mode (ATM) switches, in the form of static, high-capacity pipes (e.g., 2.5 Gbps, 10 Gbps). In this traditional architectural view, the network "intelligence" has resided primarily at the electrical layer, with the optical transport infrastructure essentially serving as a collection of "dumb," inflexible pipes. These static optical link connections are typically hand-stitched across DWDM (dense wavelength-divison multiplexing)-expanded SONET/SDH (Synchronous Optical Network/Synchronous Digital Heirarchy) networks. Internetworking between the electrical layer and optical layer in this traditional model has proven to be slow, complex, manually intensive, and error prone, resulting in an architecture that is limited in terms of flexibility, functionality, and scalability.

The introduction of intelligent optical networking changes this paradigm. Intelligent optical networking technology enables dynamic and automated control of optical layer topology reconfiguration. The ability to switch optical channels automatically introduces increased functionality that electrical layer devices can now use to enhance and optimize end-user network services. At the same time, developments in traffic engineering at the electrical layer, such as multiprotocol

IP over WDM: Building the Next-Generation Optical Internet, Edited by Sudhir Dixit.
ISBN 0-471-21248-2 © 2003 John Wiley & Sons, Inc.

label switching (MPLS), DiffServ, and so on, are bringing new features and functionality to the equation. These technological advances have heightened industry focus on efforts to define standards for an open interface between the electrical and optical boundary. During the past year there has been considerable industry debate surrounding the development of this standard, the optimal model to employ, and how to efficiently carry fast-growing data traffic such as Ethernet-over-optical transport.

In this chapter we start with an intelligent optical network overview, followed by a description of two dominant optical internetworking models: the overlay and peer schemes. Then we describe the latest standards progress regarding optical internetworking, Ethernet and Ethernet optics technology advancement, and how to map layer 2 traffic efficiently onto the optical layer.

16.2 TRADITIONAL NETWORKING VIEW

The provisioning of high-speed bandwidth within the traditional optical transport network is basically static, an approach that is referred to as the *provisioned bandwidth service business model* [1]. For network operators, it is a slow and painstaking operation that requires a considerable amount of manual configuration. In many cases, provisioning a new service requires a costly redesign of optical network internals. In this model, electrical devices such as IP routers are connected to optical networks with standard OC-n/STM-n interfaces. In general, the control plane of the traditional optical transport network is implemented via network management. This approach suffers from the following limitations [2]:

1. It leads to relatively slow convergence following failure events. Typical restoration times are measured in minutes, hours, or even days and weeks, especially in systems that require explicit manual intervention. The only way to expedite service recovery in such environments is to pre-provision dedicated protection channels.

2. It complicates the task of interworking equipment from different manufacturers, especially at the management level. Generally, a customized umbrella network management system or operations support system (OSS) is required to integrate otherwise incompatible element management systems from different vendors.

3. It precludes the use of distributed dynamic routing control in such environments.

4. It complicates network growth and service provisioning.

5. It complicates internetwork provisioning (due to the lack of electronic data interchange between operator network management systems).

16.3 INTELLIGENT OPTICAL NETWORKS

Today's high-speed networking applications now require a new way to provision services. In response, the industry is adding more automation to the optical layer to create new, high-value broadband services. Two simultaneous developments in networking technology make the dynamic request of high-speed bandwidth from the optical network feasible. First, a new generation of dynamically reconfigurable optical systems is enabling point-and-click bandwidth provisioning by network operators. These optical systems, which include optical cross-connects (OXCs) and optical add–drop multiplexers (OADMs), use existing data network control protocols [e.g., MPLS, open shortest-path forwarding (OSPF)] to determine routing within the control channels. Second, traffic engineering [3] and constraint-based routing enhancements to IP routers [4,5] and/or ATM switches are allowing these devices to determine dynamically when and where bandwidth is needed. This intelligent optical network view is also referred to as the *bandwidth-on-demand service business model* [1].

Automatic control and switching of optical channels are beneficial for the following reasons [6]:

- *Reactive traffic engineering.* This is a prime attribute, which allows the network resources to be managed dynamically according to a client's system needs.

- *Restoration and recovery.* These attributes can maintain graduated preservation of service in the presence of network degradation.

Linking optical network resources to data traffic patterns in a dynamic and automated fashion will result in a highly responsive and cost-effective transport network. It will also help pave the way for new types of broadband network services.

16.4 INTERNETWORKING MODELS TO SUPPORT OPTICAL LAYER INTELLIGENCE

A new internetworking boundary between the electrical layer service networks and the high-speed optical switched core is needed to achieve the signaling and routing capabilities needed for faster provisioning of bandwidth. The internetworking models described in this chapter consist of electrical layer devices (e.g., IP routers, IP edge devices, ATM switches, etc.) that are attached to an optical core network and connected to their peers via dynamically established switched link connections.

For network architects, the control plane essentially encompasses the following two functions:

1. *Signaling*: the process of control message exchange using a well-defined protocol to achieve communication between the controlling functional entities connected through specified communication channels. It is often used for dynamic connection setup across a network.

2. *Routing*: the process that supports adjacency discovery and ongoing health monitoring, propagation of reachability and resource information, as well as traffic engineering, path calculations, and the path selection process.

To enable dynamic and rapid provisioning of an end-to-end service path across the optical network, the model should provide two automatic discovery functions [1]:

1. *Service discovery*: the process of information exchange between two directly connected pieces of end-system equipment: the user edge device and the network edge device. The objective is for the network to obtain essential information about the end system and thereby provide the information about available services to the end system.

2. *End-system discovery*: the process of information exchange between two directly connected pieces of end-system equipment: the user edge device and the network edge device. The objective is for each side to obtain essential information about the network element at the other side of the optical link, and to understand the connection between them.

The optical network essentially provides point-to-point connectivity between electrical devices in the form of fixed-bandwidth optical paths. The collection of optical paths therefore defines the topology of the virtual network interconnecting electrical layer devices [7]. With the intelligence (signaling and routing capabilities) built into the optical layer, not only can the interconnection topology be established rapidly, it can be maintained in an autonomous manner and be reconfigured dynamically as triggered by either restoration or traffic engineering needs.

Note that there are several key differences between the electrical layer control plane (packet switched) and the optical layer control plane (circuit switched), as shown in Table 16.1. From the traffic engineering perspective shown in the table, we can see that the optical layer intelligence is different from the intelligence residing in the electrical layer. However, instead of competing or overlapping with each other, these two layers complement one another. With optical layer traffic engineering, we can create dynamically the most effective/optimized physical topology Ψ. Based on Ψ, electrical layer traffic engineering can be used to achieve load balancing, maximum degree statistical multiplexing, etc. A good internetworking model strives to marry the intelligence residing in both the optical and electrical layers.

Note that the optical layer control plane has more dependence on the physical layer technology than does the electrical layer control plane. Optical layer technology is expected to evolve rapidly, with a very real possibility of additional "disruptive" advances. The analog nature of optical layer technology compounds the problem for the corresponding control plane, because these advances are likely to be accompanied by complex, technology-specific control plane constraints. Hence, an internetworking model is highly desirable if it can allow gradual and seamless introduction of new technologies into the network without time-consuming and costly changes to the embedded control planes [10].

It is also worth noting that there have been instances in which people in the industry have tried to make an analogy between "electrical layer over optical

TABLE 16.1   Control Plane Differences

	Control Plane	
	Electrical Layer	Optical Layer
Physical connectivity	Static physical topology	Dynamically reconfigurable physical topology
Traffic engineering focus	Efficiently fill in the given physical connectivity; fully exploit statistical multiplexing potential	Efficiently reconfigure (create/tear down) the physical connectivity
Control channel (carrying signaling and routing messages)	In-band: control traffic mixed with data traffic (payload) Implication: control channel's health status can be used to determine the corresponding data channel's health status	Out-of-band: because the data traffic (payload) is not processed at the optical layer, control traffic is carried separately from the payload (via overhead bytes in the same channel or via a separated channel) Implication: control channel's health status may be independent of the corresponding data channel's health status
Interior gateway Protocol (IGP) solution: adjacency discovery	In-band, bidirectional digital "hello" mechanism is used	Out-of-band, coordination-based digital or analog "hello" may be needed
IGP solution: link representation	The main link characteristics considered are link bandwidth and/or administrative cost	In addition to bandwidth and administrative cost, several other critical link characteristics also need to be considered, including transparency level, protection level, shared risk link group (SRLG)[a] diversity [8], etc.
Exterior gateway protocol (EGP) solution	Existing EGP solution [e.g., border gateway protocol (BGP)] assumes preestablished physical interdomain connectivity	EGP solution assumes dynamic interdomain physical connectivity
Quality of service (QoS) differentiation	Since the processing granularity is down to the packet level, QoS differentiation can be supported at each packet level, which includes weighted fair queuing (WRQ), random early detection (RED)–based queue management, etc. The delay incurred by each packet includes both deterministic transmission delay and (generally) nondeterministic queuing delay.	Since the processing granularity is channel level, there is no packet-level QoS differentiation. The main QoS differentiation is the protection level that differs at the channel level. Since there is no mature optical layer buffer technology, the only delay incurred is deterministic transmission delay.

(Continued)

TABLE 16.1   (*Continued*)

	Control Plane	
	Electrical Layer	Optical Layer
Protection/restoration	Granularity: packet level In general, fault detection, fault propagation and fault processing (rerouting, etc.) is based on layer 2 schemes and above.	Granularity: channel/path level In general, fault detection, fault propagation, and fault processing (switching, etc.) is based on layer 1 scheme.
Path computation	Physical characteristics in general do not need to be considered during path computation.	In addition to SRLG diversity, other physical characteristics, such as polarization mode dispersion (PMD), amplifier spontaneous emission (ASE), etc.[9], also need to be considered during path computation.

[a]An SRLG is an identifier assigned to a group of optical links that share a physical resource. For instance, all optical channels routed over the same fiber could belong to the same SRLG. Similarly, all fibers routed over a conduit could belong to the same SRLG [15].

layer" and "IP over ATM." As shown in Table 16.1, IP routers and ATM systems share electrical layer control plane characteristics, and both the IP router and ATM system control plane differ significantly from the optical layer control plane. From this perspective, we can see that the "IP over ATM" analogy is misleading.

Given these recognized control plane differences, two dominant models have been proposed to date to internetwork the electrical layer and the intelligent optical layer: the overlay model and the peer model. In the *overlay model*, electrical devices (user/client) operate more or less independent of optical devices (network) and have no visibility into the optical network. The *peer model* is more complicated. It encompasses a single IP control plane that exists at the client layer and at the optical layer, giving a global view of the combined network. The networks then act as one network, as opposed to two networks signaling to each other. There is some debate in the industry as to which architecture to use. Some data network managers want their routers to control the network; others would prefer not to allow outsiders to view their network topology and resources.

In the following sections we describe the peer and overlay approaches and how they differ in the areas of signaling and routing.

16.5   OVERLAY MODEL

The overlay model is commonly referred to as the user–network interface (UNI) model. As we mentioned above, the electrical layer devices (user/client) operate more or less independently of the optical devices (network). In this client–server network architecture, the layers of the network remain isolated from each other, but dynamic provisioning of bandwidth is made possible. The client device (e.g., a router or switch) requests the connection and the "server" either grants or denies the request for bandwidth. General characteristics of the overlay model include:

- User–client devices signal the optical network over the UNI to request an optical circuit.
- Routing protocols, topology distribution, signaling protocols, and control plane are independent.
- UNI is used only at the edge of the network; optical layer devices use an internal signaling protocol between optical network switches.
- The model supports multiple services, such as IP, ATM, and time-division multiplexing (SONET/SDH).
- The model implies that one side of the connection is a client, such as an IP router, and the other side is a network, such as an optical switch/OXC.
- The network hides optical layer topology and management from user/client devices.
- Analogies include public switched telephone network, ATM switched virtual circuits, and integrated services digital network.
- The model supports both public and private intradomain signaling.

As described in Table 16.1, the control plane differences are resolved by deploying a separate and different control plane in the electrical and optical layers, respectively. In other words, the electrical layer address scheme, routing protocol, and signaling scheme run independent of the address scheme, routing protocol, and signaling scheme deployed in the optical layer. With this independence, the overlay model can be used to support network scenarios in which an operator owns a multiclient optical transport network and each client network employs individual address, routing, and signaling schemes. In this model, the optical layer essentially provides point-to-point connections to the electrical layer. The electrical layer device, in this respect, acts as a client to the optical layer.

One of the benefits of the overlay model is that it does not require the electrical layer devices to store both the electrical and optical layer topologies. This separation provides failure isolation, domain security, and independent evolution of technologies in both the electrical and optical layers. As the networks scale in size and scope, this hierarchical approach provides further benefits in the area of network scaling and transport layer survivability.

Other benefits of this model include:

- The optical layer comprises subnetworks with well-defined interfaces to client layers.
- Each subnetwork can evolve independently, taking advantage of the various technological innovations.

- Older infrastructures are not "stranded."
- Optical network topology and resource information is kept secure.

Interworking (signaling) between these two layers, including control coordination, can be established through static configuration or through dynamic procedures. Based on interworking differences and whether routing information is exchanged between these two layers, the overlay model can be further divided into two submodels: the static overlay model and the dynamic overlay model.

16.5.1   Static Overlay Model

In the static overlay model, optical path endpoints are specified statically through the network management system (NMS). Therefore, no UNI signaling is needed. Additionally, in this model there is no exchange of routing information between the client domain and the optical domain. This scheme is similar to ATM permanent virtual circuits (PVCs) [7].

16.5.2   Dynamic Overlay Model

Certain industry documents [11] also refer to the dynamic overlay model as an augmented model. In this model, the path endpoints are specified via UNI signaling (Fig. 16.1). Paths must be laid out dynamically, since they are specified dynamically by signaling, similar to ATM switched virtual circuits (SVCs). In this model, electrical layer devices residing on the edge of the optical network can signal and request bandwidth dynamically. The resulting bandwidth connection will look like a leased line. In addition, certain exchanges of routing information may occur between the electrical and optical layers. For example, externally attached electrical

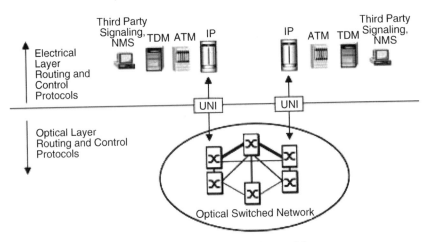

Figure 16.1   Dynamic overlay model.

end-system addresses could be carried within the optical routing protocols, and disseminated further via UNI signaling, to allow reachability information to be passed to all the other attached electrical layer devices.

Most internetworking models discussed in various standards bodies can be classified as dynamic overlay models, and the model has several advantages. First, as described in Table 16.1, there are significant differences between the optical and electrical layer control planes. The dynamic overlay model allows the development of a control plane mechanism for the optical layer independent of the electrical layer control plane. This is particularly important because the optical network is still evolving; more and more network characteristics are being discovered, and most have a direct impact on both signaling and routing schemes. For example, as technology advances from the opaque stage into the all-optical stage, where all the processing (including performance monitoring, channel switching, wavelength conversion, etc.) is done in the optical domain, a set of physical transmission impairments will have a significant impact on wavelength routing. These physical constraints include polarization mode dispersion (PMD) and amplifier spontaneous emission (ASE) [9]. By using the dynamic overlay model, rapidly evolving optical layer technological advances can be contained in the optical layer itself and the relatively mature electrical layer control plane scheme left intact.

Note that in the dynamic overlay model, unlike most conventional IP-over-ATM overlay models, we don't need to assume that any electrical attaching device is connected to all the other electrical devices attached to the optical network. In fact, from a routing perspective, the electrical layer attaching device is only interested in electrical-level reachability. This is achieved in this model via the exchange of routing information between the electrical and optical layers, facilitated by enhancing the optical layer internal topology dissemination mechanism (piggyback) and UNI interface. This model eliminates the need to expose the client layer to the full optical layer topological details; hence, there is no N^2 scalability issue assumed for the conventional IP-over-ATM overlay model.

Requiring the electrical layer system to store only the electrical layer topological reachability information significantly improves the overall scalability of the network. In addition, any optical layer topology change, such as a change triggered by the addition/removal of optical elements, or a breakdown of certain optical layer nodes, fiber, or wavelengths, will not have any effect on the electrical layer unless it affects the corresponding electrical layer connectivity. This benefit contributes significantly to network stability and performance.

Take Figure 16.2 as an example. In this case, after routers A and B have formed the physical connectivity, their primary concern is this connectivity. With this model, as long as optical system E, F, and G and all the related links A $\leftrightarrow$ E, E $\leftrightarrow$ F, F $\leftrightarrow$ G and G $\leftrightarrow$ B are functioning normally, all the other optical network details should be irrelevant from the router network's perspective. Scenarios in which the number of optical systems in the optical network increases from 6 to 100, or the number of optical links increases from 10 to 350, or the link between optical systems C and D is down, will affect only the optical layer internal topology; they will have no effect on the router network topology.

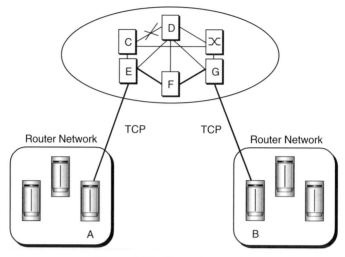

Figure 16.2   Network example.

Using this dynamic overlay model, independent protection/restoration schemes can be deployed in both layers. This addresses the difference in the protection-level requirements of the two layers, including protection granularity, diversity, and so on, as described in Table 16.1. In addition, a multilayer survivability scheme can also be developed, if needed, based on the enhancement of the UNI interaction between these two layers. From the operator's perspective, this model addresses a critical business requirement common to network deployment: operators typically run the operator network independently.

Whereas the UNI provides a signaling mechanism between the user domain and the service provider domain in the context of the optical switched network, the network-to-network interface (NNI) refers to a connection between any of the following (Fig. 16.3):

- Different service provider networks
- Subnetworks belonging to the same provider
- Same subnetworks with connections between different vendors' switches

The definition of the NNI in the optical network remains in the very early stages of development. NNI interface routing options under consideration include static routing, default routing (applicable only to the single homed scenario), and dynamic routing. NNI interface signaling options include constraint-based routing label distribution protocol (CR-LDP) and resource reservation protocol with traffic engineering extensions (RSVP-TE), both of which are Internet Engineering Task Force (IETF) drafts. Generalized multiprotocol label switching (GMPLS) extensions to both CR-LDP and RSVP-TE have been proposed to adapt the respective

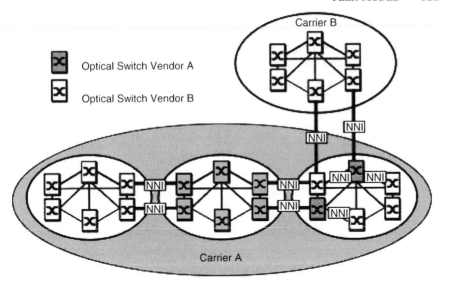

Figure 16.3   Network-to-network interface implementations.

protocols within the context of the optical domain. The use of these protocols in the context of optical switched networks is examined in greater detail below. As the standards organizations continue their progress on defining NNI requirements in the optical domain, control plane reliability and network scalability will form two critical areas of focus and concern.

16.6   PEER MODEL

In the peer model, the electrical layer systems (such as IP/ATM devices) and optical layer systems act as peers, and a uniform control plane is assumed for both the optical and electrical layers (Fig. 16.4). The idea behind this model is to exploit the existing control plane technology deployed in the electrical layer to foster rapid development and deployment of the control plane in the optical domain. The main leveraging target is the electrical layer control plane technology developed for MPLS traffic engineering [2]. The control plane differences between these two layers, as described in Table 16.1, are addressed in this model by extension of the existing electrical layer control plane. Note that with the uniform control plane deployed across both the electrical and optical layers, no special boundary layer scheme (such as UNI interaction) is needed.

Characteristics of the peer model include:

- Single/common signaling and routing protocol scheme
- Common addressing scheme used for optical and IP networks

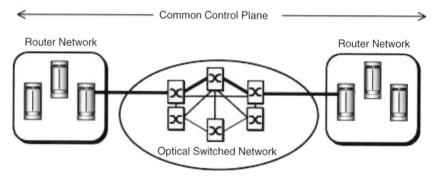

Figure 16.4   Peer model.

- Routers/layer 3 devices and optical switches/OXCs act as peers and share a common view of the entire network topology
- Optimized for IP
- Optical connections established using IP routing/signaling protocols
- NNI not required in the peer model
- Single administrative domain (does not isolate failures at the optical layer [7])

In the peer model, unlike the overlay model, routers are required only to keep adjacency with the next hop rather than keeping track of both the source address (SA) and destination address (DA) IP addresses. The peer model in this sense is a flat networking approach, whereas the overlay model is a hierarchical approach.

There are several advantages to the peer model [2]:

1. It exploits recent advances, such as MPLS control plane technology, and leverages accumulated operational experience with IP distributed routing control.

2. It obviates the need to reinvent a new class of control protocols for optical transport networks and allows reuse of software artifacts originally developed for the MPLS traffic engineering application. Subsequently, it fosters rapid development and deployment of a new class of versatile OXCs.

3. It facilitates the introduction of control coordination concepts between data network elements and optical network elements.

4. It simplifies network administration in facilities-based service provider networks by providing uniform semantics for network management and control in both the data and optical domains.

With the same control plane deployed in both electrical and optical layers, this approach implies that the same routing scheme must be used in both layers. With the peer model approach, an electrical layer device (such as an IP router) obtains both electrical and optical layer topology. Although this permits an electrical layer device to compute one complete end-to-end path to another electrical layer system across the optical network, this is done at the cost of sacrificing network scalability. Each electrical layer system needs to maintain an enormous link-state database, significantly affecting overall network scalability since most of the optical layer link-state information may potentially have no meaning in the electrical layer. In addition, any optical network outage, even if irrelevant to the electrical layer system, may need to be leaked to the client layer, introducing unnecessary topological instability. For example, after a router has formed the physical connectivity, any optical network internal topology changes (such an optical device losing connectivity, or the number of optical systems increasing from 6 to 100, etc.) will cause all the routers to update their topology, although none of these changes has any effect on the inter-connectivity between this pair of routers.

Application of the peer model would necessitate updating the existing control plane. Although these updates are introduced purely to accommodate optical layer intelligence, this common control plane needs to be applied to the electrical layer as well. Potentially, any control plane update triggered by advances in either optical layer or electrical layer technology will always have implications on both domains. While there are several advantages to using one common control plane instead of two for both the optical and the electrical layer, from the perspective of electrical layer control plane stability, the peer model approach is best viewed as a long-term rather than short-term solution.

16.7   OPTICAL INTERNETWORKING AND ETHERNET STANDARDS ACTIVITY STATUS REPORT

In order to achieve automatic optical switching, a certain degree of global standardization is required because neither rapid provisioning nor the operational improvements are likely if each vendor generates a proprietary control plane. As a consequence, a wide range of standards organizations and industry forums are addressing various aspects of the Intelligent Optical Network with a goal of creating open interfaces. Included below are brief summaries of activities from some of the major groups that are currently developing proposals for Internetworking schemes, as well as the latest standard developments regarding Ethernet and Ethernet over optics. Although the list below is by no means exhaustive (that would fill an entire book in and of itself), it provides a good overview of the standards organizations at the time this book was published.

16.7.1   Internet Engineering Task Force

Generally, the Internet Engineering Task Force (IETF) is viewed as home for protocol development. One of the ongoing IETF protocol efforts related to the optical network interworking issue is the development of generalized MPLS (GMPLS). GMPLS extends the MPLS protocol family to encompass time-division systems (e.g., SONET ADMs), wavelength switching (e.g., optical lambdas), and spatial switching (e.g., incoming port or fiber to outgoing port or fiber) [12]. MPLS is a set of protocols developed within the IETF as a means of enabling IP to behave in a connection-oriented fashion. The basic idea behind MPLS is to serve as a circuit-switching system at the network layer, setting up label-switched paths (LSPs) across a network based on bandwidth or other policy constraints. The components in this scheme include a signaling protocol for establishing the LSP, a routing protocol, such as OSPF, with appropriate extensions for advertising the network topology and available link resources, and a mechanism for forwarding packets independent of their IP header and payload contents.

MPLS was proposed initially as a means to move beyond the limitations of the next-hop forwarding scheme for IP packets through the network. Rather than examining IP headers and routing packets based on the destination IP address, the MPLS scheme lets IP routers look at short labels attached to the packets, enabling them to do a quick table lookup to switch them through the network more efficiently. MPLS is similar to cell or packet switching in virtual circuit (VC)–oriented networks, such as ATM or X.25 networks. These networks use a small VC number for switching from hop to hop.

In essence, MPLS gave network operators better control over the path that packets took through the network. One of the primary benefits of the MPLS scheme is that it separates the control and forwarding planes. This functionality creates a traffic engineering and service-provisioning framework that operates independent of the network devices themselves. With an independent forwarding plane, routing protocols such as open shortest path first (OSPF) can be used to establish optimal routing and restoration of lightpaths in the network.

By extending MPLS to the transport domain, GMPLS intends to cover both the peer model and the dynamic overlay model as described above. This approach proposes constructing an optical network control plane based on the IGP extensions for MPLS traffic engineering with additional enhancements to distribute relevant optical transport network state information, including topology state information. In the peer model case, the full optical layer internal topology is disseminated to the electrical layer. This state information is subsequently used by a constraint-based routing system to compute paths for point-to-point optical channels. The proposed optical network control plane also uses an MPLS signaling protocol to establish point-to-point optical channels between access points in the optical transport network. Note that in the peer model case of the GMPLS approach, no specific UNI signaling is needed, whereas in the dynamic overlay model case, UNI signaling is required.

GMPLS specification is in the final IETF draft stage. In addition to the GMPLS specification, there have been several IETF contributions similar to GMPLS-based signaling extensions, such as RSVP-TE and CR-LDP extensions, and routing extensions, including both OSPF and IS-IS extensions. All of the contributions noted above are still in the draft stage. However, this work has gained momentum not only in the IETF, but also in the OIF. The OIF UNI 1.0 as described below, for example, is closely aligned with the GMPLS draft proposals. This alignment within the standards organizations is an important step toward industry consolidation around an open interface to the intelligent optical network.

16.7.2   Optical Internetworking Forum

By definition, the Optical Internetworking Forum (OIF) has a charter of developing interoperability agreements between the emerging optical layer of the network and other layers already defined in the open systems interconnection (OSI) model. The OIF has also been tasked with developing interoperability agreements between different vendors within the optical network layer. The initial focus of OIF is to define the requirements of UNI and the network services offered across the UNI, particularly with respect to the requirements of Internetworking IP with the optical network layer through SONET framed and rate circuits. Although the group needs to create an open architecture that can internetwork with a variety of clients, such as IP, SONET, gigabit Ethernet, and frame relay, the working group must also prioritize which clients will be defined first. It has been decided that the initial focus should be on IP as a client to the optical network.

Currently, OIF has finished UNI 1.0 specification [13]; UNI signaling is based on the GMPLS extension of both RSVP and CR-LDP. During Supercomm 2001, OIF successfully demonstrated UNI interoperability among over 20 system vendors, including Sycamore and Cisco. OIF's UNI 1.0 functions as a dynamic overlay model, with the electrical domain signaling the optical network to request bandwidth (Fig. 16.5). The protocol has five basic components:

1. Neighbor discovery
2. Service discovery
3. Address registration
4. UNI signaling
5. Policy and security

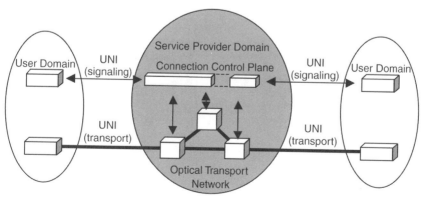

Figure 16.5   UNI used between service provider and user domains. (Courtesy of OIF.)

As the next step, OIF is going to expand UNI features while resolving intradomain and interdomain NNI issues.

16.7.3   International Telecommunications Union

Automatic switched optical network (ASON) is an emerging International Telecommunications Union (ITU-T) standard in optical networking. It seeks to define the control and management architecture for automatic switched optical transport networks (Fig 16.6). In March 2000, contributions to the standard from the U.S. and British Telecom (BT) groups were presented at the ITU-T meeting in Kyoto, Japan, at which time Study Group 13 decided to initiate work on ASON. The outcome of discussions at that time was a new recommendation, G.ASON, the architecture for the automatic switched optical network. Further, it was agreed that the U.S. and British contributions would be combined to produce the first draft of G.ASON. Contributions were requested to further the work, and these contributions continue to be submitted. The ASON architecture provides a useful framework for understanding the issues and requirements of the NNI interface protocol.

The ASON UNI and NNI interface, formally defined in T1X1.5/2001-044, "Clarification of Interface Definitions in G.ASTN" [14], are as follows:

1. *UNI:* the interface between an optical connection controller and a request agent. This interface is concerned predominantly with the connection setup service. The main information carried is the service request, such as create/delete/modify/query.

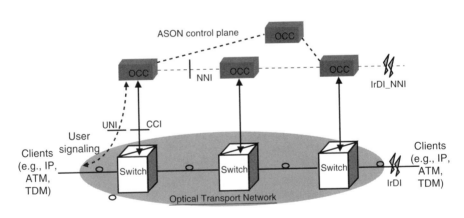

OCC: Optical Connection Controller
UNI: User Network Interface
CCI: Connection Control Interface
NNI: ASON Control Node–Node Interface
IrDI: Interdomain Interface

Figure 16.6   ASON control plane and architecture. (Courtesy of ITU-T.)

2. *NNI:* the interface between two optical connection controllers. This interface can be further clasified into interior NNI (I-NNI) or exterior NNI (E-NNI), as follows:

(a) *I-NNI:* A bidirectional interface between optical connection controllers with the control plane belonging to the same administrative domain. This interface is concerned predominantly with routing information interchanged within an operator domain. The main information carried is link-state information.

(b) *E-NNI:* bidirectional interface between optical connection controllers with the control plane belonging to different administrative domains. This interface is concerned predominantly with address reachability information, specifying the range of network address that can be reached over this link. The main information carried is the summarized network address.

Both NNI types carry the signaling information associated with connection setup. NNI signaling is used primarily for service provisioning and restoration. It allows the exchange of various types of information and affects the network state at many points in the network, all with the end goal of maintaining the appropriate state of the network.

During 2001, ITU made significant progress in this area. In addition to finishing the architecture recommendation for ASON [15], it also finished the following three related recommendations:

• Architecture and specification of data communication network [16]
• Distributed call and connection management (DCM) [17]
• Generalized automatic discovery techniques [18]

From the network requirement's perspective, ITU-T has established a solid basis for the wide deployment of intelligent optical networks.

To facilitate the flexibility of mapping higher-layer traffic into the right-sized SONET/SDH circuits, the ITU-T standardized virtual concatenation, a byte-level inverse-multiplexing technique, along with a generic framing procedure (GFP) for encapsulating different layer 2 data units into GFP frames for transport over SONET/SDH networks. At this stage, the ITU-T GFP standard has passed the consensus stage; the scope of this GFP standard includes both frame-based GFP and a transparent GFP scheme, which specifies how GFP can be used to carry transparently 8B/10B codes commonly used in gigabit Ethernet, fiber channel, ESCON, and FICON.

16.7.4   IEEE 802: Emerging Ethernet Standards

While the work being done at the core of the optical network is in signaling and routing control, other work is focused on physical interoperability and expanding data types that can be transported over optical networks. One such development involves Ethernet and Ethernet over optics. Until recently, Ethernet has been confined to local area networks (LANs), due to distance and performance limitations inherent in the original protocol. Over the past two years, however, refinements and advances in Ethernet protocols and the emergence of an intelligent optical layer have given birth to an alternative transport architecture that extends Ethernet to metropolitan, regional, and long-haul backbone networks.

Based on Ethernet-over-optics, this new architecture enables carriers/service providers to capitalize on today's market demand for high-performance business and residential data services. Examples of these services include:

- LAN-to-LAN interconnects
- Broadband corporate VPNs
- High-speed Internet access
- Web hosting

Optimized for IP applications, Ethernet-over-optics provides a common end-to-end transport protocol from the user's premise through the backbone of the network. Leveraging these networking advances, carriers/service providers can simplify their network architecture and reduce the complexity and cost associated with managing overlay networks.

16.8   GIGABIT ETHERNET

The initial focus of gigabit Ethernet (GbE) was that of a switched full-duplex operation over fiber-optic cabling, as opposed to shared-media copper. GbE would be used to connect different backbones or superservers and workstations. This combination of increased demand, increased performance, and lower prices will continue throughout the history of Ethernet.

Supporting this architecture at the optical layer are intelligent optical transport and switching network devices capable of mapping GbE streams directly over managed wavelengths. Carrier-class Ethernet solutions over a fiber infrastructure offer the flexibility and capacity to support broadband applications. Easy to provision and easy to adjust, Ethernet-over-optics solutions give carriers/service providers the ability to offer customers a range of scalable bandwidth services optimized for the needs of their diverse data applications:

- 1-2-, and 10-Mbps access for private households
- 10/100-Mbps and GbE connections to business users
- Migration to 10-GbE and 100-GbE business services

As users demand more bandwidth, we are seeing a gradual shift from GbE to 10-gigabit Ethernet, described below.

TABLE 16.2   Comparison of GbE and 10 GbE

Characteristic	GbE	10 GbE
Physical media	Optical and Copper media	Optical media only
Distance	Local area networks (LANs) up to 5 km	LANs to 40 km Direct attachment to SONET/ SDH equipment for wide-area networks (WANs)
Physical media-department sublayer (PMD)	Leverages fiber channel PMDs	3 new serial PMDs 1 new WWDM PMD
Physical coding sublayer (PCS)	Reuses 8B/10B coding	64B/66B coding for serial PMDs Reuses 8B/10B coding for WWDM PMD
Medium access control (MAC) protocol	Half-duplex (CSMA/CD) + full duplex	Full duplex only
Additions	Carrier extension for half-duplex	Throttle MAC speed

16.8.1   10-Gigabit Ethernet

There are several important restrictions that apply to 10 GbE: the traditional CSMA/CD protocol will not be used, half-duplex operation will not be supported and copper wiring will not be an option (at least not for end-station links). Table 16.2 provides a brief comparison of GbE and 10 GbE. The 10-GbE specifications include all of the major logical elements, as shown in Fig. 16.7. In the medium access control (MAC) layer, the specification will preserve the Ethernet frame format, including both minimum and maximum frame sizes. It will also support full-duplex operation only and thus has no inherent distance limitations. Because 10 GbE is still Ethernet, it minimizes the user's learning curve by maintaining the same management tools and architecture. The reconciliation sublayer (RS) maps the physical layer signaling primitives, provided by and for the MAC layer, to the 10G media independent interface (XGMII) signals.

To address different application requirements, the task force has determined a number of appropriate physical layers (PHYs). A PHY defines the electrical and optical signaling, line states, clocking and timing requirements, data encoding, and circuitry needed for data transmission and reception. Within a PHY, there are several sublayers, which perform functions including the physical coding sublayer (PCS), physical medium attachment (PMA), and optical transceiver or physical media-dependent (PMD) sublayer for fiber media.

In addition to traditional LAN connectivity (LAN PHY at 10 Gbps), the standard for 10 GbE is being developed with an option for connection across MAN and WAN links at an expected data rate compatible with OC192c/SDH VC-4-64c (WAN PHY at 9.58 Gbps). The MAC must perform rate adaptation to ensure a data

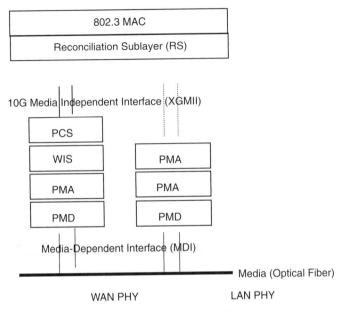

Figure 16.7   Gigabit layer architecture.

rate of 9.58 Gbps by using an open loop mechanism and appropriately adjusting the interpacket gap. The WAN interface sub layer (WIS) performs SONET/SDH framing, scrambling, and error detection to allow minimal compatibility with the requirement of SONET/SDH networks. This will enable the benefits of Ethernet technology to be extended to the construction of MANs and WANs and provide cost-effective solutions for larger geographic areas using a single technology end to end. These MAN/WAN connections can be over dark fiber, dark wavelengths, or SONET/SDH networks.

16.8.2   IEEE 802.3ab Gigabit Ethernet over Copper

Ethernet technology is currently progressing in many fronts to make end-to-end Ethernet a reality. An example is the growing popularity of 1000Base-T GbE [19]. GbE-over-copper products are now becoming available. Although standards were finished by late 1999, suppliers said that silicon shortages and faulty chips for GbE NICs hindered the technology's emergence. Vendors are now shipping 1000Base-T products, and this will accelerate the deployment of gigabit Ethernet, particularly for short-distance applications.

16.8.3   IEEE 802.3af Task Force

The IEEE 802.3af Task Force (DTE Power via MDI) [20] was formed in December 1999 to address the need to provide electrical power over standard

Ethernet cabling to power devices that include IP telephones, Web cams, wireless LAN base stations, premises access control devices, and security sensors. This eliminates the need for connecting each Ethernet device to both an electrical socket and a data outlet, thereby reducing installation and maintenance costs. It also enhances system reliability and overall cost-effectiveness by maintaining a dial tone during power interruptions through the use of a single, centralized, uninterruptible power supply (UPS).

16.8.4   IEEE 802.3 Ethernet to the First Mile Study Group

Because of the many benefits provided by Ethernet, such as high bandwidth, ease of installation and maintenance, and low cost, Ethernet technology is now being considered for point-to-point subscriber access application. For enterprise users, fiber to the premises using Ethernet has been advocated by a number of startup equipment vendors and greenfield carriers/service providers. However, it is more important to provide a low-cost access solution that does not require any protocol conversion and can deliver the major benefits of Ethernet to residential users. The IEEE 802.3 Ethernet to the First Mile Study Group [21] was recently formed in November 2000 to tackle this challenging problem of providing a low-cost Ethernet solution to the "first mile," perhaps over voice-grade copper cables. The study group had its first technical meeting in the IEEE P802.3 interim meeting in January 2001. The ubiquitous end-to-end Ethernet will truly emerge if the study group can come up with a cost-effective solution to address this market segment.

16.8.5   IEEE 802.17 Working Group on Resilient Packet Rings

Recognizing that many optical fiber infrastructures in the metro area are ring-based and that network equipment is TDM-based and not optimized for packet transport, a number of packet equipment vendors and startups have proposed proprietary schemes to meet the need for IP-optimized, ring-based packet transport networks. These schemes try to provide a higher priority to real-time traffic, such as voice and video, while guaranteeing the fair allocation of bandwidth between lower-priority data traffic. They also promise the capability of sub-50-ms protection switching, similar to that available in SONET/SDH networks. In the IEEE 802 plenary meeting held in November 2000, the IEEE 802 committee approved the forming of the new IEEE 802.17 working group [22] to develop the MAC and physical layer standards for a resilient packet ring.

16.9   SUMMARY

Enabled by a new generation of optical network elements, the optical network is transitioning from a static, dumb, unaware transport layer to a more dynamic, flexible, self-aware layer. The emergence of intelligence at the optical layer—and continuing advances in the data layer—has introduced new features and functionality to next-generation networks. Network operators have been forced to reevaluate

TABLE 16.3   Standards Status Summary

Standard	Internetworking Model	Focus	Status
ITU-T	G.ASON: dynamic overlay model	Architecture requirement GFP: layer 2 mapping scheme	G.ASON requirement document finished; G.GFP finished
IETF	GMPLS: peer model and dynamic overlay model	Protocol development	Ongoing
OIF	Dynamic overlay model	Interoperability	UNI 1.0 document finished
IEEE 802	—	Ethernet technology	GbE finished 10GE; RPF will be finalized in 2002

some fundamental assumptions. One of the biggest challenges for network operators today is to understand how to exploit and couple the intelligence residing at both the electrical and optical layers, so that they can optimize their network resources and deploy the most advanced services to their end-user customers. The internetworking models described above are examples of efforts under way within the industry to harness the powerful features and functionality that emerging optical layer intelligence brings to next-generation networks. In addition to optical internetworking, we have also summarized the latest standards activity regarding various Ethernet technologies and schemes, which efficiently maps Ethernet and layer 2 frame onto the optical layer. Table 16.3 summarizes the status of these standard activities at the beginning of 2002.

REFERENCES

1. OIF Carrier Group, *Carrier Optical Services Framework and Associated Requirements for UNI*, OIF2000.155.1.

2. D. Awduche, Y. Rekhter, J. Drake, and R. Coltun, Multi-protocol lambda switching: Combining MPLS traffic engineering control with optical crossconnects, draft memo, *http://www.watersprings.org/links/mlr/id/draft-awduche-mpls-te-optical-03.txt*, Internet Engineering Task Force, 2001.

3. D. Awduche, J. Malcolm, J. Agogbua, M. O'Dell, and M. McManus, Requirements for traffic engineering over MPLS, network memo, *http://www.watersprings.org/links/mlr/rfc/rfc2702.txt*, Internet Society, 1999.

4. T. Li and H. Smit, IS-IS extensions for traffic engineering, Internet draft, *http://www.watersprings.org/links/mlr/id/draft-ietf-isis-traffic-03.txt*, Network Working Group, IETF, 2001.

5. D. Katz, D. Yeung, and K. Kompella, Traffic engineering extensions to OSPF, Internet draft, *http://www.watersprings.org/links/mlr/id/draft-katz-yeung-ospf-traffic-06.txt*, Network Working Group, IETF, 2002.

6. *First Draft of G.ASON*, T1X1.5/2000-128.

7. D. Pendarakis, B. Rajagopalan, and D. Saha, Routing information exchange in optical networks, Internet draft, *http://www.watersprings.org/links/mlr/id/draft-prs-optical-routing-01.txt*, IETF, 2001.

8. D. Papadimitriou, et al., *Lightpath Parameters*, OIF 2000.267.

9. A. Chiu et al., *Wavelength Routing in All-Optical Network*, OIF 2000.251.

10. A. Chiu, J. Strand, R. Tkach, and J. Luciani, Unique features and requirements for the optical layer control plane, Internet draft, *http://www.watersprings.org/links/mlr/id/draft-chiu-strand-unique-olcp-02.txt*, IETF, March 2001.

11. B. Rajagopalan, J. Luciani, D. Awduche, B. Cain, B. Jamoussi, and D. Saha, IP over optical networks: A framework, Internet draft, *http://www.watersprings.org/links/mlr/id/draft-ietf-ipo-framework-00.txt*, IETF,

12. P. Ashwood-Smith, D. Awduche, A. Banerjee, D. Basak, L. Berger, G. Bernstein, S. Dharanikota, J. Drake, J. Fan, D. Fedyk, G. Grammel, D. Guo, K. Kompella, A. Kullberg, J. P. Lang, F. Liaw, T. D. Nadeau, L. Ong, D. Papadimitriou, D. Pendarakis, B. Rajagopalan, Y. Rekhter, D. Saha, H. Sandick, V. Sharma, G. Swallow, Z. B. Tang, J. Yates, G. R. Young, J. Yu, and A. Zinin, Generalized multi-protocol label switching (GMPLS) architecture, *http://www.watersprings.org/links/mlr/id/draft-ietf-ccamp-gmpls-architecture-01.txt*.

13. OIF Architecture, OAM&P, PLL, and Signaling Working Groups, *User Network Interface (UNI) 1.0 Signaling Specification*, OIF2000.125.7.

14. *Clarification of Interface Definitions in G.ASTN*, T1X1.5/2001-044, Feb. 2, 2001.

15. ITU-T G.8080, "*Architecture for the Automatically Switched Optical Network (ASON)*."

16. *Architecture and Specification of Data Communication Network*, ITU-T G.7712.

17. *Distributed Call and Connection Management (DCM)*, ITU-T G.7713.

18. *Generalized Automatic Discovery Techniques*, ITU-T G.7714.

19. IEEE P802.3ab 1000BASE-T Task Force, *http://grouper.ieee.org/groups/802/3/ab/index.html*.

20. IEEE P802.3af DTE Power via MDI Task Force, *http://grouper.ieee.org/groups/802/3/af/index.html*.

21. IEEE 802.3ah Ethernet in the First Mile Task Force, *http://grouper.ieee.org/groups/802/3/efm/index.html*.

22. IEEE 802.17 Resilient Packet Ring Working Group (RPRWG), *http://grouper.ieee.org/groups/802/17/*.

INDEX